HANDBOOK OF
ECOTOXICOLOGY

HANDBOOK OF
ECOTOXICOLOGY

EDITED BY
PETER CALOW
DSc, PhD, CBiol, FIBiol
Department of Animal and Plant Sciences
The University of Sheffield

IN TWO VOLUMES
VOLUME TWO

OXFORD
BLACKWELL SCIENTIFIC PUBLICATIONS
LONDON EDINBURGH BOSTON
MELBOURNE PARIS BERLIN VIENNA

© 1994 by
Blackwell Scientific Publications
Editorial Offices:
Osney Mead, Oxford OX2 0EL
25 John Street, London WC1N 2BL
23 Ainslie Place, Edinburgh EH3 6AJ
238 Main Street, Cambridge
    Massachusetts 02142, USA
54 University Street, Carlton
    Victoria 3053, Australia

Other Editorial Offices:
Librairie Arnette SA
1 rue de Lille
75007 Paris
France

Blackwell Wissenschafts-Verlag GmbH
Kurfürstendamm 57
10707 Berlin
Germany

Blackwell MZV
Feldgasse 13
A-1238 Wien
Austria

First published 1994

Set by Setrite Typesetters, Hong Kong
Printed and bound in Great Britain at the University Press,
Cambridge

DISTRIBUTORS

  Marston Book Services Ltd
  PO Box 87
  Oxford OX2 0DT
  (Orders: Tel: 0865 791155
            Fax: 0865 791927
            Telex: 837515)

USA
  Blackwell Scientific Publications, Inc.
  238 Main Street
  Cambridge, MA 02142
  (Orders: Tel: 800 759-6102
            617 876-7000)

Canada
  Oxford University Press
  70 Wynford Drive
  Don Mills
  Ontario M3C 1J9
  (Orders: Tel: 416 441-2941)

Australia
  Blackwell Scientific Publications Pty Ltd
  54 University Street
  Carlton, Victoria 3053
  (Orders: Tel: 03 347-5552)

A catalogue record for this title is available from the
British Library

ISBN 0-632-02989-7   Po 4940

Library of Congress Cataloging-in-Publication Data
(Revised for vol. 2)

Handbook of ecotoxicology.
    Includes bibliographical references and indexes.
    1. Pollution — Environmental aspects.   2. Pollutants
— Toxicology.   3. Ecological risk assessment — Handbooks,
manuals, etc.   I. Calow, Peter.
QH545.A1H36 1993   574.5'222   92-39663
    ISBN 0-632-03573-0 (v. 1)
    ISBN 0-632-02989-7 (v. 2)

Contents

List of Contributors

N. ADAMS, *Monsanto, Technical Center Europe, Rue Laid Burniat, Parc Scientifique, B1348 Louvain-la-Neuve, Belgium*

D. BEALING, *WRc, Medmenham Laboratory, PO Box 16, Medmenham, Marlow, Bucks SL7 2HD, UK*

C.I. BETTON, *Environmental Affairs Department, Burmah Castrol Trading Ltd, Technology Centre, Whitchurch Hill, Pangbourne, Reading, Berks RG8 7QR, UK*

P. BJERREGAARD, *Ecotoxicology Group, Institute of Biology, Odense University, Dk 5230 Odense M, Denmark*

P. CALOW, *Department of Animal and Plant Sciences, School of Biological Sciences, PO Box 601, University of Sheffield, Sheffield S10 2UQ, UK*

D.W. CONNELL, *Faculty of Environmental Sciences, Griffith University, Nathan Campus, Kessels Road, Brisbane, Queensland, Australia*

C.P. CUMMINS, *The Institute of Terrestrial Ecology, Monks Wood, Abbots Ripton, Huntingdon, PE17 2LS, UK*

N.T. DE OUDE, *Procter and Gamble, European Technical Center, Temselaan 100, B-1853, Strombeek-Bever, Belgium*

M.H. DEPLEDGE, *Department of Biological Sciences, University of Plymouth, Drake Circus, Plymouth, PL8 4AA, UK*

J. DOI, *Roy F. Weston Inc., Environmental Fate and Effect Laboratory, 254 Welsh Pool Road, Lionville, PA 19341–1345, USA*

P. DONKIN, *Plymouth Marine Laboratory, Prospect Place, The Hoe, Plymouth, PL1 3DH, UK*

S. HEDGECOTT, *WRc, Medmenham Laboratory, PO Box 16, Medmenham, Marlow, Bucks SL7 2HD, UK*

E.C. HENNES-MORGAN, *Procter and Gamble, European Technical Center, Temselaan 100, B-1853, Strombeek-Bever, Belgium*

C.N. HEWITT, *Institute of Environmental and Biological Sciences, Lancaster University, Lancaster LA1 4YQ, UK*

W.J. LANGSTON, *Plymouth Marine Laboratory, Citadel Hill, Plymouth PL1 2PB, UK*

D. MACKAY, *Institute for Environmental Studies, University of Toronto, Toronto, Ontario M5S 1A4, Canada*

T.A. MANSFIELD, *Institute of Environmental and Biological Sciences, Lancaster University, Lancaster LA1 4YQ, UK*

L.C. McEWEN, *Department of Fishery and Wildlife Biology, Colorado State University, Fort Collins, Colorado 80523, USA*

A.J. NIIMI, *Department of Fisheries and Oceans, Canada Centre for Inland Waters, Burlington, Ontario L7R 4A6, Canada*

D.R. NIMMO, *National Biological Survey, US Department of the Interior, Colorado State University, Fort Collins, Colorado 80523, USA*

L. SALTER, *Institute of Environmental and Biological Sciences, Lancaster University, Lancaster LA1 4YQ, UK*

S.K. SPENCE, *Plymouth Marine Laboratory, Citadel Hill, Plymouth PL1 2PB, UK*

M.J. WALDOCK, *Ministry of Agriculture, Fisheries and Food, Fisheries Laboratory, Remembrance Avenue, Burnham on Crouch, Essex CMO 8HA, UK*

J.M. WEEKS, *The Institute of Terrestrial Ecology, Monks Wood, Abbots Ripton, Huntingdon, PE17 2LS, UK*

Preface

The addition of one further substantial volume to this work makes the description of *handbuch* — small book or treatise that may be held in the hand — yet more dubious (see Preface to Volume 1). Nevertheless, we continue to hope that this volume makes the *Handbook* even more 'handy' for the ecotoxicological practitioners in the academic, governmental and business sectors for whom it is intended. It is designed to complement the techniques emphasis of Volume 1 by providing, in the first part, a review of ecotoxicological results from a host of chemicals that have been grouped into convenient categories. In general, each of these chapters contains information on how to measure the chemicals under consideration, as well as a review of their effects. A final part of the volume collects together a more diverse group of chapters that address issues of fate, partitioning between environmental phases, quantitative structure—activity relationships, complex mixtures and prioritization systems.

Again, to cover such a broad range of subject areas has involved bringing together a large number of authors; and with some surprise I discover that the whole *Handbook* (both volumes) has now involved more than 50 authors. Without their hard work, care and patience in attending to editorial detail, the final outcome would not have been so thorough, so it is a pleasure to thank them for their efforts. Similarly, the support team behind the authors — Editorial Board, Deputy Editor, secretarial staff, publishers — has made what has been quite a complex operation less chaotic than it could have been, and so I also take this opportunity to offer thanks to them.

Finally, I would like to repeat what I wrote in the Preface to Volume 1 — at the end of the day a handbook must be judged on how 'handy' it is for its users — so if you have any comments that might help improve future editions please let me have them.

P.C.
Sheffield, 1993

1: Overview with Observations on Risk Assessment and Management

P. CALOW

1.1 INTRODUCTION

Volume 1 opened by presenting ecotoxicology as being concerned with protecting ecological systems from adverse effects caused by chemicals derived from human activity. This involves assessing the ecological risks associated with chemicals, as an important consideration in their management. The key elements in this involve the anticipation/assessment of where chemicals go in the environment (fate) and hence what levels ecosystems and their components are exposed to (exposure), and the anticipation/assessment of what responses these elicit (effects). Volume 1 described the standard methodology involved in measuring effects — the ecotoxicity tests. This volume is largely concerned with the results from these for general categories of chemicals, together with a description of the principles of fate modelling in Chapter 15. However, before describing the detailed structure of this volume, brief overviews will be given of risk assessment and management as being important goals of the ecotoxicological effort.

1.2 HAZARDS AND RISKS

The potential that chemicals have for causing adverse effects to humans or ecological systems — assessed through the laboratory tests described in Volume 1 — depends upon their *intrinsic properties*, and characterizing these is sometimes known as *hazard identification*. The probability of this potential being realized is often referred to as *risk assessment* or *characterization*. It depends not only on hazard but also on the likelihood and level of exposure (Suter, 1993). Ideally, therefore,

it will be expressed in terms of the proportion (probability) of populations or communities expected to be affected by a particular level of manufacture, and marketing and use regime (this is analogous to how risk is defined in toxicology, e.g. Rodricks, 1992). However, rarely, if ever, do we have enough information and understanding to make such a thorough analysis of ecological risks associated with any chemical. Instead, we usually have a predicted environmental concentration (PEC) and predicted no-effect concentration (PNEC), from ecotoxicity tests, and risk analysis consists simply of comparing the one with the other (i.e. PEC with PNEC). The presumption is that, as PEC recedes from PNEC, the probability of an adverse effect reduces. This, then, is a kind of risk assessment as defined above, but without the probability values being explicitly defined (Calow, 1993). This has led some to the view that PEC/PNEC comparisons are not true risk assessments and so they should be described, instead, as hazard assessments (ECETOC, 1993). It is certainly true that words should be clearly defined, but it is even more important that the processes they describe are clearly understood so that the assumptions and weaknesses associated with them can be taken into account when they are used.

PEC

This depends upon levels of manufacture and ways of use, on the ways substances are transported in the environment, and on the properties of these substances that determine the extent to which they partition between water and air, the extent to which they stick to soil particles and sediments, the extent to which they break down in light and microbially-mediated processes, and the extent to which they accumulate in organisms. Fate models are addressed in Chapter 15. Biodegradation and bioaccumulation processes are treated in Chapters 18 and 19 respectively of Volume 1, and the partition of chemicals between water and organic phases in Chapter 13 of this volume. Even in the absence of detailed modelling of fate and hence exposure, biodegradation rates, octanol : water partition coefficients (Chapter 13), and bioaccumulation rates can give useful insight into the persistence and potential exposure regimes associated with chemicals that can be used in the first stages of risk assessment (e.g. Chapter 16).

PNEC

It is obvious, yet nevertheless important, that whether or not a particular concentration of chemical has an effect on biological systems depends upon what effects are being observed. For individuals and populations they might range from molecular responses, through physiological ones, to changes in survival, reproduction and development. The ones of ecological relevance here will be those likely to significantly influence population dynamics and Darwinian fitness. In multi-species systems the measures of effects might range from structural ones such as species composition, diversity, dominance and equitability (Magurran, 1988) to functional ones such as energy flows (e.g. Odum, 1985) and material cycles and spirals (Newbold, 1992). Here it is more difficult to be categorical about ecological relevance, since we know less about the structure and functioning of normal ecosystems — if, indeed, they exist (Calow, 1992). A major point to note, though, from Volume 1 and the reviews on chemicals in this volume, is that the majority of ecotoxicity data in the literature is on the survivorship (or behavioural) responses of single species measured at high concentrations over short time intervals, i.e. on data from acute tests expressed in terms of $L(E)C_{50}$ values (see also Maltby & Calow, 1989). These, therefore, are often the only data available as a basis for estimating PNECs, and, in doing that, due allowance

has to be made for the extrapolation involved in moving from acute to chronic effects, single- to multi-species effects, and laboratory to field effects. One way of doing this is simply to divide $L(E)C_{50}$s by appropriate application or safety factors. These are often factors of 10 (10, 100, 1000, depending on the quality and quantity of the data) and are, to say the least, somewhat arbitrary. More systematic approaches have been followed, in which the factors have been derived from careful correlation analysis on chemicals for which data exist from acute, chronic and multi-species tests (e.g. ECETOC, 1993). But still, it always has to be borne in mind that a correlation based on one dataset does not guarantee that all chemicals will always comply (this is a fundamental problem with induction — Maltby & Calow, 1989). Yet another approach is to define concentrations likely to be safe for a given percentage of population or community (often 95%) by reconstruction of sensitivity distributions from a few observations (e.g. Kooijman, 1987; Van Straalen & Denneman, 1989; Volume 1, Chapter 13). But it has been claimed that, given our ignorance about the form of these distributions and the computational effort involved with them, the more complicated approach is unlikely to lead to any more environmental protection than the simpler one involving safety factors (Forbes & Forbes, 1993, 1994). In either case, what we need, to be able to put more scientific confidence in these approaches, is more understanding of the mechanistic basis of the relationship between concentrations and effects; similar to that described in Chapter 14.

PEC vs PNEC

The ecological risk assessments described above are, as already noted, more or less pragmatic and generally involve some kind of comparison of PEC with PNEC. This is the basis of the way that they are carried out in the context of chemical legislation in the USA and Europe (Norton *et al.*, 1992; Calow, 1993). In doing this, the two volumes of this handbook amply demonstrate the need not only for more data *per se* but for more useful data (e.g. from chronic experiments) and for more understanding.

1.3 RISK MANAGEMENT

The handbook has not set out to explicitly review all the existing regulatory frameworks applied to chemicals. Rather it has been concerned with the techniques behind the risk assessments that are used by regulators in managing the risks associated with chemicals. There has, nevertheless, been a need to discuss regulatory issues in passing, especially in Volume 1, since the regulatory requirements act as an important driving force behind the methodology and approaches of ecotoxicology (see Volume 1, Chapter 1). Here, we consider further some general principles of risk management and regulation.

In considering regulations it is useful to make a distinction between those concerned with gathering information for the classification and labelling of hazardous substances and those concerned with controls. In principle, the information gathering (risk assessment) should provide a logical basis for the application of controls. In practice, the vast number of chemicals already in use (alluded to in several chapters) means that initial prioritization processes are required to identify those chemicals in need of urgent attention. These, by necessity, are often based on what is known (however little) about the intrinsic ecotoxicological properties of substances (Chapter 16). More rigorous risk assessments can then be carried out on the prioritized substances; and this is indeed required by EC legislation (Calow, 1993).

A distinction can also be made between controls applied on the marketing and use of chemicals and those applied to emissions from stacks into atmosphere and from pipes into water. Marketing and use involves a distributed problem, in which controls often have to be either all (banning) or nothing on the basis of risk assessment, or even sometimes just the suspicion of a problem. Controlling releases from point sources, on the other hand, is relatively more straightforward. Limit values and environmental standards are established by hazard- and risk-assessment procedures (Chapter 16), and ecotoxicity criteria might be incorporated into consents and authorizations for releases so that monitoring (Volume 1, Chapter 20), *in-situ* assays (Volume 1, Chapter 21) and systematic approaches to the analysis of complex

effluents (Chapter 12) becomes of considerable importance.

Finally, controls can also be applied through so-called market instruments that provide the consumer with information upon which decisions about purchasing can, in principle, be made on the basis of environmental criteria. Good examples are product labels. These can be either of a negative kind (warning consumers about environmental hazards) or of a positive kind (indicating products that, within a class, have least adverse effects on the environment.) In the EC there are examples of both. Directive 67/548/EEC, in its seventh amendment (92/32/EEC), requires that substances classified as dangerous to the environment should be labelled as such. For new chemicals, labels are allocated on the basis of a notification that specifies intrinsic properties (hazard identification), but the legislation requires that notifying authorities carry out a risk characterization with a possible view to the need for risk management (Calow, 1993). Regulation (EEC) No. 880/92 establishes a scheme for labelling of various products with reduced environmental impact. This involves a life-cycle assessment (SETAC, 1990) that should, at least in principle, incorporate an assessment of the risk of environmental impact of the product from raw materials, through manufacturing and on to ultimate disposal.

1.4 STRUCTURE OF VOLUME 2

This volume is separated into two main parts. Part 1 provides a systematic review of major chemical categories. Part 2 is a general section which consists of a range of chapters that have, as indicated explicitly in the foregoing, important parts to play in the risk-assessment procedure.

The pressing conclusion that emerges, from both this volume and Volume 1, is that we currently know a lot about ecotoxicological responses, without necessarily understanding them very thoroughly. Moreover, though we know a lot, there is a lot yet to be known. Repeatedly the point is made that there are so many existing chemicals that we shall never be able to carry out detailed risk assessments of them all. We need to have short-cut methods for extrapolating from what we know to what we do not know; and our scientific confidence in them will be increased by a better understanding of basic principles. Several chapters in Part 2 are important from this point of view, but the issue is explicitly addressed by Chapter 14 on QSARs. We also have to have methods of prioritization that are clearly defined and transparent in their mode of operation, and these are reviewed in Chapter 16.

REFERENCES

Calow, P. (1992) Can ecosystems be healthy? Critical consideration of concepts. *J. Aquat. Ecosyst. Health* **1**, 1–5.

Calow, P. (1993) Hazards and risks in Europe: challenges for ecotoxicology. *Environ. Toxicol. Chem.* **12**, 1519–1520.

ECETOC (1993) Environmental hazard assessment of substances. Technical Report No. 51. European Centre for Ecotoxicology and Toxicology of Chemicals, Brussels, Belgium.

Forbes, T.L. & Forbes, V.E. (1993) A critique of the use of distribution-based extrapolation models in ecotoxicology. *Funct. Ecol.* **7**, 249–254.

Forbes, V.E. & Forbes, T.L. (1994) *Ecotoxicology in Theory and Practice*. Chapman and Hall, London.

Kooijman, S.A.L.M. (1987) A safety factor for LC_{50} values for differences in sensitivity between species. *Water Res.* **21**, 269–276.

Magurran, A.E. (1988) *Ecological Diversity and its Measurement*. Croom Helm, London.

Maltby, L. & Calow, P. (1989) The application of bioassays in the resolution of environmental problems: past, present and future. In: *Environmental Bioassay Techniques and their Applications* (Eds M. Munawar, G. Dixon, C.I. Mayfield, T.R. Reynoldson & M.H. Sadar), pp. 65–67. Kluwer, London.

Newbold, J.D. (1992) Cycles and spirals of nutrients. In: *The Rivers Handbook*, Vol. 1 (Eds P. Calow & G.E. Petts), pp. 379–408. Blackwell Scientific Publications, Oxford.

Norton, S.B., Rodier, D.J., Gentile, J.H., van der Schalie, W.H., Wood, W.P. & Slimak, M.W. (1992) A framework for ecological risk assessment at the EPA. *Environ. Toxicol. Chem.* **11**, 1663–1679.

Odum, E.P. (1985) Trends in stressed ecosystems. *BioScience*, **35**, 419–422.

Rodricks, J.V. (1992) *Calculated Risk*. Cambridge University Press, Cambridge.

SETAC (1990) *Technical Framework for Life-Cycle Assessments*. SETAC and SETAC Foundation, Washington, DC.

Suter, G.W. (Ed.) 1993 *Ecological Risk Assessment*. Lewis Publishers, Boca Raton.

Van Straalen, N.M. & Denneman, G.A.J. (1989) Ecotoxicological evaluation of soil quality criteria. *Ecotoxicol. Environ. Safety*, **18**, 241–251.

PART 1
CHEMICAL CATEGORIES

The chapters in this part review the ecotoxicological properties of a number of major chemical categories. The classification is not a chemically rigorous one, nor is it claimed to be complete. Rather it collects chemicals into major groupings that are likely to be useful in risk assessment and management activities. In general, each describes methods of chemical analysis and reviews results from the ecotoxicity tests described in detail in Volume I. An exception involves the metals; in this case the analytical procedures are so diverse and yet so well established that they have been put in a chapter (Chapter 4) separate from the ecotoxicological review (Chapter 5). Clearly there would not be sufficient space in this volume either to provide detailed descriptions of all available analytical techniques or to provide detailed lists of $L(E)C_{50}s$ and NOECs, etc. Instead, authors have provided reviews of reviews, key references and useful summary tables and figures. To link with Part 2 it is important to draw attention to the fact that most of the ecotoxicity data are from tests on individual chemicals. More often in nature, though, chemicals occur in complex mixtures. Hence, the first chapter of the next part addresses the philosophy and methodology of handling this complexity.

2: Gaseous Compounds

C.N. HEWITT, L. SALTER AND T.A. MANSFIELD

2.1 INTRODUCTION

The Earth's atmosphere is composed almost entirely of four gases — nitrogen, oxygen, argon and water vapour. The first three of these comprise 78.084%, 20.946% and 0.934% by volume, respectively, of dry unpolluted air, concentrations which are remarkably invariable. The concentration of water vapour, on the other hand, is very variable and may approach 4% by volume. All other gases, however, are present, in trace amounts only, even in polluted air, at concentrations of parts per million, parts per billion or parts per trillion by volume (10^{-6} v/v = 1 ppm; 10^{-9} v/v = 1 ppb; 10^{-12} v/v = 1 ppt). Of these, carbon dioxide is the most abundant, its global average concentration now exceeding 350 ppm. It is into this background matrix of relatively unreactive and persistent gases that primary air pollutants are emitted, and in which the complex chemistry of the atmosphere takes place. As a result of reactions involving primary pollutants and, often, energy from solar radiation, other secondary pollutants are formed in the atmosphere. Of these, ozone is the most important in the present context.

In this chapter, the physical and chemical properties of the most important gaseous pollutants are introduced, as are their sources in the atmosphere, the analytical methods used for their determination and their typical ambient concentrations. Some of their chemical and biochemical effects on humans, animals and plants are then reviewed.

2.2 PHYSICOCHEMICAL PROPERTIES AND SOURCES OF GASEOUS AIR POLLUTANTS

Several textbooks provide reviews of the physicochemical properties and sources of gaseous air pollutants; Guderian (1985), Findlayson-Pitts & Pitts (1986), Seinfeld (1986) and Warneck (1988) being among the most useful.

2.2.1 Sulphur dioxide

Sulphur dioxide (SO_2) has a molecular weight of 64 and is a colourless gas with a very pungent smell. It is quite soluble in water, having a Henry's law constant K^H of 5.4 M/atm at 288 K. It is emitted into the atmosphere by the combustion of fossil fuels, predominantly coal and fuel oil, which may contain up to 3% sulphur by weight. For example, current emissions of SO_2 in the UK are around 3.5 Mt/year, having decreased from 5 Mt/year in 1976. In common with many other compounds which are usually thought of as pollutants, there are several important natural sources of SO_2 in the atmosphere, including emissions from volcanoes and the oxidation of naturally produced dimethyl sulphide and hydrogen sulphide. While these may be important to the global budget of the gas, on local or regional scales it is the industrial point sources of the pollutant that are of most ecological concern.

Oxidation of SO_2 in the atmosphere may take place in both the gas and liquid phases, and proceeds at widely varying rates, depending on the local concentrations of oxidants. Both gas and liquid phase oxidations ultimately result in the formation of sulphuric acid and sulphate, so it is important to consider the ecotoxicity of both the primary pollutant and these oxidation products.

2.2.2 Nitric oxide

Nitric oxide (NO) is formed in combustion processes by the oxidation of nitrogen in both the combustion air (thermal NO) and the fuel itself (fuel NO)

$$O + N_2 \rightarrow NO + N$$
$$N + O_2 \rightarrow NO + O$$

$$\overline{N_2 + O_2 \rightarrow 2NO}$$

At ambient temperatures, this equilibrium lies far to the left, but at elevated temperatures, significant quantities of NO are formed (the equilibrium constants at 300 K and 2500 K being $\sim 10^{-30}$ and 3.5×10^{-3} respectively).

NO has a molecular weight of 30 and is a colourless gas, weakly soluble in water ($K^H = 0.0023$ M/atm at 288 K). In the atmosphere it is oxidized to nitrogen dioxide (NO_2) by reaction with ozone

$$NO + O_3 \rightarrow NO_2 + O_2$$

or with peroxy radicals

$$NO + RO_2 \rightarrow RO + NO_2$$

It is thus an important precursor to NO_2 formation, and hence to many photochemical reactions (see below).

2.2.3 Nitrogen dioxide

Nitrogen dioxide (NO_2) has a molecular weight of 46 and is a brown gas with a characteristic odour. It exists in equilibrium with its colourless dimer N_2O_4, the concentration of which is negligible in the atmosphere. The solubility of NO_2 in water is low ($K^H = 7.5 \times 10^{-3}$ M/atm at 295 K) and in aqueous solution it is slowly converted to nitrous and nitric acids.

NO_2 strongly absorbs solar radiation and is readily dissociated to give atomic oxygen

$$NO_2 + h\nu \rightarrow NO + O$$

which then goes on to form ozone

$$O + O_2 + M \rightarrow O_3 + M$$

where M is another molecule, usually N_2. NO_2 is therefore an important air pollutant in its own right, being an oxidant, but it also takes part in photochemical reactions in the atmosphere that produce ozone.

2.2.4 Ozone

Pure ozone (O_3) is a pale blue gas which liquefies at 161 K to give an explosive blue liquid. At a concentration of ~ 0.5 ppm it is not visible but has a characteristic irritating odour. It is reasonably soluble in water ($K^H = 0.02$ M/atm at 288 K), a property that is of importance both in atmospheric cloud chemistry and in its effects on plants and animals.

O_3 strongly absorbs ultraviolet (UV) radiation, and so effectively shields the Earth's surface from light of wavelengths less than ~ 300 nm. Its occurrence at high concentrations (around 10 ppm) in a layer in the stratosphere is therefore essential to above-ground present-day life forms. In the stratosphere it is formed by the photolysis of oxygen

$$O_2 + h\nu \rightarrow 2O$$
$$O + O_2 + M \rightarrow O_3 + M$$

and is removed by photolysis and reaction with H_2, N_2 and halogen-containing species. Industrial emissions of chlorine compounds are now known to be causing severe depletion of O_3 from the stratosphere, allowing enhanced fluxes of UV radiation to penetrate the atmosphere, and hence having an indirect detrimental effect on the biosphere.

O_3 is present in the troposphere following its downward transport from the stratosphere. Additionally, and more importantly, it is formed in the lower atmosphere by reactions involving oxides of nitrogen, hydrocarbons and sunlight.

2.2.5 Hydrocarbons

An enormous variety of hydrocarbon compounds is present in the atmosphere, resulting from the incomplete combustion of fossil fuels, evaporation of solvents, fuel and chemical feedstock, and emissions from industrial processes (see also Chapter 10). The three most important classes of hydrocarbons are the alkanes (C_nH_{2n+2}), alkenes (C_nH_{2n}) and aromatic or cyclic compounds (the simplest being benzene, C_6H_6).

Table 2.1 summarizes some physical properties of a few representative hydrocarbon compounds. The solubility of hydrocarbons in water is extremely low, and they exhibit a very wide range of chemical properties. Those of interest as air pollutants fall into two classes: those that are toxic, mutagenic or carcinogenic (e.g. benzene and the polynuclear aromatic hydrocarbons such as benzo(*a*)pyrene) and those that are photochemically reactive in the atmosphere and take part in reactions that lead to ozone and peroxy radical formation.

Methane, by far the most abundant hydrocarbon in the atmosphere, with a global mean concentration of ~ 1.7 ppm, does not fall into either of these classes, and is therefore not considered further here. It should be noted, however, that methane is a radiatively active gas, and so may have indirect effects on the biosphere because of its contribution to global warming.

Hydrocarbons play a crucial role in ozone formation because they are oxidized to form organic peroxy radicals (e.g. peroxymethyl, CH_3O_2) which can efficiently oxidize NO to NO_2. This latter species may then be photolysed and form ozone. These reactions may be shown simplistically for an alkane RCH_3, although similar sequences can be written for any hydrocarbon:

$$OH + RCH_3 \rightarrow H_2O + RCH_2$$
$$RCH_2 + O_2 \rightarrow RCH_2O_2$$
$$RCH_2O_2 + NO \rightarrow RCH_2O + NO_2$$
$$RCH_2O + O_2 \rightarrow RCHO + HO_2$$
$$HO_2 + NO \rightarrow OH + NO_2$$

then

$$NO_2 + h\nu \rightarrow NO + O$$
$$O + O_2 + M \rightarrow O_3 + M$$

Chain reactions of this type may produce considerable quantities of ozone from the primary pollutants in the presence of sunlight, with a stoichiometry that is hard to predict.

2.3 METHODS OF ANALYSIS FOR THE MAJOR GASEOUS AIR POLLUTANTS

Findlayson-Pitts & Pitts (1986), Harrison & Perry (1986) and Hewitt (1991) all provide useful reviews of the principal methods of analysis used for the determination of air pollutants, but brief descriptions of the techniques used for the major pollutant gases are given below.

2.3.1 Sulphur dioxide

The recommended procedures for the wet chemical, manual determination of SO_2 are the West–Gaeke colorimetric technique (widely used in

Table 2.1 Physical properties of selected hydrocarbons

Name and chemical formula	Molecular weight	Melting point (°C)	Boiling point (°C)
Alkanes			
Ethane, C_2H_6	30	−183	−89
Propane, C_3H_8	44	−190	−42
n-Butane, C_4H_{10}	58	−138	−1
Alkenes			
Ethene, C_2H_4	28	−169	−104
Propene, C_3H_6	42	−185	−47
trans-2-Butene, C_4H_8	56	−105	1
Aromatic hydrocarbons			
Benzene, C_6H_6	78	6	80
Toluene, $C_6H_5CH_3$	92	−95	111
o-Xylene, $C_6H_4(CH_3)_2$	106	−25	144

the USA) and the hydrogen peroxide technique (widely used in the UK). However, they have now been very largely superseded by the development and commercial availability of a pulsed gas-phase fluorescence method which has a detection limit of ~ 0.5 ppb and a response time of ~ 2 min.

It is also possible to collect SO_2 by reaction to sulphate on an impregnated filter paper, with a Teflon pre-filter used to exclude aerosol-phase sulphate. After aqueous extraction the sulphate may be readily determined by high-pressure liquid chromatography (HPLC). Such a filter-pack method is cheap, sensitive, relatively portable and free of interferences.

2.3.2 Oxides of nitrogen

NO_2 is hydrolysed in aqueous solution to nitrite, and this ion may be determined colorimetrically by the formation of an azo-dye complex (the Griess–Isolvay reaction, as modified by Saltzmann). However, there are uncertainties in the stoichiometry of the conversion of NO_2 to nitrite ion, and the method is now little used. NO may also be determined by this method if NO_2 is first removed from the sampled air stream and the NO oxidized to NO_2 by chromic acid. The stoichiometric problem remains.

A modification of the Griess–Saltzmann reaction is now widely used for the determination of NO_2 in ambient air using passive diffusion tubes (e.g. Hewitt, 1992). Triethanolamine-impregnated stainless-steel discs are used to absorb NO_2 as it diffuses into an acrylic tube; the NO_2 is then determined colorimetrically as nitrite. This is converted to an air concentration of NO_2 using the diffusion coefficient, the dimensions of the tube and the length of exposure. Detection limits down to a few ppb are achievable with sampling times of 2–4 weeks, providing care is taken to minimize the blank signal.

The most reliable, accurate and precise method for determining oxides of nitrogen concentrations utilizes the chemiluminescent reaction of NO and O_3.

$$NO + O_3 \rightarrow NO_2^* + O_2$$
$$NO_2^* \quad \rightarrow NO_2 + h\nu$$
$$\text{(max. intensity} = 1200\,\text{nm)}$$

The response is linear over a very wide range of NO concentrations, is rapid and gives a detection limit of 1–2 ppb. There is little interference from other pollutants. NO_2 may be quantitatively reduced to NO over a heated catalyst, allowing the determination by one instrument of NO, NO plus NO_2, and hence of NO_2 by difference. Several commercial instruments based on this system are available, and they are now widely used.

2.3.3 Ozone

The determination of O_3 by wet chemical means (reaction with potassium iodide with subsequent titration of the liberated iodine) has been rendered obsolete by two instrumental methods. Of these, the chemiluminescence method has itself been very largely superseded by the UV absorption technique. This relies on the absorption of the 253.7 nm mercury resonance line by ambient unfiltered and filtered (O_3-free) air and the Beer–Lambert relationship:

$$\log \frac{I_0}{I} = \Sigma_{253.7} \times \ell \times [O_3]$$

where I_0 is the intensity of the incident light of wavelength 253.7 nm, I the intensity of the transmitted light, Σ the extinction coefficient, ℓ the path length and $[O_3]$ the ozone concentration. The method is sensitive, with a detection limit of ~ 1 ppb, rapid and selective. By using measurements of ambient air with and without the removal of O_3, it is possible to correct for absorption by mercury, hydrocarbons and other interfering species.

2.3.4 Hydrocarbons

Hydrocarbons may be determined by flame ionization detection (see also Chapter 10). However, the great complexity of the hydrocarbon mixture in ambient air means that a separation step is required, so that each compound may be identified and quantified individually. Excellent separation of many species is achievable using capillary gas chromatography (GC). In addition, a pre-concentration step may be required, since ambient hydrocarbon concentrations are low (<1–100 ppb). Both adsorption onto a polymer surface and cryogenic trapping are now routinely used for this purpose. Detection by mass spectrometry (GC−MS) may also be used to confirm peak identifications.

Polynuclear aromatic hydrocarbons (PAHs) occur in both the gas and aerosol phases. The former may be collected by adsorption onto polyurethane foam and the latter by filtration. Extraction into dichloromethane then allows their separation by HPLC with fluorescence and/or UV absorption detection.

Peroxyacetyl nitrate (PAN: $CH_3COO_2NO_2$) may be determined by electron capture detection following its separation by GC. A detection limit of < 1 ppb is achievable.

2.4 TYPICAL AMBIENT CONCENTRATIONS OF GASEOUS AIR POLLUTANTS

Ambient concentrations of gaseous air pollutants vary widely both spatially and temporally. It is therefore necessary to distinguish between natural background concentrations, 'typical' ambient concentrations and peak concentrations observable near to a primary pollutant source or found during pollution episodes. Table 2.2 provides a summary of these concentration ranges.

2.5 ECOTOXICOLOGY OF GASEOUS COMPOUNDS

2.5.1 Sulphur dioxide

Wellburn (1988) provides a recommended introduction to the toxic properties of SO_2 and the other gases discussed below. Inhalation of SO_2 mainly affects the upper respiratory tract and bronchi — unlike ozone and nitrogen oxides it is mostly absorbed in airways above the larynx. Clinical signs include conjunctivitis, cough, rhinitis, bronchoconstriction, pneumonia and thickening of the respiratory mucous layer — the sulphate, sulphite and bisulphite anions formed on contact with moisture at physiological pH stimulate histamine release. At hundreds of ppm it dissolves in water in upper respiratory tract tissues but at 1 ppm very little penetrates. Changes in nasal dynamics have been observed at levels as low as 2.0 ppm (Bedi & Horvath, 1989). People with respiratory weaknesses, e.g. asthmatics, are particularly susceptible (Freudenthal *et al.*, 1989; Horstman & Folinsbee, 1989). Combined exposure to SO_2 and particulates is more damaging to lung tissue than either alone. Neonatal development can be altered by maternal SO_2 exposure (Singh, 1989).

Sulphite and bisulphite can either react directly with disulphide linkages, cyclic compounds and DNA or, in the presence of metal ions (e.g. Mn^{2+}), form free radicals (O_2^-, OH, SO_2^-, HSO_3). In humans, sulphite and bisulphite are usually quickly removed by sulphite oxidases but high concentrations in and around the lungs may cause problems. Cells (alveolar macrophages and lymphocytes) show high levels of adenosine triphosphatase (ATPase) and lysozyme, which suggests cell function degradation. Some gas could be bound to (or react with) protein to give *S*-sulphonate from the disulphide/sulphite reaction. If sulphuric acid is formed, cell dehydration occurs and (at low pH) proteins could be denatured. Animal responses parallel those of humans.

Table 2.2 Background, typical ambient and peak concentrations of gaseous air pollutants

Pollutant	Background (ppb)	Typical ambient concentrations (ppb)	Peak (ppb)
SO_2	< 0.1	1−10	10−100
$NO_x = NO + NO_2$	< 1	5−50	> 100
O_3	~ 25	25−60	> 100
Hydrocarbons	< 1	1−20	50−100
PAHs	< 1	1−10	> 10
PAN	< 1	< 1−10	> 10

In most land plants access to reaction sites is dominated by entry via the stomata and, subsequently, resistance to diffusion offered by the mesophyll. In this context it should be noted that exposure to low concentrations of SO_2 at high humidities interferes with stomatal action, and the pores open more widely than usual. Plants exposed to SO_2 become more sensitive to frost injury (Mansfield & Lucas, 1990). Necrosis can result from acute exposure and chronic exposure causes chlorosis. Buffering capacity can be altered in extra- and intracellular fluid (Chauhan & Mehra, 1989) and both CO_2 fixation efficiency and photophosphorylation activity are reduced (Bennett, Lee & Heggestad, 1990). Damage to membranes occurs — for example the thylakoid membranes of chloroplasts and the phloem tissue appear to be very sensitive to SO_2 (Mansfield & Lucas, 1990). Sulphur is an amino acid component and it is used in a number of metabolic pathways. At low levels these would incorporate sulphite by normal metabolic mechanisms, but at higher levels sulphite-induced free radical activity increases. Reductions in productivity and other gross effects are observed (Yadav, Chand & Singh, 1988; Kropff *et al.*, 1989a,b) but small amounts of sulphur deposition can be beneficial (Mansfield & Lucas, 1990). SO_2 enhances PAH carcinogenicity. Wellburn (1988) provides a useful tabular summary of synergistic interactions of SO_2 with other pollutant gases, and comments on SO_2 dose–response relationships.

The interpretation of ecotoxicological effects of SO_2 is difficult because concentrations of SO_2 that have little direct impact on plants can sometimes have substantial secondary effects. SO_2 can alter amino acid composition in phloem sap, and in consequence its nutritional value to aphids can increase. Figure 2.1 shows how the mean relative growth rate of the pea aphid increases with SO_2 concentration up to about 100 ppb, and thereafter declines. The effects of higher concentrations are thought to indicate a direct action of SO_2 on the aphids (Warrington, 1987).

2.5.2 Oxides of nitrogen

The effects of the oxides of nitrogen are dominated by those of the dioxide — partly because forma-

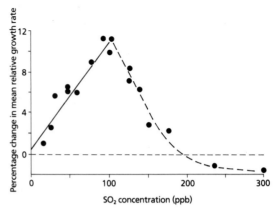

Fig. 2.1 Effects of SO_2 on mean relative growth rate of pea aphids (*Acyrthosiphon pisum*). The aphids were allowed to feed uninterrupted for 4 days on garden peas grown for 20–22 days in the various SO_2 concentrations. Changes in growth rates are percentage changes compared with controls in clean air (< 5 ppb SO_2). After Warrington (1987).

tion of nitric acid gives it the largest aqueous solubility. A useful short review of the physiological significance of NO is available (Collier & Vallance, 1991) but generally its toxicity to human beings is much less than the dioxide at comparable concentrations (Stavert & Lehnert, 1990). In the case of plants, recent evidence suggests that there are circumstances when NO can display high toxicity, in a manner different from that of NO_2 (Wellburn, 1990).

At low levels of dioxide (0.3 ppm for 3 h) few effects are observed (Avol *et al.*, 1989) but acute exposure to NO_2 is especially harmful because symptoms are not necessarily immediate. Coughing, headache and tight chest develop 1–24 h after exposure to 50–100 ppm for several minutes, and the consequent inflammation of lung tissue can last for 6–8 weeks. There are often recurrent chest complaints after such exposure. At 150–200 ppm for less than 1 h death occurs 3–5 weeks later, from circulatory collapse, congestion and/or pulmonary oedema (Manahan, 1990).

Physiologically, the most sensitive regions are the junctions between the bronchioles and alveolar regions. Type I pneumocytes are the most sensitive cells (destroyed after 2 h at 15 ppm of NO_2), and during exposure they are replaced by Type II cells which, together with replacement ciliary cells, can cause areas of hypertrophy in the lung. Due to inhibition of prostaglandin E_2, vasoconstriction can occur up to 60 h after exposure (Wellburn, 1988). Acute NO_2 uptake in pulmonary air spaces is limited by reaction with epithelial surfaces rather than solubility (Postlethwaite & Bidani, 1990).

Many of the toxic effects of NO_2 can be related to cell membrane damage. Lipid autoxidation is initiated by addition of NO_2 to double bonds. Hydroxyl radicals diminish lung elasticity by reacting with methionine in a_1-proteinase. As with all autoxidation reactions, the effects are increased by vitamin E_1 deficiency. Thiols are also easily oxidized by NO_2. Some effects on DNA have been observed (Georsdorf *et al.*, 1990) and increased dissemination of circulating cancer cells has been reported (Richter, 1988).

The access of nitrogen oxides into plant tissues will not be fully understood without further research into the basic processes involved. In the case of NO_2 it appears that uptake through stomata is dominant, and if the stomata are closed (as they usually are in the dark), entry into internal tissues of the leaf is very small (Fowler, Duyzer & Baldocchi, 1991). In the case of NO, uptake in the dark can be as high as in the light, and this may mean that cuticles are much more permeable to NO than to NO_2, or that most of the NO is sorbed into the cuticle and does not enter the leaf via stomata (Lendzian & Kerstiens, 1991).

A further complication is that solubility in the complicated aqueous mix of organic and inorganic species in the extracellular fluid is greater than expected. The detailed interrelationship between anions (nitrate and nitrite) and undissociated acids (nitrous and nitric) varies greatly with the nature of the extracellular fluid. Hence, plant responses to oxides of nitrogen are strongly species-dependent, and beneficial effects can occur when exposure increases nitrogen availability to an appropriate level. Nitrate accumulation is much less harmful than that of nitrite. It has been suggested (Wellburn, 1988) that it is the increase in nitrous acid concentration (and its consequent effect on the pH of the chloroplast stroma) which is a major contributor to nitrite toxicity — particularly since, as well as nitrite from oxides

of nitrogen in the extracellular fluid, nitrite could build up in the cytoplasm from the action of nitrate reductase on nitrate anions (Mansfield & Lucas, 1990). In the presence of SO_2, raised nitrite levels are encouraged by the inhibition of nitrite reductase which would normally convert nitrite to ammonia (or ammonium) prior to amino acid synthesis. Increased acidification (due to nitrous acid) damages the ability of chloroplasts to form adenosine triphosphate (ATP) (0.3 ppm NO_2 raised levels of nitrite in chloroplasts), and both protein synthesis and CO_2 fixation may be reduced (respiration and photorespiration are also inhibited). Lipid peroxidation is relatively unimportant.

NO_2 acts synergistically with ozone and SO_2 in both animals (Chang *et al.*, 1988) and plants (Engelbach & Fangmeier, 1988).

2.5.3 Ozone

For humans, peak levels of ozone (0.15–0.30 ppm as 8 h averages) cause coughing and bronchial constriction. During exercise there is a decline in tidal volume and an increase in breathing frequency (McDonnell, 1988). Low-level long-term exposure is considered to be particularly important (Wright, Daniel & Wheeler, 1990). Exposure to higher levels (50 ppm — much higher than currently observed in polluted air) for periods of hours causes death from pulmonary oedema. Goldstein (1979) remarks: 'of all agents currently of concern to environmental medicine, ozone perhaps has the narrowest range between presumed acute lethal concentration and the known human exposure.'

Bearing these comments in mind, it is not surprising that ozone toxicity in humans is regularly reviewed in the literature. Ozkaynak *et al.* (1988), Bates (1989), Boushey (1989), Lippmann (1989), Schlesinger (1989), Tilton (1989), Witschi & Sagai (1989) and Wright, Daniel & Wheeler (1990) are all relevant. The effects of elevated concentrations of ozone on human health have been authoritatively reviewed by the Department of Health's Advisory Group on the Medical Aspects of Air Pollution Episodes (HMSO, 1991).

Pulmonary effects dominate in animals, with Type I pneumocytes being more sensitive than Type II cells — replacement of I by II during exposure changes lung morphology. At a chemical level, many of the effects of ozone can be interpreted in terms of reaction with the cell membrane. Ozone attacks polyunsaturated fatty acids via the Criegee reaction (Criegee, 1975) forming hydroxy or peroxy radicals (via a hydroperoxide) or forming more stable (but reactive) ozonides (1,2,4-trioxolanes). Radical formation causes autoxidation. Ozone (and ozonides) also oxidize thiols, which may affect enzyme activity. Increase in lung glutathione peroxidase activity follows lipid peroxidation and vitamin E levels are important in determining sensitivity to ozone. Extrapulmonary effects are due in part to carbonyl compounds produced as end-products from membrane oxidation processes (US National Academy, 1977). Heinz bodies are observed, as are direct effects on DNA and RNA. Tolerance to ozone is thought to occur by a decrease in lung microsomal cytochrome P-450 concentration (Goldstein, 1979) though increased sensitivity has been reported (Brookes, Adams & Schelegle, 1989). Immunological changes have also been observed (Orlando *et al.*, 1988; Burleson, Keye & Stutzman, 1989; Fujimaki, 1989; Goodman *et al.*, 1989).

Studies on animals are mostly limited to the laboratory (Costa, Stevens & Tepper, 1988) but some large-animal research has been performed (for example, Hornof *et al.*, 1989; Mariassy *et al.*, 1989; Riedel, 1989). Species differences can be profound (Ichinose & Sagai, 1989) and extrapolation from one species to another is difficult (Hatch *et al.*, 1988). The effects of ozone on farm animals deserve closer attention.

Plants are very sensitive to ozone (Olszyk, 1988). A characteristic brown or white flecking appears on plant leaves at high ozone concentrations (Mansfield & Lucas, 1990). Differences in ozone sensitivity are partly determined by stomatal conductance. Exposure to 0.06 ppm can halve photosynthesis rates (Manahan, 1990) and crop losses are significant (Heagle, 1989). Although there is a general parallel between the biochemistry of oxidative stress in plants and animals, there are significant differences (Schulte-Hostede *et al.*, 1988; Hewitt *et al.*, 1990). In particular, the production of unsaturated hydrocarbons such as 'stress ethylene' (Mehlhorn & Wellburn, 1987) and isoprene in the presence

of ozone has been linked to the formation of hydroperoxides (Hewitt, Kok & Fall, 1990) which almost certainly contribute to the gross effects of ozone on plants.

2.5.4 Hydrocarbons and other gases

Hydrocarbons

In the non-industrial, non-occupational ambient atmosphere, two types of hydrocarbon species are important: the polycyclic (or polynuclear) aromatic hydrocarbons (PAHs) and those hydrocarbons contributing to and/or produced from photochemical smog. These are the compounds chiefly considered here.

PAHs

Finlayson-Pitts & Pitts (1986) offer a superb introduction to this subject and surveys the literature up to 1985. PAHs are present in motor vehicle exhaust emissions (especially diesel; see Chapter 10), wood combustion emissions and all other sources of incompletely combusted organic matter — usually in particulate associated form (respirable ambient urban aerosols, combustion-generated respirable particles, POM — particulate organic matter) (DeFlora *et al.*, 1989). Most of the biological activity is in small particles (< 2 µm diameter), and is usually due to frameshift-type chemical mutagens. Larger particles are held in the nasal cavity and larynx, and smaller particles can enter the blood system via the lung or (in the ciliated region of the respiratory system) be carried by mucus to the gastrointestinal tract. Mostly, direct effects on DNA are observed (Georgellis, Parvinen & Rydstroem, 1989) though immunotoxic effects (Ginsberg, Atherholt & Butler, 1989; Ginsberg & Atherholt, 1990) and antioxidant induction (Chen & Shiau, 1989; Ha, Storkson & Pariza, 1990) have been reported. In urban environments, co-pollutants (ozone, NO_2, SO_2, nitric acid, PAN, etc.) are present. Conclusions from different researchers on specific effects can vary dramatically depending on source, sampling, the details of the bioassay and analytical techniques used, etc., and because of the lack of a complete range of pure reference substances.

The nitrogen analogues of PAHs (azo-arenes), benzacridines, dibenzacridines, nitro-PAHs (from reaction with NO_2/nitric acid) (Pegram & Chou, 1977) and oxy-PAHs (from reaction with ozone) occur. A degree of uncertainty exists concerning the exact health risks associated with PAHs, but, for instance, bioassays on diesel POM demonstrate genotoxic activity in every particle extract.

Benzo[*a*]pyrene is the archetypal PAH. First characterized in 1932, it is present in air pollution, cigarette smoke and food sources. It is metabolically activated by the cytochrome P-450-dependent mono-oxygenase system to form arene oxides which are hydrolysed (epoxide hydrolase) and then (via cytochrome P-450) metabolized to the ultimate carcinogen benzo[*a*]pyrene-7,8-transdihydrodiol-9,10-epoxide. Mono-oxygenase is induced by binding of benzo[*a*]pyrene to a cytostolic receptor protein after it has crossed the membrane, and this receptor complex stimulates transcription and enzyme production (Ayrton *et al.*, 1990).

For other PAHs the general pattern is the same. A good level of correlation exists between the detailed PAH structure and its mutagenic activity, and several structure−function relationships have been proposed (see, for instance, Klopman & Raychaudhury, 1990; Klopman, Frierson & Rosenkrantz, 1990). PAH metabolites are available for bioaccumulation (McElroy & Sisson, 1989; McCain *et al.*, 1990), and their activity is modulated by diet (O'Neill *et al.*, 1990). The specific details of their interactions with plants remain to be revealed.

Components of photochemical smog

Photochemical smog consists of primary pollutants emitted by motor vehicles (CO, NO, unburnt hydrocarbons: alkanes, alkenes, aromatics and some ketonic compounds). In the oxidizing milieu of photochemical smog, secondary pollutants (NO_2, ozone, more aldehydes and ketones, peroxyacyl nitrates and alkyl nitrates) are formed (Findlayson-Pitts & Pitts, 1986). The complex mix of chemicals making up photochemical smog makes predictions concerning its precise toxicological effects difficult. As well as forming toxic photoproducts, innocuous compounds may become toxic through synergistic effects with other chemicals.

The peroxyacyl nitrates are one group of toxic photoproducts produced from hydrocarbons, oxides of nitrogen, oxygen and UV light. Peroxyacetyl nitrate (PAN) is the best known of these compounds, and levels of 50 ppb have been detected in polluted air. The other members of the series (e.g. PPN: peroxypropionyl nitrate, PBN: peroxybutyl nitrate) become progressively more harmful as carbon chain lengths increase, but are present in lower concentration than PAN (e.g. 6 ppb PPN) (Mudd, 1976). As a group, these compounds make a large contribution to the eye irritation which is the most noticeable feature of photochemical smog (possibly by reaction with protein sulphydryl groups) and to the other effects (e.g. irritation of the mucous membrane and respiratory distress). As with NO_2, exposure of laboratory animals to high levels has resulted in death 2−3 weeks later. Biochemically the behaviour of the peroxyacyl nitrates parallels that of ozone (except that with PAN acetylation may occur), but plants must be illuminated during fumigation to produce a toxic response and (by an unrelated mechanism) photosynthesis is also inhibited. 'Bronzing' and 'glazing' of leaf surfaces (particularly younger plants) occur after several hours exposure to 0.02−0.05 ppm.

PAN may react with lipids to form epoxides, and is known to react with a large range of SH compounds (enzymes, glutathione, etc.), possibly forming thiyl radicals. It can thus inactivate enzymes such as glucose-6-phosphate dehydrogenase, isocitric dehydrogenase, malic dehydrogenase and also those enzymes that synthesize cellulose and lipids in plants. *In-vivo* studies also suggest that PAN interferes with normal indole metabolism in plant cells. NADH and NADPH are oxidized by PAN. Nucleic acids also react (Mudd, 1976).

Other hydrocarbons

A number of other volatile organic chemicals (VOC) are categorized as toxic air pollutants (TAP) or toxic air contaminants (TAC) (O'Neill *et al.*, 1990), e.g. benzene, vinyl chloride, ethylene dibromide, trichloroethane, acrolein, pesticides and herbicides. Most VOCs have lung absorption rates of 30–50% at high concentrations and short exposures (Harkov, 1987). The effect on living systems of ambient concentrations of VOCs and their contribution to the toxic products formed in the troposphere is multifactorial, and generally a detailed knowledge of the specific compounds involved is required (Dowd, 1985; Harkov, Kebbekus & Bozzelli, 1987). Research on the contribution of VOCs to the toxicity of tropospheric pollution is expanding to accommodate these requirements.

Other gases

1 *Carbon monoxide*: reacts with haemoglobin to produce carboxyhaemoglobin and hence symptoms are similar to those of reduced oxygen availability, *viz.* decreased awareness then (with higher exposures) drowsiness, coma, pulmonary and cardiac effects followed by death. Biochemically protein, nucleic acid and lipid synthesis are inhibited—anaerobic glycolosis causes pyruvic and lactic acids to increase, acidosis occurs and, amongst other effects, denaturation of proteins.
2 *Hydrogen fluoride*: although present in small amounts, fluoride is ranked fourth after ozone, SO_2 and nitrogen compounds in terms of economic damage to crops (Wellburn, 1988). Access to plants is via the stomata, and many metabolic processes are inhibited through (it is thought, but not clearly demonstrated) Ca/Mg/P−F interactions. In animals, acute poisoning (fluorosis)

results in immediate death, and chronic intoxication results in lethargy, imperfect calcification and wasting—sometimes also followed by death.
3 *Hydrogen sulphide*: at low levels irritation of the eyes and the respiratory tract occurs. At 1000 ppm respiratory failure occurs due to paralysis of the central nervous system (CNS). Effects are similar to those of cyanide. Hydrogen sulphide is slightly phytotoxic. Plants can be stimulated (30 ppb) or killed (300–3000 ppb) by continuous exposure. It can often occur with carbon disulphide and related compounds.

2.6 EXTERNAL FACTORS AFFECTING ECOTOXICOLOGY OF GASEOUS COMPOUNDS

Prediction of the biological effects of pollutants is made difficult by the changes in response that occur in different environments. The problem is illustrated by two examples arising from studies of the effects of air pollutants on plants. Low concentrations of NO_2 often stimulate plant growth, but when the same concentrations are applied in the presence of SO_2 they can be extremely toxic. Table 2.3 shows the results of an experiment lasting 140 days at near-ambient temperatures over winter, in which the grass *Phleum pratense* (Timothy) was exposed to clean air (the control), 68 ppb SO_2, 68 ppb NO_2 or 68 ppb of each gas (Ashenden & Williams, 1980).

It is clear that the effects of the two pollutants together could not have been predicted from

Table 2.3 Percentage reductions (relative to control) of growth parameters for *Phleum pratense* after being exposed for 140 days to atmospheres containing: (a) 0.068 ppm NO_2, (b) 0.068 ppm SO_2, (c) 0.068 ppm NO_2 + 0.068 ppm SO_2. After Ashenden & Williams (1980)

	NO_2	SO_2	$NO_2 + SO_2$	Effect
Number of tillers	6	33	55	
Number of leaves	10[a]	29	68	S
Leaf area	30[a]	11	82	S
Dry weight of green leaves	14[a]	25	84	S
Dry weight of dead leaves and 'stubble'	12	47	64	
Dry weight of roots	1[a]	58	92	

[a] Increase.
S, synergistic effects of pollutants (i.e. a statistically significant interaction).

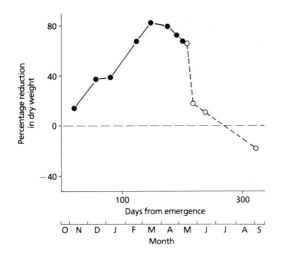

Fig. 2.2 Percentage reductions (compared with controls in clean air) in the growth of the grass *Poa pratensis* in 0.062 ppm SO_2 + 0.062 ppm NO_2. ● Represents harvest of the whole plant (i.e. root and shoot). After Whitmore & Freer-Smith (1982), reproduced by permission of Macmillan Magazines Ltd.

experimental fumigations involving SO_2 and NO_2 separately. This data set does not, however, permit us to come to simple conclusions about the effects of SO_2 and NO_2 on production of grasses. Further research showed that the deleterious effects of the two pollutants occurring during the winter rapidly disappeared in spring and early summer (Fig. 2.2). It is now known from studies in controlled environments that the effects of SO_2 and NO_2 on grasses are very dependent on environmental factors such as light intensity and temperature. It is believed that the damaging effects of the pollutants are associated with slower metabolism under winter conditions (Mansfield & McCure, 1988).

Information such as this clearly indicates that simplistic approaches to defining critical loads of air pollutants are not acceptable. Much more research will be necessary to establish dose–response relationships under a wide range of environmental conditions.

REFERENCES

Ashenden, T.W. & Williams, I.A.D. (1980) Growth reductions in *Lolium multiflorum* and *Phleum pratense* as a result of SO_2 and NO_2 pollution. *Environ. Pollut. A* **21**, 131–139.

Avol, E.L., Linn, W.S., Peng, R.C., Whyot, J.D., Shamoo, D.A., Little, D.E., Smith, M.N. & Hackney, J.D. (1989) Experimental exposures of young asthmatic volunteers to 0.3 ppm nitrogen dioxide and to ambient air pollution. *Toxicol. Ind. Health* **5**(6), 1025, 1–34.

Ayrton, A.D., McFarlane, M., Walker, R., Neville, S., Coombs, M.M. & Ioannides, C. (1990) A relationship appears to exist between carcinogenicity of PAH's and their ability to induce hepatic P450 I activity. *Toxicology* **60**(1–2), 173–186.

Bates, D.V. (1989) Ozone—myth and reality. *Environ. Res.* **50**(2), 230–237.

Bedi, J.F. & Horvath, S.M. (1989) Inhalation route effects on exposure to 2.0 ppm sulphur dioxide in normal subjects. *JAPCA* **39**(11), 1448–1452.

Bennett, J.H., Lee, E.H. & Heggestad, H.E. (1990) Inhibition of photosynthesis and leaf conductance interactions induced by sulphur dioxide, nitrogen dioxide and sulphur dioxide + nitrogen dioxide. *Atmos. Environ, A* **24A**(3), 557–562.

Boushey, H. (1989) Ozone and asthma. *ASTM Spec. Tech. Publ.* **1024**, 214–217.

Brookes, K.A., Adams, W.C. & Schelegle, E.S. (1989) 0.35 ppm ozone exposure induces hyperresponsiveness on 24-hour reexposure to 0.20 ppm ozone. *J. Appl. Physiol.* **66**(6), 2756–2762.

Burleson, G.R., Keye, L.L. & Stutzman, J.C. (1989) Immunosuppression of pulmonary natural killer activity by exposure to ozone. *Immunotoxicology* **11**(4), 715–735.

Chang, L.Y., Mercer, R.R., Stockstill, B.L., Miller, F.J., Graham, J.A., Ospital, J.J. & Grapo, J.D. (1988) Effects of low levels of nitrogen dioxide on terminal bronchiolar cells and its relative toxicity compared to ozone. *Appl. Pharmacol.* **96**(3), 451–464.

Chauhan, A. & Mehra, P. (1989) Effect of sulphur dioxide on leaf buffering capacity in some crop plants. *Indian J. Exp. Biol.* **27**(11), 992–995.

Chen, H.L. & Shiau, C.C.A. (1989) Induction of glutathione-s-transferase activity by antioxidants in hepatocyte culture. *Anticancer Res.* **9**(4), 1069–1072.

Collier, J. & Vallance, P. (1991) Physiological importance of nitric oxide. *BMJ* **30**, 1289–1300.

Costa, D.L., Stevens, M.S. & Tepper, J.S. (1988) Repeated exposure to ozone and chronic lung disease: recent animal data. *Proc. APCA Annu. Meet. 81st*(7), Paper 88/122.3.

Criegee, R. (1975) Mechanisms of oxonolysis. *Angew. Chem. Int. Edn.* **14**, 745–752.

DeFlora, S., Bagnasco, M., Izzotti, A., D'Agostini, F., Pala, M. & Valerio, F. (1989) Mutagenicity of polycyclic aromatic hydrocarbon fractions extracted from urban air particulates. *Mutat. Res.* **224**(2), 305–318.

Dowd, R.M. (1985) Proposed air toxics legislation. *Environ. Sci. Technol.* **19**, 580.

Engelbach, G. & Fangmeier, A. (1988) Effect of sulphur dioxide, nitrogen dioxide and ozone on the seasonal growth of ground vegetation in a melick-beech forest near Giessen, Germany. *Verh-Ges. Oekol.* **18**, 379–385.

Findlayson-Pitts, B.J. & Pitts, J.N. Jr (1986) *Atmospheric Chemistry*. John Wiley, New York.

Fowler, D., Duyzer, J.H. & Baldocchi, D.D. (1991) Inputs of trace gases, particles and cloud droplets to terrestrial surfaces. *Proc. R. Soc. Edinb.* **97B**, 35–59.

Freudenthal, P.C., Roth, D.H., Hammerstrom, T. & Lichtenstein, C. (1989) Health risks of short-term sulphur-dioxide exposure to exercising asthmatics. *JAPCA* **39**(6), 831–835.

Fujimaki, H. (1989) Impairment of humoral immune responses in mice exposed to nitrogen dioxide and ozone mixtures. *Environ. Res.* **48**(2), 211–217.

Georgellis, A., Parvinen, M. & Rydstroem, J. (1989) Inhibition of stage-specific DNA synthesis in rat spermatogenic cells by polycyclic aromatic hydrocarbons. *Chem-Biol. Interact.* **72**(1–2), 79–92.

Georsdorf, S., Appel, K.E., Engeholm, C. & Obe, G. (1990) Nitrogen dioxide induces DNA single-strand breaks in cultured Chinese hamster cells. *Carcinogenesis* **11**(1), 37–41.

Ginsberg, G. & Atherholt, T.B. (1990) DNA adduct formation in mouse tissue in relation to serum levels of benzo[a]pyrenediol epoxide after injection of benzo[a]pyrene or the diol epoxide. *Cancer Res.* **50**(4), 1189–1194.

Ginsberg, G.L., Atherholt, T.B. & Butler, G.H. (1989) Benzo[a]pyrene-induced immunotoxicity: comparison to DNA adduct formation *in vivo*, in cultured splenocytes, and in microsomal systems. *J. Toxicol. Environ. Health* **28**(2), 205–220.

Goldstein, B. (1979) The pulmonary and extrapulmonary effects of ozone. In: *Oxygen Free Radicals and Tissue Damage*, Ciba Foundation Symposium; new series 65, pp. 295–319. Excerpta Medica, Amsterdam.

Goodman, J.W., Peter-Fizaine, F.E., Shinpock, S.G., Hall, E.A. & Fahmie, D.J. (1989) Immunologic and hematologic consequences in mice of exposure to ozone. *J. Environ. Pathol. Toxicol. Oncol.* **9**(3), 243–252.

Guderian, R. (1985) *Air Pollution by Photochemical Oxidants*. Springer-Verlag, Berlin.

Ha, Y.L., Storkson, J. & Pariza, M.W. (1990) Inhibition of benzo[a]pyrene-induced mouse forestomach neoplasia by conjugated dienoic derivatives of linoleic acid. *Cancer Res.* **50**(4), 1097–1101.

Harkov, R. (1987) The New Jersey ATEOS program: an overview of its importance and health/regulatory implications. In: *Toxic Air Pollution: A Comprehensive Study of Non-Criteria Air Pollutants* (Eds P.J. Lioy & J.M. Daisey), pp. 251–282. Lewis, Michigan.

Harkov, R., Kebbekus, B. & Bozzelli, J.W. (1987) Volatile organic compounds at urban sites in New Jersey. In: *Toxic Air Pollution: A Comprehensive Study of Non-Criteria Air Pollutants* (Eds P.J. Lioy & J.M. Daisey), pp. 69–90. Lewis, Michigan.

Harrison, R.M. & Perry, R. (1986) *Handbook of Air Pollution Analysis*, 2nd edn. Chapman & Hall, London.

Hatch, G.E., Wiester, M.J., Overton, J.H. & Aissa, M. (1988) Respiratory tract dosimetry of oxygen-18-labeled ozone in rats: implications for a rat–human extrapolation of ozone dose. Report EPA/600/D-88/174.

Heagle, S.A. (1989) Ozone and crop yield. *Annu. Rev. Phytopathol.* **27**, 397–423.

Hewitt, C.N. (1991) *Instrumental Analysis of Pollutants*. Elsevier, London.

Hewitt, C.N. (1992) Spatial and temporal variations in urban NO_2 concentrations. *Atmos. Environ. B: Urban Atmos.* **25B**, 429–434.

Hewitt, C.N., Kok, G.L. & Fall, R. (1990) Hydroperoxides in plants exposed to ozone mediate air pollution damage to alkene emitters. *Nature* **344**, 56–58.

Hewitt, C.N., Lucas, P., Wellburn, A.R. & Fall, R. (1990) Chemistry of ozone damage to plants. *Chem. Ind.* 478–481.

HMSO (1991) *First Report of the Advisory Group on the Medical Aspects of Air Pollution Episodes, 'Ozone'*. HMSO, London.

Hornof, W.J., Schelege, E., Kammerman, M., Gunther, R.A., Fisher, P.E. & Cross, C.E. (1989) Ozone-induced accelerated lung clearance of Tc-DTPA aerosol in conscious sheep. *Respir. Physiol.* **77**(3), 277–290.

Horstman, D.H. & Folinsbee, L.J. (1989) Sulphur dioxide-induced bronchoconstriction in asthmatics exposed for short durations under controlled conditions: a selected review. *ASTM Spec. Tech. Publ.*, 1024 (*Susceptibility Inhaled Pollutants*), 195–206.

Ichinose, T. & Sagai, M. (1989) Biochemical effects of combined gases of nitrogen dioxide and ozone. III. Synergistic effects on lipid peroxidation and antioxidative systems in the lungs of rats and guinea pigs. *Toxicology* **59**(3), 259–270.

Klopman, G., Frierson, M.R. & Rosenkrantz, H.S. (1990) The structural basis of the mutagenicity of chemicals in *Salmonella typhimurium*; the Gene–Tox data base. *Mutat. Res.* **228**(1), 1–50.

Klopman, G. & Raychaudhury, C. (1990) Vertex indexes of molecular graphs in structure-activity relationships: a study of the convulsant-anticonvulsant activity of barbiturates and the carcinogenicity of unsubstituted polycyclic aromatic hydrocarbons. *J. Chem. Inf. Comput. Sci.* **30**(1), 12–19.

Kropff, M.J., Mooi, J., Goudrian, J., Smeets, W., Leemans, A. & Kliffen, C. (1989a) The effects of long-term open-air fumigation with sulphur dioxide on a field of broad bean (*Vicia faba* L.). II. Effects on growth components, leaf area development and elemental composition. *New Phytol.* **113**(3), 345–351.

Kropff, M.J., Mooi, J., Goudrian, J., Smeets, W., Leemans, A., Kliffen, C. & Van der Zalm, A.J.A. (1989b) The effects of long-term open-air fumigation with sulphur dioxide on a field of broad bean (*Vicia faba* L.). I. Depression of growth and yield. *New Phytol.* **113**(3), 337–344.

Lendzian, K.J. & Kerstiens, G. (1991) Sorption and transport of gases and vapors in plant articles. *Rev. Environ. Contam. Toxicol.* **121**, 65–128.

Lippmann, M. (1989) Health effects of ozone: a critical review. *JAPCA* **39**(5), 672–695.

Manahan, S.E. (1990) *Environmental Chemistry*, 4th edn., Lewis, Ann Arbor.

Mansfield, T.A. & Lucas, P.W. (1990) Effects of gaseous pollutants on crops and trees. In: *Pollution: Causes, Effects and Control*, 2nd edn (Ed. R.M. Harrison), pp. 237–259. Royal Society of Chemistry, London.

Mansfield, T.A. & McCure, D.C. (1988) Problems of crop loss assessment when there is exposure to two or more gaseous pollutants. In: *Assessment of Crop Loss from Air Pollutants* (Eds W.W. Heck, O.C. Taylor & D.T. Tingey), pp. 317–344. Elsevier, London.

Mariassy, A.T., Sielczak, M.W., McCray, M.N., Abraham, W.M. & Wanner, A. (1989) Effects of ozone on lamb tracheal mucosa. Quantitative glycoconjugate histochemistry. *Am. J. Pathol.* **135**(5), 871–879.

McCain, B.B., Malins, D.C., Krahn, M.M., Brown, D.W., Gronlund, W.D., Moore, L.K. & Chan, S.L. (1990) Uptake of aromatic and chlorinated hydrocarbons by juvenile chinook salmon (*Oncorhynchus tshawytscha*) in an urban estuary. *Arch. Environ. Contam. Toxicol.* **19**(1), 10–16.

McDonnell, W.F. (1988) Do functional changes in humans correlate with the airway removal efficiency of ozone. Report EPA/600/D-88/165.

McElroy, A.E. & Sisson, J.D. (1989) Trophic transfer of benzo[*a*]pyrene metabolite between benthic marine organisms. *Mar. Environ. Res.* **28**(1–4), 265–269.

Mehlhorn, H. & Wellburn, A.R. (1987) Stress ethylene formation determines plant sensitivity to ozone. *Nature* **327**, 417–418.

Mudd, J.B. (1976) The role of free radicals in toxicity of air pollutants (peroxyacyl nitrates). In: *Free Radicals in Biology*, Vol. II, (Ed. W.A. Pryor), pp. 203–211. Academic Press, London.

Olszyk, D.M. (1988) Documentation of ozone as the primary phytotoxic agent in photochemical smog. Report ARB-R-88/349, US EPA.

O'Neill, I.K., Povey, A.C., Bingham, S. & Cardis, E. (1990) Systematic modulation by human diet levels of dietary fibre and beef on metabolism and disposition of benzo[*a*]pyrene in the gastrointestinal tract of Fischer F344 rats. *Carcinogenesis* **11**(4), 609–616.

Orlando, G.S., House, D., Daniel, E.G., Koren, H.S. & Becker, S. (1988) Effect of ozone on T-cell proliferation and serum levels of cortisol and beta-endorphin in exercising males. *Inhal. Toxicol.* (premier issue), 53–63.

Ozkaynak, H., Burbank, B., Kinney, P.L. & Thurston, G.D. (1988) Review of epidemiological studies on ozone exposures. *Proc. APCA Annu. Meet. 81st*(7), Paper 88/122.7.

Pegram, R.A. & Chou, M.W. (1977) Effect of nitro-substitution of environmental polycyclic aromatic hydrocarbons on activities of hepatic phase II enzymes in rats. *Drug Chem. Toxicol.* **12**(3–4), 313–326.

Postlethwaite, E. & Bidani, A. (1990) Reactive uptake governs the pulmonary air space removal of inhaled nitrogen dioxide. *J. Appl. Physiol.* **68**(2), 594–603.

Richter, A. (1988) Effects of nitrogen dioxide and ozone on blood-borne cancer cell colonization of the lungs. *J. Toxicol. Environ. Health* **25**(3), 383–390.

Riedel, F. (1989) Animal experiments on the relationship between bronchial damage by irritant gases and bronchial sensitization. *Allergologie* **12**(3), 112–113.

Schlesinger, R.B. (1989) Comparative toxicity of ambient air pollutants: some aspects related to lung defense. *Environ. Health Perspect.* **81**, 123–128.

Schulte-Hostede, S., Darrall, N.M., Black, L.W. & Wellburn, A.R. (Eds) (1988) *Air Pollution and Plant Metabolism*. Elsevier, London.

Seinfeld, J.H. (1986) *Atmospheric Chemistry and Physics of Air Pollution*. Wiley, New York.

Singh, J. (1989) Neonatal development altered by maternal sulphur dioxide exposure. *Neurotoxicology* **10**(3), 523–527.

Stavert, D.M. & Lehnert, B.E. (1990) Nitric oxide and nitrogen dioxide as inducers of acute pulmonary injury when inhaled at relatively high concentrations for brief period. *Inhal. Toxicol.* **2**(1), 53–67.

Tilton, B.E. (1989) Health effects of tropospheric ozone. *Environ. Sci. Technol.* **23**(3), 257–263.

US National Academy of Sciences Committee on Medical and Biological Effects of Environmental Pollutants (1977) *Ozone and Other Photochemical Oxidants*. Washington, DC.

Warneck, P. (1988) *Chemistry of Natural Atmosphere*. Academic Press, San Diego, CA.

Warrington, S. (1987) Relationship between SO$_2$ dose and growth of the pea aphid, *Acyrthosiphon pisum*, on peas. *Environ. Pollut.* **43**, 155–162.

Wellburn, A. (1988) *Air Pollution and Acid Rain — the Biological Input*. Longman, London.

Wellburn, A.R. (1990) Why are atmospheric oxides of nitrogen usually phytotoxic and not alternative fertilizers? *New Phytol.* **115**, 395–429.

Whitmore, M.E. & Freer-Smith, P.H. (1982) Growth effects of SO$_2$ and/or NO$_2$ on woody plants and grasses during winter and summer. *Nature* **300**, 55–57.

Witschi, H. & Sagai, M. (1989) Ozone, nitrogen dioxide and lung cancer: A review of some recent issues and problems. *Takai Osen Gakkaishi* **24**(1), 1–20.

Wright, E.S., Daniel, D. & Wheeler, C.S. (1990) Cellular, biochemical and functional effects of ozone: new research and perspectives on ozone health effects. *Toxicol. Lett.* **51**(2), 125–145.

Yadav, N.K., Chand, S. & Singh, V. (1988) Growth stress — reductions in growth, chlorophyll content and productivity. *Biovigyanam* **14**(2), 89–94.

3: Acid Solutions

C.P. CUMMINS

3.1 INTRODUCTION

This chapter is concerned with solutions in which the hydrogen ion (H^+) is the predominant determinant of ecotoxicity, either as a toxicant in its own right or through its effects on the speciation, mobility and bioavailability of other toxicants such as Al, Cd, Hg and Pb.

Examples of acid solutions of possible ecotoxicological interest include rain, snow, mist and cloud droplets, acid mine drainage and acid industrial effluents, soil water, and a wide range of surface waters (pools, streams, lakes and rivers). In this chapter I shall pay particular attention to studies of freshwaters sensitive to acidification by atmospheric pollutants.

The range of studies of organisms and ecological processes in acid waters is enormous, and it is both impractical to attempt to summarize them all and invidious to select examples. Therefore, my aim is to identify ecotoxicological principles and suggest methods appropriate to studies in acid waters. Where examples are cited, these should be regarded not necessarily as model studies, although some may be, but rather as points of reference in the literature from which a greater range of related work may be identified.

It has long been recognized that the acid status of surface waters is not uniform, and that the pattern of presence and absence of certain species can be accounted for by acid status. Waters can become acidified by a variety of means. Many of the processes are 'natural', in the sense that they are features of biological and biogeochemical processes, and take place in the absence of human interference. Nevertheless, human activities have often increased the rate and the scale on which these processes impinge on the environment. While there is a legitimate scientific rationale for enquiring into the evolution of acid tolerance and the role of acid conditions in limiting the distribution and abundance of organisms, it is largely in the field of investigation of impacts of anthropogenic acidification that ecotoxicology has evolved. As progress has been made, the nature of the questions being addressed has changed, from 'has there been a change?', through 'what has changed?', 'is the change due to acid rain?' and 'what can we do to reverse the change?' to 'if we do this, what will be the result?' and 'what must we do to get back to that position?'. Accordingly, the burden of responsibility on ecotoxicologists has shifted from demonstrating or quantifying effects of acid conditions to making predictions about the ecological consequences of changing acid deposition.

3.2 CAUSES OF ACIDIFICATION OF FRESHWATERS

3.2.1 Atmospheric pollution and acid deposition

Research on the acidification of lakes and rivers in Scandinavia and North America has bourgeoned since the 1970s (e.g. Almer et al., 1974; Seip & Tollan, 1978, Drabløs & Tollan, 1980; Overrein et al., 1980; Haines, 1981; Renberg & Hellberg, 1982). Areas where surface waters are at risk are those where the rate of input or generation of acids outstrips the neutralizing capacity of the matrix or substrate with which the water is, or has been, in contact. These tend to be areas of high relief, high precipitation, base-poor soils and slowly weathering bedrock. The water chemistry in such areas may be influenced substantially by snowmelt and precipitation.

The most familiar anthropogenic contributors to large-scale surface water acidification are atmospheric pollutants derived from the burning of fossil fuels. Smelters are also notorious sources of SO_2, and their emissions may also contain a variety of toxic heavy metals. Airborne acids and acid precursors may be transported over large distances before being deposited in dry or wet form. Historically, SO_2 has been the principal acidifying pollutant, but oxides of nitrogen (NO_x) have increased in importance in recent years. It has also become apparent that ammonia, of which much derives from domestic livestock, can contribute significantly to soil acidification. For a brief summary of emissions, transport and deposition of major acidifying pollutants see Irwin & Williams (1988).

Early evidence of recent, large-scale acidification was derived from historical records of pH measurements (see Haines, 1981, for a review). The reliability of such evidence has been debated, however, as methods for measuring pH have changed over the period of concern (see Howells, 1982; Haines, 1982). More recently, several approaches have been applied to investigations of the acid status of lakes and rivers.

Regional correlations have been found between sulphate deposition and sulphate in surface waters in Scandinavia (Henriksen *et al.*, 1988) and the eastern USA (Sullivan *et al.*, 1988). Kaufmann *et al.* (1992) recognized three categories of acid waters sampled in the US National Surface Water Survey, according to their probable sources of acidity: (i) organic-dominated (organic anions $> SO_4^{2-} + NO_3^-$); (ii) watershed SO_4^{2-}-dominated (watershed $SO_4^{2-} >$ deposited SO_4^{2-}); (iii) deposition-dominated (anion chemistry dominated by deposited SO_4^{2-} and NO_3^-). Kaufmann *et al.* (1992) estimated that 75% of acid lakes and 47% of acid streams in the eastern USA were deposition-dominated, and were probably acid because of acid deposition; about a quarter of acid lakes and streams were organic-dominated; in the 26% of acid streams where watershed SO_4^{2-} predominated, acid mine drainage was a major source.

Examination of fossil diatoms from dated sections of lake sediment cores, and knowledge of current associations between lake pH and species composition of the diatom flora, has allowed pH-histories of lakes to be reconstructed (Renberg & Hellberg, 1982; Battarbee *et al.*, 1990). For example, a series of palaeoecological studies has shown that several lochs in Galloway, southwestern Scotland, have become acidified since the early or middle nineteenth century, and that acid precipitation is the most likely cause of this acidification (Flower & Battarbee, 1983; Battarbee *et al.*, 1985; Jones *et al.*, 1986).

An important line of evidence that anthropogenic sulphur emissions have contributed significantly to surface water acidification lies in the recovery of lakes as emissions have declined. The best-known example is that of lakes close to the Sudbury smelter in Ontario (Gunn & Keller, 1990). However, Skeffington & Brown (1992) considered that the Sudbury lakes became acidified only because of severe acid loading from local point sources, and that their catchment geology provides a much greater capacity for recovery than that of lakes in southern Norway or on the Canadian Precambrian Shield. Recovery in the Sudbury lakes may not provide a model for recovery in more acid-sensitive waters. Dillon & LaZerte (1992) found that in the catchment of Plastic Lake, on the Precambrian Shield, the quality of upland runoff responded rapidly to a reduction in sulphur deposition, but recovery in Plastic Lake was delayed, apparently by reoxidation of stored sulphur in wetlands in the catchment. Nevertheless, Allott *et al.* (1992) reported a rise in pH from 4.7 to 4.9 in a lake in Galloway between 1978 and 1989 — a period of sharply declining sulphur emissions. Diatom remains in sediments also indicated a rise in pH in Galloway lakes (Battarbee *et al.*, 1988; Allott *et al.*, 1992).

3.2.2 Acid mine drainage

Probably the most severe instances of anthropogenic surface water acidification are due to the oxidation of pyrite (Fe_2S) and other sulphides exposed by mining activities (Kelly, 1988). The chemical pathway by which H^+ is generated from sulphides has been described by Singer & Stumm (1970); the oxidation process is catalysed by bacteria which use pyrite as a source of energy, and the rate of oxidation may be increased thus by a factor of 10^6. The extremely acid conditions generated at the site of oxidation lead to rapid dissolution of minerals, and very high conductivities can be attained. In this respect, the chemical characteristics of acid mine drainage are generally very different from those of waters whose acidity derives from natural organic acids or waters acidified by atmospheric pollutants (Kelly, 1988). Mining activities often focus on ores rich in heavy metals such as Pb, Cu and Zn, and mobilization of such toxicants can cause severe and long-lasting environmental damage, at least on a local scale. Inevitably, acid mine drainage is neutralized downstream of the source; as this happens, flocculation and precipitation, especially of iron hydroxides, may produce a substrate inimical to many benthic plants and animals. Potentially toxic metals may be remobilized from such deposits if there is a subsequent drop in pH, as may occur during periods of high flow.

3.2.3 Biological acidification

Sphagnum has a capacity for exchanging H^+ for other cations in solution, thereby acidifying the surrounding water, and pH values of 3.2−4.0 are common in ombrotrophic sphagnaceous bogs. It has been estimated that the contribution of such cation exchange to total acidity could be on a par with that deriving from wet deposition of airborne acid where the measured pH of rain was 4.0−4.2 (Clymo, 1984).

Afforestation has been implicated as a factor in the acidification of soils, and hence surface waters (Nilsson *et al.*, 1982; Reynolds *et al.*, 1988; Ormerod *et al.*, 1989): trees may scavenge pollutants from the air, remove base cations from the soil and exacerbate nitrification (Johnson *et al.*, 1991). Precipitation emerging as throughfall and stemflow may be affected by leaching of organic acids or deposited strong acids, and by ion exchange. The influence of trees on soil nutrient dynamics is complex (Reynolds *et al.*, 1988), and varies according to the species (broadly, coniferous or deciduous), season and many other interacting factors (Schulze & Freer-Smith, 1991). Although soil acidification may result from the

growth and harvesting of trees (Hornbeck, 1992), transport of H^+ from forest soils to surface waters depends on the availability of mobile anions, such as organic anions or SO_4^{2-}. Further changes in the aluminium chemistry of soil and stream water may follow clearfelling (Reynolds *et al.*, 1992).

Where peatlands have been afforested, disturbances of the soil/vegetation and changes in patterns of drainage associated with planting and cultivation may have had a significant influence on the acid status of surface waters (Waters & Jenkins, 1992).

Thus, atmospheric pollution and land management have combined to exacerbate surface water acidification.

3.3 ECOTOXICOLOGICAL STUDIES IN ACID WATERS

3.3.1 Chemical analysis of acid waters

In ecotoxicological studies, chemical analysis of waters serves two purposes: one is to define the organisms' proximate chemical environment, both in the field and in experimental systems; the other is to characterize water bodies in the field, thereby allowing some assessment of the likely nature and duration of variations in the chemical environment.

In the former case, one is concerned principally with the concentrations of: (i) hydrogen ions (H^+); (ii) potentially toxic metals such as Al, Pb, Cd, etc., (iii) ligands which may reduce the bioavailability of toxic metals, such as dissolved organic matter (DOM); (iv) major cations such as Na^+, K^+, Ca^{2+}, Mg^{2+}; (v) organic acids which may be toxic in their own right; (vi) any other aspect of water quality which may influence metal speciation, e.g. alkalinity.

Two aspects of the chemistry of acid freshwaters require special mention. First, the great majority of surface waters that are naturally acid or which have been acidified by human activities other than mining or point discharge of acid effluents are of low ionic strength, and this presents special difficulties in measuring pH (Davison, 1990). Second, the speciation of several potentially toxic metals is pH-dependent within the range of pH occurring in the field.

3.3.2 Measurement of pH

The pH of a solution is a measure of the activity, or effective concentration, of H^+ therein; it is temperature dependent. The principles underlying the measurement of pH are described in detail by Galster (1991) and Midgley & Torrance (1991). For most practical purposes, pH should be measured electrochemically, i.e. with a pH meter and glass electrode. Some suitable pH electrodes are described by Davison (1990), who recommends avoiding gel-filled or permanently sealed designs. Electrodes of a particular design may vary in their performance in low-conductivity waters; criteria for accepting an electrode for such use are given by Davison & Harbinson (1988).

Several precautions can be taken to improve accuracy and precision of measurement in waters of low conductivity (Davison, 1990). Electrodes with a high liquid junction potential should be avoided: this was found to be the major source of error in interlaboratory comparisons (Koch *et al.*, 1986), the error increasing with increasing difference between the ionic strengths of the calibration and sample solutions. Proprietary calibration buffer solutions are typically of significantly greater ionic strength than acid natural waters, and it may be worthwhile calibrating electrodes against standard dilute solutions, such as dilute strong acid or reference solutions designed to mimic rainwater (Galloway *et al.*, 1979; Covington *et al.*, 1983; Koch *et al.*, 1986). Galster (1991) suggests that precision of measurement may be increased by adding equal amounts of KCl to the sample, and by diluting standard solutions used for calibration (but see Galloway *et al.*, 1979).

Calibration and measurement should, as far as possible, be carried out at the same temperature; failing this, temperature correction should be applied (Midgley & Torrance, 1991). Large or sudden changes in temperature during measurement should be avoided.

Stirring often produces a lower pH reading: this effect is thought to be due largely to loss of reference electrolyte from the boundary layer outside the electrode, and consequent expression of the junction potential (Galster, 1991). A recommended procedure is to stir briefly, speeding equilibration of the electrode with the sample, then make the reading when the sample has become quiescent.

When considering the reliability of pH measurements made in the field and in the laboratory one must weigh the greater precision likely to be achievable in the laboratory against the risk of errors arising through change in the sample during transportation and storage. As a rule, it is advisable to take measurements in the field whenever possible, whether or not measurements are to be made in the laboratory. Interlaboratory standard deviations (SDs) for river waters (pH 4.8–5.6, specific conductivity $\simeq 35\,\mu S/cm$) measured in the laboratory were 0.05 pH, whereas SDs of measurements on river waters and dilute acids in the field were 0.1–0.5 pH (Davison & Gardner, 1986; Davison & Woof, 1985). Temporal and spatial variations in pH of natural waters can be of a similar magnitude, arising from variations in biological activity, particularly photosynthesis, and changes in the pattern and magnitude of drainage. Water draining from soil may have an elevated CO_2 content, and degassing on equilibration with the atmosphere may lead to a rise in pH. As polyethylene may be permeable to CO_2, borosilicate glass (not soda glass) is recommended if such water samples are to be collected for pH determination later. However, for acid surface waters with pH < 5.6, the influence of CO_2 may be sufficiently small as to permit the use of polyethylene bottles (Davison, 1990). Bottles should be rinsed, filled, then capped under water. If samples must be stored for subsequent chemical analysis, this should be in the dark at 4°C, and pH should be checked periodically.

An important part of most ecotoxicological studies is the comparison of observations made in experiments under controlled conditions with observations made in the field. Therefore, in spite of the practical limitations on accuracy of measurement of pH under field conditions, there is a case for using the same equipment in both field and laboratory components of a study. This need not preclude more accurate laboratory measurements, and as long as laboratory media are chemically similar to waters in the field, confidence in the applicability of laboratory results should be enhanced.

3.4 MEASUREMENT OF METALS

Two major approaches have been taken in estimating the concentrations of metal species in natural waters (Campbell & Tessier, 1987a,b). One approach is to calculate equilibrium concentrations of species, using published values of stability constants, measured concentrations of metals and ligands, pH, temperature, etc. A widely used example is the MINEQL-1 model of Westall *et al.* (1976). Campbell & Stokes (1985) have noted the sensitivity of such models to the values of stability constants and the paucity of data for many chemical species involving organic matter. Another major concern over the applicability of this approach to field studies of toxicity is its assumption of equilibrium conditions — transient conditions, particularly those associated with changes in flow (e.g. during snowmelt or heavy rainfall), may be critical.

The other approach to measuring chemical speciation in natural waters is to partition the metal of interest into fractions defined operationally by their physical properties (e.g. size, partitioned by filtration or dialysis) or chemical reactivity, as determined by ion-exchange (Driscoll, 1984), potentiometry (Midgley & Torrance, 1991) or polarography (O'Shea & Mancy, 1978).

The very low aqueous concentration of Cd, Pb and Hg in most acidified soft waters presents difficulties of analysis, and demands the use of trace metal-free methods of sampling (Coale & Flegal, 1989; Fitzgerald & Watras, 1989). Nelson & Campbell (1991) suggest that concentrations of these metals reported in earlier studies which did not employ such methods should be regarded as suspect.

Al and Cd are both mobilized in acid soil water, and are exported readily from acidified mineral soils to surface waters; Pb and Hg are associated closely with organic matter, and their mobility is limited accordingly — see LaZerte (1986) and Nelson & Campbell (1991) for reviews of effects of acidification on the geochemistry of Al, Cd, Pb and Hg. Whereas most metals bind less readily

with organic matter as pH decreases, Hg shows an opposite tendency, as demonstrated in acidification experiments in Canada (Schindler *et al.*, 1980; Jackson *et al.*, 1980).

The speciation of heavy metals and problems of measurement are discussed elsewhere in this volume (Chapter 4). However, as Al is recognized as the principal metal toxicant in freshwaters influenced by atmospheric acid deposition, some further discussion of this metal is warranted.

3.5 MEASUREMENT OF ALUMINIUM

Aluminium is an important potential toxicant in acid soil and surface waters, but it can also ameliorate the toxic effects of H^+ under certain conditions. Al is ubiquitous in mineral soils; its solubility and speciation are pH-dependent, and the bioavailability of its various forms is influenced further by a tendency to sorb onto particles and bind to organic matter.

Both direct analysis and calculation of equilibrium partitioning have been used in studies of the toxic effects of Al. However, the assumption of equilibrium conditions may be unrealistic in many cases of toxicological interest. Indeed, much attention has focused on mortality among fish in conditions where disequilibrium exists (Rosseland *et al.*, 1992).

The most widely used methods for partitioning Al fractions are based on that described by Driscoll (1984). In this method, the sample is passed through an ion-exchange column which retains 'labile' aluminium. The Al content of the column effluent (Al_{org}) is measured and compared with the Al content of the original sample (Al_{tot}). The concentration of labile, or 'inorganic', Al is then obtained by subtraction:

$$Al_{inorg} = Al_{tot} - Al_{org}$$

The column needs to be conditioned to a pH close to that of the sample before use, in order to avoid changes in speciation during passage of the sample. The colorimetric method of Dougan & Wilson (1974) has been modified in various respects for measuring $[Al_{tot}]$ and $[Al_{org}]$ (e.g. Seip *et al.*, 1984; Bull & Hall, 1986; Goenaga & Williams, 1988; Tipping *et al.*, 1989).

Further references on speciation and the

measurement of aluminium include LaZerte (1984); Cronan *et al.* (1986); Goenaga *et al.* (1987); Lovgren *et al.* (1987); Schecher & Driscoll (1987); Tipping *et al.* (1988); Bertsch & Anderson (1989); for a recent summary see Nelson & Campbell (1991).

3.6 DESIGN OF EXPERIMENTS

Various approaches have been used in the investigation of effects of acidification on animals and plants. A typical sequence might be:

1 (a) observation of change in abundance, or
   (b) *a priori* expectation of change in abundance;
2 field survey to identify correlations between acid status of environment and status of species of concern;
3 laboratory tests of exposure−response relationships;
4 field experiments:
   (a) introduction or translocation into existing acid waters, or
   (b) experimental acidification and/or neutralization.

Progression from stage 1 to stage 4 may not be linear−some stages may run concurrently, and there may be a process of iteration between stages 3 and 4, for example, as hypotheses of cause and effect are tested and refined. While it is these latter stages that are truly ecotoxicological, it is important to recognize the value and the limitations of information from field observations and surveys.

The aim of a study may be to explain past events, assess current hazard or risk to a population or system, or predict the ecological consequences of a projected change in acid status. In each case it is important to establish the nature, duration and intensity of organisms' exposure to toxicants associated with acid conditions. For example, the pH and ionic composition of surface waters susceptible to acidification by atmospheric pollutants may exhibit considerable temporal and spatial change on a variety of scales. These range from diurnal fluctuations associated with photosynthetic uptake of CO_2 to seasonal changes reflecting broader climatic effects on catchment dynamics and chemistry, e.g. spring snowmelt

(Hagen & Langeland, 1973; Haapala *et al.*, 1975; Jeffries *et al.*, 1979; Hendershot *et al.*, 1986). Superimposed on these more-or-less cyclical changes are irregular events such as rainstorms (Goenaga & Williams, 1988; Kahl *et al.*, 1992; Wigington *et al.*, 1992). Spatial variations in H^+ concentration $[H^+]$ may reflect the local distribution of photosynthesising vegetation, heterogeneity of the substratum, or patterns of drainage − surface runoff, lateral flow through surface soil horizons, or baseflow (Bull & Hall, 1986; Reynolds *et al.*, 1986; Sullivan *et al.*, 1986; Soulsby & Reynolds, 1993).

Such variations in the environment need to be viewed in the context of the life history and behaviour of the organisms of concern. For example, the breeding seasons of many fish and amphibians coincide with seasonal depressions of pH. Morever, specific spawning events may be triggered by heavy rain, which may have significant short-term effects on pH and other aspects of water chemistry, particularly the concentration of toxic aluminium species. The most immobile and sensitive life stages, eggs and embryos, may thus be exposed to the most acid conditions.

A toxicant may exhibit more than one mechanism of toxicity, and susceptibility to each may vary from one life stage to another. For example, embryonic development is arrested in many amphibian species during chronic exposure to a pH of about 4.2 or less (reviewed by Pierce, 1985 and Freda, 1986); at higher levels of pH, mortality may still occur due to the failure of embryos to hatch, either because of a change in the vitelline membranes or because the hatching enzyme is ineffective at low pH (Dunson & Connell, 1982). Several reports of laboratory studies have not distinguished between the contributions of these two mechanisms to overall prehatching mortality. As the two mechanisms of toxic effect are differentially sensitive to the concentration of calcium in solution (Freda & Dunson, 1985a), and a wide variety of media have been used in experiments, the applicability of such experimental results to either the prediction of field responses or comparisons of sensitivity among populations or species is questionable.

Given that organisms may be exposed to various potential toxicants at different times, in varying concentrations and in many different combinations, it is clearly important to tailor experimental studies to specific objectives. Studies of mechanisms of toxic action may best be carried out under tightly controlled conditions in the laboratory, whereas impacts of acid conditions at population level may require field studies which allow a more or less full range of biotic and abiotic interactions. Aspects of experimental design to be considered include duration, concentrations of toxicants, stability of conditions, system characteristics (static, semistatic, flowthrough, mesocosm or field manipulation) and characteristics of the test organisms (age, life stage, size). These should be selected with reference to circumstances in the field, such as expected exposure to chronic or episodic acid conditions.

3.7 EXPERIMENTAL METHODS

3.7.1 Laboratory experiments

Experimental media

How closely should experimental conditions correspond to conditions observed in the field? Clearly, field conditions vary, and it is impossible to conduct toxicity tests under all possible conditions. We therefore need to discriminate between environmental variables of greater and lesser significance. For example, the sensitivity of most aquatic animals to low pH is highly dependent on the ambient Ca concentration. It would therefore be inappropriate to use a medium based on hard tapwater with a [Ca] of (say) 40 mg/l if the interest were in effects of low pH on survival in acidified lakes with a [Ca] < 5 mg/l. In contrast, although it is known that mortality of fish due to Al toxicity is reduced in waters rich in dissolved organic matter, it may not be necessary to include organic matter in media for testing the sensitivity of fish to Al. The essential difference is that calcium modifies the sensitivity of the test organism, whereas organic matter modifies the bioavailability of the toxicant (in this case, Al). As long as the conditions in the laboratory can be related to those in the field *as perceived by the organism*, then there is scope for simplifying

experimental conditions without compromising the applicability of the results. Nevertheless, experimenters should remain aware of the possibility of interactive effects of toxicants (e.g. Rueter *et al.*, 1987). This applies particularly in the case of organic acids, which may be toxic in their own right but at the same time reduce the availability of other toxicants (Petersen & Persson, 1987; Freda *et al.*, 1990).

The selection of acid(s) for use in acidifying test solutions may take account of field conditions. For example, sulphuric acid would be appropriate in studies of acid mine drainage, which is typically very high in sulphate from the oxidation of sulphides. The contribution of NO_x to atmospheric acid deposition is increasing relative to that of SO_2 in many developed countries, and studies which incorporate or address ion-exchange processes in plant canopies and soil should allow for the status of N as a plant nutrient. In experiments with simulated rainwaters, Smith *et al.* (1984) adjusted pH with sulphuric, nitric and hydrochloric acids to mimic the reported ionic composition of New Hampshire rain. However, in spite of the increasing role of NO_x in the process leading to soil and surface water acidification, sulphate remains the predominant mobile anion in surface waters (Wright & Haughs, 1991). Therefore, sulphuric acid will be an appropriate choice in most direct, experimental acidifications of solutions, be they laboratory media, mesocosms, streams or lakes.

Exposure systems

Laboratory exposure systems may be categorized as static, semistatic or flowthrough (see Volume 1). The terminology reflects the dynamics of adjustment of the experimental medium, not just the flow. A static system is one in which no adjustment of the medium is made during the period of exposure. In a semistatic system the medium is adjusted or replaced at intervals during the experiment. In a flowthrough system the medium is renewed or replaced continuously, necessitating a flow of water. In all cases the aim is to achieve a known, and usually constant, level of exposure throughout the experiment.

Whatever the type of system, it is essential to

measure the concentrations of toxicants and other aspects of the environment that may influence their effects on the test organisms.

In waters of low ionic strength, an upward drift of pH is often observed in static systems. Uptake of H^+ by test organisms may have a significant effect on pH of the medium, particularly if the volume of water is relatively small. Exposure to other potential toxicants whose bioavailability is pH-dependent may then deviate from that intended. For example, the availability of labile aluminium may change, owing to changes in speciation with changing pH, uptake by test organisms, or sorption onto the test chamber, food, excreta, etc. This characteristic of static and semistatic systems limits their usefulness for certain types of ecotoxicological study, particularly where investigation of chronic and sublethal effects requires long periods of exposure. Nevertheless, there are undoubtedly circumstances in which the chemical conditions in an organism's microenvironment differ greatly from those in the water column, where most measurements of 'field conditions' are made (Table 3.1), and a static system may allow such microenvironments to develop.

Semistatic systems have been used where a static system has proved too unstable over the required period of exposure, and where technical difficulties or specific requirements of the test organisms have rendered a flowthrough system impractical or unsuitable. Almost inevitably, the periodic adjustment or replacement of the medium results in a series of gradual changes in water quality, punctuated by stepwise changes back to some standard condition. It can then be difficult to determine the contributions of chronic and acute components of exposure to the overall effect.

For large and active organisms such as fish it is almost essential to use a flowthrough system to maintain reasonable control over exposure. If a suitable source of water is available, it may be appropriate to modify this before feeding it into the system (e.g. by acidifying, neutralizing or adding Al salts), then run the outflow to waste. Consideration should be given to the likelihood of change in Al speciation, and it may be appropriate to allow a period for chemical equilibration.

Table 3.1 Field measurements of pH within a single clump of spawn of the common frog, *Rana temporaria*, and in the surrounding water at a site near Cairnsmore of Fleet, Scotland (Cummins & Ross, 1986). The spawn was laid on 20 April. Note the variation in exposure to low pH within a single clump of spawn (typically 600–1000 eggs), and the potential for overestimating risk from episodic depressions of pH on the basis of results from laboratory experiments using small batches of embryos[*]. pH was persistently < 4.0 at this site, and 99% of embryos in this spawn clump were dead by 3 May.

Site no.	Date (1986)	pH		
		Water	Spawn edge	Spawn centre
12	25 April	3.43	3.83	4.36
	26 April	3.61	3.82	4.24
	27 April	3.67	3.98	4.30
	28 April	3.79	3.89	4.00

[*] Mortality of *Rana temporaria* embryos in laboratory tests of H^+-toxicity in soft water (no Al added): 95% to hatch at pH 4.0, 8% to hatch at pH 4.5 (2.6 mg Ca/l)[a]; 100% in ≤ 120 h at pH 4.0[b]; 100% in 48 h at pH 3.75, 0% in 48 h at pH ≥ 4.25 (0.5–4.0 mg Ca/l)[c]; 0–100% in 90 h at pH 3.8 ([Ca]-dependent), 0% in 90 h at pH ≥ 4.25 (0.5–4.0 mg Ca/l)[c]; < 10% to hatch at pH 4.5 (1.6 mg Ca/l)[d].
References: [a] Leuven *et al.* (1986); [b] Gebhardt *et al.* (1987); [c] Cummins (1988); [d] Tyler-Jones *et al.* (1989).

Raw domestic water supplies are often too hard to form the basis of a realistic experimental medium; they can be pretreated by filtering, reverse osmosis, resin-bed ion exchange, etc. before realistic concentrations of important ions are restored. The cost of such treatment can be significant, and it may be economically expedient to recirculate used water, albeit with further treatment to remove waste products or other chemical mediators of stress. If control of pH is to be automated, it is worth noting that a pH electrode will give a lower reading in a moving solution than in a static one of identical composition. In solutions of low ionic strength there may also be a significant loss of electrolyte during the course of an experiment, leading to errors of measurement.

Examples of flowthrough systems used for

exposing various life stages of fish and amphibians are described by Brown & Lynam (1981), Leuven *et al.* (1986) and Tyler-Jones *et al.* (1989).

Temperature

Severity of effect is often a function of duration of exposure, as well as intensity. Temperature may therefore be of great importance because of its influence on rates of development and growth, and hence the duration of more-or-less-sensitive life stages, such as embryos or small larvae or the frequency of moult in invertebrates.

Selection of organisms

Important considerations include not only the selection of species, which clearly should correspond to those at risk in the field, but also the source of material, its condition (life stage, age, size) and its history. For example, differences in acid tolerance between populations have been reported in brown trout, *Salmo trutta* (McWilliams, 1982) and brook trout, *Salvelinus fontinalis* (Robinson *et al.*, 1976), wild fish from acid waters being more tolerant of acid conditions. There is some evidence that fish can acclimatize to Al in acid water (Wood *et al.*, 1988), but generally prior exposure to acid conditions does not enhance resistance to further acid stress (Audet & Wood, 1988). Fish may acclimatize to other aspects of water chemistry, such as ionic strength and $[Ca^{2+}]$, also. For experiments on effects of chronic exposure to acid conditions it may be appropriate to begin with the earliest life stage that will be at risk of exposure in the field. Or, if later life stages must be used, apply acid conditions gradually. If the experiment is concerned with effects of episodic acidification, the test organism should at least be allowed to acclimatize to the baseline conditions before a treatment is applied.

Feeding

Too many 'toxicity tests' have been carried out on animals without access to food. The most significant toxic effect of H^+ on most aquatic animals is the disruption of ionic regulation, and loss of body sodium has been identified as a proximate cause of death in invertebrates, fish and amphibians. Apart from any effect of starvation on hormonal status (Tam *et al.*, 1990), depriving animals of food is likely to increase the significance of any loss of sodium due to low pH.

A disadvantage of feeding during experiments is the potential of food or additional waste products to modify the experimental milieu in undesirable and possibly unexpected ways. Adsorption or binding of metals to organic matter is a likely problem. Spry & Wiener (1991) have suggested that microbial methylation of inorganic mercury can occur in exposure systems during long-term experiments.

The need to sample or monitor experimental conditions during the experiment should never be overlooked.

3.7.2 Field experiments

Experiments carried out in existing acid waters have the advantage that baseline conditions are realistic, but suffer in that the conditions may not be fully quantifiable and probably cannot be replicated. It may be difficult to generalize from events in one system to predict events in another. Nevertheless, experience suggests that biotic interactions play a major part in the structuring and functioning of aquatic systems, and therefore cannot be ignored if useful predictions are to be made.

Field translocations

Field experiments in which organisms are introduced into an existing acid site, and their performance is observed, have been used commonly to identify and quantify toxic effects. Typically, they are followed by laboratory tests with candidate toxicants.

Microcosms and mesocosms

Laboratory systems generally facilitate strict control over major environmental variables that might otherwise introduce unwanted variance into the experiment. While this allows effects of specific treatments to be quantified under the experimental conditions, it often excludes the possibility of significant interactive effects of the treatment variable with other components of the natural environment, and may not allow the effects of treatments to be put in a proper perspective against the background of other environmentally induced variation. Micro- and mesocosms are intended to be more realistic than simple laboratory systems by allowing some such interactions to take effect (see Chapters 7 and 12 in Volume 1). This is achieved by constructing simple systems or by isolating parts of larger, more complex systems. They may be separate bodies of water — pools or streams — or enclosures within lakes. The size, biotic and abiotic complexity, and degree of control that may be applied are infinitely variable, but one might aim for a series of experiments in which control is relaxed or complexity is increased in steps.

Examples of mesocosm studies are those of DeLisle *et al.* (1984), Havens & Heath (1989), Havens & DeCosta (1987), Warner *et al.* (1991) and Rowe *et al.* (1992).

Experimental acidification

Streams, rivers or whole lakes have been acidified or neutralized and responses observed at organismic, population or ecosystem levels. This method allows the target system to be characterized biotically and abiotically before perturbation, and the rate of acidification or neutralization can be controlled (Schindler, 1990; Schindler *et al.*, 1991). This whole-system approach appears to offer the most realistic means of assessing the role of acid conditions in the structuring and functioning of aquatic systems. However, there are certain constraints. It is rarely possible to find enough natural systems of sufficient similarity to allow replication of treatments and controls. One approach to this problem was used at Little Rock Lake, where two very similar basins were isolated from one another; one basin was acidified and the other acted as a control (Brezonik *et al.*, 1986). It may be impracticable to treat a complete watershed, so important influences of soil, such as leaching of metals, are neglected (Schindler *et al.*, 1991). In the case of lakes, the rate of experimental

acidification or neutralization may be unrealistically high.

Experimental acidification is, perhaps, most applicable to the simulation of acid events in streams (e.g. Hall *et al.*, 1985, 1987; Ormerod *et al.*, 1987; Merrett *et al.*, 1991), which are typically of the order of hours to days. However, unless experiments are carried out in the headwaters, where naturally occurring acid episodes may be most severe, invertebrates lost from the experimental area through mortality or drifting may be replaced from untreated sources upstream. This may give an unrealistic picture of the opportunity for recovery. Despite such limitations, this methodology is clearly of great value.

3.8 EFFECTS OF ACID CONDITIONS ON FRESHWATER ORGANISMS

In principle, ecotoxicology is about explaining or predicting changes at population level on the basis of toxic effects on survival, growth and reproduction of individual members of the population (see Chapter 1, Volume 1). In practice, however, certain factors in the overall equation linking environmental concentration to population response receive more attention than others — there have been many more studies of acute toxicity, using death as the endpoint, than of sublethal effects on growth or reproduction. There is a growing requirement for scientific input to policy-making on the control of transboundary pollutants, particularly as emission targets are set on the basis of tolerable or acceptable impacts on ecosystems. There will be a need for better integration of existing studies on toxic mechanisms and interactions of toxicants with other environmental variables, and further studies pertinent to conditions during recovery from acidification.

3.8.1 Physiological effects of acid conditions on animals

Disruption of ionic regulation, and loss of sodium in particular, is the principal symptom exhibited by crustaceans (Havas *et al.*, 1984; Økland &

Økland, 1986; Appelberg, 1987), fish (Muniz & Leivestad, 1980; McDonald *et al.*, 1989; Potts & McWilliams, 1989; Wood, 1989; Wood & McDonald, 1987) and amphibians (Freda & Dunson, 1984; McDonald *et al.*, 1984) on exposure to realistic, low levels of pH. Measurements of sodium fluxes in and out of organisms, using ^{22}Na, have shown that increasing $[H^+]$ increases the rate of passive loss of sodium and reduces the rate of active uptake from the medium (Freda & Dunson, 1984; McDonald *et al.*, 1984; Havas & Likens, 1985; Wood & McDonald, 1987). The rate of passive loss of sodium may decline with duration of exposure. In fish this may be due partly to the decline in concentration gradient across the gill surface, but also to mucification, lamellar cell swelling, hormonal effects or other mechanisms (Wood & McDonald, 1987). In brook trout (*Salvelinus fontinalis*), which is one of the more acid-tolerant salmonids, mortality at pH 4.4 was preceded by a 35% reduction in plasma Na^+ (Wood & McDonald, 1987). Duration of survival of amphibian larvae at low pH has been equated with the length of time taken for body sodium content to be reduced by 50% (Freda & Dunson, 1984, 1985b). At elevated, but sublethal $[H^+]$, body sodium concentrations are typically lower than in controls, and sodium status may therefore be an indicator of 'acid stress'.

The detrimental influence of low pH on ionic regulation can be ameliorated by calcium, which is known to modulate the permeability of cell membranes. Acid resistance of aquatic vertebrates is particularly sensitive to calcium concentration in the range typical of acid natural waters, i.e. $< 5\,mg/l$ (Cummins, 1988). Concentrations of other major cations (Na^+, K^+, Mg^{2+}) can also influence acid resistance, albeit to a much lesser extent (Freda & Dunson, 1985a). For an interesting review of reported effects of acid conditions on calcium metabolism and reproductive physiology in fish, see Munkittrick (1991).

Campbell & Stokes (1985) proposed a theoretical framework for predicting pH-dependent changes in metal toxicity, based on effects of pH on metal speciation and the relative affinity of binding sites on cell membranes for H^+ and free metal ions or metal-ligand species. Thus, for metals whose speciation changes significantly within an ecologically relevant range of pH (e.g. Al, Cu, Mn and Pb), the net toxic effect of decreasing pH may depend on the changing relative activities of H^+ and the metal species, the outcome of competition of H^+ and the metal species for binding sites on cell membranes, and the relative toxicities of H^+ and the metal species at those sites. Therefore, decreasing pH may be expected to increase the binding of some metals by cell membranes, but to reduce the binding of other metals. Campbell & Stokes (1985) suggested other mechanisms by which pH may influence the toxic effect of metals, such as changing the cell membrane potential, or changing the conformation of binding sites, and thus their affinities for toxicants.

At least two toxic modes of action of aluminium in freshwaters have been suggested. An effect of Al on fish at pH levels upwards of 5.5 has been attributed to the formation of polymeric precipitates on the gills (Skogheim *et al.*, 1986). This has been shown to occur under conditions of chemical disequilibrium, and particularly when there is a rapid change in pH. Such conditions can occur in the mixing zone below the confluence of limed and unlimed acid streams (Rosseland *et al.*, 1992). At a pH less than about 5.0, Al has been shown to interfere with ion exchange in invertebrates, fish and amphibians, and the magnitude of this effect has been found to be related to the concentration of labile or 'inorganic monomeric' Al in solution.

Besides the disruption of ionic regulation manifested in changed plasma concentrations or whole-body concentrations of Na^+ and Cl^-, high $[H^+]$ can arrest the development of fish and amphibian embryos, and delay or prevent hatching (Nelson, 1982; Dunson & Connell, 1982). The functioning of the hatching enzyme exuded by the embryo is known to be pH-dependent, but there may also be some structural change in the membranes which renders them less susceptible to the hatching enzyme.

In a number of laboratory studies in which late-term amphibian embryos were exposed to low pH, the perivitelline space was reduced relative to controls. In severe cases this led to mechanical deformation of the embryos—the 'curling defect' (Gosner & Black, 1957; Freda 1986)—and hatching was delayed or failed altogether. Although the curling defect has been observed in many laboratory studies of anurans and urodeles, reports of its occurrence in spawn *in situ* in the field have been rare for urodeles (Pough, 1976; Freda & Dunson, 1986) and lacking for anurans. The curling defect is exacerbated by high $[Ca^{2+}]$ (Freda & Dunson, 1985a), and some laboratory media used in studies of amphibians and fish have been unrealistic in this respect. The proximate cause of death during embryogenesis in acid water may be the same as in later life stages, but in amphibian embryos at least it is preceded (sublethally) by sloughing of cells from the outer surface of the embryo.

Deformities have been described in fish and amphibian embryos developing in acid water (Daye & Garside, 1980; Tyler-Jones *et al.*, 1989). Tyler-Jones *et al.* (1989) described teratological effects on tadpoles of *Rana temporaria* following non-retraction of the embryonic yolk plug; this occurred in water of realistic ionic composition. Mechanical deformities have been observed in a number of laboratory studies of amphibian embryos, apparently due to restriction of the perivitelline space. Skeletal deformities reported among adult fish in acid waters (Fraser & Harvey, 1982; Campbell *et al.*, 1986) have been attributed to disturbances in Ca metabolism, which was also suggested as a factor in failure of some fish to attain reproductive condition (Beamish *et al.*, 1975; cf. Munkittrick, 1991). A suite of symptoms, or syndrome, has been observed among salmonid fish exposed to acid conditions (low pH ± Al): in addition to changed plasma ion concentrations this includes elevated levels of plasma glucose and cortisol (Audet *et al.*, 1988). Elements of this syndrome may not be specific to acid stress.

Within species, resistance to acute acid exposure has been found to be correlated with body size and/or developmental stage (e.g. France & Stokes, 1987). It is, perhaps, to be expected that an organism's capacity for maintaining its *milieu interieur* will increase progressively during early ontogenesis, as organs differentiate and function in a coordinated manner. Nevertheless, 'sensitive stages' have been reported at various points in the life cycle. For example, the susceptibility of aquatic arthropods to water with low pH and low ionic concentration is said to be greatest at moult, when permeability to water and ions increases (Sutcliffe & Hildrew, 1989). Moult may occur more frequently in young or small animals. Allometric relationships such as the ratio of gill area to body mass may influence the rate of disruption of ionic composition of body compartments, such as the plasma, in fish (Laurén & McDonald, 1986).

Evidence of acclimatization to high $[H^+]$, i.e. compensatory increases in active uptake of Na during chronic exposure, is sparse. Indeed, fish and amphibians pre-exposed to sublethal acid conditions subsequently suffered greater net losses of sodium than 'naive' conspecifics when challenged with a further increase in $[H^+]$ (McDonald *et al.*, 1984; Audet & Wood, 1988). However, Wood *et al.* (1988) have presented evidence that fish can acclimatize to aluminium in acid water.

Intraspecific differences in acid tolerance have also been found at the population level. Acid-tolerant strains have been identified among *Salmo trutta* (McWilliams, 1982) and *Salvelinus fontinalis* (Robinson *et al.*, 1976). Potts & McWilliams (1989) attributed greater acid tolerance of *S. trutta* from acid streams at least in part to the lower permeability of their gills to H^+.

Interspecific differences in acid tolerance may depend on differences in the sensitivity of sodium loss and sodium uptake to pH, aluminium and other metals. Havens (1990) reported that aluminium bound to ion exchange sites much less in acid-tolerant cladocerans than in acid-sensitive species.

3.8.2 Organism-level responses to acid conditions

Sublethal effects of acid conditions on animals in laboratory experiments have included reduced efficiency of prey capture by zooplankton (reviewed by Locke, 1991) and fish (Hill, 1989). Reduced growth was reported in zooplankton (reviewed in Locke, 1991), fish (Mount *et al.*, 1988; cf. Sadler & Lynam, 1986), and amphibians (Freda & Dunson, 1985b; Cummins, 1986, 1989).

In the field it becomes increasingly difficult to account for the role of the numerous and inter-acting factors impinging on animals and plants. This problem, of course, has always presented one of the greatest challenges to ecotoxicologists. While the disappearance of certain species may well be accounted for by their sensitivity to acid conditions *per se*, changes in the abundance and performance of other species may owe more to shifts in biotic interactions than to direct sub-lethal effects of abiotic conditions.

One type of response which can be studied at organism level in the field, however, is behaviour. For example, acid spawning sites are reported to be avoided by brook trout (Johnson & Webster, 1977) and natterjack toads, *Bufo calamita* (Beebee *et al.*, 1990). Free-ranging trout carrying trans-mitters dropped downstream during episodic acidification of Linn Run, Pennsylvania and suf-fered lower mortality than caged conspecifics (Carline *et al.*, 1992), and Merrett *et al.* (1991)

reported an increase in numbers of invertebrates drifting out of an experimentally acidified reach of a stream in Wales (Figs. 3.1 and 3.2).

3.9 SOME CHANGES OBSERVED IN ACIDIFIED FRESHWATERS

3.9.1 Decomposition

Decomposition of detritus is slowed under acid conditions (Grahn *et al.*, 1974; Friberg *et al.*, 1980; Traaen, 1980). After liming, the rate of decomposition of accumulated organic material usually increases (Hultberg & Andersson, 1982), and bacterial decomposition may exceed that in similar, non-acidified waters (Gahnström *et al.*, 1980). Invertebrate decomposers such as *Asellus aquaticus* may increase in abundance after liming, and the abundance of several insect orders may change significantly (Hasselrot & Hultberg, 1984).

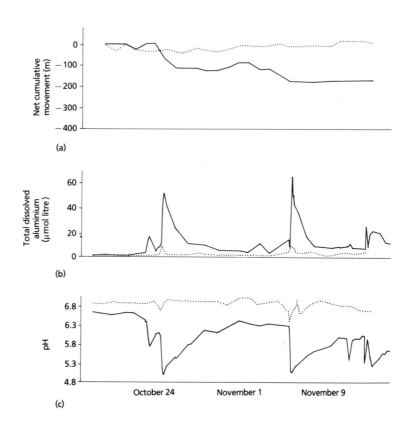

Fig. 3.1 Net cumulative movement of radio-tagged brook trout and changes in stream pH and total dissolved aluminium in Linn Run and Baldwin Creek, autumn 1988. (——), Linn Run; (·····), Baldwin Creek, Pennsylvania, USA. After Carline *et al.* (1992).

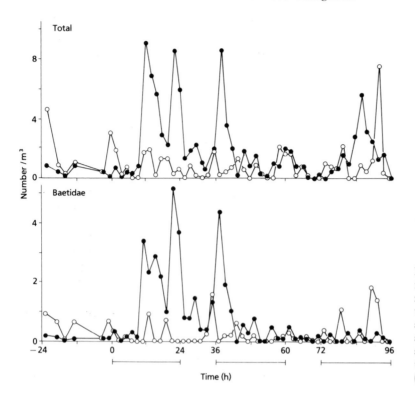

Fig. 3.2 Drift densities of the invertebrate fauna in reference zone A (○) and zone B (●) in which three acid episodes were simulated by addition of sulphuric acid and aluminium sulphate (shown by bars below the x-axis). After Merrett *et al.* (1991).

3.9.2 Primary production

Photosynthesis may be depressed by low pH (Nalewajko & O'Mahoney, 1989). As pH decreases, the availability of bicarbonate as a source of carbon for algae declines.

Acidification is often accompanied by a decline in planktonic species of algae and an increase in benthic and attached forms. A small range of algal species occurs in a wide range of acid waters, particularly *Euglena mutabilis*, *Chlamydomonas acidophila* and diatoms of the genera *Eunotia*, *Nitzschia* and *Pinnularia* (Kelly, 1988).

Schindler *et al.* (1991) have reviewed biological changes in softwater lakes during experimental acidification and recovery. Species diversity in the phytoplankton declined with decreasing pH, owing to a reduction in relative abundance of a few formerly dominant species: chrysophyceans declined while dinoflagellates and some cyanophytes increased. These changes were in accordance with those reported from atmospherically acidified lakes. However, the number of species

in the phytoplankton showed little change with pH during the experimental acidification (Schindler *et al.*, 1991), whereas surveys have revealed 30–80 species in circumneutral oligotrophic lakes, 10–20 species in acid lakes, and as few as three species in some lakes at pH 4 (Muniz, 1991). Formerly sparse, periphytic, filamentous green algae grew to form large mats (metaphyton) at pH 5.6 and below. An increase in transparency of the water column, attributed largely to changes in DOC (Davis *et al.*, 1985, Schindler *et al.*, 1991), allowed benthic and planktonic algae to grow at greater depths, and there was a small increase in production. However, a four-fold increase in respiration of the littoral community as pH declined from >6 to <5 led to a decline in production by the periphyton.

In some atmospherically acidified lakes, littoral mats dominated by cyanophytes have an internal pH of about 6, and may provide a refuge for acid-sensitive species (Schindler *et al.*, 1991).

Effects of experimental acidification have been tested on periphyton in artificial stream channels (Hendrey, 1976, cited in Muniz, 1991) and on phytoplankton in mesocosms (Yan & Stokes, 1978).

3.9.3 Zooplankton and limnetic insects

Laboratory studies of acid effects on zooplankton have been reviewed by Locke (1991). The zooplankton includes grazers, which are susceptible to changes in the phytoplankton, and predators that depend on the grazers. Both groups may be affected directly by low pH and aluminium (e.g. Havas & Likens, 1985; Havens & DeCosta, 1987), and by fish predation.

Following liming and the introduction of brook trout to two formerly fishless acid lakes, Schaffner (1989) reported large changes in the zooplankton: he attributed a very rapid decline in the acid-tolerant rotifer *Keratella taurocephala* to a direct effect of liming on water chemistry; longer-term shifts in relative abundance were attributed to changes in the dynamics of individual species (see also Zimmer, 1987) and effects of predation. In the same lakes, Evans (1989) observed a rapid reduction in numbers of notonectids, corixids and *Chaoborus americanus* in the epilimnion. Notonectids, which are vulnerable to predation by fish because they must surface for air, are often abundant in fishless acid lakes (Bendell, 1986; Brett, 1989).

In a review, Muniz (1991) concluded that the disappearance of acid-sensitive species, such as daphnids, could be explained on the basis of water chemistry. The complexity of trophic interactions involving zooplankton species suggests that the composition and structure of the zooplankton during recovery from acidification may be less predictable.

3.9.4 Zoobenthos

Effects of aluminium and low pH on benthic macroinvertebrates have been reviewed by Herrman (1987). In addition to direct effects of water chemistry and predation pressure, Sutcliffe & Hildrew (1989) suggested that changes in the microbial decomposer community in acid streams could affect benthic invertebrates which rely heavily on allochthonous material. Taxonomic differences in acid sensitivity have been summarized by Muniz (1991): characteristic disturbances of ionic and osmotic regulation account for many such differences, though trophic interactions may be influential (see below).

3.9.5 Fish

Because of their economic importance, fish have received much more attention than other taxa. A review of current knowledge of the physiological effects of acid conditions is beyond the scope of this chapter. Recent, relevant reviews include those by Wood & McDonald (1987), Wood (1989) and Spry & Wiener (1991).

Acidification is reported to cause changes in the age-structure of fish populations, largely as a result of poor recruitment (reviewed by Rosseland *et al.*, 1980; Haines, 1981, 1986; Harvey, 1982; Muniz & Walløe, 1990). At pH less than 5.0–5.5 mortality of eggs and fry is a major cause of reproductive failure (Andersen *et al.*, 1984), but reduced reproductive output has been observed in experiments, in association with reduced food intake and lowered gonadotrophic hormone levels (Tam *et al.*, 1990).

Fish populations are susceptible to, and may contribute to, gross changes in invertebrate species abundance (Eriksson *et al.*, 1980; Schindler *et al.*, 1985; Brett, 1989; Evans, 1989). Removal of fish from non-acidified lakes caused changes in invertebrate communities, which then resembled those of fishless acidified lakes: large invertebrate predators increased in abundance, and there was a shift in the grazing zooplankton from daphnids to larger calanoid copepods (Eriksson *et al.*, 1980).

3.9.6 Amphibians

Field studies in acid waters have revealed widespread and sometimes extensive mortality among embryos of anurans and urodeles that lay their eggs in clumps (e.g. Pough, 1976; Strijbosch, 1979; Hagström, 1981; Cummins & Ross, 1986; Leuven *et al.*, 1986; Gebhardt *et al.*, 1987). The formation of chemical microenvironments may be important in the field: pH has been observed to be higher at the centre of frog spawn clumps than in the surrounding water, and damage is often greatest at the periphery of spawn clumps (Cummins & Ross, 1986; Gebhardt *et al.*, 1987). Species which lay eggs singly, in small clumps or in strings may suffer greater exposure to low pH than those which produce large clumps and which spawn communally. The use of abnormally small groups of eggs and/or flowing water in experiments may therefore lead to overestimates of susceptibility of certain species to acid water in the field.

Field studies showed that *Rana temporaria* tadpoles hatched in acid waters where substantial embryo mortality had occurred (Cummins & Ross, 1986; Gebhardt *et al.*, 1987). The principal physical symptoms of acid stress in tadpoles are loss of sodium and dehydration (McDonald *et al.*, 1984; Freda & Dunson, 1985b); Linnenbach *et al.* (1987) also described changes in the epidermis and increased secretion of mucus in *Rana temporaria* tadpoles at pH 4.2.

Laboratory studies of the sensitivity of anuran embryos to low pH were reviewed by Freda (1986) and Pierce (1985). Tadpoles of most species tested are very tolerant of low pH: long-term exposure of *Rana temporaria* tadpoles to realistic [H$^+$] (nominal pH 3.6−6.5, actual mean pH 4.0−7.0) in a semistatic system resulted in progressive reductions in rates of growth and development with decreasing pH, but all tadpoles achieved metamorphosis (Cummins, 1986). It is widely agreed that, after hatching, amphibians' acid tolerance increases with size and developmental stage.

A mesocosm study by Warner *et al.* (1991) demonstrated significant interactions between effects of acid conditions and biotic factors (inter- and intraspecific interactions).

3.9.7 Birds

Nyholm (1981) suggested that defective eggshells and low reproductive success among insectivorous birds breeding close to acidified lakes resulted from the ingestion of aluminium-rich insects from the lakes. However, other studies have not revealed similar eggshell defects under comparable circumstances (Glooschenko *et al.*, 1986; Ormerod *et al.*, 1988). Scheuhammer (1991), reviewing studies involving the addition of aluminium to the diet of birds, considered that aluminium was unlikely to have any toxic effects

unless dietary phosphorus was in short supply (see Sparling, 1990); although insects from acid waters contained large quantities of aluminium (Sadler & Lynam, 1985; Ormerod *et al.*, 1988) they also contained sufficient phosphorus for birds eating them not to suffer significant biological effects. Possible causes of a correlation between stream pH and breeding success in dippers has been discussed by Tyler & Ormerod (1992).

3.10 MITIGATION OF ANTHROPOGENIC ACIDIFICATION

An important objective of ecotoxicological studies of acidified waters is the provision of a sound scientific basis for mitigation. Consequently, the nature and course of ecological change taking place as waters become less acid is of concern.

Three types of study have addressed the effects of amelioration or reversal of large-scale acidification: (i) addition of neutralizing agents where long-term acidification has occurred — so-called 'liming' (Hasselrot & Hultberg, 1984; Raddum *et al.*, 1986; Howell *et al.*, 1991; several authors in Morrison, 1992); (ii) reduction or cessation of (relatively) short-term experimental acidification (Schindler *et al.*, 1991); (iii) observation of change against a background of declining inputs of acid, e.g. in areas where deposition of acidifying air pollutants has declined (Schindler, 1987; Gunn *et al.*, 1988; Gunn & Keller, 1990; Keller *et al.*, 1992).

3.10.1 Liming — addition of CaCO$_3$

Liming has been carried out on a large scale in lakes and streams in Scandinavia since the late 1960s. It is a costly process, and much research has been carried out into its effectiveness. It is not a sustainable cure for acifidication, but it may preserve certain valued components of acid-sensitive systems until acid deposition is reduced to a tolerable level (Hasselrot & Hultberg, 1984). As the restoration of fish populations is a common motive for liming, the introduction of fish has had a major influence on subsequent changes in many systems relieved from acid stress by liming (Hultberg & Andersson, 1982; Eriksson *et al.*,

1983). Furthermore, liming may produce calcium concentrations higher than those found in such acid-sensitive areas prior to acidification, and indigenous flora (e.g. *Sphagnum*) and fauna may be damaged as a result. For these reasons, studies of limed waters may be poor models for predicting the recovery of systems where the acidifying input is declining. Nevertheless, many observations made on systems after liming have provided opportunities to test our understanding of mechanisms of toxic effect at various levels, from organism to community.

3.10.2 Recovery from experimental acidification

Schindler *et al.* (1991) described changes in Lake 223 in the Experimental Lakes Area (ELA) of Canada during a phase of declining additions of acid. Following a 7-year period in which pH was reduced from 6.6–6.8 to 5.0–5.1, pH was allowed to rise to 5.4–5.5 for 4 years and to 5.8 for a further 3 years. Responses of the various types of biota were said to be 'mixed': (i) the phytoplankton continued to be dominated by acid-tolerant forms, despite the presumed dispersive abilities and pool of resting stages of acid-sensitive species; (ii) several zooplankton species returned, after delays of 1–3 years, but biomass and species diversity were lower than during the acidification phase; (iii) the mats of filamentous green algae characteristic of acidified lakes disappeared rapidly as pH reached 5.8; (iv) large fluctuations in chironomid emergence were thought to be due to variable fish recruitment (and predation); (v) all surviving fish species resumed reproduction, but changing availability of prey led to a decline in condition in some species and recruitment was variable (Schindler *et al.*, 1991).

3.10.3 Recovery from acidification caused by acid deposition

Analysis of fossil algae in lake sediments showed that diatoms and chrysophyceans recovered quickly during recovery from acidification (Dixit *et al.*, 1992); benthic filamentous algae declined in Swan Lake as pH rose from 4.8 to 5.4 (Vandermeulen *et al.*, cited in Keller *et al.*, 1992). As in Lake 223, numbers of zooplankton species increased with rising pH in several lakes in the Sudbury area, but remained below those expected on the basis of nearby reference lakes (Keller *et al.*, 1992). In Whitepine Lake the number of taxa in the zoobenthos increased from 39 in 1982–83 (pH 5.4) to 72 in 1988 (pH 5.9); some acid-sensitive species increased in abundance during this period (Keller *et al.*, 1992). Residual populations of lake trout *Salvelinus namaycush* resumed recruitment after pH increased in two Sudbury lakes; as lake trout populations increased, the acid-tolerant yellow perch declined greatly; small residual populations of *Catostomus commersoni* and *Lota lota* in Whitepine Lake continued their decline to extinction (Keller *et al.*, 1992).

3.10.4 Conclusions from observed recovery from acidification

It is already evident that freshwater systems can exhibit hysteresis during the process of recovery from acidification, and at least three factors limit our ability to predict the rate and route of recovery. First, species differ in their ability to persist in the face of acid conditions which are ultimately inimical to survival — the lifespan of reproductive adults or resting stages relative to the time-course of acidification and recovery is critical in determining whether a species has to recolonize a water or may simply resume reproduction *in situ*. Relevant data are available for many species, but these need to be applied on a case-by-case basis in each water. Occasional reproductive success, due to local or short-term relaxation of acid stress, for example, may be difficult to predict but highly influential. Second, dispersal can be an important factor. For example, although the acid-sensitive amphipod *Hyalella azteca* was able to recolonize the experimentally acidified Lake 223 during its recovery phase (Schindler *et al.*, 1991), apparently from an unacidified lake 300 m upstream, it had failed to return to lakes recovering from long-term acidification up to 7 years after pH levels matched the lower limit of the species' occurrence (Keller *et al.*, 1992). Aside from differences in species' powers of dispersal, actual colonization events may be highly stochastic.

Both of the above illustrate limitations of acidification experiments, even when they are carried out on a whole-lake or whole-catchment scale: (i) the experiment may be too short to effect the extinction of some acid-sensitive species; (ii) close proximity of unacidified waters may allow recolonization to take place much more quickly than in areas where acidification is extensive.

A third limitation on our ability to predict rates and routes of recovery is the predominant role of interactions in the structuring of communities in stable and transitional conditions. Many toxicological studies have sought to explain or predict the decline of species in the face of increasingly deleterious physical conditions (H^+, Al, etc.), but how useful will they be at predicting recovery? Many of the species which predominate in acidified waters were present prior to acidification, albeit in small numbers, and it is not unreasonable to suppose that they were kept in check primarily by biotic interactions, such as competition or predation. The course of recovery may therefore be governed by such interactions. As acidifying emissions are brought under control, at least in the West, and attention shifts towards the process of recovery, future ecotoxicological studies will need to tend more towards the 'eco' than the 'tox'.

REFERENCES

Allott, T.E.H., Harriman, R. & Battarbee, R.W. (1992) Reversibility of lake acidification at the Round Loch of Glenhead, Galloway, Scotland. *Environ. Pollut.* 77, 219–226.

Almer, B., Dickson, W., Eckström, C., Hörnström, E. & Miller, U. (1974) Effects of acidification on Swedish lakes. *Ambio* 3, 30–36.

Andersen, R., Muniz, I.P. & Skurdal, J. (1984) Effects of acidification on age class composition in Arctic char (*Salvelinus alpinus* (L.)) and brown trout (*Salmo trutta* L.) in a coastal area, SW Norway. *Rep. Inst. Freshwat. Res. Drottningholm* 61, 5–15.

Appelberg, M. (1987) Some factors regulating the crayfish *Astacus astacus* L. in acid and neutralized waters. In: *Ecophysiology of Acid Stress in Aquatic Organisms* (Eds H. Witters & O. Vanderborght), pp. 167–179. *Ann. Soc. R. Zool. Belg.* 117, Suppl. 1.

Audet, C. & Wood, C.M. (1988) Do rainbow trout (*Salmo gairdneri*) acclimate to low pH? *Can. J. Aquat. Sci.* 45, 1399–1405.

Audet, C., Munger, R.S. & Wood, C.M. (1988) Long-term sublethal acid exposure in rainbow trout (*Salmo*

gairdneri) in soft water: effects on ion exchanges and blood chemistry. *Can. J. Fish. Aquat. Sci.* 45, 1387–1398.

Battarbee, R.W., Flower, R.J., Stevenson, A.C., Jones, V.J., Harriman, R. & Appleby, P.G. (1988) Diatom and chemical evidence for reversibility of acidification of Scottish lochs. *Nature* 332, 530–532.

Battarbee, R.W., Flower, R.J., Stevenson, A.C. & Rippey, B. (1985) Lake acidification in Galloway: a palaeoecological test of competing hypotheses. *Nature* 314, 350–352.

Battarbee, R.W., Mason, J., Renberg, I. & Talling, J.F. (1990) *Palaeolimnology and Lake Acidification*. Royal Society, London.

Beamish, R.J., Lockhart, W.L., Van Loon, J.C. & Harvey, H.H. (1975) Long-term acidification of a lake and resulting effects on fishes. *Ambio* 4, 98–102.

Beebee, T.J.C., Flower, R.J., Stevenson, A.C., Patrick, S.T., Appleby, P.G., Fletcher, C., Marsh, C., Natkanski, J., Rippey, B. & Battarbee, R.W. (1990) Decline of the natterjack toad *Bufo calamita* in Britain: palaeoecological, documentary and experimental evidence for breeding site acidification. *Biol. Conserv.* 53, 1–20.

Bendell, B.E. (1986) The effects of fish and pH on the distribution and abundance of backswimmers (Hemiptera: Notonectidae). *Can. J. Zool.* 64, 2696–2699.

Bertsch, P.M. & Anderson, M.A. (1989) Speciation of aluminum in aqueous solutions using ion chromatography. *Anal. Chem.* 61, 535–539.

Brett, M.T. (1989) The distribution of free-swimming macroinvertebrates in acidic lakes of Maine: the role of fish predation. *Aqua Fennica* 19, 113–118.

Brezonik, P.L., Baker, L.A. Eaton, J.R., Frost, T.M., Garrison, P., Kratz, T.K., Magnuson, J.J., Rose, W.J., Shephard, B.K., Swenson, W.A., Watras, C.J. & Webster, K.E. (1986) Experimental acidification of Little Rock Lake, Wisconsin. *Water, Air Soil Pollut.* 31, 115–121.

Brown, D.J.A. & Lynam, S. (1981) The effect of sodium and calcium on the hatching of eggs and on the survival of the yolk sac fry of brown trout, *Salmo trutta* L. at low pH. *J. Fish Biol.* 19, 205–211.

Bull, K.R. & Hall, J.R. (1986) Aluminium in the rivers Esk and Duddon and their tributaries, Cumbria. *Environ. Pollut. B* 12, 165–193.

Campbell, R.N.B., Maitland, P.S.M. & Lyle, A.A. (1986) Brown trout deformities: an association with acidification? *Ambio* 15, 244–245.

Campbell, P.G.C. & Stokes, P.M. (1985) Acidification and toxicity of metals to aquatic biota. *Can. J. Fish. Aquat. Sci.* 42, 2034–2049.

Campbell, P.G.C. & Tessier, A. (1987a) Current status of metal speciation studies. In: *Metals Speciation, Separation and Recovery* (Ed. J.W. Patterson &

R. Passino), pp. 201–224. Lewis, Chelsea.

Campbell, P.G.C. & Tessier, A. (1987b) Metal speciation in natural waters: influence of environmental acidification. In: *Sources and Fates of Aquatic Pollutants* (Eds R.A. Hites & S.J. Eisenreich), pp. 185–207. American Chemical Society, Washington, DC.

Carline, R.F., DeWalle, D.R., Sharpe, W.E., Dempsey, B.A., Gagen, C.J. & Swistock, B. (1992) Water chemistry and fish community responses to episodic stream acidification in Pennsylvania, USA. *Environ. Pollut.* **78**, 45–48.

Clymo, R.S. (1984) *Sphagnum*-dominated peat bog: a naturally acid ecosystem. *Phil. Trans. R. Soc. Lond B* **305**, 487–499.

Coale, K.H. & Flegal, A.R. (1989) Copper, zinc, cadmium and lead in surface waters of Lakes Erie and Ontario. *Sci. Tot. Environ.* **87/88**, 297–304.

Covington, A.K., Whalley, P.D. & Davison, W. (1983) Procedures for the measurement of pH in low ionic strength solutions including freshwater. *Analyst* **108**, 1528–1532.

Cronan, C.S., Walker, W.J. & Bloom, P.R. (1986) Predicting aqueous aluminium concentrations in natural waters. *Nature* **324**, 140–143.

Cummins, C.P. (1986) Effects of aluminium and low pH on growth and development in *Rana temporaria* tadpoles. *Oecologia* **69**, 248–252.

Cummins, C.P. (1988) Effect of calcium on survival times of *Rana temporaria* embryos at low pH. *Funct. Ecol.* **2**, 297–302.

Cummins, C.P. (1989) Interaction between the effects of pH and density on growth and development in *Rana temporaria* tadpoles. *Funct. Ecol.* **3**, 45–52.

Cummins, C.P. & Ross, A. (1986) Effects of acidification of natural waters upon amphibians. *Final report to the Commission of the European Communities, CEC/NERC Contract EV3V.0907.UK(H).*

Davis, R.B., Anderson, D.S. & Berge, F. (1985) Palaeolimnological evidence that lake acidification is accompanied by loss of organic matter. *Nature* **316**, 436–438.

Davison, W. (1990) A practical guide to pH measurement in freshwaters. *Trends Analyt. Chem.* **9**, 80–83.

Davison, W. & Gardner, M.J. (1986) Interlaboratory comparisons of the determination of pH in poorly buffered fresh waters. *Anal. Chim. Acta,* **182**, 17–31.

Davison, W. & Harbinson, T.R. (1988) Performance testing of pH electrodes suitable for low ionic strength solutions. *Analyst* **113**, 709–713.

Davison, W. & Woof, C. (1985) Performance tests for the measurement of pH with glass electrodes in low ionic strength solutions including natural waters. *Anal. Chem.* **57**, 2567–2570.

Daye, P.G. & Garside, E.T. (1980) Structural alterations in embryos and alevins of the Atlantic salmon, *Salmo salar* L., induced by continuous or short-term exposure to acidic levels of pH. *Can. J. Zool.* **58**, 27–43.

DeLisle, C.A., Roy, L., Bilodeau, P. & Andre, P. (1984) Effets d'une acidification artificielle in situ sur le phytoplancton et le zooplancton lacustre. *Verh. Int. Ver. Limnol.* **22**, 383–387.

Dillon, P.J. & LaZerte, B.D. (1992) Response of the Plastic Lake catchment, Ontario, to reduced sulphur deposition. *Environ. Pollut.* **77**, 211–218.

Dixit, A.S., Dixit, S.S. & Smol, J.P. (1992) Algal microfossils provide high temporal resolution of environmental change. *Water, Air Soil Pollut.* **62**, 75–87.

Dougan, W.K. & Wilson, A.L. (1974) The absorptiometric determination of aluminium in water. A comparison of chromogenic reagents and the development of an improved method. *Analyst* **99**, 413–430.

Drabløs, D. & Tollan, A. (Eds) (1980) *Proceedings of the International Conference on the Ecological Impacts of Acid Precipitation.* SNSF Project, Oslo, Norway.

Driscoll, C.T. (1984) A procedure for the fractionation of aqueous aluminium in dilute acidic waters. *Int. J. Environ. Analyt. Chem.* **16**, 267–283.

Dunson, W.A. & Connell, J. (1982) Specific inhibition of hatching in amphibian embryos at low pH. *J. Herpetol.* **57**, 435–443.

Eriksson, F., Hörnström, E., Mossberg, P. & Nyberg, P. (1983) Ecological effects of lime treatment of acidified lakes and rivers in Sweden. *Hydrobiologia* **101**, 145–164.

Eriksson, M.O.G., Henrikson, L., Nilsson, B.I., Nyman, G., Oscarson, H.G., Stenson, A.E. & Larsson, K. (1980) Predator–prey relations important for the biotic changes in acidified lakes. *Ambio* **9**, 248–249.

Evans, R.A. (1989) Response of limnetic insect populations of two acidic, fishless lakes to liming and brook trout (*Salvelinus fontinalis*) introduction. *Can. J. Fish. Aquat. Sci.* **46**, 342–351.

Fitzgerald, W.F. & Watras, C.J. (1989) Mercury in surficial waters of rural Wisconsin lakes. *Sci. Tot. Environ.* **87/88**, 223–232.

Flower, R.J. & Battarbee, R.W. (1983) Diatom evidence for recent acidification of two Scottish lochs. *Nature* **305**, 130–133.

France, R.L. & Stokes, P.M. (1987) Life stage and population variation in resistance and tolerance of *Hyalella azteca* (Amphipoda) to low pH. *Can. J. Fish. Aquat. Sci.* **44**, 1102–1111.

Fraser, G.A. & Harvey, H.H. (1982) Elemental composition of bone from white sucker (*Catostomus commersoni*) in relation to lake acidification. *Can. J. Fish. Aquat. Sci.* **39**, 1289–1296.

Freda, J. (1986) The influence of acidic pond water on amphibians. *Water, Air Soil Pollut.* **30**, 439–450.

Freda, J. (1991) The effects of aluminium and other metals on amphibians. *Environ. Pollut.* **71**, 305–328.

Freda, J., Cavdek, V. & McDonald, D.G. (1990) Role of

organic complexation in the toxicity of aluminum to *Rana pipiens* embryos and *Bufo americanus* tadpoles. *Can. J. Fish. Aquat. Sci.* **47**, 217–224.

Freda, J. & Dunson, W.A. (1984) Sodium balance of amphibian larvae exposed to low environmental pH. *Physiol. Zool.* **57**, 435–443.

Freda, J. & Dunson, W.A. (1985a) The influence of external cation concentration on the hatching of amphibian embryos in water of low pH. *Can. J. Zool.* **63**, 2649–2656.

Freda, J. & Dunson, W.A. (1985b) Field and laboratory studies of ion balance and growth rates of ranid tadpoles chronically exposed to low pH. *Copeia 1985*, 415–423.

Freda, J. & Dunson, W.A. (1986) Effects of low pH and other chemical variables on the local distribution of amphibians. *Copeia 1986*, 454–466.

Friberg, F., Otto, C. & Svensson, B. (1980) Effects of acidification on the dynamics of allochthonous leaf material and benthic invertebrate communities in running water. In: *Proceedings of the International Conference on the Ecological Impacts of Acid Precipitation* (Eds D. Drabløs & A. Tollan), pp. 304–305. SNSF Project, Norway.

Gahnström, G., Andersson, G. & Fleischer, S. (1980) Decomposition and exchange processes in acidified lake sediments. In: *Proceedings of the International Conference on the Ecological Impacts of Acid Precipitation* (Eds D. Drabløs & A. Tollan), pp. 306–309. SNSF Project, Norway.

Galloway, J.N., Cosby, B.J. Jr & Likens, G.E. (1979) Acid precipitation: measurement of pH and acidity. *Limnol. Oceanogr.* **24**, 1161–1165.

Galster, H. (1991) *pH Measurement: Fundamentals, Methods, Applications, Instrumentation.* VCH, Weinheim.

Gebhardt, H., Kreimes, K. & Linnenbach, M. (1987) Untersuchungen zur Beeinträchtigung der Ei- und Larvalstadien von Amphibien in sauren Gewässern. *Natur Landschaft* **62**, 20–23.

Gloosschenko, V., Blancher, P., Herskowitz, J., Fulthorpe, R. & Rang, S. (1986) Association of wetland acidity with reproductive parameters and insect prey of the eastern kingbird (*Tyrannus tyrannus*) near Sudbury, Ontario. *Water, Air Soil Pollut.* **30**, 553–567.

Goenaga, X. & Williams, D.J.A. (1988) Aluminium speciation in surface waters from a Welsh upland area. *Environ. Pollut.* **52**, 132–149.

Goenaga, X., Bryant, R. & Williams, D.J.A. (1987) Influence of sorption processes on aluminum determinations in acidic waters. *Anal. Chem.* **59**, 2673–2678.

Gosner, K.L. & Black, I.H. (1957) The effects of acidity on the development and hatching of New Jersey frogs. *Ecology* **38**, 256–262.

Grahn, O., Hultberg, H. & Landner, L. (1974) Oligotrophication—a self-accelerating process in lakes subjected to excessive supply of acid substances. *Ambio* **3**, 93–94.

Gunn, J.M. & Keller, W. (1990) Biological recovery of an acid lake after reductions in industrial emissions of sulphur. *Nature* **345**, 431–433.

Gunn, J.M., McMurty, M.J., Casselman, J.M., Keller, W. & Powell, M.J. (1988) Changes in the fish community of a limed lake near Sudbury, Ontario: effects of chemical neutralization or reduced atmospheric deposition of acids? *Water, Air Soil Pollut.* **41**, 113–136.

Haapala, H., Sepponen, P. & Meskus, E. (1975) Effect of spring floods on water acidity in the Kiiminkijoki area, Finland. *Oikos* **26**, 26–31.

Hagen, A. & Langeland, A. (1973) Polluted snow in southern Norway and the effect of the meltwater on freshwater and aquatic organisms. *Environ. Pollut.* **5**, 45–57.

Hagström, T. (1981) Reproductive strategy and success of amphibians in waters acidified by atmospheric pollution. In: *Proceedings of the European Herpetological Symposium, 1980*, pp. 55–57. Cotswold Wildlife Park, Oxford.

Haines, T. (1981) Acidic precipitation and its consequences for aquatic ecosystems: a review. *Trans. Am. Fish. Soc.* **110**, 669–707.

Haines, T. (1982) Interpretation of aquatic pH trends. *Trans. Am. Fish. Soc.* **111**, 779–786.

Haines, T. (1986) Fish population trends in response to surface water acidification. In: *Acid Deposition: Long Term Trends*, pp. 304–334. National Academy Press, Washington, DC.

Hall, R.J., Driscoll, C.T., Likens, G.E. & Pratt, J.M. (1985) Physical, chemical and biological consequences of episodic aluminium additions to a stream. *Limnol. Oceanogr.* **30**, 212–220.

Hall, R.J., Driscoll, C.T. & Likens, G.E. (1987) Importance of hydrogen ions and aluminium in regulating the structure and function of stream ecosystems: an experimental test. *Freshwater Biol.* **18**, 17–43.

Harvey, H.H. (1982) Population responses of fishes in acidified waters. In: *Acid Rain/Fisheries* (Ed. R.E. Johnson), pp. 103–121. American Fisheries Society, Bethesda, MD.

Hasselrot, B. & Hultberg, H. (1984) Liming of acidified Swedish lakes and streams and its consequences for aquatic ecosystems. *Fisheries* **9**, 4–9.

Havas, M. & Likens, G.E. (1985) Changes in ^{22}Na influx and outflux in *Daphnia magna* (Straus) as a function of elevated Al concentrations in soft water at low pH. *Proc. Natl. Acad. Sci. USA* **82**, 7345–7349.

Havas, M., Hutchinson, T.C. & Likens, G.E. (1984) The effects of low pH on sodium regulation in two species of *Daphnia*. *Can. J. Zool.* **62**, 1965–1970.

Havens, K.E. (1990) Aluminium binding to ion exchange sites in acid-sensitive versus acid-tolerant cladocerans. *Environ. Pollut.* **64**, 133–141.

Havens, K.E. & DeCosta, J. (1987) Freshwater plankton community succession during experimental acidification. *Arch. Hydrobiol.* **111**, 27–65.

Havens, K.E. & Heath, R.T. (1989) Acid and aluminium effects on freshwater zooplankton: an *in situ* mesocosm study. *Environ. Pollut.* **62**, 195–211.

Hendershot, W.H., Dufresne, A., Lalande, H. & Courchesne, F. (1986) Temporal variation in aluminium speciation and concentration during snowmelt. *Water, Air Soil Pollut.* **31**, 231–237.

Henriksen, A., Lien, L., Traaen, T.S., Sevaldrud, S. & Brakke, D.F. (1988) Lake acidification in Norway — present and predicted chemical status. *Ambio* **17**, 259–266.

Herrmann, J. (1987) Aluminium impact on freshwater invertebrates at low pH: a review. In: *Speciation of Metals in Water, Sediment and Soil Systems. Lecture Notes in Earth Sciences*, Vol. 11 (Ed. L. Landner), pp. 157–175. Springer-Verlag, Berlin.

Hill, J. (1989) Analysis of six foraging behaviors as toxicity indicators, using juvenile smallmouth bass exposed to low environmental pH. *Arch. Environ. Contam. Toxicol.* **18**, 895–899.

Hornbeck, J.W. (1992) Comparative impacts of forest harvest and acid precipitation on soil and streamwater acidity. *Environ. Pollut.* **77**, 151–155.

Howell, E.T., Coker, G., Booth, G.M., Keller, W., Neary, B., Nicholls, K.H., Tomassini, F.D., Yan, N.D., Gunn, J.M., Reitveld, H. & Wales, D. (1991) Ecosystem responses of a pH 5.9 trout lake to whole-lake liming. In: *International Lake and Watershed Liming Practices* (Eds H. Olem, R.K. Schreiber, R.W. Brocksen & D.B. Porcella), pp. 61–95. Terrene Institute, Washington, DC.

Howells, G. (1982) Interpretation of aquatic pH trends. *Trans. Am. Fish. Soc.* **111**, 779–786.

Hultberg, H. & Andersson, I. (1982) Liming of acidified lakes: induced long-term changes. *Water, Air Soil Pollut.* **18**, 311–331.

Irwin, J.G. & Williams, M.L. (1988) Acid rain: chemistry and transport. *Environ. Pollut.* **50**, 29–59.

Jackson, T.A., Kipphut, G., Hesslein, R.H. & Schindler, D.W. (1980) Experimental study of trace metal chemistry in soft-water lakes at different pH levels. *Can. J. Fish. Aquat. Sci.* **37**, 387–402.

Jeffries, D.S., Cox, C.M. & Dillon, P.J. (1979) Depression of pH in lakes and streams in central Ontario during snowmelt. *J. Fish. Res. Bd Can.* **36**, 640–646.

Johnson, D.W. & Webster, D.A. (1977) Avoidance of low pH in selection of spawning sites by brook trout (*Salvelinus fontinalis*). *J. Fish. Res. Bd Can.* **34**, 2215–2218.

Johnson, D.W., Cresser, M.S., Nilsson, S.I., Turner, J.,

Ulrich, B., Binkley, D. & Cole, D.W. (1991) Soil changes in forest ecosystems: evidence for and probable causes. In: *Acidic Deposition: its Nature and Impacts* (Eds F.T. Last & R. Watling), 97B:81–116. Proceedings of the Royal Society of Edinburgh.

Jones, V.J., Stevenson, A.C. & Battarbee, R.W. (1986) Lake acidification and the land use hypothesis: a mid-post-glacial analogue. *Nature* **322**, 157–158.

Kahl, J.S., Norton, S.A., Haines, T.H., Rochette, E.A., Heath, R.H. & Nodvin, S.C. (1992) Mechanisms of episodic acidification in low-order streams in Maine, USA. *Environ. Pollut.* **78**, 37–44.

Kaufmann, P.R., Herlihy, A.T. & Baker, L.A. (1992) Sources of acidity in lakes and streams of the United States. *Environ. Pollut.* **77**, 115–122.

Keller, W., Gunn, J.M. & Yan, N.D. (1992) Evidence of biological recovery in acid-stressed lakes near Sudbury, Canada. *Environ. Pollut.* **78**, 79–85.

Kelly, M. (1988) *Mining and the Freshwater Environment*. Elsevier, Amsterdam.

Koch, W.F., Marinenko, G. & Paule, R.C. (1986) Development of a standard reference material for rainwater analysis. *J. Res. Natl. Bur. Stand.* **91**, 33.

Laurén, D.J. & McDonald, D.G. (1986) Influence of water hardness, pH and alkalinity on the mechanisms of copper toxicity in juvenile rainbow trout, *Salmo gairdneri. Can. J. Fish. Aquat. Sci.* **43**, 1488–1496.

LaZerte, B.D. (1984) Forms of aqueous aluminum in acidified catchments of central Ontario: a methodological analysis. *Can. J. Fish. Aquat. Sci.* **41**, 766–776.

LaZerte, B.D. (1986) Metals and acidification: an overview. *Water, Air Soil Pollut.* **31**, 569–576.

Leuven, R.S.E.W., den Hartog, C., Christiaans, M.M.C. & Heijligers, W.H.C. (1986) Effects of water acidification on the distribution pattern and the reproductive success of amphibians. *Experientia* **42**, 495–503.

Linnenbach, M., Marthaler, R. & Gebhardt, H. (1987) Effects of acid water on gills and epidermis in brown trout (*Salmo trutta* L.) and in tadpoles of the common frog (*Rana temporaria* L.). In: *Ecophysiology of Acid Stress in Aquatic Organisms* (Eds H. Witters & O. Vanderborght), pp. 365–374. *Ann. Soc. R. Zool. Belg.* **117**, Suppl. 1.

Locke, A. (1991) Zooplankton responses to acidification: a review of laboratory bioassays. *Water, Air Soil Pollut.* **60**, 135–148.

Lovgren, L., Hedlund, T., Ohman, L.-O. & Sjoberg, S. (1987) Equilibrium approaches to natural water systems — 6. Acid-base properties of a concentrated bog water and its complexation reactions with aluminium(III). *Water Res.* **21**, 1401–1407.

McDonald, D.G., Ozog, J.L. & Simons, B.P. (1984) The influence of low pH environments on ion regulation in the larval stages of the anuran amphibian (*Rana clamitans*). *Can. J. Zool.* **62**, 2171–2177.

McDonald, D.G., Reader, J.P. & Dalziel, T.R.K. (1989) The combined effects of pH and trace metals on fish ionoregulation. In: *Acid Toxicity and Aquatic Animals* (Eds R. Morris, E.W. Taylor, D.J.A. Brown & J.A. Brown), pp. 221–242. Cambridge University Press, Cambridge.

McWilliams, P.G. (1982) A comparison of physiological characteristics in normal and acid-exposed populations of the brown trout, *Salmo trutta*. *Comp. Biochem. Physiol.* **72A**, 515–522.

Merrett, W.J., Rutt, G.P., Weatherley, N.S., Thomas, S.P. & Ormerod, S.J. (1991) The response of macro-invertebrates to low pH and increased aluminium concentrations in Welsh streams: multiple episodes and chronic exposure. *Arch. Hydrobiol.* **121**, 115–125.

Midgley, D. & Torrance, K. (1991) *Potentiometric Water Analysis*. John Wiley & Sons Ltd, Chichester.

Morrison, B. (Ed.). (1992) Effects of acidic pollutants on freshwater plants and animals. *Environ. Pollut.* **78** (special issue).

Mount, D.R., Ingersoll, C.G., Gulley, D.D., Fernandez, J.D., LaPoint, T.W. & Bergman, H.L. (1988) Effect of long-term exposure to acid, aluminum and low calcium on adult brook trout (*Salvelinus fontinalis*). 1. Survival, growth, fecundity, and progeny survival. *Can. J. Fish. Aquat. Sci.* **45**, 1623–1632.

Muniz, I.P. (1991) Freshwater acidification: its effects on species and communities of freshwater microbes, plants and animals. In: *Acidic Deposition: its Nature and Impacts* (Eds F.T. Last & R. Watling), Proceedings of the Royal Society of Edinburgh. **97B**:227–254.

Muniz, I.P. & Leivestad, H.R. (1980) Acidification effects on freshwater fish. In: *Ecological Impact of Acid Precipitation*. Proceedings of an international conference, Sandefjord, Norway, 11–14 March 1980 (Eds D. Drabløs & A. Tollan), pp. 84–92. SNSF Project, Oslo-Ås, Norway.

Muniz, I.P. & Walløe, L. (1990) The influence of water quality and catchment characteristics on the survival of fish populations. In: *The Surface Waters Acidification Programme* (Ed. J. Mason), pp. 327–339. Cambridge University Press, Cambridge.

Munkittrick, K.R. (1991) Calcium-associated reproductive problems of fish in acidified environments: evolution from hypothesis to scientific fact. *Environ. Toxicol. Chem.* **10**, 977–979.

Nalewajko, C. & O'Mahoney, M.A. (1989) Photosynthesis of algal cultures and phytoplankton following an acid pH shock. *J. Phycol.* **25**, 319–325.

Nelson, J.A. (1982) Physiological observations on developing rainbow trout, *Salmo gairdneri* (Richardson), exposed to low pH and varied calcium ion concentrations. *J. Fish Biol.* **20**, 359–372.

Nelson, W.O. & Campbell, P.G.C. (1991) The effects of acidification on the geochemistry of Al, Cd, Pb and Hg in freshwater environments: a literature review. *Environ. Pollut.* **71**, 91–130.

Nilsson, S.I., Miller, H.G. & Miller, J.D. (1982) Forest growth as a possible cause of soil and water acidification. *Oikos* **39**, 40–49.

Nyholm, N.E.I. (1981) Evidence of involvement of aluminium in causation of defective formation of eggshells and of impaired breeding in wild passerine birds. *Environ. Res.* **26**, 363–371.

Økland, J. & Økland, K.A. (1986) The effects of acid deposition on benthic animals in lakes and streams. *Experientia* **42**, 471–486.

Ormerod, S.J., Boole, P., McCahon, C.P., Weatherley, N.S., Pascoe, D. & Edwards, R.W. (1987) Short-term experimental acidification of a Welsh stream: comparing the biological effects of hydrogen ions and aluminium. *Freshwater Biol.* **17**, 341–356.

Ormerod, S.J., Bull, K.R., Cummins, C.P., Tyler, S.J. & Vickery, J.A. (1988) Egg mass and shell thickness in dippers *Cinclus cinclus* in relation to stream acidity in Wales and Scotland. *Environ. Pollut.* **55**, 107–121.

Ormerod, S.J., Donald, A.P. & Brown, S.J. (1989) The influence of plantation forestry on the pH and aluminium concentration of upland Welsh streams: a re-examination. *Environ. Pollut.* **62**, 47–62.

O'Shea, T.A. & Mancy, K.H. (1978) The effect of pH and hardness metal ions on the competitive interaction between trace metal ions and inorganic and organic complexing agents found in natural waters. *Water Res.* **12**, 703–711.

Overrein, L.N., Seip, H.M. & Tollan, A. (Eds) (1980) *Acid Precipitation—Effects on Forest and Fish*. Final report, SNSF project 1972–1980, Oslo-Ås, Norway.

Petersen, R.C. & Persson, U. (1987) Comparison of the biological effects of humic materials under acidified conditions. *Sci. Total Environ.* **62**, 387–398.

Pierce, B.A. (1985) Acid tolerance in amphibians. *BioScience* **35**, 239–243.

Potts, W.T.W. & McWilliams, P.G. (1989) The effects of hydrogen and aluminium ions on fish gills. In: *Acid Toxicity and Aquatic Animals* (Eds R. Morris, E.W. Taylor, D.J.A. Brown & J.A. Brown), pp. 201–220. Cambridge University Press, Cambridge.

Pough, F.H. (1976) Acid precipitation and embryonic mortality of spotted salamanders, *Ambystoma maculatum*. *Science* **192**, 68–70.

Raddum, G.G., Brettum, P., Matzow, D., Nilssen, J.P., Skov, A., Sveälv, T. & Wright, R.F. (1986) Liming the acid lake Hovvatn, Norway: a whole-lake study. *Water, Air Soil Pollut.* **31**, 721–763.

Renberg, I. & Hellberg, T. (1982) The pH history of lakes in southwestern Sweden, as calculated from the subfossil diatom flora of the sediments. *Ambio* **11**, 30–33.

Reynolds, B., Neal, C., Hornung, M., Hughes, S. &

Stevens, P.A. (1988) Impact of afforestation on the soil solution chemistry of stagnopodzols in mid-Wales. *Water, Air Soil Pollut.* **38**, 55–70.

Reynolds, B., Neal, C., Hornung, M. & Stevens, P.A. (1986) Baseflow buffering of streamwater acidity in five mid-Wales catchments. *J. Hydrol.* **87**, 167–185.

Reynolds, B., Stevens, P.A., Adamson, J.K., Hughes, S. & Roberts, J.D. (1992) Effects of clearfelling on stream and soil water aluminium chemistry in three UK forests. *Environ. Pollut.* **77**, 157–165.

Robinson, G.D., Dunson, W.A., Wright, J.E. & Mamolito, G.E. (1976) Differences in low pH tolerance among strains of brook trout (*Salvelinus fontinalis*). *J. Fish. Biol.* **8**, 5–17.

Rosseland, B.O., Blakar, I.A., Bulger, A., Kroglund, F., Kvellstad, A., Lydersen, E., Oughton, D.H., Salbu, B., Staurnes, M. & Vogt, R. (1992) The mixing zone between limed and acidic river waters: complex aluminium chemistry and extreme toxicity for salmonids. *Environ. Pollut.* **78**, 3–8.

Rosseland, B.O., Sevaldrud, I., Svalastog, D. & Muniz, I.P. (1980) Studies of freshwater fish populations—effects on reproduction, population structure, growth and food selection. In: *Proceedings on the International Conference on the Ecological Impacts of Acid Precipitation* (Eds D. Drabløs & A. Tollan), pp. 336–337. SNSF Project, Oslo, Norway.

Rowe, C.L., Sadinski, W.J. & Dunson, W.A. (1992) Effects of acute and chronic acidification on three larval amphibians that breed in temporary ponds. *Arch. Environ. Contam. Toxicol.* **23**, 339–350.

Rueter, J.G., O'Reilly, K.T. & Petersen, R.R. (1987) Indirect aluminium toxicity to the green alga *Scenedesmus* through increased cupric ion activity. *Environ. Sci. Technol.* **21**, 435–438.

Sadler, K. & Lynam, S. (1985) The mineral content of some freshwater invertebrates in relation to stream pH and calcium concentration. *Technical Report TPRD/L/2781/84*, Central Electricity Research Laboratories, Leatherhead, UK.

Sadler, K. & Lynam, S. (1986) Some effects of low pH and calcium on the growth and tissue mineral content of yearling brown trout, *Salmo trutta. J. Fish Biol.* **29**, 313–324.

Schaffner, W.R. (1989) Effects of neutralization and addition of brook trout (*Salvelinus fontinalis*) on the limnetic zooplankton communities of two acidic lakes. *Can. J. Fish. Aquat. Sci.* **46**, 295–305.

Schecher, W.D. & Driscoll, C.T. (1987) An evaluation of uncertainty associated with aluminum equilibrium calculations. *Water Resour. Res.* **23**, 525–534.

Scheuhammer, A.M. (1991) Effects of acidification on the availability of toxic metals and calcium to wild birds and mammals. *Environ. Pollut.* **71**, 329–375.

Schindler, D.W (1987) Recovery of Canadian lakes from acidification. In: *Reversibility of Acidification* (Ed. H. Barth), pp. 2–13. Elsevier Applied Science, London.

Schindler, D.W. (1990) Experimental perturbations of whole lakes as tests of hypotheses concerning ecosystem structure and function. *Oikos* **57**, 25–41.

Schindler, D.W., Frost, T.M., Mills, K.H., Chang, P.S.S., Davies, I.J., Findlay, L., Malley, D.F., Shearer, J.A., Turner, M.A., Garrison, P.J., Watras, C.J., Webster, K., Gunn, J.M., Brezonik, P.L. & Swenson, W.A. (1991) Comparisons between experimentally- and atmospherically-acidified lakes during stress and recovery. In: *Acidic Deposition: its Nature and Impacts* (Eds F.T. Last & R. Watling), Proceedings of the Royal Society of Edinburgh. **97B**:193–226.

Schindler, D.W., Hesslein, R.H., Wagemann, R. & Broecker, W.S. (1980) Effects of acidification on mobilization of heavy metals and radionuclides from sediments of a lake. *Can. J. Fish. Aquat. Sci.* **37**, 373–377.

Schindler, D.W., Mills, K.H., Malley, D.F., Findlay, D.L., Shearer, J.A., Davies, I.J., Turner, M.A., Linsey, G.A. & Cruikshank, D.R. (1985) Long-term ecosystem stress: the effects of years of experimental acidification of a small lake. *Science* **228**, 1395–1401.

Schulze, E.-D. & Freer-Smith, P.H. (1991) An evaluation of forest decline based on field observations focused on Norway spruce, *Picea abies*. In: *Acidic Deposition: its Nature and Impacts* (Eds F.T. Last & R. Watling), Proceedings of the Royal Society of Edinburgh. **97B**: 155–168.

Seip, H.M. (1980) Acidification of freshwaters—sources and mechanisms. In: *Ecological Impact of Acid Precipitation* (Eds D. Drabløs & A. Tollan), pp. 358–366. SNSF Project, Ås, Norway.

Seip, H.M. & Tollan, A. (1978) Acid precipitation and other possible sources for acidification of rivers and lakes. *Sci. Total Environ.* **10**, 253–270.

Seip, H.M., Muller, L. & Naas, A. (1984) Aluminium speciation: comparison of two spectrophotometric analytical methods and observed concentrations in some acidic aquatic systems in southern Norway. *Water, Air Soil Pollut.* **23**, 81–95.

Singer, P.C. & Stumm, W. (1970) Acidic mine drainage: the rate-determining step. *Science* **167**, 1121–1123.

Skeffington, R.A. & Brown, D.J.A. (1992) Timescales of recovery from acidification: implications of current knowledge for aquatic organisms. *Environ. Pollut.* **77**, 227–234.

Skogheim, O.K., Rosseland, B.O., Hoell, E. & Kroglund, F. (1986) Base addition to flowing acidic water: effects on smolts of Atlantic salmon (*Salmo salar* L.). *Water, Air Soil Pollut.* **30**, 587–592.

Smith, W.H., Geballe, G. & Fuhrer, J. (1984) Effects of acidic deposition on forest vegetation: interaction with insect and microbial agents of stress. In: *Direct and Indirect Effects of Acidic Deposition on Veg-*

etation (Ed. R.A. Linthurst), pp. 33–50. Butterworth, Stoncham, MA.

Soulsby, C. & Reynolds, B. (1993) Influence of soil hydrological pathways on stream aluminium chemistry at Llyn Brianne, mid-Wales. *Environ. Pollut.* **81**, 51–60.

Sparling, D.W. (1990) Acid precipitation and food quality: inhibition of growth and survival in black ducks and mallards by dietary aluminum, calcium and phosphorus. *Arch. Environ. Contam. Toxicol.* **19**, 457–463.

Spry, D.J. & Wiener, J.G. (1991) Metal bioavailability and toxicity to fish in low-alkalinity lakes: a critical review. *Environ. Pollut.* **71**, 243–304.

Strijbosch, H. (1979) Habitat selection of amphibians during their aquatic phase. *Oikos* **33**, 363–372.

Sullivan, T.J., Christopherson, N., Muniz, I.P., Seip, H.M. & Sullivan, P.D. (1986) Aqueous aluminium chemistry response to episodic increases in discharge. *Nature* **323**, 324–327.

Sullivan, T.J., Eilers, J.M., Church, M.R., Blick, D.J., Eshleman, K.N., Landers, D.H. & DeHaan, M.S. (1988) Atmospheric wet sulphate deposition and lakewater chemistry. *Nature* **331**, 607–609.

Sutcliffe, D.W. & Hildrew, A.G. (1989) Invertebrate communities in acid streams. In: *Acid Toxicity and Aquatic Animals* (Eds R. Morris, E.W. Taylor, D.J.A. Brown & J.A. Brown), pp. 13–29. Cambridge University Press, Cambridge.

Tam, W.H., Fryer, J.N., Valentine, B. & Roy, R.J.J. (1990) Reduction in oocyte production and gonadotrope activity, and plasma levels of estrogens and vitellogenin, in brook trout exposed to low environmental pH. *Can. J. Zool.* **68**, 2468–2476.

Tipping, E., Woof, C., Walters, P.B. & Ohnstad, M. (1988) Conditions required for the precipitation of aluminium in acidic waters. *Water Res.* **22**, 585–592.

Tipping, E., Ohnstad, M. & Woof, C. (1989) Adsorption of aluminium by stream particulates. *Environ. Pollut.* **57**, 85–96.

Traaen, T.S. (1980) Effects of acidity on decomposition of organic matter in aquatic environments. In: *Proceedings of the International Conference on the Ecological Impacts of Acid Precipitation* (Eds D. Drabløs & A. Tollan), pp. 340–341. SNSF Project, Oslo-Ås, Norway.

Tyler, S.J. & Ormerod, S.J. (1992) A review of the likely causal pathways relating the reduced density of breeding dippers *Cinclus cinclus* to the acidification of upland streams. *Environ. Pollut.* **78**, 49–55.

Tyler-Jones, R., Beattie, R.C. & Aston, R.J. (1989) The effects of acid water and aluminium on the embryonic development of the common frog, *Rana temporaria*. *J. Zool.* **219**, 355–372.

Warner, S.C., Dunson, W.A. & Travis, J. (1991) Interaction of pH, density and priority effects on the survivorship and growth of two species of hylid tadpoles. *Oecologia* **88**, 331–339.

Waters, D. & Jenkins, A. (1992) Impacts of afforestation on water quality trends in two catchments in mid-Wales. *Environ. Pollut.* **77**, 167–172.

Westall, J.C., Zachary, J.L. & Morel, F.M.M. (1976) MINEQL, a computer program for the calculation of the chemical equilibrium composition of aqueous systems. *MIT Dep. Civ. Eng. Tech. Rep. 18*.

Wigington, P.J. Jr, Davies, T.D., Tranter, M. & Eshleman, K.N. (1992) Comparison of episodic acidification in Canada, Europe and the United States. *Environ. Pollut.* **78**, 29–35.

Wood, C.M. (1989) The physiological problems of fish in acid waters. In: *Acid Toxicity and Aquatic Animals* (Eds R. Morris, E.W. Taylor, D.J.A. Brown & J.A. Brown), pp. 125–152. Cambridge University Press, Cambridge.

Wood, C.M. & McDonald, D.G. (1987) The physiology of acid/aluminium stress in trout. In: *Ecophysiology of Acid Stress in Aquatic Organisms* (Eds H. Witters & O. Vanderbergh). *Ann. Soc. R. Zool. Belg.* **117**, suppl. 1, 399–410.

Wood, C.M., Simons, B.P., Mount, D.R. & Bergman, H.L. (1988) Physiological evidence of acclimation to acid/aluminium stress in adult brook trout (*Salvelinus fontinalis*). 2. Blood parameters by cannulation. *Can. J. Fish. Aquat. Sci.* **45**, 1597–1605.

Wright, R.F. & Haughs, M. (1991) Reversibility of acidification: soils and surface waters. In: *Acidic Deposition: its Nature and Impacts* (Eds F.T. Last & R. Watling), *Proc. R. Soc. Edinb.* **97B**, 169–191.

Yan, N.D. & Stokes, P.M. (1978) Phytoplankton in an acid lake, and its responses to experimental alterations of pH. *Environ. Conserv.* **5**, 93–100.

Zimmer, D.J. (1987) Effects of low pH acclimatization on cladocerans: clues to the interaction of physiology and ecology of acid lake zooplankton. In: *Ecophysiology of Acid Stress in Aquatic Organisms* (Eds H. Witters & O. Vanderborght), pp. 139–149. *Ann. Soc. R. Zool. Belg.* **117**, Suppl. 1.

4: Metal Analysis

W.J. LANGSTON AND S.K. SPENCE

The purpose of this chapter is not to provide the reader with a comprehensive list of methods, for this would run into volumes as a result of advances in chemical analysis made over the past decade. Rather it is intended as a guide to the criteria that must be considered in order to obtain meaningful data on metals in ecotoxicological studies.

For the newcomer to the field, the choice of possible ways in which to determine and define metal contamination in the environment must seem daunting. When initiating studies, careful consideration of the precise reasons for, and objectives of the research, together with painstaking selection of the most appropriate sampling regime, is time well spent. Having decided on the most appropriate means of achieving these objectives and acquiring information, it is then essential that the method of sampling and analysis, and hence the quality of the data, is fully validated. Modern analytical equipment can generate numbers at an impressive rate, but without reasonable verification such data may be meaningless or, at very least, may restrict their impact if other members of the scientific community are uncertain of their precision.

It is now acknowledged that measuring total metal concentrations in environmental samples (sediments and water) can sometimes lead to an overestimation of potential impact on biota. Throughout the chapter we have tried to emphasize the importance of speciation in ecotoxicological studies, and to provide some perspective on why and how the environmental scientist should determine those chemical/physical forms which are biologically relevant (bioavailable).

4.1 SAMPLING

The reasons for analysing metals in an ecotoxicological context are usually related to one or more of the following:

1 to determine sources of metals in the environment;

2 to measure dispersal and release over distance and time;

3 to study routes of transfer along food chains;

4 to investigate human health consequences and more general biological effects (thereby assessing the need for emission controls or equivalent directives on environmental quality).

Depending on the purpose of the research, measurements of metal contamination may involve collection and analysis of air, water, sediment/soil or organisms. Biological monitoring is potentially capable of addressing points 1–4, while standard environmental sampling addresses only points 1 and 2. Arguably then, measurement of metals in a single phase is of limited value, on its own, to the ecotoxicologist without some understanding of how organisms respond. Ideally, representatives of all components should be analysed, perhaps several times a year, to measure the health of a particular ecosystem. Since the cost of this would be prohibitive, the sampling strategy must be tailored to available funds. If physical and chemical sampling fulfil the criteria of the research quickly and cheaply, biological monitoring may not be justified. Clearly then, the all important first stage is to establish as precisely as possible the purpose of the exercise and to define which metals should be measured. The second stage involves decisions as to where, when and by what methods these objectives can be achieved. We first consider sampling requirements for each type of medium.

4.1.1 Air

Metal analysis of air samples is principally the domain of those involved in determining fluxes between the atmosphere and land/sea surfaces, public health scientists concerned with industrial exposures, and toxicologists studying the impact of airborne metals in the vicinity of·fossil-fuel-burning power plants and metal smelters.

Atmospheric processes involved in metal transport are still poorly understood, and estimates of inputs to the sea, for example, may vary widely due to difficulties in sampling and analysis. Nevertheless, for some metals the importance of aerial deposition is clearly demonstrated; thus the major proportion of lead entering the North Sea is derived from the atmosphere, and for other metals (Cu, Zn and Hg) inputs from this source are also highly significant (North Sea Conference, 1987). Furthermore, the important influence of anthropogenic activity on atmospheric cycling of elements has been highlighted by comparisons of fluxes from natural and human sources (Nriagu and Pacyna, 1988). Metal pollution is largely a result of industrial activity including coal-burning (MaoMei *et al.*, 1989) and smelting (Tan, 1988), and as a result inputs may be localized to the areas adjacent to the main emission sources (e.g. the North Sea — Krell & Roeckner, 1987). Gradient sampling is therefore required to obtain an overall estimate of the total atmospheric input of trace metals into a particular body of water.

Average anthropogenic emissions of As, Cd, Cu, Ni and Zn exceed inputs from natural sources by a factor of two or more, while this ratio rises to approximately 17 for lead. Although beyond the scope of the majority of ecotoxicologists, some description of sampling methods and potential problems in this important subject area are warranted.

Atmospheric elements may exist predominantly as gases (Hg, Se) or as aerosols (most metals) and may reach the Earth's surface due to fallout of particles (dry deposition) or by precipitation (wet deposition) following scavenging of the air column. Sampling must therefore take into account these various forms. Sampling which includes the study of size distributions for each element may be used to identify the source(s) of individual metals (Flament *et al.*, 1987). Mechanically generated particles (e.g. rock weathering, composed predominantly of Ca, Al, Si and Fe) are larger than $2\,\mu m$ in diameter, while those from anthropogenic sources (Pb $0.65\,\mu m$, Cd $1.15\,\mu m$, Cu $1.48\,\mu m$, Zn $1.74\,\mu m$) are between 0.1 and $2\,\mu m$ in diameter (Aalst, 1988; Kersten *et al.*, 1988). Measurements are often taken at coastal land sites due to the difficulties of the collection of atmospheric data at sea (Aalst, 1988). However, development of systematic sampling which eliminates the influence of sea spray and the effects of the sampling equipment itself is required to increase the reliability of current estimates of atmospheric input.

Trace metals in seawater may be lost to the atmosphere by the formation of marine aerosols. Metals are scavenged by bubbles which burst at the surface and release their contents into the air, as demonstrated in the field for plutonium and americium in the Irish Sea (Walker *et al.*, 1986). Maximum heavy-metal concentrations, in marine aerosols, may be particularly dependent on air flow and temperature (Flament *et al.*, 1987; Otten *et al.*, 1989), but may more accurately represent local contamination of the water than direct measurement of atmospheric input. As above, direct measurements of pollutant fluxes over the sea are limited, and do not yet allow comparison of dry deposition and emission fluxes (Aalst, 1988).

Due to difficulties in the measurement of atmospheric deposition, various plant materials have been used as surrogates. The use of bryophytes and lichens is discussed by Glooschenko (1986).

Atmospheric input of heavy metals is discussed in more detail by Nriagu & Davidson (1986) and Pacyna *et al.* (1991).

4.1.2 Sediments

The physical and chemical partitioning of heavy metals between particulates and water, and an appreciation of the modes of action of these contaminants, allows the collection of information to protect against acute lethality, chronic effects (including reproductive impairment, susceptibility to disease, development) and the accumulation of contaminants. Research has moved from initial surveys of sources and pathways of heavy metals to detailed investigation of the factors controlling the mobility and behaviour of different metal species in sediment and water (Calmano *et al.*, 1990) allowing a greater understanding of the mechanisms involved in metal availability to biota.

Collection, storage, characterization and manipulation of sediments

The methods used to collect, store, characterize and prepare sediments can greatly influence the results of any sediment toxicity or bioaccumulation study (Chapters 6 and 11 in Volume 1). A greater consideration of the variables involved would allow comparisons to be made between the many studies that have been done, providing a route to the development of standard methods for assessing and predicting toxicity of, and bioaccumulation from, a given sediment.

The exposure of organisms to sediment contaminants is influenced by a number of factors which control contaminant speciation (Forstner *et al.*, 1990). These are iron and manganese oxide, organic carbon, grain size and shape, ion-exchange capacity, redox chemistry and acid volatile sulphide (AVS) concentrations. Information is therefore required on the biological and physicochemical nature of the sediment and the content of potentially available contaminants, accompanied by chemical analysis. Chemical and physical analyses, however, are unlikely to fully characterize the actual bioavailable fraction, and currently biological effects testing is required in almost all cases.

Correct sampling and sediment collection are seen to play the most fundamental role in metal speciation work (NATO, 1990). These are discussed below.

Sediment collections

Methods used to collect sediment include: grab samplers, dredge samplers, corers or hand collection (DoE, 1980; Parker & McCann, 1988; ASTM, 1990). All disturb the integrity of the sediment to some extent, and the use of corers is suggested to be the best if intact samples are required for toxicity assessment (ASTM, 1990). All samplers and equipment used for transporting sediment must be carefully cleaned to avoid contamination between sites; any metal surfaces on sampling equipment should ideally be coated with a suitable plastic or replaced with Teflon, and any rubber parts replaced by silicone (Forstner, 1989). The sampler used should be operated to prevent the loss of the metal-rich fine surface sediments. To ensure homogeneity of samples a suitable sampler should be used following careful selection of the collection sites (Ap Rheinallt *et al.*, 1989). Replication is recommended to prevent subsampling error when a site is visited, although revisit error is not usually significant.

Changes that may occur in the sediment during collection include alterations in chemical speciation due to oxidation and reduction, chemical equilibrium disruption, changes in biological activity, the degree of mixing and collection contamination. Such changes may affect the results of toxicity experiments in a way that makes extrapolation to the field inaccurate, and should therefore always be taken into consideration.

If sediment is to be spiked with heavy metals for laboratory experiments, enough sediment should be collected on each occassion to allow a blank control to be set up. Control samples should be collected as a reference against contaminated samples, allowing seasonality within natural sediments to be taken into account. If contaminated sediments from the field are used in toxicity tests, clean sediments of similar physical, chemical and biological characteristics, particularly particle size and organic content, should be used as a control. Particle size may be of particular importance if a specific test species, which selectively feeds on sediment, is to be used.

Sediment storage

Ideally, sediment for use in experiments or analysis should be processed as quickly as possible after collection. However, this may not always be possible and storage may be required.

Freezing and cold storage both affect sediment toxicity (Stemmer *et al.*, 1990), and for bioassays freezing of sediments is generally not recommended (Jones & Rae, 1989; Meador *et al.*, 1990). Freezing has been shown to decrease the release of copper into the overlying water with a subsequent reduction in toxicity to daphnids (Malueg *et al.*, 1986). In addition, changes in the chemical composition of the interstitial water occur during freezing, particularly in ammonia and phosphate levels. Drying of sediment (air-drying, N_2-drying, freeze-drying) should also be avoided (Campbell & Tessier, 1991; Murdoch & Bourbonniere, 1991).

The recommended limits for cold storage of metal-spiked sediments range from 2 to 7 days (ASTM, 1990). Ideally, sediment should be stored at 4°C, under anoxic conditions if necessary, with tests being carried out within 2 weeks. However, storage at 4°C for up to 12 months has been shown not to significantly affect toxicity (DeWitt *et al.*, 1988; ASTM, 1990).

If the sediment is anoxic, and assessment of its toxicity is required, then it should be kept under nitrogen or argon to prevent oxidation and pre-cipitation of metal species or the loss of acid volatile sulphide — thought to be the reactive solid-phase pool that binds metals in reduced sediment (Forstner, 1989; Di Toro *et al.*, 1990). Oxidation processes during the handling of sedi-ments may effect an enhanced consumption of oxygen and a reduction of bonding strength of metals (Forstner *et al.*, 1987). Failure to exclude oxygen from anoxic sediment results in hydrogen sulphide evolution in addition to a decrease in AVS, changes in the amount and nature of organic matter in the pore water and loss of iron and phosphate (Orem, 1982). Stabilization of sediments under inert gases also minimizes microbial trans-formations of critical pollutants during storage, thus maintaining the integrity of the sediment.

A discussion of methods for handling and storing sediments, assessing the use of the auto-clave, storage at 10°C and sieving is given by Deans *et al.* (1982).

Sediment characterization

Factors known to influence toxicity and bioavail-ability of metals include temperature, organic carbon, particle size distribution, pH, Eh, cation exchange capacity, AVS, ammonia and interstitial water content (Calmano *et al.*, 1990). Bioavail-ability of a given metal may differ by as much as 1000-fold among different sediment types (Campbell & Tessier, 1991); sediment character-ization is therefore important if comparisons are to be made between the results obtained in differ-ent experiments, within and between laboratories.

Temperature. The temperature in natural systems (usually 4–20°C) can differ appreciably from lab-oratory tests (Tessier & Campbell, 1987), and should therefore be considered when comparing results from the field and artificial systems.

Organic carbon. One method to estimate the organic matter content of sediment is to deter-mine the loss in weight of dry sediment (80°C) heated at 400°C for 6 h, the sample being cooled for 1 h in a desiccator after each stage prior to weighing. A correction factor for marine/ estuarine sediment, to take into account the loss of seawater salts at 400°C, may be required (Bryan *et al.*, 1985). The importance of the proportion of particulate organic carbon has been shown in sandy sediments, when a small variation in the quantity of organic-rich fines can result in greatly differing particle-bound cadmium concentrations, though interstitial water cadmium concen-trations remain relatively constant (Kemp & Swartz, 1988).

Other, more accurate methods for organic carbon include the use of an elemental (CHN) analyser, but analysis should be carried out on a carbonate-free basis (Craft *et al.*, 1991). A cali-bration of organic carbon determined using a CHN analyser and dry combustion is included by Craft *et al.* (1991).

Particle size. Organic content and particle size have been shown to modify amphipod survival (DeWitt *et al.*, 1988), and positive correlations between heavy-metal concentrations and decreas-ing grain size have been reported (Langston, 1982; Forstner, 1989). Particle size may be determined by a number of methods which include: wet sieving, hydrometers, settling techniques, X-ray

fine
It is
ates
ters,
TM,
ded.
hape
sons
any
nent

n or
for
inly
on is
sus-
not,
for
ater
rous

pollutant studies, particularly with resp t to heavy metals, have already been performed on the $< 63\,\mu m / < 100\,\mu m$ fraction, and further studies using similar fraction sizes would allow comparisons to be made (Forstner, 1989).

The $< 63\,\mu m$ fraction is often used to normalize data (ASTM, 1990). However, geochemical normalization is superior to granulometric methods to compensate for the natural variability of trace metals in sediments, because it compensates for the mineralogical, as well as the granular, variability of the sediment (Loring, 1991).

pH. The distribution of metals between solution and sediment is related to pH (Hunt, 1986; Fu & Allen, 1992). The lowering of pH, for example, enhances the release of heavy metals from sediments into solution (Forstner, 1989).

The potential release of heavy metals from sediments can be assessed in detail using buffer capacity and acid potential. These are determined respectively by (i) assessing the difference in pH of natural sediment and that of the sediment following one hour shaken with O.1 N acid and (ii) determination of the release of hydrogen ions when anoxic sediment is oxidized (Calmano *et al.*, 1990).

Eh. The Eh of a system is at best a semiquantitative expression of its oxidizing or reducing intensity. Redox potential may be defined as the electron-escaping tendency of a reversible oxidation−reduction system. However, the pH and temperature at which measurements are made, together with sediment composition, influence the redox potential of sediment samples (Zobell, 1946). Positive Eh values are characteristic of well-oxygenated sediments and negative redox potentials of deposits rich in organic matter and consisting of mainly fine sediment. Organic matter and micro-organisms create reducing conditions.

Eh meters are readily available which use, for example, a platinum rod electrode coupled with a calomel reference half-cell (Wharfe *et al.*, 1985). Redox methodology is outlined in detail by Zobell (1946) and Fenchel (1969). Alternative methods of measuring redox status include pS electrodes, $Fe^{2+} : Fe^{3+}$ ratios and AVS.

Acid volatile sulphide (AVS). AVS is a reactive pool of solid-phase sulphide that is available to bind metals and may render that portion unavailable and non-toxic to biota (Di Toro *et al.*, 1990). Metals that are frequently associated with the sulphide fraction of suspended matter and sediments in anaerobic environments include Zn, Pb, Cu, Co, Ni, Cd, As, Sb, Hg, Mo and occasionally Mn (Morse *et al.*, 1987). The relative association of the different trace metals between sulphide and organic complexes varies in different environments. The quantity and distribution of AVS and pyrite-S in marine sediments depends on a large number of variables including carbon input rate, sediment burial rate, sulphate diffusion, bioturbation, iron input and reactivity and sediment temperature (Morse *et al.*, 1987). The determination of AVS is therefore important in studies of trace metal cycling in the aquatic environment (Cutter & Oatts, 1987).

To determine AVS concentration, dilute acid is added to the sediment, forming AVSs and pyrite. Hydrogen sulphide produced is stripped and trapped, followed by analysis of the sulphide formed by gravimetry or iodometric titration (see Di Toro *et al.* 1990; Morse *et al.*, 1987 for a review of the extraction conditions used to determine AVS content). The acid may be warmed or utilized at room temperature (Jorgensen & Fenchel, 1974; Hsieh & Yang, 1989) with precautions being taken when anaerobic sediment is assessed for AVS content to prevent loss due to oxidation (Wang & Barcelona, 1983).

AVS is thought to be of importance in limiting the bioavailability and toxicity of trace metals, and AVS normalization has been suggested as a reasonable means for assessing the hazard of some sediment-associated metals in aquatic ecosystems. Those assessed to date include Cd and Ni (Ankley *et al.*, 1991, Di Toro *et al.*, 1990, Carlson *et al.*, 1991). Zn, Pb, Cu, Hg, Cr, As and Ag may displace Fe in the same way from the monosulphide complex and subsequently form insoluble and biologically unavailable sulphides in sediments, modulating metal bioavailability (Ankley *et al.*, 1991b).

Cation exchange capacity. Cation exchange capacity is measured using anodic stripping voltammetry (ASV). This measures metal ion concentration at the levels of the hydrated cation and labile complex present in solution. It is a sensitive method (microgram per gram detection level), and minimizes errors due to resorption of displaced metal ions (Waller & Pickering, 1990). However, the need to filter samples to remove colloidal matter, prior to analysis, limits its use (Waller & Pickering, 1990).

An alternative method (which may be used to determine bioavailable fractions of metals in sediments) is the equilibration of ion-exchange resins with aqueous suspensions of sediments, used as a means of subdividing the metal content of estuarine sediments (Pb, Cu, Cd, Zn, Ca, Mg, Mn, Fe, Al) into different 'labile' or 'available' fractions (Beveridge *et al.*, 1989). The tendency for readsorption on the sediment or other surfaces is minimized during extraction. The adsorbed cations are then extracted into ethylenediaminetetraacetic acid (EDTA) which is analysed by atomic absorption spectrometry (AAS). The value found with any particular exchange resin or chemical extractant, however, may depend on the chemical form of the species to be extracted, and requires further study.

Manipulation of sediments for experiments

Manipulation of sediments is often required to obtain consistent material for toxicity testing and laboratory experiments. Methods include: sieving, spiking, mixing, dilution and sterilization.

Sieving is used to obtain a maximum particle size and to remove unwanted biota. Sieving should be done on all samples, including the control, obtaining homogeneity and allowing for contamination from the sieves to be taken into account.

Spiking of sediments with pollutants may be done by addition to the overlying water, or directly into the sediment by mixing or injection. The pollutant is mixed into the sediment by shaking, stirring, tumbling or simple diffusion. Contact time between toxicant and sediment can affect partitioning and bioavailability due to rapid labile sorption followed by toxicant movement and stabilization (ASTM, 1990). Preliminary studies may therefore be required to determine the most suitable protocol for the type of sediment and toxicant under investigation. Care must be taken to mix sediment adequately so as to distribute the contaminants evenly. This is a complex process during which the sediment should be fluid enough for the particles to come into contact with the added contaminant. Tumbling has been shown to be a particularly effective method of mixing (Ditsworth *et al.*, 1990) and shaking may also be used (Stemmer *et al.*, 1990).

Contaminated sediment may be diluted with clean sediment to obtain the required toxicant concentrations. However, it is recommended that heavy metals should be added concurrently, at the final concentrations required, to allow the same equilibration time between the toxicant and all the sediment to be used, resulting in greater homogeneity (Apte *et al.*, 1989). If a solvent is used to add the pollutant, it should also be added to the control in a comparable concentration. The sediment should then be washed in the water type to be used in the experiment to remove any excess solvent.

If required, various methods may be used for sediment sterilization to inhibit biological activity. Autoclaving kills any live material present, but organic matter remains which may act as a sink for added pollutants and food for the test species. Use of the furnace to overcome this problem has its own drawback of releasing bound metals from the sediment being used. The addition of antibiotics (ASTM, 1990) may be a compromise, reducing the chemical and physical changes due to autoclaving or combustion, although the dead cell material remains.

A review of the physical, chemical and microbiological changes that occur during collection and preparation of sediment for laboratory experiments is given by Deans *et al.* (1982).

Extraction procedures

During any sediment extraction procedures the acids and other reagents used must be of the highest purity available (AR grade) to prevent contamination. Blanks should be prepared according to the same procedures used for heavy-metal extraction, to allow any sources of error introduced during such procedures to be taken into account when the final analyses are made.

Total metal

The total metal present in sediment is traditionally determined using alkali fusion or hydrofluoric acid/mineral acid mixtures (Campbell *et al.*, 1988). However, comparison of the safer nitric acid digestions with these more vigorous procedures indicates that it usually extracts most of the metal, allowing it to be used routinely for many elements. It seems unlikely that metals remaining insoluble after this treatment are ever likely to become bioavailable (Bryan *et al.*, 1985).

To determine total sediment metals by this latter method, about 1 g of oven-dried (80°C) sediment is digested with 20 ml of HNO_3, in a flask covered with a glass bubble, by heating on a hot plate for 2–3 days at 80°C until a pale yellow solution is obtained. The acid is then slowly evaporated, the residue dissolved in concentrated AR grade HCl (1 ml per 1 g of dry sediment), with warming if necessary, and then diluted to give 10% HCl (Bryan *et al.*, 1985). The extracts are decanted into test tubes and insoluble particles allowed to settle prior to analysis of the extract-

ant. Total Hg and other volatile elements such as As and Se, are determined by refluxing 2–3 g slurry with 20 ml HNO_3 in a Kjeldahl flask fitted with a condenser for 2 h, making up to 100 ml when cool (Langston, 1982).

These methods of digestion are time-consuming and have led to the development of rapid microwave digestion techniques (Littlejohn *et al.*, 1991). Samples are placed in polytetrafluoroethylene (PTFE) bombs, nitric acid added and the vessels heated in a domestic microwave (650 W at 90°C for 3 min). This has been shown to be a satisfactory alternative to conventional reflux extraction (Nieuwenhuize *et al.*, 1991).

Bioavailable fractions

Many studies have shown that the trace metal levels in benthic organisms are often best related not to the total metal content of the sediment but rather to relatively easily extracted fractions (Luoma & Bryan, 1978; Langston, 1980; Diks & Allen, 1983; an overview of interactions between contaminated sediments and benthic invertebrates in marine and freshwater systems is given by Reynoldson, 1987 and by Bryan & Langston, 1992). Of the many methods developed to define such fractions, however, no single technique is clearly superior (ASTM, 1990).

The total metal content of sediments is the sum of the fractions present in different chemical forms or bonded to different sediment components. The fraction of greatest significance to the biota is the 'biologically available' metal (Beveridge *et al.*, 1989). Since modelling of sediment-bound metals is far less advanced than that of dissolved species (primarily because the thermodynamic data needed for sediment–interstitial water systems are not yet available – Tessier & Campbell, 1987) chemical extraction is therefore required. The most widely used approach to determine the biologically available fraction is selective chemical extraction; various schemes are summarized by Campbell *et al.* (1985) and Tessier & Campbell (1987). Care must be taken during processing to preserve sediments in their natural state, to allow such selective extraction methods to be effectively used (Jenne *et al.*, 1980; Kersten & Forstner, 1987; Jordao & Nickless,

1989). Selectivity deteriorates markedly in sulphide-rich anoxic sediment, but is acceptable for sulphide-poor, oxic sediments (Luoma & Bryan, 1981; Tessier & Campbell, 1987; Campbell & Tessier, 1991). However, despite numerous limitations, such methods probably represent the best approach to determining biologically important fractions, currently available.

Dilute (10%) hydrochloric acid has been used with some success to determine the easily extractable or 'bioavailable' fraction of heavy metals present in sediments (Bryan & Langston, 1992), employing methods such as the following.

1 Aliquots (2 ml) of sediment slurry are pipetted into glass liquid-scintillation-counting vials and extracted for 2 h with 20 ml of 10% HCl, stirring continuously.

2 The extract is separated from the sediment by filtration under pressure through a 0.45 µm membrane filter using a glass syringe connected to a plastic membrane filter holder (Millipore, Swinnex 25 mm). The apparatus and filters are all precleaned with 10% HCl.

3 Other aliquots of slurry are dried, allowing calculation of heavy-metal concentration on a dry-weight basis. This is a relatively quick and simple method which may be used to obtain a guide to the bioavailability of heavy metals in a particular sediment (Bryan *et al.*, 1985).

Sequential extraction

The use of sequential extraction subdivides the 'easily extractable' metal into a number of fractions, in addition to dividing the more strongly bound metal into operationally defined fractions. Sequential extraction procedures remove specific metal forms from the sediment matrix based on the assumptions that selective partial extraction is feasible and neglible readsorption of solubilized metals occurs (Campbell & Tessier, 1991).

The chemical extractants are selective rather than specific in their action. However, if used in an appropriate sequence going from least to most aggressive, the distribution of the trace metals among their principal sedimentary sinks may be assessed (Campbell *et al.*, 1985). Sequential extractions cannot emulate biological systems, but may be of considerable indirect use when assessing biological availability of metals by: (i) predicting the most probable response to changes in physical (e.g. the effects of bioturbation (Krantzberg & Stokes, 1985) and chemical conditions and (ii) establishing the metal-binding capacity of operationally defined components of the sediment (Campbell *et al.*, 1985). When the results from such extractions are obtained, care must be taken in interpreting relationships between metal levels in tissues and the individual fractions due to the occurrence of significant intercorrelations between the fractions themselves (Tessier *et al.*, 1984).

The following is an example of the various sequential extraction procedures. To determine metal partitioning, a volume of slurry equivalent to about 1 g dry weight is pipetted into a centrifuge tube. After centrifuging, the aqueous layer is removed and the retained solids extracted sequentially as follows (based on schemes by Tessier *et al.*, 1979 and Rapin & Forstner, 1983).

1 Exchangeable fraction — shaken at room temperature with 30 ml of 1 M ammonium acetate/0.25 M calcium chloride at pH 6 (acetic acid) for 3 h.

2 Carbonate fraction — shaken at room temperature for 5 h with 20 ml of 1 M sodium acetate adjusted to pH 5 with acetic acid.

3 Easily reducible fraction — shaken with 20 ml 1 M sodium acetate, 0.25 M hydroxylammonium chloride at pH 5 for 30 min.

4 Moderately reducible — digested at 96°C for 6 h with 20 ml 0.25 M hydroxylamine hydrochloride in 25% (v/v) acetic acid.

5 Residual organics and sulphides — extracted at 85°C for 2 h with 5 ml acidified 30% hydrogen peroxide (pH 2), then continuing extraction for a further 3 h after addition of another 5 ml of 30% hydrogen peroxide. After cooling, sample shaken for 30 min at room temperature extracting with 10 ml of 3.2 M ammonium acetate in 20% (v/v) nitric acid.

6 Acid-extractable (residual) fraction — digested for 4 h at 120°C with 20 ml of concentrated nitric acid.

7 Humic associated — on a separate aliquot of sediment from the sequential scheme above; sample shaken at room temperature for 6 h with 20 ml 0.1 M tetrasodium pyrophosphate.

After each stage the residues are separated from

the extracts by centrifugation and washed with distilled water. The combined extracts and washings are filtered (25 mm, 0.45 µm membrane (Millipore) filter) then acidified with 1 M HCl (AR grade) and made up to volume prior to analysis (50 ml for extracts 1–5 and 7 200 ml for extract 6).

The importance of partitioning of trace metals in sediments has been highlighted in a number of laboratory and field studies (Luoma & Bryan, 1982). However, at present there is a need for a method to determine the actual, and not merely the operationally defined, partitioning of the metals. The latter gives valuable information about the important sinks in natural sediments, but, if generalizations are to be made, theoretical models need to be developed and verified in predicting the behaviour of trace metals under natural conditions (Tessier & Campbell, 1987).

· Bioavailability of trace metals to aquatic organisms is markedly influenced by the metal speciation. The conduct of tests using only simple inorganic metal species to spike 'clean' sediment may therefore lead to overestimation of the impact of many trace metals because other species, important in nature (for example organic complexes and adsorbed forms – Lion *et al.*, 1982), are usually less bioavailable than simple inorganic forms (Hunt, 1986). Organic complexation of many heavy metals may also reduce their toxicity to aquatic organisms, and residual toxicity is often related to the 'free' metal ion concentration (review Hunt, 1986; Lake *et al.*, 1989). Comparisons of the impact of natural sediments, loaded with contaminants, and spiked sediments (with similar physical and chemical characteristics, prepared in the laboratory) are required to clarify this important topic.

Prediction of metal concentrations in organisms is often improved when metal concentrations in the sediment are normalized with respect to Fe, Mn and/or organic content of the sediments (Langston, 1982; Luoma & Bryan, 1981, 1982; Tessier *et al.*, 1984), highlighting the importance of complexation and adsorption. The extractable (10% HCl) phase of Fe shows a stronger association with Ag, Cd, Cu, Pb and Zn in oxidized sediments than does total Fe, and the humic fraction of organic materials is highly important in binding Ag and Cu (Luoma & Bryan, 1981). Sequential extraction procedures allow the initial development of models to predict the bioavailability of trace metals to benthic organisms (Tessier & Campbell, 1990) but, as already mentioned, the subject of speciation in sediments – and fluxes between the particulate, aqueous and living matter – requires further detailed investigation; in particular our understanding of the geochemical processes that control the scavenging of trace metals to sediments, and their release to overlying and interstitial water, needs to be improved while a knowledge of the route of entry of trace metals into benthic organisms should help in the selection of predictive variables (free-ion concentration, pH, organic matter). Biological factors that affect metal accumulation (age, sex, etc.) must also be quantified if meaningful models are to be constructed (Tessier & Campbell, 1990).

4.1.3 Water

Because of the low levels of metals encountered in the majority of water (and air) samples, strenuous efforts must be made to avoid the introduction of contamination. Particular attention should be given to cleaning glassware, filters and reagents; the use of clean-room facilities and laminar flow hoods may be essential in many instances.

Sampling of water

Bioavailable metal species may be present either in the interstitial water, within the bulk of the sediment itself, loosely associated with the sediment particles or in the water column overlying the sediment.

Interstitial water

Metal bioavailability from interstitial water (IW) is of interest for organisms that live in intimate contact with the sediment but do not ingest particulate material, as well as to those capable of ingesting sediment and taking up metals from the particulate phases (Campbell & Tessier, 1991). Digestive chemistry is of importance in the latter organisms since metals must normally pass into solution before they are assimilated in the gut.

IW has been collected for a variety of reasons: to determine potential availability of heavy metals to biota (Campbell & Tessier, 1991); to study depth profiles (Brumsack & Gieskes, 1983; Tolonen & Merilaeinen, 1983) and metal partitioning in relation to sediment type (Sawlan & Murray, 1983); to assess the effects of biota on nutrient profiles (Herndl *et al.*, 1989; Huttel, 1990; Jorgensen *et al.*, 1990) and nutrient cycling (Blackburn & Henriksen, 1983; Giblin & Howarth, 1984; Edenborn *et al.*, 1985).

Fluid extraction from sediments dates back over 100 years, and until recently has mainly involved centrifugation, gas displacement and pressure filtration. The development of *in-situ* collections allows more standard methods to be used, and serves to check and refine more traditional techniques (Kriukov & Manheim, 1982). The collection of IW may be achieved by a number of procedures.

1 Centrifugation (Jorgensen *et al.*, 1981).

2 Squeezing, which ideally requires a controlled temperature and redox environment (Carr *et al.*, 1989; Jahnke, 1988).

3 Dialysis, preferably using cellulose-based membranes rather than polycarbonate membranes to avoid the problem of iron precipitation within the apparatus (Carignan, 1984). The use of dialysis is limited by the long equilibration time that is often required (Bottomley & Bayly, 1984).

4 *In-situ* samplers, which allow regular temporal sampling without disturbing the sediment or the biota (Gaillard *et al.*, 1989; Buddensiek *et al.*, 1990; Huttel, 1990; Watson & Frickers, 1990). Samples may be taken along a depth profile without interference between the levels, providing accurate assessment of redox, pH, nutrient and contaminant changes. Simple, one-level, *in-situ* samplers have been developed using fritted glass samplers which fill by hydrostatic pressure. These are prepared from glass tubing and the 15 µm pore size fritted section from a gas dispersion tube (Pittinger *et al.*, 1988) or from nylon (10–15 µm) mesh and glass tubing (Spence & Bryan, 1990). The tubes are wrapped with 0.7 µm glass-fibre filter paper (Whatman GF/F) to prevent clogging.

Centrifugation and squeezing allow the recovery of large volumes of IW for use in aqueous testing, but the remaining methods are more useful in laboratory or short-term, *in-situ*, field studies. The use of IW or pore water for bioassays and toxicity tests for heavy metals is, however, limited due to the manipulations required prior to their use (Carr *et al.*, 1989). Dissolved oxygen must be added, the pH increased and, often, the salinity adjusted, for biota to survive. The collection of IW is therefore recommended for direct analyses but not for general use in toxicity studies. The composition of IW in sediments is perhaps the most sensitive indicator of the types and the extent of reactions that take place between pollutant-loaded sediment particles and the aqueous phase in contact with them (Calmano *et al.*, 1990). The immediate processing of IW from field sediments is recommended since chemical changes may occur even when sediments are stored for a short time period. Once removed from the sediment, IW should be acidified or deoxygenated. Storage of IW for arsenic (III), for example, is recommended by acidification to pH 2 and refrigeration at 0°C, or by deoxygenation in the absence of refrigeration (Aggett & Kriegman, 1987).

Preliminary results suggest that measurements of adsorbed metal concentrations in sediment may be a valuable alternative to those attempts aimed at predicting the correlation of the free-metal ion concentration in IW with metal levels in organisms. The use of Teflon sheets, to collect sorbed metals, would reduce the errors that may occur when IW is collected and analysed, and may be more accurate than direct determination of metal concentrations in IW (Campbell & Tessier, 1991).

Elutriate tests

Elutriate tests are valuable in assessing the mobility and impact of sediment-associated contaminants, such as those released during sludge disposal and dredging operations (Brannon *et al.*, 1980; Craig, 1984; Palermo, 1986), notably on those organisms living in the water column (Munawar *et al.*, 1985). The tests involve mixing the sediment with clean water, settling of the solids and filtration of the water for use in toxicity tests. It is assumed that a determined proportion of adsorbed contaminants will transfer to the aqueous phase (Ludwig *et al.*, 1988; Samoiloff, 1989). Elutriate bioassays can provide guidance when chemical analysis is insufficient to assess the potential impact of contamination (Chapman, 1987; Athey *et al.*, 1989), and may be followed by sediment bioassays studies as a reproducible measure of *in situ* toxicity.

The predictive ability of a particular test is often related to the test species used. A study comparing the predictive ability of pore water (IW) and elutriate tests for the toxicity of bulk sediments suggested pore water to be the most effective fraction for predicting the presence of bulk sediment toxicity to benthic organisms (Ankley *et al.*, 1991b). In contrast, elutriate tests are more suitable for toxicity assessment with pelagic rather than benthic species.

Overlying water

Overlying water (OW) is collected using open containers or, if required from depth, special containers are available which allow a discrete sample to be obtained from a chosen depth. The containers should be thoroughly cleaned to prevent contamination and chemical and/or biological reactions between the water and its components and the vessel walls. Glass is advisable, but plastic is often suitable for heavy-metal determination (DoE, 1980), and PTFE bottles are being used increasingly, for example for mercury. If collections are being made from a research vessel, the contaminated surface layer should be avoided by passing the sampler through it in closed configuration, opening it at the required depth by messenger. Alternatively, if uncontaminated samples from the surface are required, samples should be collected away from the vessel, sampling from the bow of a dinghy moving at slow speed.

Preparation of water samples for analysis

The determination of heavy metals in estuarine and marine waters causes difficulties due to interference effects arising from the high salt content of the matrix. Initial interferences may be reduced by filtration of the water through 0.45 µm, removing undissolved components (Hunt, 1986). The sample may then be acidified to maintain solubility of the metals, a standard procedure prior to heavy-metal analysis. Particulate trace-metal concentrations may be determined by flame or furnace AAS after complete digestion of solids with nitric acid and hydrofluoric acid in Teflon bombs (Littlejohn *et al.*, 1991). To reduce interference further, a matrix separation/analyte preconcentration step prior to analysis may be used (Apte & Gunn, 1987; Mantoura *et al.*, 1987; Smith & Windom, 1980). Thus, for example, preconcentration may be achieved by chelation of the metals with ammonium pyrrolidine dithiocarbamate (APDC) and subsequent extraction into 1,1,1-trichloroethane. The organic solvent can be injected directly into the graphite furnace, eliminating the need for back-extraction into acid, used in more time-consuming methods (Danielsson *et al.*, 1978, 1982; Campbell *et al.*, 1985). The extraction time for 25 samples and standards is estimated to be 90 min, with detection limits by atomic absorption allowing routine monitoring of filterable metals in many coastal and estuarine waters (Cu 0.3 µg/l, Cd 0.02 µg/l, Pb 0.7 µg/l, Ni 0.5 µg/l). The following method, based on a scheme by Apte & Gunn (1987) uses a matrix separation/analyte preconcentration step.

The metals to be analysed in the water samples are complexed as their pyrrolidine dithiocarbamates and extracted into a small volume of 1,1,1-trichloroethane in which these complexes are very soluble, thus achieving a concentration factor of $\times 5$, as well as salt removal.

1 A sample of 2.50 ml of filtered seawater is pipetted into a 5 ml conical Teflon PFA (Cole-Parmer) extraction vial. To each vial 15 µl of 1 M HCl is added per 1 ml of sample.

2 To each vial 250 µl of neutralizing/chelating agent is added. This is prepared by dissolving 0.40 g APDC and 5.00 g sodium hydrogen carbonate in water—final volume of 100 ml. To purify the reagent it should be transferred to a precleaned PTFE container and shaken with 20 ml 1,1,1-trichloroethane for 5 min, allowing the phases to separate before use and pipetting from the upper aqueous layer.

3 The extractant, 500 µl of 1,1,1-trichloroethane, is added (pipetting is easier if the solvent is cold), the caps placed on all the vials and secured in a shaking rig.

4 The rig is shaken vigorously for 10 min and samples allowed to stand for 5 min for the phases to separate.

5 The lower 1 ml of solution is transferred into an autosampler cup. Some of the upper aqueous layer must be transferred along with the 0.5 ml of organic extract, to prevent evaporation losses.

6 Analysis of 1,1,1-trichloroethane is done using furnace AAS (or equivalent), ensuring that the sampling probe is adjusted to sample from the lower organic layer only. Other methods of preconcentration include: 8-hydroxyquinoline (Broekaert, 1990).

The 'free metal' concentration is considered most important when assessing toxic effects of dissolved metals in water. This refers either to the free metal ion concentration (M^{2+}) or the total inorganic metal concentration (free ion plus inorganic complexes). ASV may be used for measurement of the free metal ion concentration (Hunt, 1986). Various methods of extraction may be used to divide heavy metals present in water, e.g. into particulate, labile, dissolved and acid-leachable fractions (Danielsson *et al.*, 1983), or partitioned between solution, ion exchange, organic materials, metallic coatings and crystalline solids (Gibbs, 1973). Speciation of trace metals in natural waters (Hunt, 1986) is easily viable for Al, Cd, Cu, Pb, Ni and Zn, but other metals and metalloids (for example As, Cr, Hg, Se), have a different chemistry involving the occurrence of well-defined kinetically inert (non-labile) organic compounds of different oxidation states, making speciation more complex. 'Speciation' of mercury is achieved sometimes by distinguishing methyl-mercury, and 'labile' inorganic mercury (that which is reduced by $SnCl_2$ without further treatment), from 'total' mercury. The latter includes organically bound mercury, and is made available following ultraviolet (UV) radiation (Schmidt, 1992). In fresh waters the hydroxide and carbonate ions are the dominant inorganic ligands for most metals, and in saline waters the chloride ion is also often of importance with respect to Al, Cd, Cu, Pb, Ni, Zn (Hunt, 1986).

4.1.4 Organisms

Analysis of organisms has many obvious advan-

tages over air, water, sediments and soils when determining mechanisms and consequences of metal uptake, since tissue burdens are often a direct manifestation of biologically available metal in the environment. Living organisms also tend to provide an integrated picture of metal contamination rather than the 'snapshot' values determined by sampling various media.

Methods for digestion and analyses of tissues are similar to those described for sediments (see, for example, Bryan *et al.* (1985), Langston (1982) and other publications cited in this chapter). Recently the use of microwaves has led to considerable reduction in processing times (Littlejohn *et al.*, 1991). Rather than reiterate these schemes here, it is more valuable to discuss the difficulties involved in achieving representative samples, when faced with the diversity and complexity of organisms in animal and plant kingdoms.

General problems in the use of organisms

Biological monitoring has assumed increasing significance in recent years, particularly when used alongside biological effects studies. As a general, though by no means universal, rule (relationships can be disturbed by a number of physiological and environmental factors), concentrations of contaminants in tissues reflect concentrations in the environment. However, variability in tissue burdens is seen as a potentially major problem in the interpretation of results, and it is important to identify (and quantify) the major causes of variability in order to 'visualize' the magnitude of change which might reasonably be detected when comparing samples from different sites or metal exposures. Causes of biological variation in metal determinations fall generally into the following categories:

1 differential metal-handling abilities between individuals (and between tissues/organs of the same individual);

2 physiological status of the organism: size, age, growth rate, reproductive stage, sex;

3 variation due to microhabitat;

4 local variation in climatic or physical/edaphic factors;

5 seasonal changes;

6 'contamination' by inclusion for analysis of gut

contents or extraneous material which is not biologically incorporated.

Pilot studies may greatly assist in reducing these potential sources of variation from the outset. For example, in the initial selection of species (where this is an option in the research programme) it may be preferable to choose an organism/organ/tissue that shows stability and responds to changes in one or more environmental phase. The work of Bryan and co-workers (1980) provides an illustration of such an approach in evaluating three burrowing species (the polychaete *Nereis diversicolor*, and deposit-feeding bivalves *Scrobicularia plana* and *Macoma balthica*) as potential indicators of sediment—metal contamination in estuaries. All three species proved useful for the majority of metals studied, though *Scrobicularia* appeared to be the best accumulator, most of the metal burden residing in its digestive gland. The disadvantage of dissecting out this tissue is balanced by its ease of analysis and lack of dilution effects caused by gonadal development. Similar conclusions were drawn by Eganhouse & Young (1978) in their studies with mussels, *Mytilus californianus*. Analysis of individual tissues may be of additional value in determining the significance of metal burdens in terms of toxicity, storage sites and routes of uptake. The analysis of digestive gland (relative to whole organism) in molluscs, for example, may indicate the comparative importance of sediment/diet as the major source of metal, while high concentrations in gills are often thought to reflect the significance of metals in the dissolved phase. For monitoring purposes, however, different tissues may often indicate common trends. Thus, multi-site comparisons of metal concentrations in digestive gland and whole soft tissues of *Scrobicularia* show that responses of these two tissue types are broadly similar. Analysis of whole organisms has subsequently been adopted as standard in this context (Bryan *et al.*, 1980).

Size

Of all the physiological factors that influence metal burdens in organisms the effect of size/growth rate is probably the most important in the majority of species. Evaluation of the effect of size on metal concentrations in *S. plana*, for example, reveal substantial differences between metals and, in contaminated estuaries, concentrations of Cd, Co, Cr, Ni, Pb and Zn tend to increase and Ag, Mn, Cu and Sn decrease, as the animal grows. Hg and As concentrations changed little over the entire size range (Bryan *et al.*, 1980). It should be recognized, however, that size—metal relationships can vary between locations (Langston, 1982). In order to make comparisons of contamination between sites using clams, Bryan *et al.* (1980) and Langston (1982) recommend standardizing samples on animals of 4 cm shell length (0.3–0.4 g dry weight soft tissues). Such a simplified procedure may not be satisfactory for all organisms, however, and alternative assessments of the effect of size are discussed for individual species below.

Boyden (1977) examined the dependence of metal levels on size in a variety of shellfish species by plotting values for metal content (Y) against body weight (X) according to the (power function) form:

$$Y = aX^b$$

Logarithmic transformation of data produced straight-line relationships in most cases with the *majority* of slopes conforming to one of two groups; those of approximately 0.77 (metal concentrations highest in small individuals) and those metal—species combinations with a slope of 1 (concentrations independent of size). However, other relationships were apparent, including examples of increasing concentrations with size (slope greater than 1), as described above for *S. plana*.

By utilizing these and other similar relationships, various methods can be devised to overcome problems of variability in metal data caused by size differences and, though not perfect, are often preferable to making no attempt at correction. One approach commonly used is to normalize metal concentrations in the animal to a standardized weight/size/condition. Since metal content and concentration are the same value for a 1 g individual, the selection of this size has attractions for many groups. The amount of metal in an animal of 1 g dry weight, a, can be derived by modifying the equation described above, such that it represents the weight-specific (metal concentration) relationship to body weight:

$$\text{metal concentration } (\mu g/g) = aX^{(b-1)} \quad \text{or}$$
$$a = \text{metal concentration}/X^{(b-1)}$$

where X is the actual dry weight and (b^{-1}) is the slope of the metal *concentration*–tissue weight relationship (as opposed to b, the slope of the metal *content*–tissue weight relationship). Negative and positive values for (b^{-1}) indicate decreasing and increasing concentrations with size respectively. It is important to note, however, that if it is wished to base collections on length stratification, rather than selecting a standard size organism, the validity of this approach hinges on the reliability of the slope and hence large size ranges (over an order of magnitude) are essential. Whether animals are analysed individually, or pooled into different size groups, similar size range distributions should ideally be selected for each sampling location/time.

Alternative methods to compensate for growth-related variability in metal concentrations in mussels, based on normalization to a constant flesh condition index (mg dry weight flesh per 1 g shell weight) are proposed by Lobel & Wright (1982), and a similar cadmium/shell weight index (the amount of cadmium present in the soft parts of mussels having a shell weight of 1 g) is described by Fischer (1983). Pre-examination of mussels to ensure similar growth rates and condition indices has subsequently featured among suggestions aimed at improving protocols for collection of shellfish in biological monitoring programmes (Lobel *et al.*, 1991).

It has been recommended that, for fish, checks and appropriate adjustments might be made for various biological variables, including age, body weight, liver weight, sex and degree of maturation to improve precision of analytical data (ICES, 1989). Size is widely perceived as being the most important of all parameters influencing contaminant burdens, and models describing length–metal relationships may be essential for monitoring trends. Fortunately, since size and a number of the above parameters covary, adjust-

ment of metal burdens for length, for example, has the effect of unifying data with respect to several of the other variables. ICES guidelines indicate collection of at least 25 individuals per sample (whose sizes should span as wide a length range as possible) and division into equal numbers (e.g. five) per length group. Length intervals between these groups should be of roughly equal size after logarithmic transformation (i.e. not clustered). The precision of such an approach will, as with invertebrates, depend on the accuracy of models defining length–concentration relationships, bearing in mind that these may vary not only between species, but at different locations and at different times.

Careful consideration of the data structure is clearly of paramount importance: where length–concentration relations are significant it will be necessary to compare slopes; where samples do not need adjustments for length or other variables, trends may be ascertained by comparisons of means. However, it is frequently worth considering additional information on conditions at the sampling sites, particularly those that might cause outliers in the data. Tissue residue data alone may not always be sufficient evidence on which to base conclusions regarding contaminant trends, and sometimes call for comparisons with environmental (water, sediment) samples.

Season

It is common that weights and contributions of different tissues to the total body weight alter throughout the year. Seasonal variations are therefore to be expected to influence slopes relating metal content to body size, and there are enough examples in the literature to provide convincing arguments for the need for preliminary surveillance before the timing of sampling is decided. Clearly the most suitable period and frequency of collection will be unique to the species under investigation, though if there is one common requirement it is likely to be advantageous to sample when body concentrations are least likely to be subject to variations caused by weight changes which are related to food supply and position in the spawning cycle. 'Dilution' effects caused by rapid development of gonadal tissues in mussels, for example, can result in apparent reductions in element concentrations in excess of 50% over relatively short periods of time (Langston, 1984).

Variability between individuals

Variability in metal concentrations between individuals is usually overcome by collecting sufficient individuals to provide normally distributed values about the mean (naturally or after transformation) such that coefficients of variance tend towards a minimum. For *S. plana* these range from 13% for chromium to 46% for silver (Bryan *et al.*, 1980). The case for bulking/pooling samples should seriously be considered in order to reduce analytical effort and costs, and comparisons between different sites/treatments should be made to indicate the levels at which differences in concentration become statistically significant. For most metals in *S. plana* a difference between values of 30–40% is likely to be detectable ($p < 0.05$, comparisons made using three pooled samples of six individuals, at each of five sites). A similar exercise with mussels, *Mytilus edulis*, revealed that by analysing homogenates of 25 pooled animals, differences of between 20% and 40% could be detected (Topping, 1983). Although the use of pooled samples is often adopted as common practice, particularly among invertebrates, such procedures may introduce bias, though if sampling and pooling methods are consistent (regarding numbers and size) this is not generally the case. If for any reason different sample numbers have to be pooled, calculation of mean values should perhaps include appropriate comparisons using weighted and unweighted procedures in order to determine whether adjustments are needed in statistical analyses. Some validation of the consequences of pooling would therefore seem desirable in the initial set-up of the sampling protocol.

For field studies, tests should then be made to assess variation within a site. Again using clams as an example, variability has been quantified along an area of mudflat by analysing six pooled samples within an area of 10×20 m. Coefficients of variance lie between 8% (Pb) and 32% (Ag), and differences of about 20% would appear to be significant at the 95% level of significance for most metals (Bryan *et al.*, 1980). This type of study is also useful in that results show that sampling of pooled animals from a single site may be representative of a fairly large area of shore.

Contamination

Attention to 'cleanliness' is usually important for tissues, especially where dissection is involved. Stainless-steel scalpels are suitable for the majority of applications, though if contamination is suspected the use of glass or Teflon dissection instruments is advisable. Homogenization of whole organisms and tissues is often required, particularly where samples are pooled. Trials should be conducted to ensure that contamination is not introduced from materials in the head of the homogenizer, particularly if bushes are not of PTFE; Sn Cr, and Ni analyses appear to be most vulnerable to this procedure.

Contamination from sedimentary material present in the digestive system may be significant particularly in detritivores, suspension and deposit-feeding aquatic organisms, and can lead to serious overestimation of metal body burdens, most notably at sites where sediment–metal concentrations are highest. Up to 25% of the total zinc in oysters, *Crassostrea virginica*, for example, may be due to contributions from gut contents and other particulate material (Mo & Nielson, 1991), so that removal of residual sediment in the intestinal tract is essential if the aim of the study is to measure metal which is truly biologically incorporated into tissues. In the above study, gut contents of oysters were flushed with distilled water to remove particulates, whilst other researchers favour holding live animals in clean conditions for a period of time to encourage natural elimination of metal-rich particulates: for clams, *S. plana*, a period of a week in contaminant-free seawater (of fixed salinity to ensure that calculations based on dry weights are not affected by variations in the salt content) appears sufficient for most metals. For some burrowing species, notably polychaetes such as *N. diversicolor*, the depuration processes may be assisted by holding animals in acid-washed sand or some similar metal-free substrate for much of this period (Bryan *et al.*, 1985). Depending on the species, however, care should be taken to avoid too long a period of clean-up, particularly if the presence of metal forms with short biological half-lives is suspected.

Analysis of tissues/organs which contain no sediment—such as gills, kidneys, liver and muscle—may be an alternative route to avoiding the problem of residual contamination or, if this proves to be unacceptable, the application of a retrospective correction has been proposed as a means of estimating the truly biologically incorporated metal in the organism. For example, the Y-intercept of a graph showing the relationship between total body concentration and amount of sediment in the gut may represent the true biological concentration in the animal. This approach is discussed by Lobel *et al.* (1991) with reference to metals in capelin, *Mallotus villosus*, netted from an unpolluted site off the coast of Newfoundland.

Surface contamination with metal-rich particulates can prove an equally important problem in terms of overestimating biologically incorporated body loadings, and is probably most pertinent with regard to the use of plants, both terrestrial and aquatic, as monitors of environmental contamination. Marine algae such as *Fucus vesiculosus*, for example, are widely regarded as ideal indicators of dissolved metals, since the dietary route for metal uptake is not involved (though some scavenging of particulate metals such as Cu, As, Pb, Zn and Ag by algal tissues probably takes place—Luoma *et al.*, 1982). Fine particulates adhering to fronds may cause errors if not cleaned by gentle brushing. Even then not all particulate contamination may be eliminated. However, it is possible to check the maximum likely contribution from particulates: thus if all the iron in the *Fucus* is assumed to be the result of sediment contamination, then the maximum contribution of particulate metal to the weed can be calculated by multiplying the seaweed:sediment ratio for iron by the sediment metal concentration. If this calculated value is equivalent in magnitude to the concentration of metal actually measured in the *Fucus*, then much of the burden in the weed is likely to be in particulate form. Similar procedures may be used to help determine relative contributions of particulates towards metal burdens of terrestial plants.

Miscellaneous

In addition to the more general sources of variability described above, it is important to be aware of relatively specific influences in the organism under study. For example, infestations of pea crabs, *Pinnotheres ostreum*, ostensibly reduce concentrations of zinc in oysters, *C. virginica* (Mo & Nielson, 1991). This is due to the low zinc levels in crab tissues, relative to their molluscan hosts, producing a 'dilution' effect analogous to that sometimes caused by variation in condition, or brought about during rapid development of gonadal tissue. Crabs and other parasites which represent a significant proportion of the host's body weight must therefore be removed where there is a large discrepancy in tissue metal burdens. For some intertidal species, comparisons of the effect of tidal height may show important differences in metal burdens in indicator species (Phillips, 1980; Bryan *et al.*, 1980; Roberts *et al.*, 1986); standardization of sampling with respect to shore height may then be necessary. Lobel *et al.* (1991) have indicated that collection of mussels from subtidal locations is preferable to shore sampling, since it is likely that growth rates are more uniform in the former and hence metal burdens will be less variable between individuals.

Finally, as a general rule, variability tends to be considerably reduced by expressing results on a dry weight basis rather than a wet weight basis. The use of dry tissue weights will therefore improve comparisons of data between different samples and also between different laboratories. However, despite the high variability associated with wet weight determination (dependent on the condition of the sample and time between dissection and weighing), metal concentrations in food organisms (e.g. fish and shellfish) are still frequently quantified in terms of fresh weights because of the obvious need to assess intake figures in relation to critical doses in humans. If it is essential to express results like this, additional information on wet/dry weight ratios should be provided so that the necessary conversions can be made.

Selection of species/standardization of methods

Having outlined how some of the problems of metal determination in organisms might be overcome (or at least reduced to a minimum), there are still further variables to consider. These stem from the fact that different metals have their own individual characteristics, and also organisms accumulate metals from a variety of sources, often at different rates. Thus there is considerable variation in the extent to which plants and animals regulate tissue concentrations with respect to environmental levels: some organisms are versatile and reflect contamination with a wide range of metals, others are limited to a few specific metals, while it is not uncommon to find certain species capable of exceptional metal accumulation (notably, for example, among polychaetes). Consequently, although it may be possible to make some generalized observations and comparisons regarding metal contamination and its consequences, there is clearly no universal organism for use either in experimental conditions or as an indicator in the environment. The most useful programmes are therefore likely to include the analysis of several species of different ecological type (primary producer, detritivore, herbivore, carnivore) to try to assess different forms of contamination and to obtain a more holistic appraisal of impact to the system.

Furthermore, although for obvious reasons there may be an inclination to study only the specific element of interest, the case for complementary analysis of other relevant metals/environmental factors may be compelling. There is an increasing body of literature confirming the importance of metal−metal/metal−ligand competition and interaction in determining metal burdens in organisms, some of which may be predictable (Bryan & Langston, 1992). For example, inhibition of cadmium uptake due to high levels of zinc in the environment appears to be a fairly common phenomenon, though occasionally elevated cadmium levels may result in con-

comitant rapid accumulation of zinc. Therefore, just because a sample type may be an excellent indicator at one location, this does not guarantee its success elsewhere if competing elements are sufficiently high. It is here where programmes incorporating several different types of sample are at an advantage over the single-species approach. Clearly, where competition is suspected as being of significance, there is a strong case for studying the behaviour of the elements in the laboratory and in the field in order to refine monitoring techniques. Indeed, there is insufficient known about the representativeness of *most* metal−organism combinations. This can only be judged in the environment being studied, taking into account local interactions and variable conditions.

Sampling should therefore ideally be evaluated for each individual case. It is worthwhile summarizing some of the steps that need to be taken to arrive at a satisfactory protocol:
• Careful definition of sampling site and time; sampling should have as little impact on the environment as possible.
• Collection of suitable numbers and appropriate size ranges (or surface area measurements in the case of vegetation). It may be necessary to ensure there are adequate numbers to provide a normal distribution of data about the mean value before statistical treatment (otherwise transformation will be needed).
• Transport to the laboratory in an undamaged state using contaminant-free containers and utensils. If storage is necessary, consideration of the effects of this procedure on metal content will be needed and should be preceded by thorough clean-up of samples where necessary. It is a waste of time and effort to collect organisms and simply freeze them if sediment/soil contamination is suspected. Excess material may be stored for future comparison.
• Standard (validated) methods of preparation and analysis should be adhered to, and if this is changed comparisons of old and new methods should be run in tandem until satisfactory agreement is achieved. Variability due to analysis may be reduced by splitting the sample into subsamples for replicate determinations.

• Accurate and complete reporting of information is essential if the information is to be used with confidence and to maximum advantage by the scientific community.

Examples of organisms as environmental monitors

The selection of appropriate organisms as indicators of environmental contamination is based on satisfaction of a number of criteria, including: widespread geographical distribution; costs of collection and analysis; organisms should ideally be sedentary and possess a reasonably high tolerance to metals (the ability to concentrate metals to levels several orders of magnitude higher than, for example, surrounding water makes analysis considerably easier); concentration factors should be similar at all sites (however, in practice this is clearly not always the case — a feature which may sometimes be used to identify factors influencing bioavailability); populations should be relatively stable in order that they may be resampled throughout the year (for information on temporal trends); specimens should be transplantable to areas which they may not inhabit normally (to investigate kinetics, which may in turn assist in establishing the frequency of sampling).

There is no universal indicator for metals in any environmental phase (above), so it would seem necessary to add to this list the fact that the selection of indicator(s) should be appropriate to the chemistry and form of metal of concern. For example, for assessing aerial deposition, uptake of metals via other routes should be relatively insignificant in the chosen organism, while for determinations of sediment–metal bioavailability, deposit-feeders might be most appropriate. The inclusion of several feeding types would clearly be essential for studies of trophic transfer. Included in Table 4.1 are some appropriate references offering guidance in the use of indicators in terrestrial, freshwater and marine environments. While not being comprehensive, the table gives examples of the relative advantages and limitations of biological monitoring for typical species and how they are used to provide information as to the sources and dispersal, temporal trends, routes of transfer and, sometimes, ecological

effects of metallic contaminants. Many of these references also discuss sampling protocols and methods of data treatment.

4.2 ANALYSIS

4.2.1 Techniques and instrumentation

There is insufficient space here to review in detail all the methods available to the analyst. Instead, a brief comparison of the major techniques which are widely used in metals determinations is shown. Inspection of the manufacturers' manuals will in most cases provide at least the essentials of methods, together with key references appropriate to the instrumentation available. In addition, a number of specialist series and journals are now devoted to this topic. The series of 'blue' booklets on 'methods for the examination of waters and associated materials', produced by the Department of the Environment (UK), and published by HMSO*, for example, provides a reasonably complete and up-to-date collection of methods for metals, applicable to the different analytical facilities available in the majority of laboratories. The topics covered in each of these booklets include principles and procedures of the method, performance characteristics (limits of detection, sensitivity), sample storage, interferences, reagents, apparatus, instrument settings and hazards.

A similar style manual, covering detection of 35 analytes in a variety of environmental sample types and involving a wide range of analytical instrumentation, has recently been produced by the United States Environmental Protection Agency (US EPA, 1991). Other organizations providing standard methods of analysis include the American Society for Testing and Materials (ASTM, 1916 Race Street, Philadelphia), the United Nations Environment Programme (UNEP, Palais des Nations, Geneva) and the International Standard Organization (ISO, Geneva).

Table 4.2 lists the most commonly used techniques for metal analysis, together with their

---

* A list of current publications may be obtained from the Secretary, Standing Committee of Analysts, Department of the Environment, 2, Marsham Street, London.

Table 4.1 Examples of the use of organisms in monitoring studies on metal contamination

Type	Typical species	References
Terrestrial		
Bryophytes	*Sphagnum* spp., *Pleurocarpus* spp. (*Hylocomium splendens, Pleurozium schreberi, Hypnum cupressiforme*)	Lee *et al.* (1977); Pakarinen & Rinne (1979); Thomas & Herrmann (1980); Pakarinen (1981); Martin & Coughtrey (1982); Gydesen *et al.* (1983).
Lichens	*Leconora conizaeoides, Hypnogymnia physodes*	Nieboer *et al.* (1972); Pilegaard (1979); Arafat & Glooschenko (1982); Martin and Coughtrey (1982)
Fungi	*Agaricus* spp., *Gasteromycetes*	Kuusi *et al.* (1981); Martin & Coughtrey (1982)
Angiosperms	Virtually all types of vegetation used from grasses to vegetables and trees. Includes analysis of whole plants or component parts (leaves, shoots, roots, fruit, bark, etc.)	Tyler (1976); Sharma & Shupe (1977); Flanagan *et al.* (1980); Martin & Coughtrey (1982)
Earthworms	*Lumbricus* spp., *Allolobophora caliginosa, Diplocardia* spp., *Eisenia* spp.	Gish & Christensen (1973); Hartenstein *et al.* (1980); Ma (1982); Ma *et al.* (1983); Stafford & McGrath (1986)
Isopods	*Oniscus ascellus, Tracheoniscus rathkei, Porcellio* spp.	Martin *et al.* (1976); Wieser *et al.* (1977); Martin & Coughtrey (1982); Hopkin *et al.* (1986)
Gastropods	*Helix aspersa, Cepaea nemoralis*	Martin & Coughtrey (1982)
Birds	*Passer domesticus, Sturnus vulgaris, Bonasa umbellus*	Osborn (1979); Pinowska *et al.* (1981); Martin & Coughtrey (1982); Rose & Parker (1982); Honda *et al.* (1986)
Mammals	*Spermophilus variegatus, Eptesicus fuscus, Myotis lucifugus, Microtus agrestis, Apodemus sylvaticus, Sorex araneus*, various domestic stock and biological monitoring in humans (blood, urine, hair, etc.)	Sharma & Shupe (1977); Roberts *et al.* (1978); Underwood (1980); Martin & Coughtrey (1982); Froslie *et al.* (1984); CEP (1985); Wren (1986); Hunter *et al.* (1987); Samiullah (1990)
Freshwater		
Microalgae	(Composite)	Foster (1982)
Macroalgae	*Lemanea fluviatilis, Cladophora glomerata, Nitella flexilis*	Harding & Whitton (1981); Whitton *et al.* (1981)
Bryophytes	*Rhynchostegium riparoides, Cinclodotus danubicus, C. riparius*	Empain (1976); Say *et al.* (1981); Whitton *et al.* (1982); Wehr & Whitton (1983)
Angiosperms	*Potamogeton* spp., *Nuphar* spp., *Myriophyllum exalbescens, Elodea canadensis, Phragmites* spp., *Typhalatifolia* spp.	Mudroch & Capobianco (1979); Welsh & Denny (1980); Taylor & Crowder (1983); Campbell *et al.* (1988)
Bivalves	*Dreissena polymorpha, Corbicula* spp., *Anodonta* spp.	Manly & George (1977); Tessier *et al.* (1985)
Gastropods	*Valvata piscinalis, Planorbis corneus, Lymnaea* spp.	Zhulidov *et al.* (1980); Everard & Denny (1984)

Continued on page 64

Table 4.1 *continued*

Type	Typical species	References
Insects	*Baetis* spp., *Ecdyonurus venosus, Brachyptera risi, Leuctra* spp., *Perlodes microcephala, Perla bipunctata, Rhyacophila dorsalis, Dicranota* spp., *Simulium* spp., *Limnius* spp.	Burrows & Whitton (1983)
Fish	*Anguilla anguilla, Rutilus rutilus, Esox lucius, Micropterus* spp., *Ictalurus* spp., *Salmo gairdnerii, Pomoxis micromaculatus, Lepomis* spp., *Cyprinus carpio*	May & McKinney (1981); Wilson *et al.* (1981); Mason (1987); Stratton *et al.* (1987); Hakanson *et al.* (1988)
Birds	Various ducks, e.g. *Anas platyrhynchos, Aythya marila*; gulls, coot, grebe, herons, eagles	Fimreite (1974); Forsyth & Marshall (1986); Ohlendorf *et al.* (1986); Gochfeld & Burger (1987)
Marine		
Plankton	(Composite usually)	Davies (1983); Fowler (1990)
Macroalgae	*Fucus* spp., *Ascophyllum nodosum, Enteromorpha* spp.	Eide *et al.* (1980); Julshamn (1981); Langston (1984); Bryan *et al.* (1985); Say *et al.* (1986)
Polychaetes	*Nereis diversicolor, Nephtys hombergi*	Bryan *et al.* (1985)
Bivalves	*Scrobicularia plana, Macoma balthica, Mytilus* spp., *Cerastoderma edule, Ostrea* spp., *Crassostrea* spp.	Boyden (1977); Bryan *et al.* (1980, 1985); Phillips (1980); Lobel & Wright (1982); Lauenstein *et al.* (1990); Lobel *et al.* (1991)
Gastropods	*Littorina* spp., *Patella vulgata, Nucella lapillus*	Bryan *et al.* (1985)
Barnacles	*Semibalanus balanoides, Balanus eburneus*	Rainbow *et al.* (1980); Ireland (1974); Barber & Trefry (1981)
Fish	*Platichthys flesus, Gadus morrhua, Thunnus thynnus, Squalus acanthias, Carcharinus obscurus, Isurus oxyrhiinchus, Makaira* spp.	van den Broek *et al.* (1981); Buffoni *et al.* (1982); Julshamn *et al.* (1982); Watling *et al.* (1982); Barber & Whaling (1983); Bryan *et al.* (1985); Jensen & Cheng (1987); ICES (1989); Nicholson & Fryer (1992)
Birds	*Calonectris diomodea, Calidris canutus, Limosa lapponica, Tringa totanus* (including the use of feathers for analysis)	Furness & Hutton (1979); Goede & de Bruin (1984); Applequist *et al.* (1985); Goede (1985); Furness *et al.* (1986); Renzoni *et al.* (1986); Bryan & Langston (1992)
Mammals	*Phoca* spp., *Halichoerus grypus, Leptonychotes weddellii, Phagophillus groenlandicus, Erignathus barbatus, Trichecus manatus, Phocoena phocoena, Stenella coeruleoalba*	Martoja & Viale (1977); Gaskin *et al.* (1979); Honda *et al.* (1983); O'Shea *et al.* (1984); Tohyama *et al.* (1986); Hoelzel & Amos (1988)

assets and limitations (Nurnberg, 1984; Shendrikar & Ensor, 1986; Dabeka & Ihnat, 1987; Falkner & Edmond, 1990; Henze, 1990; van den Berg *et al.*, 1991; Waller & Pickering, 1991; Robson, 1992). It also defines accepted abbreviations for the techniques — and this will be used in what follows.

Other techniques include isotope dilution mass spectrometry (Heumann, 1990). When comparing AAS, ICP and ICPMS, the best detection limits are attained using ICPMS and graphite furnace AA. For those elements that form volatile hydrides (As, Se), sensitivity is improved using hydride generation-flameless techniques, while

Table 4.2 Common techniques used for metal analysis

Method	Advantages	Disadvantages
AAS (FAAS/GFAAS)	Relatively interference-free Good precision and accuracy Automatic samplers available Flame detection range of $1-1000\,\mu g/l$, $3-10\,s$/sample per element Graphite furnace detection range of $0.01-1\,\mu g/l$, $2-3\,min$/sample per element Easy to use	Time-consuming, individual detection of elements Requires sample destruction Needs relatively large sample sizes for analysis $(25\,mg-1\,g)$
ICPE	Simultaneous detection of elements in range $1-100\,\mu g/l$, $10-40$ elements/min per sample Intermediate operator skill	Less sensitive than graphite furnace AAS Requires sample destruction Sample sizes of $100\,mg-1\,g$.
ICPMS	Simultaneous detection of elements, at detection range of $0.01-1\,\mu g/l$, at $5-20$ elements/min per sample	More operator skill required than ICP or FAAS
EDXRF	Detection limit $500-2000\,\mu g/l$ for mid-range transition metals, e.g. Fe, Co, Ni, Cu and elements such as Pb, As and Hg Useful for detection of Cl, difficult by AA/ICP Fast detection of halogens Multi-element, non-destructive	Sample should be monolayer
NAA	Non-destructive No sample size limitation Minimal sample preparation	Time-consuming and expensive Unable to provide data on some elements, e.g. Pb, S, Si, etc.
Voltammetry	Differentiate between labile and non-labile forms of heavy metals Sensitive to $\mu g/l$	Not all metals are accessible
e.g. ASV/CSV	The metal concentrates at an electrode before being stripped, thus increasing sensitivity in comparison with analysis of species directly in solution	Effect of pH on the stability of the complexes is more important than for, e.g., AAS Interference from organics
Polarography	Measures the concentration of the electroactive species, which behave independently	The technique is very temperature-sensitive; good temperature control is therefore required

AAS, atomic absorption spectrometry; FAAS, flame atomic absorption spectrometry; GFAAS, graphite furnace atomic absorption spectrometry; ICPE, inductively coupled plasma emission; ICPMS, inductively coupled plasma mass spectrometry; EDXRF, energy-dispersive X-ray fluorescence; NAA, neutron activation analysis; ASV, anodic stripping voltammetry; CSV, cathodic stripping voltammetry.

for Hg, reduction with $SnCl_2$ is usually preferrable (cold vapour).

Flame AAS is rapid, but not all the sample reaches the flame, thus reducing sensitivity: for samples too concentrated for graphite furnace analysis but too dilute for conventional flame analysis, determinations may be carried out by flame using an atom concentrator tube placed over the burner. In contrast, all the sample is atomized in graphite furnace AA, allowing smaller and more dilute samples to be analysed. Determination time is, however, much longer. A stabilized temperature platform in the graphite tube may be used to compensate for physical and chemical interference when using GFAAS.

ICP can reach temperatures of $10\,000\,K$, allowing complete atomization of elements and minimizes interference effects. The development of ICPMS, which links an ICP with a quadrupole mass spectrometer, combines the multi-element capability of ICP with the sensitivity of the graphile furnace. Both AA and ICPE use the principle of separating light according to wavelength to determine elemental concentration, while ICPMS separates the ions according to their mass/charge ratio. Costs rise with increasing complexity of the equipment and accessories required, and in general increase in order FAAS > GFAAS > ICPE > ICPMS. Where the selection of the technique to be used is based on analyte concentrations, FAAS and ICPE are suitable for moderate to high levels, while GFAAS and ICPMS are favoured for much lower levels. ICPE and ICPMS are multi-element techniques and are most advantageous where large numbers of samples are to be analysed for a range of metals.

Various interference difficulties arise when analysing for heavy metals. Background problems encountered with the flame and graphite furnace AAS, can in general be compensated for by using Zeeman effect or deuterium background correction. Deuterium methods do not correct for structural background, though this may not be a major problem in all samples.

Chemical interference during AAS analysis can be reduced by:
1 use of standard solutions with similar matrix content;
2 standard additions;

3 use of nitrous oxide/acetylene flame instead of air/acetylene flame;
4 chemical pretreatment such as matrix isolation/modification.

The simplest and cheapest methods for improving AAS sensitivity rely on increasing the concentration of the sample solution. These include: evaporation (Goerlach & Boutron, 1990), solvent extraction using e.g. APDC/MIBK, ion-exchange, e.g. Chelex-100 (Vermierman *et al.*, 1990; Waller & Pickering, 1991), and co-precipitation. On-line preconcentration may be available with a graphite furnace (Baeckstroem & Danielsson, 1990), although solvent extraction is the most commonly used.

The high salt content of seawater dictates a clean-up of the matrix and possible concentration of the element by ion exchange or solvent extraction. Acidification of seawater samples to be analysed by graphite furnace should be carried out with nitric or sulphuric acid rather than a halogen acid such as hydrochloric. Many metal chlorides are volatile and can be lost from the graphite tube before or during atomization. Modifiers may be used to decrease the volatility of the analyte element or to increase the volatility of a bulk matrix component that is causing interference. The element under test may then be stabilized at a higher temperature, increasing the accuracy of determination. Ammonium oxalate, for example, is used to reduce background signals from seawater samples when direct cadmium determination is required. Other modifiers include palladium and ammonium dihydrogen phosphate (Kunwar *et al.*, 1990; Littlejohn *et al.*, 1991; Quevauviller *et al.*, 1992). Manufacturers' literature should be consulted for choice of appropriate modifier.

Hydride generation/cold vapour AAS usually avoids the problem of interferences, by separation of the analyte from the matrix prior to analysis. However, it is essential that digestion techniques are validated to ensure that the analyte is in a suitable form for reduction, otherwise serious errors may be introduced.

4.2.2 Definitions of precision/detection limits

Increasingly, collaborative interlaboratory studies

are being used to estimate systematic and random error characteristics of methods of analysis, and some of the various international organizations listed above appear to be reaching consensus for interpreting results. Examples of such studies are to be found for most metals in a variety of media including air (Ross, 1992), water (Berman & Boyko, 1988), particulate matter (Hovind & Skei, 1992) and biota (Berman & Boyko, 1986). Within laboratories, analytical error is usually expressed as the standard deviation about the mean value of a series of measurements of a single sample. However, interlaboratory variability presents a more complex set of variables, and is often the largest component error in intercalibration exercises (though this should not deter laboratories from taking part in such experiments). It is only by scrutiny of the results of intercalibration studies (by organizations such as IUPAC — International Union of Pure and Applied Chemistry) that method reliability and harmonization will be achieved.

Detection limits for the instrument are normally expressed as the concentration of analyte that is equivalent to three times the standard deviation of (10) replicate measurements of a reagent blank signal. Alternatively, the method detection limit may be determined as the minimum concentration of the analyte which can be measured with 99% confidence that the analyte concentration is greater than zero. Elemental concentration in the sample must be at least five times the detection limit for satisfactory determination.

Method validation and quality control is, rightly, becoming established as a fundamental aspect by which the veracity of environmental data is judged, and for most metals significant improvements in analysis have occurred during the past two decades. Thus, because of reduced levels of sample contamination and the advent of much lower detection limits, recently reported concentrations of metals in oceanic water are in some cases between one and three orders of magnitude lower than values accepted prior to 1975 (Bryan *et al.*, 1985; Bruland, 1983).

4.2.3 Standard reference materials

Method validation is now made easier by the increasing number of agencies and institutions involved in the production of reference materials. There are, for example, some 900 types, from 13 producers, which are relevant to marine sciences alone. These, together with a short discussion of their use, have been catalogued by Cantillo (1989), and a range of examples is included in Table 4.3, along with examples from freshwater and terrestrial systems. It is recommended that analysts study the specifications produced by the various agencies, in order to select materials that correspond closest to their own samples in terms of matrix, origin and metal content.

Reference materials, or 'in-house' standards, should be analysed at regular intervals to ensure consistency and to check that no systematic errors have entered the analytical scheme. It is also valuable to keep records of parameters which indicate that analysis is proceeding satisfactorily; those using AA, for example, can observe absorbances of known standard solutions and compare these with expected values, while for users of ICP and other types of emission spectroscopy, ratios of peak intensities for various elements may be used to indicate satisfactory performance. The quality control sample/standard reference material is taken through the entire analytical scheme. If the measured concentrations are not within a specified range (usually ± 5% or 10%) of the stated value, laboratory performance may be deemed unacceptable and the source of the error should be identified before continuing analyses.

Among parameters that require validation during method testing are analyte recoveries (particularly where digestion of samples is needed, 'spikes' should be taken through the whole procedure), interference studies and applicability to different matrices, together with appropriate calibration procedures (external standards, standard additions, bracketing, calibration ranges). Thus, if, when using 'real' samples or standard reference materials during method development, analyte recovery falls outside the workable range, yet performance characteristics for that element are otherwise satisfactory, the problem is likely to be matrix-related not system-related. Analysis using

Table 4.3 Examples of reference materials available for metals

Type of material	Matrix	Reference number	Source[a]
Water	River	SLRS-1	NRCC
	Estuarine	SLEW-1	NRCC
	Nearshore seawater	CASS-2	NRCC
	Open ocean seawater	NASS-2	NRCC
Rock	Numerous powdered types	AGV-1, BHVO-1, G-2, etc.	USGS
	Manganese nodules	NOD-A-1	USGS
Ash	City incineration ash	CRM 176	BCR
	Exhaust particles	NIES No8	NIES
	Coal fly ash	SRM 1633a	NIST
Sediment	Stream sediment	SARM 46 to 52	SABS
	Lake sediment	SL-1	IAEA
	Pond sediment	NIES No2	NIES
	Estuarine sediment	SRM-1646	NIST
	Harbour sediment	PACS-1	NRCC
	Marine sediment	BCSS-1, MESS-1	NRCC
	Marine sediment	MAG-1	USGS
Tissue (plant)	Marine algae	CRM 279	BCR
	Citrus leaves	SRM 1572	NIST
Tissue (animal)	Oyster tissue	SRM 1566	NIST
	Mussel tissue	NIES No6	NIES
	Copepod homogenate	MA-A-1/TM	IAEA
	Shrimp homogenate	MA-A-3/OC	IAEA
	Lobster hepatopancreas	LUTS-1/TORT-1	NRCC
	Fish flesh homogenate	MA-A-2/TM	IAEA
	Dogfish liver	DOLT-1	NRCC
	Dogfish muscle	DORM-1	NRCC
	Albacore tuna	RM-50	NIST
	Bovine liver	SRM 1577a	NIST

[a] *Addresses*

BCR: Community Bureau of Reference, Commission of the European Communities, Directorate General for Science R&D, 200 rue de la Loi, B-1049, Brussels, Belgium.

IAEA: International Atomic Energy Agency, Analytical Quality Control Service Laboratory, Seibersdorf, PO Box 100, A-1400 Vienna, Austria.

NIES: National Institute for Environmental Studies, Yatabe-machi, Tsukuba, Ibaraki, 305 Japan.

NIST (formerly NBS): National Institute of Standards and Technology, Office of Standard Reference Materials, Gaithersburg, MD 20899, USA.

NRCC: National Research Council of Canada, Marine Analytical Chemistry Standards Programme, Division of Chemistry, Montreal Road, Ottawa, Ontario, Canada.

SABS: South African Bureau of Standards, Private Bag X191, Pretoria, Transvaal 0001, Republic of South Africa.

USGS: US Geological Survey, Branch of Geochemistry, 12201 Sunrise Valley Drive, Reston, VA 22092, USA (USGS provide accepted, as opposed to certified, values).

the method of standard additions, which involves preparing new standards in the sample matrix by adding known amounts of standard to one or more aliquots of the processed sample solution, may solve the problem. This technique compensates for a sample component that enhances or suppresses the analyte signal, producing a different slope from that of the standards alone.

Limitations are that the curve must be linear, the chemical form of the analyte in standards must respond similarly to that in the sample, and interferences must be constant over the range of concern. Failing this, checks on potential interferences, and techniques to remove them, will have to be made. Specifications for equipment, reagents, standards and conditions should also be established at this stage.

As mentioned earlier, reluctance to publish and make use of data produced by other workers may sometimes arise from a lack of confidence in presentation and documentation, particularly of 'quality control' data. A comprehensive review of quality assurance requirements is described by Williams (1989), though in many cases a simple statement of method precision, detection limits and recoveries (or validation against appropriate standard reference materials) may be sufficient for validation.

The ability to reflect differences in metal concentrations between samples which are caused by biological/environmental variables is dependent to some extent on variability in the chosen analytical technique. This is usually relatively small (1–2% or less) for the majority of metals, and a scale of change of the order of 10% caused by external sources would be easily detected against a background of this type of analytical noise. In cases where metal concentrations are extremely low, however, as in many aqueous and gaseous samples, analytical variability may increase to 10–20% and the relative influence of environmental variables must be considerably higher for their effects to be detectable above analytical noise. Generally, if sampling errors are more than twice analytical error, improvements in the latter will be of little consequence and emphasis should be placed on minimizing variability due to sampling rather than upgrading analytical accuracy.

4.3 CONCLUDING REMARKS

Having discussed problems in sampling and analysis of metals in an ecotoxicological context, it should, of course, be recognized that such measurements alone do not provide an assessment of impact on biological systems. Only when considered alongside effects studies can the relevance of environmental loadings and tissue burdens be appreciated. Nevertheless, accurate and appropriate determination of metals and other contaminants is a major component of ecotoxicology. Thus, our knowledge of the state of the environment, and our ability to make comparisons and decisions regarding the consequences of contamination, rely heavily on assurances regarding the quality of data. The need to select the correct sampling protocol, and to ensure (and report) high standards of 'quality control' should feature uppermost on the list of priorities of those undertaking metal analysis.

REFERENCES

Aalst, R.M. van (1988) Input from the atmosphere. In: *Pollution of the North Sea. An Assessment* (Eds W. Salomons, B.L. Bayne, E.K. Duursma & U. Forstner), pp. 275–283. Springer-Verlag, London.

Aggett, J. & Kriegman, M.R. (1987) Preservation of arsenic (III) and arsenic (V) in samples of sediment interstital water. *Analyst* 112, 153–158.

Ankley, G.T., Phipps, G.L., Leonard, E.N., Benoit, D.A., Mattson, V.R., Kosian, P.A., Cotter, A.M., Dierkes, J.R., Hansen, D.J. & Mahony, J.D. (1991a) Acid-volatile sulphide as a factor mediating cadmium and nickel bioavailability in contaminated sediments. *Environ. Toxicol. Chem.* 10, 1299–1307.

Ankley, G.T., Schubauer-Berigan, M.K. & Dierker, J.R. (1991b) Predicting the toxicity of bulk sediments to aquatic organisms with aqueous test fractions: pore water vs. elutriate. *Environ. Toxicol. Chem.* 10, 1359–1366.

Ap Rheinallt, T., Orr, J., van Dijk, P. & Ellis, J.C. (1989) Sources of variation associated with the sampling of marine sediments for metals. In: *Developments in Estuarine and Coastal Study Techniques* (Eds J. McManus & M. Elliott), pp. 121–127. EBSA 17 Symposium, Dundee, 14–18 September, 1987. International Symposium Series. Olsen and Olsen, Fredensborg, Denmark.

Applequist, H., Drabaek, I. & Asbirk, S. (1985) Variation in mercury content of guillemot feathers over 150 years. *Mar. Pollut. Bull.* 16, 244–248.

Apte, S.C. & Gunn, A.M. (1987) Rapid determination of copper, nickel, lead and cadmium in small samples of estuarine and coastal waters by liquid/liquid extraction and electothermal atomic absorption spectrometry. *Analyt. Chim. Acta* 193, 147–156.

Apte, S.C., Gunn, A.M. & Winnard, D.A. (1989) Mercury and sludge contaminants in marine sediments. Final report to DoE. WRc, Medmenham.

Arafat, N.M. & Glooschenko, W.A. (1982) The use of bog vegetation as an indicator of atmospheric deposition of arsenic in Northern Ontario. *Environ. Pollut. B,* **4**, 85–90.

ASTM (1990) *Standard Guide for Collection, Storage, Characterization and Manipulation of Sediments for Toxicological Testing.* E 1391–90. ASTM, Philadelphia.

Athey, L.A., Thomas, J.M., Miller, W.E. & Word, J.Q. (1989) Evaluation of bioassays for designing sediment cleanup strategies at a wood treatment site. *Environ. Toxicol. Chem.* **8**, 223–230.

Baeckstroem, K. & Danielsson, L.G. (1990) A mechanized continuous flow system for the preconcentration and determination of Co, Cu, Ni, Pb, Cd and Fe in seawater using graphite furnace atomic absorption. *Mar. Chem.* **29**, 33–46.

Barber, S. & Trefrey, J.H. (1981) *Balanus eburneus:* a sensitive indicator of copper and zinc pollution in the coastal zone. *Bull. Environ. Contam. Toxicol.* **27**, 654–659.

Barber, R. & Whaling, P.J. (1983) Mercury in marlin and sailfish. *Mar. Pollut. Bull.* **14**, 395–396.

Berman, S.S. & Boyko, V.J. (1986) Report on the results of the seventh intercalibration exercise on trace metals in biota Part 1. *ICES Coop. Res. Rep.* 138.

Berman, S.S. & Boyko, V.J. (1988) ICES sixth round intercalibration for trace metals in estuarine water. *ICES Coop. Res. Rep.* 152.

Beveridge, A., Waller, P. & Pickering, W.F. (1989) Evaluation of 'labile' metal in sediments by use of ion-exchange resins. *Talanta* **36**, 535–542.

Blackburn, T.H. & Henriksen, K. (1983) Nitrogen cycling in different types of sediments from Danish waters. *Limnol. Oceanogr.* **28**, 477–493.

Bottomley, E.Z. & Bayly, I.L. (1984) A sediment porewater sampler used in root zone studies of the submerged macrophyte, *Myriophyllum spicatum. Limnol. Oceanogr.* **29**, 671–673.

Boyden, C.R. (1977) Effect of size upon metal content of shellfish. *J. Mar. Biol. Assoc. UK* **57**, 675–714.

Brannon, J.M., Plumb, R.H. & Smith, I. (1980) Long-term release of heavy metals from sediments. In: *Contaminants and Sediments,* Vol. 2: *Analysis, Chemistry, Biology,* pp. 221–266. Ann Arbor, MI.

Broekaert, J.A.C. (1990) Use of ICP-spectrometry for environmental analysis. In: *Metal Speciation in the Environment,* (Eds J.A.C. Brockaert, S. Gucer & F. Adams), pp. 213–239. NATO ASI Series, Vol. G23. Springer-Verlag, Berlin.

Bruland, K.W. (1983) Trace elements in sea-water. In: *Chemical Oceanography,* Vol. 8, 2nd edn (Eds J.P. Riley & R. Chester), pp. 157–220. Academic Press, London.

Brumsack, H.J. & Gieskes, J.M. (1983) Interstitial water trace-metal chemistry of laminated sediments from

the Gulf of California, Mexico. *Mar. Chem.* **14**, 89–106.

Bryan, G.W. & Langston, W.J. (1992) Bioavailability, accumulation and effects of heavy metals in sediments with special reference to United Kingdom estuaries: a review. *Environ. Pollut.* **76**, 89–131.

Bryan, G.W., Langston, W.J. & Hummerstone, L.G. (1980) The use of biological indicators of heavy metal contamination with special reference to an assessment of the biological availability of metals in estuarine sediments from South West Britain. *Mar. Biol. Assoc. UK,* Occ. Publ. No. 1.

Bryan, G.W., Langston, W.J., Hummerstone, L.G. & Burt, G.R. (1985) A guide to the assessment of heavy-metal contamination in estuaries using biological indicators. *Mar. Biol. Assoc. UK* Occ. Publ. No. 4.

Buddensiek, V., Engel, H., Fleischauer-Roessing, S., Olbrich, S. & Waechtler, K. (1990) Studies on the chemistry of interstitial water taken from the defined horizons in the fine sediments of bivalve habitats in several northern German lowland waters. 1: Sampling techniques. *Arch. Hydrobiol.* **119**, 55–64.

Buffoni, G., Bernhard, M. & Renzoni, A. (1982) Mercury in Mediterranean tuna. Why is their level higher than in Atlantic tuna? A model. *Thalassia Jugoslavia* **18**, 231–243.

Burrows, I.G. & Whitton, B.A. (1983) Heavy metals in water, sediments and invertebrates from a metal-contaminated river free of organic pollution. *Hydrobiologia* **106**, 263–273.

Calmano, W., Ahlf, W. & Forstner, U. (1990) Exchange of heavy metals between sediment components and water. In: *Metal Speciation in the Environment.* NATO ASI Series, Vol. G23 (Eds J.A.C. Broekaert, S. Gucer & F. Adams), pp. 503–522. Springer-Verlag, Berlin.

Campbell, J.A., Cowling, S.J. & Gunn, A.M. (1985) Methods for the determination of dissolved and particulate trace metals Ni, Cu, Zn, Cd and Pb in estuarine waters. WRc Technical Report TR227. WRc, Medmenham.

Campbell, P.G.C. & Tessier, A. (1991) Biological availability of metals in sediments: analytical approaches. In: *Trace Metals in the Environment. Heavy Metals in the Environment* (Ed. J.P. Vernet), pp. 161–174. Elsevier, London.

Campbell, P.G.C., Lewis, A.G., Chapman, P.M., Crowder, A.A., Fletcher, W.K., Imber, B., Luoma, S.N., Stokes, P.M. & Winfrey, M. (1988) *Biologically Available Metals in Sediments.* Natural Research Council of Canada Report No. 27684, Ottawa, Canada.

Cantillo, A.Y. (1989) Standard and reference materials for marine science. NOAA Technical Memorandum NOS OMA 51.

Carignan, R. (1984) Interstitial water sampling by dialysis: methodological notes. *Limnol. Oceanogr.* **29**, 667–670.

Carlson, A.R., Phipps, G.L., Mattson, V.R., Kosian, P.A. & Cotter, A.M. (1991) The role of acid-volatile sulphide in determining cadmium bioavailability and toxicity in freshwater sediments. *Environ Toxicol. Chem.* **10**, 1309–1319.

Carr, R.S., Williams, J.W. & Fragata, C.T.B. (1989) Development and evaluation of a novel marine sediment pore water toxicity test with the polychaete *Dinophilus gyrociliatus*. *Environ. Toxicol. Chem.* **8**, 533–543.

CEP Consultants (1985) *Heavy Metals in the Environment* (Ed. T.D. Lekkas), Vols 1 and 2. CEP Consultants, Edinburgh.

Chapman, P.M. (1987) Sediment bioassay tests provide toxicity data necessary for assessment and regulation. In: *Proceedings of the Eleventh Annual Aquatic Toxicity Workshop*, 13–15 November 1984, Vancouver, British Columbia (Eds G.H. Green & K.L. Woodward), pp. 178–197. *Canadian Technical Report of Fisheries and Aquatic Sciences* **1480**.

Craft, C.B., Seneca, E.D. & Broome, S.W. (1991) Loss on ignition and Kjeldahl digestion for estimating organic carbon and total nitrogen in estuarine marsh soils: calibration with dry combustion. *Estuaries* **14**, 175–179.

Craig, G.R. (1984) Bioassessment of sediments: Protocol view and recommended approach. Abstracts of papers presented at 11th annual Aquatic Toxicity Workshop, Richmond, BC, 13–15 November.

Cutter, G.A. & Oatts, T.J. (1987) Determination of dissolved sulphide and sedimentary sulphur speciation using gas chromotography-photoionization detection. *Anal. Chem.* **59**, 717–721.

Dabeka, R.W. & Ihnat, M. (1987) Methods of cadmium detection. In: *Cadmium in the Aquatic Environment* (Eds J.O. Nriagu & J.B. Sprague). *Advances in Environmental Science and Technology*, Vol. 19, pp. 231–264. John Wiley & Sons, New York.

Danielsson, L.G., Magnusson B. & Westerlund, S. (1978) An improved metal extraction procedure for the determination of trace metals in seawater by atomic absorption spectrometry with electrothermal atomization. *Analyt. Chim. Acta* **98**, 47–57.

Danielsson, L.G., Magnusson, B., Westerlund, S. & Zhang, K. (1982) Trace metal determinations in estuarine waters by electrothermal atomic absorption spectrometry after extraction of dithiocarbamate complexes into Freon. *Analyt. Chim. Acta* **144**, 183–188.

Danielsson, L.G., Magnusson, B., Westerlund, S. & Zhang, K. (1983) Trace metals in the Gota River estuary. *Estuarine, Coastal Shelf Sci.* **17**, 73–85.

Davies, A.G. (1983) The effects of heavy metals upon natural marine phytoplankton populations. *Progr.*

Phycol. Res. **2**, 113–145.

Deans, E.A., Meadows, P.S. & Anderson, J.G. (1982) Physical, chemical and microbiological properties of intertidal sediments and sediment selection by *Corophium volutator*. *Int. Rev. Gesamt. Hydrobiol.* **67**, 261–269.

DeWitt, T.H., Ditsworth, G.R. & Swartz, R.C. (1988) Effects of natural sediment features on survival of the phoxocephalid amphipod, *Rhepoxynius abronius*. *Mar. Environ. Res.* **25**, 99–124.

Di Toro, D.M., Mahony, J.D., Hansen, D.J., Scott, K.J., Hicks, M.B., Mayr, S.M. & Redmond, M.S. (1990) Toxicity of cadmium in sediments: the role of acid volatile sulphide. *Environ. Toxicol. Chem.* **9**, 1487–1502.

Diks, D.M. & Allen, H.E. (1983) Correlation of copper distribution in a freshwater-sediment system to bioavailability. *Bull Environ. Contam. Toxicol.* **30**, 37–43.

Ditsworth, G.R., Schults, D.W. & Jones, J.K.P. (1990) Preparation of benthic substrates for sediment toxicity testing. *Environ. Toxicol. Chem.* **9**, 1523–1529.

DoE (1980) *General Principles of Sampling and Accuracy of Results. Methods for the examination of waters and associated materials.* HMSO, London.

Edenborn, H.M., Paquin, Y. & Chateauneuf, G. (1985) Bacterial contribution to managanese oxidation in a deep coastal sediment. *Estuarine Coastal Shelf Sci.* **21**, 801–815.

Eganhouse, R.P. & Young, D.R. (1978) *In situ* uptake of mercury by the intertidal mussel *Mytilus californianus*. *Mar. Pollut. Bull.* **9**, 214–217.

Eide, I., Myklestad, S. & Melson, S. (1980) Long-term uptake and release of heavy metals by *Ascophyllum nodosum* L. Le Jol. *Environ. Pollut. A* **23**, 19–28.

Empain, A. (1976) Les bryophytes aquatiques utilisés comme traceurs de la contamination en métaux lourds des eaux douces. [The use of aquatic bryophytes as tracers of heavy metals contamination of fresh waters.] *Mém. Soc. R. Bot. Belg.* **7**, 141–156.

Everard, M. & Denny, P. (1984) The transfer of lead by freshwater snails in Ulswater, Cumbria. *Environ. Pollut. A* **35**, 299–314.

Falkner, K.K. & Edmond, J.M. (1990) Gold in seawater. *Earth Planet. Sci. Lett.* **98**, 208–221.

Fenchel, T. (1969) The ecology of marine microbenthos. IV. Structure and function of the benthic ecosystem, its chemical and physical factors and the microfauna communities, with special reference to the ciliated protozoa. *Ophelia* **6**, 1–182.

Fimreite, N. (1974) Mercury contamination of aquatic birds in northwestern Ontario. *J. Wildl. Manage.* **35**, 293–300.

Fischer, H. (1983) Shell weight as an independent variable in relation to cadmium content of molluscs. *Mar. Ecol. Prog. Ser.* **12**, 59–75.

Flament, P., Lepretre, A. & Noel, S. (1987) Coastal aerosols in the northern Channel. *Oceanologica Acta* **10**, 49–61.

Flanagan, J.T., Wade, K.J., Currie, A. & Curtis, D.J. (1980) The deposition of lead and zinc from traffic pollution on two roadside shrubs. *Environ. Pollut. B* **1**, 71–78.

Forstner, U. (1989) *Contaminated Sediments. Lecture Notes in Earth Sciences*, Vol. 21 (Eds S. Bhattacharji, G.M. Friedman, H.J. Neugebauer & A. Seilacher). Springer-Verlag, London.

Forstner, U., Kersten, M. & Calmano, W. (1987) Exchange of heavy metals at the sediment/water interface in dredge spoil and natural waters. *Acta Hydrochim. Hydrobiol.* **15**, 221–242.

Forstner, U., Ahlf, W., Calmano, W., Kersten, M. & Schoer, J. (1990) Assessment of metal mobility in sludges and solid wastes. In: *Metal Speciation in the Environment*. NATO ASI Series, Vol. G23 (Eds J.A.C. Broekaert, S. Gucer & F. Adams), pp. 1–41. Springer-Verlag, Berlin.

Forsyth, D.S. & Marshall, W.D. (1986) Ionic alkylleads in herring gulls from the great lakes region. *Environ. Sci. Technol.* **20**, 1033–1038.

Foster, P.L. (1982) Species associations and metal contents of algae from rivers polluted by heavy metals. *Freshwater Biol.* **12**, 41–61.

Fowler, S.W. (1990) Critical review of selected heavy metal and chlorinated hydrocarbon concentrations in the marine environment. *Mar. Environ. Res.* **29**, 1–64.

Froslie, A., Norheim, G., Rambaek, J.P. & Steinnes, E. (1984) Levels of trace elements in liver from Norwegian moose, reindeer and red deer in relation to atmospheric deposition. *Acta Vet. Scand.* **25**, 333–345.

Fu, G. & Allen, H.E. (1992) Cadmium adsorption by oxic sediment. *Water Res.* **26**, 225–233.

Furness, R. & Hutton, M. (1979) Pollutant levels in the great skua, *Catharacta skua*. *Environ. Pollut.* **19**, 261–268.

Furness, R.W., Muirhead, S.J. & Woodburn, M. (1986) Using bird feathers to measure mercury in the environment: relationships between mercury content and moult. *Mar. Pollut. Bull.* **17**, 27–30.

Gaillard, J.F., Pauwels, H. & Michard, G. (1989) Chemical diagenesis in coastal marine sediments. *Oceanol. Acta* **12**, 175–187.

Gaskin, D.E., Stonefield, K.I., Suda, P. & Frank, R. (1979) Changes in mercury levels in harbour porpoises from the Bay of Fundy, Canada, and adjacent waters during 1969–1977. *Arch. Environ. Contam. Toxicol.* **3**, 733–762.

Gibbs, R.J. (1973) Mechanisms of trace metal transport in rivers. *Science* **180**, 71–73.

Giblin, A.E. & Howarth, R.W. (1984) Porewater evidence for a dynamic sedimentary iron cycle in salt marshes. *Limnol. Oceanogr.* **29**, 47–63.

Gish, C.D. & Christensen, R.E. (1973) Cadmium, nickel, lead and zinc in earthworms from roadside soil. *Environ. Sci. Technol.* **7**, 1060–1062.

Glooschenko, W.A. (1986) Monitoring the atmospheric deposition of metals by use of bog vegetation and peat profiles. In: *Toxic Metals in the Atmosphere* (Eds J.O. Nriagu & C.I. Davidson), Vol. 17, *Advances in Environmental Science and Technology*, pp. 507–534. Wiley & Sons, New York.

Gochfeld, M. & Burger, J. (1987) Heavy metal concentrations in the liver of three duck species: influence of species and sex. *Environ. Pollut.* **45**, 1–15.

Goede, A.A. (1985) Mercury, selenium, arsenic and zinc in waders from the Dutch Wadden Sea. *Environ. Pollut. A*, **37**, 287–309.

Goede, A.A. & de Bruin, M. (1984) The use of bird feather parts as a monitor for metal pollution. *Environ. Pollut. B*, **8**, 281–298.

Goerlach, U. & Boutron, C.F. (1990) Preconcentration of lead, cadmium, copper and zinc in water at the pg/g level by non-boiling evaporation. *Analyt. Chim. Acta* **236**, 391–398.

Gydesen, M., Pilegaard, K., Rasmussen, L. & Ruhling, A. (1983) Moss analyses used as a means of surveying the atmospheric heavy metal deposition in Sweden, Denmark and Greenland in 1980. *Statens naturvardsverk Report SNV PM* 1670, Stockholm, Sweden.

Hakanson, L., Nilsson, A. & Andersson, T. (1988) Mercury in fish in Swedish lakes. *Environ Pollut.* **49**, 145–162.

Harding, J.P.C. & Whitton, B.A. (1981) Accumulation of zinc, cadmium and lead by field populations of *Lemanea. Water Res.* **15**, 301–319.

Hartenstein, R., Leaf, A.L., Neuhauser, E.F. & Bickelhaupt, D.H. (1980) Composition of the earthworm *Eisenia foetida* Savigny and assimilation of 15 elements from sludge during growth. *Comp. Biochem. Physiol.* **66C**, 187–192.

Henze, G. (1990) Application of polarographic and voltammetric techniques in environmental analysis. In: *Metal Speciation in the Environment* (Eds J.A.C. Brockaert, S. Gucer & F. Adams), pp. 391–408. NATO ASI Series, Vol. G23. Springer-Verlag, Berlin.

Herndl, G.J., Peduzzi, P. & Fanuko, N. (1989) Benthic community metabolism and microbial dynamics in the Gulf of Trieste northern Adriatic Sea. *Mar. Ecol. Progr. Ser.* **53**, 169–178.

Heumann, K.G. (1990) Elemental species analyses with isotope dilution mass spectrometry. In: *Metal Speciation in the Environment* (Eds J.A.C. Brockaert, S. Gucer & F. Adams), pp. 153–169. NATO ASI Series, Vol. G23. Springer-Verlag, Berlin.

Hoelzel, A.R. & Amos, W. (1988) DNA fingerprinting and scientific whaling. *Nature* **333**, 305.

Honda, K., Nasu, T. & Tatsukawa, R. (1986) Seasonal changes in mercury accumulation in the black-eared kite *Milvus migrans linearus. Environ. Pollut. A*, **42**, 325–334.

Honda, K., Tatsukawa, R., Itano, K., Miyazaki, N. & Fujiyama, T. (1983) Heavy metal concentrations in muscle, liver and kidney tissue of striped dolphin *Stenella coeruleoalba* and their variations with body length, weight age and sex. *Agric. Biol. Chem.* **47**, 1219–1228.

Hopkin, S.P., Hardisty, G.N. & Martin, M.H. (1986) The woodlouse *Porcellio icaber* as a 'biological indicator' of zinc, cadmium, lead and copper pollution. *Environ. Pollut. B* **11**, 271–290.

Hovind, H. & Skei, J. (1992) Report of the second ICES intercalibration exercise on the determination of trace metals in suspended particulate matter. *ICES Coop. Res. Rep.* 184.

Hsieh, Y.P. & Yang, C.H. (1989) Diffusion methods for the determination of reduced inorganic sulphur species in sediments. *Limnol. Oceanogr.* **34**, 1126–1130.

Hunt, D.T.E. (1986) Trace metal speciations and toxicity to aquatic organisms — a review. WRc Technical Report, TR247, Water Research Centre, Medmenham, UK.

Hunter, B.A., Johnson, M.S. & Thompson, D.J. (1987) Ecotoxicology of copper and cadmium in a contaminated grassland ecosystem. 3. Small mammals. *J. Appl. Ecol.* **24**, 601–614.

Huttel, M. (1990) Influence of the lugworm *Arenicola marina* on porewater nutrient profiles of sand flat sediments. *Mar. Ecol. Progr. Ser.* **62**, 241–248.

ICES (1989) Statistical analysis of the ICES co-operative monitoring programme data on contaminants in fish mussel tissue 1978–1985 for determination of temporal trends. *ICES Co-op. Res.* No. 162. Copenhagen, Denmark.

Ireland, M.P. (1974) Variations in the zinc, copper, manganese and lead content of *Balanus balanoides* in Cardigan Bay, Wales. *Environ. Pollut.* **7**, 65–75.

Jahnke, R.A. (1988) A simple, reliable and inexpensive pore-water sampler. *Limnol. Oceanogr.* **33**, 483–487.

Jenne, E.A., Kennedy, V.C., Burchard, J.M. & Ball, J.W. (1980) Sediment collection and processing for selective extraction and for total trace element analyses. In: *Contaminants and Sediments*, Vol. 2: *Analysis, Chemistry, Biology*, pp. 169–190. Ann Arbor, MI.

Jensen, A. & Cheng, Z. (1987) Statistical analysis of trend monitoring data of heavy metals in flounder *Platichthys flesus. Mar. Pollut. Bull.* **18**, 230–238.

Jones, M.A. & Rae, J.E. (1989) Sediment preservation: the effects on phosphate exchange between sediment and water. *Oceanol. Acta* **12**, 87–90.

Jordao, C.P. & Nickless, G. (1989) An evaluation of the ability of extractants to release heavy metals from alga and mollusc samples using a sequential extraction procedure. *Environ. Technol. Lett.* **10**, 445–451.

Jorgensen, B.B. & Fenchel, T. (1974) The sulphur cycle of a marine sediment model system. *Mar. Biol.* **24**, 189–201.

Jorgensen, B.B., Bang, M. & Blackburn, T.H. (1990) Anaerobic mineralization in marine sediments from the Baltic Sea–North Sea transition. *Mar. Ecol. Progr. Ser.* **59**, 39–54.

Jorgensen, N.O.G., Lindroth, P. & Mopper, K. (1981) Extraction and distribution of free amino-acids and ammonia in sediment interstitial waters from the Limfjord, Denmark. *Oceanol. Acta* **4**, 465–474.

Julshamn, K. (1981) Studies on major and minor elements in molluscs in Western Norway. VII. The contents of 12 elements including copper, zinc, cadmium and lead in common mussel *Mytilus edulis* and brown seaweed *Ascophyllum nodosum* relative to the distance from the industrial sites in Sorfjorden, inner Hardangerfjord. *Fisk. Dir. Skr. Ser. Ernaering* **1**, 267–287.

Julshamn, K., Ringdal, O. & Braekkan, O.R. (1982) Mercury concentrations in liver and mussel of cod *Gadus morrhua* as an evidence of migration between waters with different levels of mercury. *Bull. Environ. Contam. Toxicol.* **29**, 544–549.

Kemp, P.F. & Swartz, R.C. (1988) Acute toxicity of interstitial and particle-bound cadmium to a marine infaunal amphipod. *Mar. Environ. Res.* **26**, 135–153.

Kersten, M. & Forstner, U. (1987) Cadmium associations in freshwater and marine sediment. *Adv. Environ. Sci. Technol.* **19**, 51–88.

Kersten, M., Dicke, M., Kriews, M., Naumann, K., Schmidt, D., Schulz, M., Schwikowski, M. & Steiger, M. (1988) Distribution and fate of heavy metals in the North Sea. In: *Pollution of the North Sea. An Assessment* (Eds W. Salomons, B.L. Bayne, E.K. Duursma & U. Foerstner), pp. 300–347. Springer-Verlag, Berlin.

Krantzberg, G. & Stokes, P.M. (1985) Benthic macroinvertebrates modify copper and zinc partitioning in freshwater-sediment microcosms. *Can. J. Fish. Aquat. Sci.* **42**, 1465–1473.

Krell, U. & Roeckner, E. (1987) Simulation of the atmospheric transport and deposition of heavy metals into the North Sea. *ICES Council Meeting, Santander, Spain, 1 October 1987, Collected Papers*, ICES, Copenhagen.

Kriukov, P.A. & Manheim, F.T. (1982) Extraction and investigative techniques for study of interstitial waters of unconsolidated sediments: a review. *Dynam. Environ. Ocean Floor*, 3–26.

Kunwar, U.K., Littlejohn, D. & Halls, D.J. (1990) Hot-injection procedures for the rapid analysis of biological samples by electrothermal atomic-absorption spectrometry. *Talanta* **37**, 555–559.

Kuusi, T., Laaksovirta, K., Liukkonen-Lilja, H., Lodenius, M. & Piepponen, S. (1981) Lead, cadmium and mercury contents of fungi in the Helsinki area and in unpolluted control areas. *Z. Lebensm. Unters.-Forsch.* **173**, 261–267.

Lake, D.L., Kirk, P.W.W. & Lester, J.N. (1989) Heavy metal solids association in sewage sludges. *Water Res.* **23**, 285–291.

Langston, W.J. (1980) Arsenic in U.K. estuarine sediments and its availability to benthic organisms. *J. Mar. Biol. Assoc. UK* **60**, 869–881.

Langston, W.J. (1982) The distribution of mercury in British estuarine sediments and its availability to deposit-feeding bivalves. *J. Mar. Biol. Assoc. UK* **62**, 667–684.

Langston, W.J. (1984) Availability of arsenic to estuarine and marine organisms: a field and laboratory investigation. *Mar. Biol.* **80**, 143–154.

Lauenstein, G.G., Robertson, A. & O'Connor, T.P. (1990) Comparison of trace metal data in mussels and oysters from a mussel watch program of the 1970's with those from a 1980's programme. *Mar. Pollut. Bull.* **21**, 440–447.

Lee, J.A., Brooks, R.R., Reeves, R.D. & Jaffre, T. (1977) Chromium-accumulating bryophyte from New Caledonia. *Bryologist* **80**, 203–205.

Lion, L.W., Altmann, R.S. & Leckie, J.O. (1982) Trace-metal adsorption characteristics of estuarine particulate matter: evaluation of contribution of Fe/Mn oxide and organic surface coatings. *Environ. Sci. Technol.* **16**, 660–666.

Littlejohn, D., Egila, J.N., Gosland, R.M., Kunwar, U.K. & Smith, C. (1991) Graphite furnace analysis — getting easier and achieving more? *Analyt. Chim. Acta* **250**, 71–84.

Lobel, P.B. & Wright, D.A. (1982) Total body zinc concentration and allometric growth ratios in *Mytilus edulis* collected from different shore levels. *Mar. Biol.* **66**, 231–236.

Lobel, P.B., Bajdik, C.D., Belkhode, S.P., Jackson, S.E. & Longerich, H.P. (1991) Improved protocol for collecting mussel watch specimens taking into account sex, size, condition, shell shape and chronological age. *Arch. Environ. Contam. Toxicol.* **21**, 409–414.

Lobel, P.B., Belkhode, S.P., Jackson, S.E. & Longerich, H.P. (1991) Sediment in the intestinal tract: A potentially serious source of error in aquatic biological monitoring programs. *Mar. Environ. Res.* **31**, 163–174.

Loring, D.H. (1991) Normalization of heavy-metal data from estuarine and coastal sediments. *J. Mar. Sci.* **48**, 101–115.

Ludwig, D.D., Sherrad, J.H. & Amende, R.A. (1988) Improvement of operations and maintenance techniques research program: An evaluation of the standard elutriate test as an estimator of contaminant release at the point of dredging. Compl. Rep. U.S. Army Eng. Waterways Exp. Stn. Hydraul. Lab.

Luoma, S.N. & Bryan, G.W. (1978) Factors controlling the availability of sediment-bound lead to the estuarine bivalve *Scrobicularia plana*. *J. Mar. Biol. Assoc. UK* **58**, 793–802.

Luoma, S.N. & Bryan, G.W. (1981) A statistical assessment of the form of trace metals in oxidised estuarine sediments employing chemical extractants. *Sci. Total Environ.* **17**, 165–196.

Luoma, S.N. & Bryan, G.W. (1982) A statistical study of environmental factors controlling concentrations of heavy metals in the burrowing bivalve *Scrobicularia plana* and the polychaete *Nereis diversicolor*. *Estuarine, Coastal Mar. Sci.* **15**, 95–108.

Luoma, S.N., Bryan, G.W. & Langston, W.J. (1982) Scavenging of heavy metals from particulates by brown seaweed. *Mar. Pollut. Bull.* **13**, 394–396.

Ma, W.C. (1982) The influence of soil properties and worm related factors on the concentrations of heavy metals in earthworms. *Pedobiologia* **24**, 109–120.

Ma, W., Edelman, T. & van Beersum, I. (1983) Uptake of cadmium, zinc, lead and copper by earthworms near a zinc-smelting complex — influence of soil pH and organic matter. *Bull. Environ. Contam. Toxicol.* **30**, 424–427.

Malueg, K.W., Schuytema, G.S. & Krawczyk, D.F. (1986) Effects of sample storage on a copper-spiked freshwater sediment. *Environ. Toxicol. Chem.* **5**, 245–254.

Manly, R. & George, W.O. (1977) The occurrence of some heavy metals in populations of the freshwater mussel *Anodonta anatina* L. from the River Thames. *Environ. Pollut.* **14**, 139–154.

Mantoura, R.F.C., Nelson, L.A. & Morris, A.W. (1987) Trace metal biochemistry of sewage sludge mixtures in seawater and sediments. Final Report on Commissioned Research Funded by WRc. IMER, NERC.

MaoMei, Z., LiuJing, Y., Bertine, K.K., Koide, M. & Goldberg, E.D. (1989) Atmospheric pollution in Beijing, China, as recorded in sediments of the Summer Palace Lake. *Environ. Conserv.* **16**, 233–236.

Martin, M.H. & Coughtrey, P.J. (1982) *Biological Monitoring of Heavy Metal Pollution, Land and Air*. Applied Science Publishers, London.

Martin, M.H., Coughtrey, P.J. & Young, E.W. (1976) Observations on the availability of lead, zinc, cadmium and copper in woodland litter and the uptake of lead, zinc and cadmium by the woodlouse, *Oniscus asellus*. *Chemosphere* **5**, 313–318.

Martoja, R. & Viale, D. (1977) Accumulation de granules de sélénium mercurique dans le foie d'odontocètes Mammifères, cétacés: Un mécanisme possible de détoxication du méthylmercure par le sélénium. *C.R. Acad. Sci. Paris* **285**, série D, 109–112.

Mason, C.F. (1987) A survey of mercury, lead and cadmium in muscle of British freshwater fish. *Chemo-*

sphere **16**, 901–906.

May, T.W. & McKinney, G.L. (1981) Cadmium, lead, mercury, arsenic and selenium concentrations in freshwater fish 1976–1977: National Pesticide Monitoring Programme. *Pestic. Monit. J.* **15**, 14–38.

Meador, J.P., Ross, B.D., Dinnel, P.A. & Picquelle, S.J. (1990) An analysis of the relationship between a sand-dollar embryo elutriate assay and sediment contaminants from stations in an urban embayment of Puget Sound, Washington. *Mar. Environ. Res.* **30**, 251–272.

Mo, C. & Nielson, B. (1991) Variability in measurements of zinc in oysters. *C. virginica. Mar. Pollut. Bull.* **22**, 522–525.

Morse, J.W., Millero, F.J., Cornwell, J.C. & Rickard, D. (1987) The chemistry of hydrogen sulphide and iron sulphide systems in natural waters. *Earth-Sci. Rev.* **24**, 1–42.

Mudroch, A. & Capobianco, J.A. (1979) Effects of mine effluent on uptake of Co, Ni, Cu, As, Zn, Cd, Cr and Pb by aquatic macrophytes. *Hydrobiologia* **64**, 223–231.

Mudroch, A. & Bourbonniere, R.A. (1991) Sediment preservation, processing and storage. In: *Handbook of Techniques for Aquatic Sediments Sampling* (Eds A. Murdoch & S.D. McKnight), pp. 131–170. CRC Press, Boca Raton, FL.

Munawar, M., Thomas, R.L., Norwood, W. & Mudroch, A. (1985) Toxicity of Detroit River sediment-bound contaminants to ultraplankton. *J. Great Lakes Res.* **11**, 264–274.

NATO (1990) Panel discussions. In: *Metal Speciation in the Environment* (Eds J.A.C. Brockaert, S. Gucer & F. Adams), pp. 633–639. NATO ASI Series, Vol. G23. Springer-Verlag, Berlin.

Nicholson, M.D. & Fryer, R.J. (1992) The statistical power of monitoring programmes. *Mar. Pollut. Bull.* **24**, 146–149.

Nieboer, E., Ahmed, H.M., Puckett, K.J. & Richardson, D.H.S. (1972) Heavy metal content of lichens in relation to distance from a nickel smelter in Sudbury, Ontario. *Lichenologist* **5**, 292–304.

Nieuwenhuize, J., Poley-Vos, C.H., van den Akker, A.H. & van Delft, W. (1991) Comparison of microwave and conventional extraction techniques for the determination of metals in soil, sediment and sludge samples by atomic spectrometry. *Analyst* **116**, 347–351.

North Sea Conference (1987) *Second International Conference on the Protection of the North Sea. Quality Status of the North Sea.* A report by the Scientific and Technical Working Group. HMSO, London.

Nriagu, J.O. & Davidson, C.I. (1986) *Toxic Metals in the Atmosphere.* Wiley & Sons, London.

Nriagu, J.O. & Pacyna, J.M. (1988) Quantitative assessment of worldwide contamination of air, water and soils by trace metals. *Nature* **333**, 134–139.

Nurnberg, H.W. (1984) Potentialities of voltammetry for the study of physiochemical aspects of heavy metal complexation in natural waters. In: *Complexation of Trace Metals in Natural Waters* (Eds C.J.M. Kramer, & J.C. Duinker), pp. 95–116. Martinus Nijhoff/Dr W. Junk, The Hague.

Ohlendorf, H.M., Hoffman, D.J., Saiki, M.K. & Aldrich, T.W. (1986) Embryonic mortality and abnormalities of aquatic birds: apparent impacts of selenium from irrigation drain water. *Sci. Total Environ.* **52**, 49–63.

Orem, W.H.V. (1982) *Organic Matter in Anoxic Pore Water from Great Bay, New Hampshire.* PhD Diss., New Hampshire University, Durham, NH.

Osborn, D. (1979) Seasonal changes in the fat and metal content of the liver of the starling *Sturnus vulgaris. Environ. Pollut.* **19**, 145–155.

O'Shea, T.J., Moore, J.F. & Kochman, H.I. (1984) Contaminant concentrations in manatees in Florida. *J. Wildl. Manage.* **48**, 741–748.

Otten, P., Storms, H., Xhoffer, C. & van Grieken, R. (1989) *Chemical composition, source identification and quantification of the atmospheric input into the North Sea.* ICES Council Meeting, The Hague, Netherlands, 5 October. Collected Papers. ICES, Copenhagen.

Pacyna, J.M., Munch, J. & Axenfeld, F. (1991) European inventory of trace metal emissions to the atmosphere. In: *Trace Metals in the Environment. Heavy Metals in the Environment* (Ed. J.P. Vernet), pp. 1–20. Elsevier, London.

Pakarinen, P. (1981) Metal content of ombrotrophic *Sphagnum* mosses in NW Europe. *Ann. Bot. Fenn.* **18**, 281–292.

Pakarinen, P. & Rinne, R.J.K. (1979) Growth rates and heavy metal concentrations of five moss species in paludified spruce forests. *Lindbergia* **5**, 77–83.

Palermo, M.R. (1986) *Long-term effects of dredging operations program. Development of a modified elutriate test for estimating the quality of effluent from confined dredged material disposal areas.* Technical Report of the US Army Engineers Waterways Experimental Station.

Parker, A. & McCann, C. (1988) Percussion corer and extrusion device for unconsolidated sediments. *J. Sediment Petrol.* **58**, 752–753.

Pastorok, R.A. & Becker, D.S. (1989) Comparative sensitivity of bioassays for assessing sediment toxicity in Puget Sound. *Oceans '89: The Global Ocean*, Vol. 2: *Ocean Pollution*, pp. 431–436. Institute of Electrical and Electronics Engineers, New York.

Phillips, D.J.H. (1980) *Quantitative Aquatic Biological Indicators.* Applied Science Publishers, London.

Pilegaard, K. (1979) Heavy metals in bulk precipitation and transplanted *Hypogymnia physodes* and *Dicranoweisia cirrata* in the vicinity of a Danish steelworks. *Water, Air, Soil Pollut.* **11**, 77–91.

Pinowska, B., Krasnicki, K. & Pinowski, J. (1981) Estimation of the degree of contamination of graniverous birds with heavy metals in agricultural and industrial landscape. *Ekol. Polska*, **29**, 137–149.

Pittinger, C.A., Hand, V.C., Masters, J.A. Davidson, L.F. (1988) Interstitial water sampling in ecotoxicological testing: partitioning of a cationic surfactant. In: *Aquatic Toxicology and Hazard Assessment*. Vol. 10, ASTM STP 971 (Eds W.J. Adams, G.A. Chapman & W.G. Landis), pp. 138–148. American Society for Testing Materials, Philadelphia, PA.

Quevauviller, P.H., Kramer, K.J.M., van der Vlies, E.M., Vercoutere, K. & Griepink, B. (1992) Improvements in the determination of trace elements in seawater leading to the certification of Cd, Cu, Mo, Ni, Pb and Zn CRM 403. *Mar. Pollut. Bull.* **24**, 33–38.

Rainbow, P.S., Scott, A.G., Wiggins, R.A. & Jackson, R.W. (1980) Effect of chelating agents on the accumulation of cadmium in the barnacle *Semibalanus balanoides*, and complexation of soluble Cd, Zn and Cu. *Mar. Ecol. Progr. Ser.* **2**, 143–152.

Rapin, F. & Forstner, U. (1983) Sequential leaching techniques for particulate metal speciation: The selectivity of various extractants. In: *Heavy Metals in the Environment*, pp. 1074–1077. CEP Consultants, Heidelberg.

Renzoni, A., Focardi, S., Fossi, C., Leonzio, C. & Mayol, J. (1986) Comparison between concentrations of mercury and other contaminants in eggs and tissues of Cary's shearwater *Calonectris diomedea* collected on Atlantic and Mediterranean Islands. *Environ. Pollut. A* **40**, 17–35.

Reynoldson, T.B. (1987) Interactions between sediment contaminant and benthic organisms. *Hydrobiologia* **149**, 53–66.

Roberts, D.F., Elliot, M. & Read, P.A. (1986) Cadmium contamination, accumulation and some effects of this metal in mussels from a polluted marine environment. *Mar. Environ. Res.* **18**, 165–183.

Roberts, R.D., Johnson, M.S. & Hutton, M. (1978) Lead contamination of small mammals from abandoned metalliferous mines. *Environ. Pollut.* **15**, 61–69.

Robson, N.S. (1992) EDXRF—a flexible solution to quality control, development, investigative and environmental analysis. *Int. Labmate*, **17**, 23–26.

Rose, G.A. & Parker, G.M. (1982) Effects of smelter emmisions on metal levels in the plumage of ruffed grouse near Sudbury, Ontario, Canada. *Can. J. Zool.* **60**, 2659–2667.

Ross, H.B. (1992) Report on the intercalibration of analytical methods for the determination of trace metals, nitrate and ammonium in atmospheric precipitation within the framework of the Baltic Marine Environment Protection Commission—Helsinki Commission. *Baltic Sea Environ. Proc.* **41**, 2–14.

Samiullah, Y. (1990) *Biological Monitoring of Environmental Contaminants: Animals*. MARC Report No. 37. King's College London.

Samoiloff, M.R. (1989) Toxicity testing of sediments: problems, trends and solutions. In: *Aquatic Toxicology and Water Quality Management* (Eds J.O. Nriagu & J.S.S. Lakshminarayana), pp. 143–152. *Advances in Environmental Science and Technology*, Vol. 22. Wiley & Sons, New York.

Sawlan, J.J. & Murray, J.W. (1983) Trace metal remobilization in the interstitial waters of red clay and hemipelagic marine sediments. *Earth Planet. Sci. Lett.* **64**, 213–230.

Say, P.J., Burrows, I.G. & Whitton, B.A. (1986) *Enteromorpha* as a monitor of heavy metals in estuarine and coastal intertidal waters. Occasional publication No. 1, Northern Environmental Consultants Ltd, Consett, Durham.

Say, P.J., Harding, J.P.C. & Whitton, B.A. (1981) Aquatic mosses as monitors of heavy metal contamination in the River Etherow, Great Britain. *Environ. Pollut. B* **2**, 295–307.

Schmidt, D. (1992) Mercury in Baltic and North Sea Waters. *Water, Air Soil Pollut.* **62**, 43–55.

Sharma, G.K. & Shupe, J.L. (1977) Lead, cadmium and arsenic residues in animal tissues in relation to their surrounding habitat. *Sci. Total Environ.* **7**, 53–62.

Shendrikar, A.D. & Ensor, D.S. (1986) Sampling and measurement of trace element emissions from particulate control devices. In: *Toxic Metals in the Atmosphere* (Eds J.O. Nriagu & C.I. Davidson), pp. 53–100. Wiley & Sons, New York.

Smith, R.G. & Windom, H.L. (1980) A solvent extraction technique for determining nanogram per liter concentrations of cadmium, copper, nickel and zinc in sea water. *Anal. Chim. Acta* **113**, 39–46.

Spence, S.K. & Bryan, G.W. (1990) *Trace Metal Accumulation in Tubificid Worms*. First annual report to the WRc, December. PML, Plymouth.

Stafford, E.A. & McGrath, S.P. (1986) The use of acid insoluble residue to correct for the presence of soil-derived metals in the gut of earthworms used as bioindicator organisms. *Environ. Pollut. A* **42**, 233–246.

Stemmer, B.L., Burton, G.A. & Liebfritz-Frederick, S. (1990) Effect of sediment test variables on selenium toxicity to *Daphnia magna*. *Environ. Toxicol. Chem.* **9**, 381–389.

Stratton, J.W., Smith, D.F., Fan, A.M. & Book, S.A. (1987) *Methylmercury in Northern Coastal Mountain Lakes*. Office of Environmental Health Hazard Evaluations, Department of Health Services, California, Berkeley, CA.

Tan, P.M. (1988) Palmerton zinc Superfund site remediation strategy. *Trace Subst. Environ. Health* **22**, 296–305.

Taylor, G.J. & Crowder, A.A. (1983) Uptake and

accumulation of heavy metals by *Typha latifolia* in wetlands of the Sudbury, Ontario region. *Can. J. Bot.* **61**, 63–73.

Tessier, A. & Campbell, P.G.C. (1987) Partitioning of trace metals in sediments: relationships with bio-availability. *Hydrobiologia* **149**, 43–52.

Tessier, A. & Campbell, P.G.C. (1990) Partitioning of trace metals in sediments and its relationship to their accumulation in benthic organisms. In: *Metal Speciation in the Environment* (Eds J.A.C. Brockaert, S. Gucer & F. Adams), pp. 545–569. NATO ASI Series, Vol. G23. Springer-Verlag, Berlin.

Tessier, A., Campbell, P.G.C., Auclair, J.C., Ardisson, P., Legrand, C., Huizenga, D. & Schenk, R. (1985) Accumulation of trace metals in a freshwater mussel: some physico-chemical and biological factors involved. In: *Heavy Metals in the Environment* (Ed. T.D. Lekkas), pp. 682–684. CEP Consultants, Edinburgh.

Tessier, A., Campbell, P.G.C., Auclair, J.C. & Bisson, M. (1984) Relationships between the partitioning of trace metals in sediments and their accumulation in the tissues of the freshwater mollusc *Elliptio complanata* in a mining area. *Can. J. Fish. Aquat. Sci.* **41**, 1463–1472.

Tessier, A., Campbell, P.G.C. & Bisson, M. (1979) Sequential extraction procedure for the speciation of particulate trace metals. *Anal. Chem.* **51**, 844–851.

Thomas, W. & Herrmann, R. (1980) Nachweis von Chlorpestiziden, PCB, PCA und Schwermetallen mittels epiphytischer Moose als Biofilter entlang eines Profils durch Mitteleuropa. *Staub-Reinhalt. Luft* **40**, 440–444.

Tohyama, C., Himeno, S., Watanabe, C., Suzuki, T. & Morita, M. (1986) The relationship of the increased level of metallothionein with heavy metal levels in the tissue of the harbour seal *Phoca vitulina*. *Ecotoxicol. Environ. Safety* **12**, 85–94.

Tolonen, K. & Merilaeinen, J. (1983) Sedimentary chemistry of a small polluted lake, Galltraesk, South Finland. *Hydrobiologia* **103**, 309–318.

Topping, G. (1983) Guidelines for the use of biological material in first order assessment and trend monitoring. *DAFFS, Scottish Fisheries Research Report* 28.

Tyler, G. (1976) Soil factors controlling metal ion absorption in the wood anemone, *Anemone nemorosa. Oikos* **27**, 71–80.

Underwood, E.J. (1980) *Trace Elements in Human and Animal Nutrition*, 4th edn. Academic Press, London.

United Nations Environment Programme (1988) Guidelines for the determination of selected trace metals in aerosols and in wet precipitation. *Reference Methods for Marine Pollution Studies, UNEP*, **42**.

US EPA (1991) *Methods for the Determination of Metals in Environment Samples*. United States Environmental Agency, EPA-600/4-91-010. Washington, DC.

van den Berg, C.M.G., Khan, S.H., Daly, P.J., Riley, J.P. & Turner, D.R. (1991) An electrochemical study of Ni, Sb, Se, Sn, U and V in the estuary of the Tamar. *Estuarine, Coastal Shelf Sci.* **33**, 309–322.

van den Broek, W.L.F., Tracey, D.M., Solly, S.R.B. & Avrahami, M. (1981) Mercury levels in some New Zealand sea fishes. *N.Z.J. Mar. Freshwater Res.* **15**, 137–146.

Vermierman, K., Vandecasteele, C. & Dams, R. (1990) Determination of trace amounts of cadmium, lead, copper and zinc in natural waters by inductively coupled plasma atomic emission spectrometry with thermospray nebulisation, after enrichment on Chelex-100. *Analyst* **115**, 17–22.

Walker, M.I., McKay, W.A., Pattenden, N.J. & Liss, P.S. (1986) Actinide enrichment in marine aerosols. *Nature* **323**, 141–143.

Waller, P.A. & Pickering, W.F. (1990) Evaluation of 'labile' metal in sediments by anodic stripping voltammetry. *Talanta* **37**, 981–993.

Waller, P.A. & Pickering, W.F. (1991) Evaluation of 'labile' metal levels in polluted creek sediments, using transfer of metal to cation exchangers and ASV analysis of chemical extracts. *Chem. Spec. Bioavail.* **3**, 47–54.

Wang, W.C. & Barcelona, M.J. (1983) The determination of acid-reactive sulphide. *Environ. Int.* **9**, 129–133.

Watling, R.J., Watling, H.R., Stanton, R.C., McClurg, T.P. & Engelbrecht, E.M. (1982) The distribution and significance of toxic metals in sharks from the Natal Coast, South Africa. *Water Sci. Technol.* **14**, 21–30.

Watson, P.G. & Frickers, T.E. (1990) A multilevel, *in situ* pore-water sampler for use in intertidal sediments and laboratory microcosms. *Limnol. Oceanogr.* **35**, 1381–1389.

Wehr, J.D. & Whitton, B.A. (1983) Accumulation of heavy metals by aquatic mosses, 2. *Rhyncostegium riparoides. Hydrobiologia* **100**, 261–284.

Welsh, R.P.H. & Denny, P. (1980) The uptake of lead and copper by submerged aquatic macrophytes in two English lakes. *J. Ecol.* **68**, 443–445.

Wharfe, J.R., Friend, H. & Dines, R.A. (1985) An evaluation of selected sediment parameters as a rapid means of assessing the impact of organic waste discharges to tidal waters. *Environ. Pollut.* **10**, 159–172.

Whitton, B.A., Say, P.J. & Wehr, J.D. (1981) Use of plants to monitor heavy metals in rivers. In: *Heavy Metals in Northern England: Environmental and Biological Aspects* (Eds P.J. Say & B.A. Whitton), pp. 135–145. University of Durham Department of Botany, Durham.

Whitton, B.A., Say, P.J. & Jupp, B.P. (1982) Accumulation of zinc, cadmium and lead by the aquatic liverwort *Scapania. Environ. Pollut. B* **3**, 299–316.

Wieser, W., Dallinger, R. & Busch, G. (1977) The flow of copper through a terrestrial food chain II. Factors influencing the copper content of isopods. *Oecol.* **30**, 265–272.

Williams, L.R. (1989) Harmonization of quality assurance—an Interagency perspective. In: *Aquatic Toxicology and Hazard Assessment*, Vol. 12. ASTM STP 1027 (Eds V.M. Cowgill & L.R. Williams), pp. 11–18. ASTM, Philadelphia, PA.

Wilson, D., Finlayson, B. & Morgan, N. (1981) Copper, zinc and cadmium concentrations of resident trout related to acid-mine wastes. *Calif. Fish. Game.* **67**, 176–186.

Wren, C.D. (1986) Mammals as biological indicators of environmental metal levels. *Environ. Monit. Assess.* **6**, 127–144.

Zhulidov, A.V., Emets, V.M. & Shevstov, A.S. (1980) Biomonitoring of river pollution with heavy metals in wilderness preserves based on a study of metal accumulation in aquatic invertebrates. Chem. Abstr. 94, 552537r., 342, *Dokl. Akad. Nauk. SSSR* **252**, 1018–1020.

Zobell, C.E. (1946) Studies on redox potential of marine sediments. *Bull. Am. Assoc. Petro. Geol.* **30**, 477–513.

5: Heavy Metals

M.H. DEPLEDGE, J.M. WEEKS
AND P. BJERREGAARD

5.1 INTRODUCTION

In the previous chapter, analytical techniques were described by which heavy metal concentrations and chemical species can be determined in water and sediment samples, and in biota. Application of these techniques in monitoring studies can provide valuable insights into the extent of metal contamination in terrestrial and aquatic environments. This not only permits identification of pollution 'hotspots', but also allows the effectiveness of legislation controlling metal releases to be assessed.

The second major application of metal analysis is in studies of the uptake, handling, storage and excretion of heavy metals by animals and plants. Such work promotes a mechanistic understanding of the biological significance of particular metal concentrations and distributions in biota.

There is a very extensive literature concerning the accumulation of heavy metals in ecosystems and their components. It is not the purpose of this chapter to compile a review of such work; see instead Eisler (1981); Lepp (1981); Fowler, 1990; Samiullah (1990); Vernet (1991) and Bryan & Langston (1992). Rather, it gives general guidance on the metal concentrations that can be expected in particular types of organisms from clean and polluted localities, in aquatic and terrestrial ecosystems. More importantly, the factors that should be considered when trying to determine the biological significance of particular metal concentrations in organisms will be discussed. Finally, current management practices applicable to heavy metal pollution will be briefly summarized.

It is worth noting at the outset that many of the terms used in the 'heavy metal' literature have never been defined accurately, or have been used inappropriately. Every effort should be made to avoid using terms ambiguously in future. For example, the term 'bioavailable' has been used to refer to that proportion of a chemical in the environment that might be taken up into an organism. Alternatively, it has been used exclusively to refer to the amount of chemical that *is* taken up into an organism. In this account, the first definition is preferred. Similarly, there appears to be confusion about the terms 'uptake' and 'accumulation'. The former refers to the entrance of the chemical into an organism, the latter to the amount of chemical that remains in an organism following exposure over a certain time period. To give an extreme example, accumulation of a chemical in an organism may be negligible if excretory mechanisms are about 100% effective; however, uptake may have been great. Other terms will be defined in appropriate sections.

While this chapter deals with heavy metals in biota in general, the majority of the examples are taken from marine organisms. This reflects their dominance in the literature. Nonetheless, the principles discussed are broadly applicable.

5.2 WHAT ARE HEAVY METALS?

The term 'heavy metal' has been used very loosely in the literature. Chemical dictionaries define them as metals with a specific gravity greater than 4 or 5; but this includes the lanthanides and actinides, which are not usually regarded as 'heavy' on the basis of their chemical properties (Nieboer & Richardson, 1980). At present, the separation of metals into class A, borderline and class B appears to be the most useful (Nieboer & Richardson, 1980; Rainbow, 1985; Hopkin, 1989; see Table 5.1). If this is accepted, then all 'heavy metals' belong to either the 'borderline' or 'B' group. Among ecotoxicologists, the term 'heavy metals' is generally used to refer to metals that have been shown to cause environmental problems. Those of major concern are: Cd, Hg, Zn, Cu, Ni, Cr, Pb, Co, V, Ti, Fe, Mn, Ag, Sn (Rainbow, 1985; Hopkin, 1989; Bryan & Langston, 1992). In addition, the metalloids, As and Se (having some properties of metals and others of non-metals) are usually included.

A brief examination of the literature reveals that some heavy metals have been studied intensively while others have been neglected (see Samiullah, 1990). Cd, Hg and Pb have probably received most attention (see Goyer, 1991). Cu, Zn, Fe and Ni come next (Nriagu, 1979, 1980a,b). The biological significance of the metalloids, As and Se, has also been examined in some detail (e.g. Phillips & Depledge, 1985; Phillips, 1990; Jackson, 1991).

5.3 SOURCES OF HEAVY METALS IN THE ENVIRONMENT

Metals are continuously released into the biosphere by volcanoes, natural weathering of rocks, and by human activities such as mining, the combustion of fossil fuels and the release of sewage. On a global scale there is now abundant evidence that humans have contaminated the environment with heavy metals (and other pollutants) from the poles to the tropics and from the mountains to the abyssal depths (Samiullah, 1990; GESAMP, 1990). It has proven more difficult to determine the extent of adverse effects on biota resulting from such contamination.

5.3.1 Marine ecosystems

Numerous international bodies have published reports on heavy metal contamination and effects in marine ecosystems (Gray, 1979; UNEP, 1983; Kullenberg, 1986; GESAMP, 1990). Bryan (1976) categorized natural heavy metal inputs into the sea as follows.

1 Coastal supply, which includes input from rivers and from erosion produced by wave action and glaciers.

2 Deep sea supply, which includes metals released from deep sea volcanism and those removed from particles or sediments by chemical processes.

3 Supply which bypasses the nearshore environment. This includes metals transported in the atmosphere as dust particles or as aerosols, and also material produced by glacial erosion in polar regions and transported by floating ice.

Domestic effluents and urban stormwater runoff are also significant sources of heavy metal input into coastal waters. Appreciable amounts of heavy metals are contributed to domestic effluents from metabolic waste, corrosion of water pipes (Cu, Pb, Zn and Cd) and consumer products (e.g. detergent formulations containing Fe, Mn, Cr, Ni, Co, Zn, B and As). Concentrations are often in the milligram per litre range (Connell & Miller, 1984).

The oceans provide a vital sink for many heavy metals and their compounds, and are intimately involved in biogeochemical cycling. Indeed, there is growing concern that the natural cycling rates of many metals are being disturbed by human activities (Ramade, 1987). For example, it has been estimated that anthropogenic releases of Hg, Pb, Zn, Cd and Cu are between one and three orders of magnitude higher than natural fluxes (see Schindler, 1991). Recent estimates indicate that up to 150 000 tonnes of mercury are released naturally per year as a result of degassing from

Table 5.1 The separation of some essential and non-essential metal ions into class A, class B and borderline elements based on the classification scheme of Nieboer & Richardson (1980)

Class A	Borderline	Class B
Calcium	Zinc	Cadmium
Magnesium	Lead	Copper
Manganese	Iron	Mercury
Potassium	Chromium	Silver
Strontium	Cobalt	
Sodium	Nickel	
	Arsenic	
	Vanadium	

the Earth's crust. This compares with about 8000–10000 tonnes released by human activities (Goyer, 1991).

Routes of anthropogenic inputs of metals into the sea are from the atmosphere and rivers (GESAMP, 1990). With regard to the atmospheric route, metal particles released into the air at ground level are then mixed vertically. A consequence of this is that contaminants may be transported many thousands of kilometres from where they were first released (see Pacyna *et al.*, 1991 for a thorough discussion). This obviously creates problems when trying to relate effects of pollutants at one locality to a particular pollution source elsewhere. Thus, the role of human activity in the redistribution of heavy metals is evident in the 200-fold increase in the lead content of Greenland ice. Beginning with a low natural level about 2700 years ago, lead concentration increased during the industrial age and has risen very rapidly since lead was added to petrol during the 1920s (Goyer, 1991). Global differences in climate also result in uneven deposition of heavy metals (Bruenig, 1989; McKay & Thomas, 1989). The relative absence of precipitation scavenging and strong atmospheric inversions have been blamed for the net accumulation of Hg, Cd, V and Mn in Arctic biota and ice. These metals are thought to have been emitted in industrialized temperate zones and have therefore undergone considerable directional transport (Rahn & McCaffrey, 1979).

On the basis of currently available information, comparisons of atmospheric and riverine inputs of heavy metals into the sea indicate that approximately 98% of the lead which eventually dissolves in seawater enters the global ocean via the atmosphere. Similarly, most of the dissolved Cd, Cu, Fe and Zn is primarily from the atmosphere. Atmospheric inputs of arsenic and nickel are also significant (GESAMP, 1990). On a regional scale, it is thought that for the heavy metals, Cd, Hg, Cu, Pb and Zn, between 80% and 60% of the input into the North Sea is via atmospheric deposition (QSRNS, 1987). Similar routes of heavy metal input have been recognized for other temperate water bodies. Nonetheless, it would be unwise to conclude that the relative importance of these routes of entry is the same for all marine

and brackish water ecosystems. Little information is available for subtropical and tropical regions, or for southern hemisphere ecosystems where different patterns of rainfall, ocean currents, prevailing winds and annual temperature cycles affect heavy metal input and persistence.

Goldberg (1989) reviewed aquatic transport of chemicals. He concluded that organic compounds in the sea play a key role in determining the extent of heavy metal transport. Particulate organics can take up metals and artificially produced radionuclides, and thereby enhance their descent through the water column.

5.3.2 Freshwater ecosystems

An extensive review of heavy metal concentrations in the water, sediment and biota of ponds and lakes was conducted from a global perspective by Nriagu (1990). Similar syntheses are available for trace metals in rivers (see, for example, Burrows & Whitton, 1983; Dixit & Witcomb, 1983; Barak & Mason, 1989). Important factors affecting heavy metal concentrations and bioavailability in freshwater were discussed by Mason (1981), Connell & Miller (1984), Hellawell (1986), Kelly (1988) and, more recently, by Schindler (1991). These authors emphasize the importance of metal speciation associated with changes in redox potential, surface complexes, concentrations of major ions, pH, organic ligand concentrations, etc. For example, metals are primarily found adsorbed to particulates in freshwaters and chemical speciation to various soluble metal–ligand complexes is regarded as being more extensive than in seawater (Connell & Miller, 1984). The situation is further complicated by acidification of lakes and rivers. Acid rain resulting from dissolved gases such as hydrogen sulphide, sulphur dioxide and oxides of nitrogen, of both natural and anthropogenic origin, has contributed to alterations in soil and freshwater acidity (Samiullah, 1990). A major consequence of this is to increase the bioavailability of many heavy metals to freshwater biota. For example, Wood *et al.* (1988) showed that ambient calcium and aluminium concentrations together with lowered pH in freshwaters combine in a complex way to cause toxicity in brook trout (*Salvelinus fontinalis*). Toxicity due to aluminium increases as calcium ion concentration and pH fall.

5.3.3 Terrestrial ecosystems

Hopkin (1989) reviewed natural and anthropogenic sources of heavy metal release into the terrestrial environment. The metal content of soils may be strongly influenced by their origins. For example, soils derived from shales are often rich in cadmium, while those deriving from serpentine rocks contain elevated concentrations of Co, Ni and Cr, which may then be passed on to plants (Hopkin, 1989). Anthropogenic sources of heavy metals in the terrestrial environment include mining, ore smelting, combustion of fossil fuels and the use of certain pesticide formulations which contain As, Cu and Hg (Orians & Pfieffer, 1970; Mathew & Al-Doori, 1976; Godan, 1983; Hopkin, 1989).

5.4 THE SPECIATION AND BIOAVAILABILITY OF HEAVY METALS

Metals exist in water in equilibrium between free metal ions, metal bound in organic and inorganic complexes, and metal bound to organic and inorganic particulate matter. The chemical composition of the water (freshwater, seawater or porewater in soil or sediment) strongly influences the speciation of metals. Thus, in turbid river water, a large proportion of the total metal load is bound in or to organic or inorganic particulate matter (Salomons & Forstner, 1984). Furthermore, pH, hardness and the dissolved organic matter concentration of freshwater also affect metal speciation. When river water mixes with seawater in estuaries, the metal speciation is altered as ionic strength increases. Dissolved organic complexes and particulate matter may undergo flocculation, and for some metals a large proportion of the load transported in the river water will end up in the sediments of the estuarine area. However, some metals, for example cadmium, are released from particulate matter under the influence of the increasing chloride ion concentration due to the formation of chloride complexes (Elbay-Poulichet *et al.*, 1987). In estuarine areas, the speciation of metals that remain in solution is affected by the increasing concentration of anions — mainly chloride — and for most metals the free ions constitute only a minor proportion of the total dissolved metal load (Zirino & Yamamoto, 1972; Long & Angino, 1977; Mantoura *et al.*, 1978).

Some metals seem to be available for uptake into organisms from solution only as free ions, whereas others are supposedly transported over biological membranes as inorganic complexes. In experiments in which the free species of copper and cadmium were either carefully controlled by organic chelators or determined by means of ion selective electrodes, the toxicity (and thus probably bioavailability) was correlated with the concentration of free metal ions rather than total dissolved metal concentration (Sunda & Guillard, 1976; Sunda *et al.*, 1978; Zamuda & Sunda, 1982; Sanders *et al.*, 1983). This is consistent with the fact that uptake and toxicity of these metals increase with decreasing salinity (and thereby free ion concentration) in most estuarine organisms studied (McLusky *et al.*, 1986). In contrast with copper and cadmium, inorganic mercury is apparently transported over lipid membranes principally as uncharged chloride complexes (Gutknecht, 1981; Bienvenue *et al.*, 1984).

Since sediment is a much more complex chemical environment than water, there are so far no generally applicable, reliable methods for assessing bioavailability of metals in such media (see Luoma, 1989; Chapter 4). With regard to heavy metals in soils, it has been shown that they are associated with several distinct geochemical phases, the most important of which are considered to be clay minerals, organic matter, carbonates and sulphides (Kheboian & Bauer, 1987; Alloway, 1990). Metals that are mobilized through weathering of parent materials during pedogenesis, or following oxidation of metal sulphides in mining waste, eventually partition among various inorganic/organic phases as they are transported from the area of weathering. Subsequent distribution of metals depends on prevailing chemical and mineral environments. The types of reactions that are likely to control the partitioning of metals in soils are: (i) adsorption and desorption, (ii) precipitation and solubilization, (iii) surface complex formation, (iv) ion exchange, (v) penetration of the crystal structure of minerals, and (vi) biological mobilization and immobilization (Chao, 1984). The relative impact of elevated metals on soil–plant–animal systems is controlled by the predominant form(s) of the metal. For example, the speciation of a heavy metal in solution and absorbed to, or incorporated in, solid phases following the application of municipal sewage sludge to a soil is of concern because of the health risks associated with potential enhancement of available metal concentrations, or the loss in productivity due to the phytotoxicity of metals such as copper and zinc. Operationally defined extraction procedures have been used to characterize metal species, assess their availability to plants, or establish their potential for movement in the soil profile (Beckett, 1989).

With regard to uptake of metals from ingested soil or sediment, very few studies have assessed the actual species of metal that are available for transport across the lining of the digestive system. Luoma (1989) speculated that the processes that would determine availability of metals from ingested sediment in the gut of deposit-feeding invertebrates would be identical to the processes that control sorption/desorption processes in the sediment.

5.4.1 Heavy metals in biota

The concentrations of heavy metals in biota are the result of complex interactions among several factors. Thus, in any particular organism, tissue metal concentrations reflect the amount of metal taken up into the organism, the proportion of that metal that is distributed to each tissue, and the extent to which the metal enters and is retained within each tissue. With regard to the latter, the metabolic requirement (if any) for the metal, the metal concentration that can be tolerated in the tissue without initiating detoxification (via excretory or sequestration mechanisms), and whether the tissue has a role in storage of detoxified metal forms, are of primary importance (Depledge, 1989a).

Table 5.2 summarizes whole-body concentrations of an essential metal (Cu) and a non-essential metal (Cd) in selected terrestrial and

Table 5.2 Whole-body concentrations (µg/g dry weight) of copper and cadmium in selected terrestrial and aquatic biota

	Terrestrial		Aquatic	
	Cu	Cd	Cu	Cd
Plants/algae	c 10	c 0.6	1–16	0.1–16.3
Annelida	c 25	c 4	20–30	0.08–3.6
Mollusca	c 46	c 7	62–200	1–2.5
Crustacea	c 80	c 15	20–25	c 10

aquatic plants, annelids, molluscs (gastropods) and crustaceans. While there are many exceptions, concentrations of the two metals usually appear to be of the same order of magnitude for species within these groups sampled from sites that have not been exposed to significant pollution.

5.4.2 Heavy metals in marine organisms

The most comprehensive reviews of heavy metals in marine organisms conducted to date are those by Eisler (1981), Bryan (1984) and Furness & Rainbow (1990). The vast majority of information on metal concentrations in whole organisms and tissues concerns molluscs and crustaceans, although there are now substantial data bases for annelids (polychaetes), coelenterates and echinoderms. Close examination of available information reveals that some species have received a great deal of attention, while others have only been investigated at one locality on one occasion. This is clearly a major deficiency in the current data base, since it may give us a distorted view of metal handling and toxicity in biota. A common motive for selecting particular species appears to be the need to investigate the potential for heavy metal transfer to humans through ingestion of seafood. Thus, not surprisingly, edible bivalves (including mussels, oysters and scallops) are especially well studied among molluscs, while edible crabs, shrimps and lobsters have received most attention among the crustaceans. Of course, this means that many of the key species in ecosystems that may have relevance from an ecotoxicological point of view (rather than human toxicological viewpoint) have been neglected.

Phytoplankton and macroalgae have also been examined with regard to their heavy metal content in clean and polluted environments (Phillips, 1980). In general, their metal contents closely reflect short-term fluctuations in metal bioavailability in seawater (Phillips, 1977; Munda & Hudnik, 1991). Uptake mechanisms have been studied in detail by, for example, Black & Mitchell (1952) and Yamamoto *et al.* (1980). Direct uptake and release of metals can occur as seawater metal concentrations fluctuate, but ligands may also be produced which have strong metal-complexing ability. Concentration factors of up to six orders of magnitude are not uncommon in macroalgae (Phillips, 1977).

Table 5.3 provides a rough guide to the metal concentrations in marine organisms.

5.4.3 Heavy metals in freshwater organisms

Concentrations of metals found in freshwater organisms were summarized by Hellawell (1986). Some interesting detailed studies have also been carried out at particular localities; for example, Timmermanns (1991) studied the Maarsseveen Lakes system in the Netherlands. This lake system is considered to be relatively free from trace metal pollution. The majority of invertebrates had cadmium concentrations that were either undetectable or up to 1.9 μg/g dry weight. This range is similar to that reported by Barak & Mason (1989) for river invertebrates. Lead concentrations were in the range 0.29–5.0 μg/g dry weight while copper and zinc concentrations were in the ranges *c* 5–110 and *c* 50–400 μg/g dry weight, respectively. Timmermanns (1991) also went on to determine which biotic and abiotic factors influenced heavy metal concentrations in the lake invertebrate fauna. The most important biotic factors were body weight, trophic position and the presence of haemocyanin in the organism. Important abiotic factors influencing metal accumulation were calcium and dissolved organic carbon concentration in lake water. Zinc concentrations in benthic macro-invertebrates were highly correlated with trophic position; predators had higher zinc concentrations than their prey. Organisms possessing haemocyanin, not surprisingly, had higher whole-body copper loads than those which

did not. With regard to body weight this was particularly important in the case of lead, where younger, smaller animals had lower lead concentrations than older, larger ones. This applied both intraspecifically and interspecifically.

5.4.4 Heavy metals in terrestrial organisms

Studies on terrestrial animals have focused on a few species of annelids (earthworms), isopod crustaceans (woodlice) and molluscs (snails and slugs). Earthworms have been used extensively as bioindicator organisms (Ireland, 1983; Hopkin, 1989). Not only are they in intimate contact with the soil, but they ingest it as well. They are also confined within quite a small area, and consequently heavy metal concentrations found within their tissues reflect bioavailability at the sampling locality.

Terrestrial isopods have an extraordinary ability to accumulate metals. Wieser & Makart (1961) measured more than 1.5% copper on a dry weight basis in *Porcellio scaber* collected from spoil tips of disused mining areas. More recently, Hopkin & Martin (1982a) have shown that the concentrations of Zn, Cd, Pb and Cu in the hepatopancreas of individuals of this species from metal-contaminated sites may exceed 1.2%, 0.4%, 2.55% and 3.4% of the dry weight, respectively, with no apparent adverse effects. The hepatopancreas has been shown to be the most important storage organ of Pb, Zn, Cd and Cu (Hopkin & Martin, 1982b), often containing more than 75% of the Zn, 95% of the Cd, 80% of the Pb and 85% of the Cu in the whole body. Despite the massive concentrations of Zn, Cu, Cd and Pb in the hepatopancreas of isopods from metal-contaminated sites, the concentrations of these metals in other tissues remain relatively constant. High concentrations of pollutant metals have been measured in isopods collected from a variety of localities and including roadside verges (Williamson & Evans, 1972), spoil tips of metal mines (Avery *et al.*, 1983), areas adjacent to smelting works (for example, Martin & Coughtrey, 1976) and sites subject to contamination with metal-containing fungicides (Wieser *et al.*, 1976, 1977). The ability of isopods to accumulate metals, and their abundance in most

Table 5.3 A guide to the heavy-metal concentrations in the major groups of marine organisms (μg/g dry weight)

	Al	As	Cd	Cr	Co	Cu	Fe	Pb	Mn	Hg	Ni	Se	Zn
Marine algae and plants	c 200	10–20	<5	<1	<15	1–16	50–3300	<15	50–300	<1	<10	0.04–0.84	<500
Porifera	3700	2.8	1–4.5	–	–	8.5–31	4040	–	–	0.33	–	–	63–180
Coelenterates	c 80	c 70	0.07–5.3	0.4–35	0.004–0.2	2–57	5–730	2–24	2–130	0.07–0.86	2–19	–	3.4–170
Annelids	–	c 5	0.08–3.6	0.1–38	–	22–74	300–500	c 45	5–10	0.01–0.35	2–5	–	78–165
Molluscs (soft parts)	50–150	5–10	1–5	0.8–24	1–3	3–12	110–5000	0.5–40	5–150	0.005–0.1	1–10	0.03–3	50–500
Crustaceans	55–70	0.6–16	0.1–5.0	0.4–2.0	0.05–0.4	10–30	c 80	3–10	c 5	0.1–1.0	2–40	c 3	20–30
Echinoderms	<100	1–10	1–10	0.5–13.0	0.09–0.7	4–30	16–400	0.5–7.0	5–30	0.2–0.9	0.3–4.0	0.8–4.4	50–250
Fish	–	1–8	0.03–4.2	0.2–0.8	0.07–1.45	2–42	5–20	8–25	5–20	0.10–7.5	c 8.5	1–5	10–100
Seabirds	–	0.2–6	<0.3–6	<1	–	c 10–20	100–500	<5	<3	<2	<5	<5	c 50

Compiled from data presented by Eisler (1981), Bryan (1984) and Furness & Rainbow (1990).

terrestrial ecosystems, has encouraged their use as indicators of metal pollution (see Hopkin, 1989).

Research dealing with heavy metals in terrestrial molluscs has tended to focus on the application of metals in suitable forms to control slug and snail pests of agriculture effectively (Hopkin, 1989) and in the control of diseases in which molluscs serve as intermediate hosts (for example, control of the pulmonate gastropod, *Biomphalaria* spp., the intermediate host in schistosomiasis). Metal (and metalloid) compounds that are particularly effective in this regard include $CuSO_4$, Na_3AsO_4, Zn compounds and organotin formulations (Godan, 1983). Snails have also been included in biomonitoring studies of metal contaminated areas; for example, Hopkin (1989) reports concentrations of metals in a variety of tissues from *Helix aspersa* from an uncontaminated site (Kynance Cove, UK) and a contaminated site (a site 1 km from a smelting works at Avonmouth). Digestive gland accumulated most metal. At the contaminated site, Zn, Cd, Pb and Cu concentrations were raised by factors of c 4, 13, 16 and 1.8, respectively, compared with the uncontaminated site. Simkiss & Watkins (1991) have been able to demonstrate differences in zinc uptake in terrestrial snails (*H. aspersa*) from metal- and bacteria-polluted sites. Bacteria are well recognized for their ability to concentrate certain heavy metals. Bacteria are, of course, ingested by snails browsing on soil. Decreasing the numbers of bacteria in soil artificially was associated with a reduced rate of zinc uptake in the snail intestine.

A limited literature is available for nematodes and bees (Samiullah, 1990). Analysis of honey from bees from around suspected metal sources has been used to map metal releases both in space and time (Tong *et al.*, 1975; Jones, 1987). Roadside heavy metal contamination has been assessed with the aid of beetles, collembolans, spiders and mites. There have also been some interesting studies in which the population dynamics of soil arthropod populations have been used to detect adverse effects of heavy metal pollution (reviewed by Samiullah, 1990).

With regard to small mammals, some information is available for bats (Zook *et al.*, 1970; Miura *et al.* 1978; Clark, 1979). Otherwise, data are largely restricted to small rodents, especially mice, voles, shrews, squirrels and rats (Samiullah, 1990). A few studies have also been performed on cattle, sheep, horses and deer (reviewed by Samiullah, 1990). One particularly interesting example is the work of Baars (1990) on sheep grazing on the salt marshes of the Scheldt Estuary, the Netherlands. Heavy metals, brought into the estuary from rivers contaminated by industrial discharges, contaminated the marsh soil and flora. Heavy metals were then taken up by sheep grazing on the marsh.

Another study involving mammals was conducted between 1972 and 1975, in the Wisconsin River watershed. Selected heavy metal concentrations were measured in beaver, muskrat, red fox, raccoon mink and otter. The study revealed an interesting pattern of mercury accumulation, with piscivorous animals accumulating more mercury than herbivorous ones (Sheffy & St Amant, 1982). This serves to highlight the influence of feeding habit and position in trophic webs on the heavy metal load that organisms accumulate (Ramade, 1987).

Heavy metals in birds have been discussed by Samiullah (1990) and Peakall (1992). The latter identifies a number of biomarkers in birds and mammals that can be used to signal exposure to heavy metal pollution (see later).

The reduced mobility of heavy metals in soils and sediments, due to their strong binding to organic and inorganic colloids, constitutes a barrier to heavy metal toxicity in land plants. However, there are areas of very marked heavy metal enrichment where damage may result (for review see Lepp, 1981). Indeed, metals have been shown to exert a selective pressure leading to the evolution of metal-resistant plants in some localities (Bradshaw & Hardwick, 1989).

5.4.5 Unusual cases of metal accumulation

Although most organisms from clean and polluted sites contain metals in the concentration ranges mentioned so far, there are cases in which extraordinarily high heavy metal concentrations have been measured, either in whole organisms or in specific tissues. Sometimes these extremes cannot be attributed to pollutant exposure, but appear to occur naturally. Obviously, it is important to distinguish these exceptions from the general trends to avoid misinterpretation of the seriousness of pollution threats.

Examples of some of the highest reported metal concentrations follow: 18 500 µg Al/g dry weight was reported for the sponge *Dysidea crawshayi* (Bowen & Sutton, 1951); 19 300–40 800 µg Zn/g dry weight were measured in the excretory organs of *Pecten maximus* and *Chlamys opercularis* (Bryan, 1976), and extraordinarily high concentrations of arsenic have been measured in the gastropod *Hemifusus ternatanus* and in the feeding palps of the polychaete *Tharyx marioni* (Gibbs *et al.*, 1983; Phillips & Depledge, 1986a,b). Similarly, barnacles from the Thames estuary contain zinc concentrations of 153 000 µg Zn/g dry weight which is equivalent to 15% of the dry weight of the animal (Rainbow, 1987). Barnacles also have the potential to accumulate high concentrations of other metals; for example, the highest concentration so far reported for cadmium in a barnacle is 156 µg Cd/g dry weight in *Chthamalus stellatus* from the Azores (Weeks *et al.*, in press). In the terrestrial isopod, *Porcellio scaber* from the vicinity of a mine, more than 1000 µg Cu/g dry weight have been recorded (Prosi & Dallinger, 1988).

The literature also contains numerous examples of plants containing extremely high heavy metal concentrations (see reviews by Petersen, 1971 and Baker & Brooks, 1989). For example, in *Aeolanthus floribundus*, concentrations of up to 230 000 µg/g ash weight of nickel have been recorded (Severne & Brooks, 1972), while in *Ipomaea alpina* (Convolvulaceae) in Zaire, copper concentrations reach 12 300 µg/g dry weight (Malaise *et al.*, 1979). The term 'hyperaccumulator' has been used to describe plants capable of accumulating > 1000 µg/g nickel in their tissues (Brooks *et al.*, 1977). Such a threshold is now thought to be appropriate for hyperaccumulators of Cu, Cr and Pb (Baker & Walker, 1990). Metal hyper-accumulation is exhibited by representatives of all sections of the Plant Kingdom. Nonetheless, it is still a relatively uncommon response restricted to endemic taxa from metalliferous substrates (Baker & Walker, 1990). The latter authors have provided an excellent review of heavy metal handling in tolerant plants,

which should be consulted for further details.

The capacity of certain plants to accumulate heavy metals has been used extensively in the search for natural ore deposits in so-called 'geobotanical prospecting' (see Martin & Coughtrey, 1982).

5.5 THE USE OF BIOINDICATOR (BIOMONITOR) ORGANISMS TO MAP METAL DISTRIBUTION IN THE ENVIRONMENT

In the past, the main purpose of measuring heavy metal concentrations in biota has been to determine the toxicological threat posed to humans from the ingestion of excessive heavy metal loads in edible species. Other effects of direct relevance to humans include the accumulation of heavy metals in soils which may then be passed on to edible crops (Peterson, 1990) and heavy metal pollution of rivers that affects fish stocks. For example, in 1984, 4000 fish were reported to have been killed in the River Mawddach in Wales following copper and zinc discharges from disused mine workings (ENDS, 1984). While these concerns remain valid, both for medical and economic reasons, there is now the additional goal of ascertaining the effects of heavy metals on individuals, populations and communities, and on ecosystem structure and function.

The implementation of bioindicator studies (more recently referred to as biomonitor studies) to assess the accumulation of metals (and other pollutants) in biota was described in Chapter 4. Monitoring surveys have been carried out in the marine ecosystem (the Global Mussel Watch Programme, Bayne, 1989), in freshwater ecosystems (for example, using zebra mussels, Kraak *et al.*, 1991) and in terrestrial ecosystems (for example, using isopods, Hopkin, 1989). The following account identifies important factors that must be considered when interpreting data obtained from such studies.

5.5.1 Abiotic and biotic factors affecting bioavailability

Heavy metals that accumulate in water, aquatic sediments and soils are not necessarily freely available for uptake into biota. A proportion of the metal, as we have seen earlier, may be strongly bound in dissolved complexes or onto sediment surfaces, or in organic films surrounding particles. This speciation of metals is extremely difficult to follow or predict either qualitatively or quantitatively (Cantillo & Segar, 1975; Turner, 1984). Alterations in the physicochemical conditions in the environment can strongly influence the relative proportions of the metal species that can be taken up. Thus, alterations in pH, redox potential, salinity, temperature, etc., can all influence the bioavailability of metals for uptake into organisms (Mantoura *et al.*, 1978). This should be borne in mind in biomonitoring studies. Just because organisms do not contain especially high heavy metal concentrations in their tissues, this does not preclude the possibility that metals may be present in the environment at high concentrations, but firmly bound to sediment/soil particles, etc., at the locality in question. This gives rise to the possibility of so-called chemical time bombs (Hertling & Kuhnt, 1992). Scenarios can easily be envisaged in which high, non-bioavailable metal concentrations occurring in, for example, soil are not reflected in high heavy metal concentrations in biota and may therefore go undetected. As physicochemical conditions change, with, for example, acidification, rapid conversion of metals to bioavailable ionic forms may result in both high concentrations and toxicity in biota (see Samiullah, 1990).

5.5.2 Routes of uptake

Metal accumulation in biota may be the result of direct uptake from the surroundings across the body wall and/or respiratory structures, via the food, or by a combination of both routes. It is often not clear which route is the most important, although this may have great significance for subsequent inter-organ distribution of metals and, indeed, toxicity.

In the terrestrial environment, uptake via the food might be expected to be of greatest importance. For example, in insects, small mammals and birds, uptake across the body surfaces is usually minimal. Similarly, in many soil arthropods, the food route represents the major route of metal

uptake. There are exceptions, though. The moist, permeable body surface of an earthworm is in direct contact with soil and pore water, and it is thought that a proportion of metal uptake occurs via this route (Ireland, 1983).

In aquatic organisms it is often assumed that uptake across the body surface is the predominant route of uptake (Rainbow, 1988). This is because this is constantly bathed in metal-containing water, and indeed, large volumes of such water often pass over respiratory surfaces for the purpose of gas exchange. However, for many invertebrates and vertebrates (fish and aquatic mammals), direct uptake from water may be minor. Bryan (1984) reviewed available evidence and concluded that for many molluscs, crustaceans and annelids, metal uptake via the food may still present the major route of metal entry into organisms. Depledge & Rainbow (1990) point out that this is often ignored in standard toxicity tests. There is abundant evidence that the route of heavy metal uptake influences both intraorganismal metal distribution among tissues and toxicity. For example, the accumulation patterns of copper in the starfish *Asterias rubens* are markedly different when animals are exposed to copper in seawater or via their food (Depledge & Payet, unpublished). Similarly, Weeks & Rainbow (1990) reported that the zinc accumulated from a dietary source by the talitrid amphipod *Orchestia gammarellus* can be re-excreted after 24 h. However, zinc accumulated from a seawater source is retained with 100% efficiency (Weeks & Rainbow, 1991).

For most heavy metals it is usually free ionic species that are most readily taken up by diffusion across the entire body surface of the organism or across specialized respiratory structures (gills or lungs). Free ions probably bind to ligands in cell membranes, and this facilitates transport across the cellular barrier. Simkiss & Taylor (1989) proposed that non-ionic, inorganic species and organic derivatives may also diffuse into organisms due to their high lipid solubility. Uptake kinetics by passive diffusion are described by Fick's Law; for a detailed description of the factors involved see Depledge & Rainbow (1990). Fick's equation may require some modification to take account of the electrical potential differences that exist across most cellular barriers. Such potentials might influence to different extents the trans-barrier distribution of metal ions of different size and valency (Depledge & Rainbow, 1990).

As well as passive diffusion, some metals enter organisms by active transport. For example, cadmium may enter a variety of crustaceans, molluscs and fish via active transport through calcium ion pumps, while pinocytosis has been shown to be involved in the uptake of metal-rich particles in the gills and pharynx of some molluscs and ascidians, respectively (Kalk, 1963; Hobden, 1967). It is important to note that metal ions entering organisms along these routes are bioavailable, but would not have entered the organism were it not for these energy-consuming biological processes. This serves to emphasize the importance of interactions between the bioavailable fraction of metals in the environment and the biological characteristics of exposed organisms in jointly determining the amount of metal that accumulates in tissues.

5.5.3 Essential versus non-essential metals

Organisms exhibit selectivity with regard to their body loads of metals (Simkiss & Mason, 1983). Thus, the so-called 'bulk metals', Na, K, Mg and Ca, occur in large amounts while 'heavy metals' are present at much lower concentrations (Simkiss & Taylor, 1989). A further distinction should be made between heavy metals that are essential for life and those that are non-essential. Examples of essential heavy metals are Cu, Zn and Fe. These are vital components of enzymes, respiratory proteins and certain structural elements of organisms (Depledge & Rainbow, 1990). Thus, carbonic anhydrase, carboxypeptidase A and B and several hydrogenases contain zinc; pyruvate carboxylase contains manganese; the metalloid selenium is a component of glutathione peroxidase; copper is present in cytochrome oxidase; cobalt is present in vitamin B_{12}; haemocyanin contains copper and haemoglobin contains iron (Bryan, 1976, Hopkin, 1989). Animals and plants must somehow deliver a variety of heavy metals to their tissues to meet these diverse metabolic and respiratory requirements whilst at the same time preventing excessive accumulation of potentially toxic metal species. Dose–response curves highlight the dependency of the well-being of organisms on an appropriate metal supply, and the fact that essential metal deficiency or excess, beyond certain threshold concentrations, give rise to detrimental effects (see Rainbow, 1985). Other heavy metals, for example Cd, Hg and Pb, have no known biological role and are highly toxic when found at metabolically active sites, even at relatively low concentrations (Rainbow, 1985). They are therefore considered non-essential.

5.5.4 Metal handling

A prerequisite in biomonitoring studies is that normal background concentrations of heavy metals in the tissues of organisms can be distinguished from elevated concentrations associated with environmental pollution. The principles involved in the selection of bioindicator species have been discussed elsewhere (Volume 1, Chapters 19 and 20) and will not be repeated here. Suffice it to say that the organisms in question should reflect changes in metal bioavailability in the environment through changes in heavy metal concentrations in tissues. It was recognized early on in monitoring studies in the marine environment that, for essential metals in many of the marine invertebrates, changes in bioavailability were not always reflected in changes in tissue metal concentrations. For example, as seawater zinc concentrations increase within a certain range, whole-body zinc concentrations in the shrimp *Palaemon elegans* remain unchanged (White & Rainbow, 1984; Nugegoda & Rainbow, 1987, 1989). Apparently, zinc uptake in the shrimp increases with increasing zinc bioavailability, but increased zinc excretion results in maintenance of a relatively constant body load. This has been termed 'metal regulation', and the organism involved is said to be a 'regulator' of zinc (Rainbow, 1988). At high environmental zinc concentrations, 'regulatory ability' apparently breaks down, and body Zn concentrations rise with increasing environmental zinc bioavailabilities. It is not clear what the significance of regulation breakdown is, since animals appear to survive long after regulatory ability is exceeded, without mishap.

Other organisms have been recognized that are apparently unable to 'regulate' body concentrations of essential metals independent of environmental concentrations. For example, the barnacle, *Elminius modestus*, simply accumulates zinc at all exposure concentrations above background concentrations. Excess zinc is not excreted, but is instead stored in granules bound to pyrophosphate (Pullen & Rainbow, 1991). It has been proposed that these barnacles (and other species which respond in the same way) should be designated as 'accumulators', and are particularly suitable for inclusion in biomonitoring studies (Rainbow, 1992). Whilst this approach helps to provide a pragmatic solution to the problem of selecting biomonitoring organisms from a host of species found in ecosystems, it should be recognized that it has limited value scientifically, and can indeed obscure understanding of the mechanisms involved in heavy metal handling (Depledge & Rainbow, 1990). For example, the designation of organisms as either 'regulators' or 'accumulators' disguises the fact that there are a range of intermediate possibilities. Furthermore, some organisms may 'regulate' one metal, but not another. Thus, it is important to specify for which metal 'regulatory' ability applies. Another important consideration is that, for a particular species, 'regulatory' ability for a given metal may vary with route of uptake.

From a mechanistic viewpoint, the 'regulator'/ 'accumulator' classification is also misleading. In a number of scientific articles, organisms have been referred to as handling heavy metals via 'regulator strategies' or 'accumulator strategies'. This terminology should be avoided in future, as 'strategy' implies an active component in which an organism chooses a particular course of action (cf. the *Oxford English Dictionary*). Clearly, this is not the case. Furthermore, 'accumulators' of heavy metals avoid excessively high intracellular concentrations at metabolically active sites by sequestering metals in granules or by eliminating them from the metabolically active tissues and storing them in inert tissues, such as shells or carapaces. Even within cells, toxicity may be prevented by storing metals in inert granules (Simkiss, 1976; Coombs & George, 1978). The ability to store metals in this way appears to be common to all the major invertebrate phyla, with a prime site of granule storage being the digestive

gland. To prevent the transfer of potentially toxic metals into the blood of an animal, the cells of the gut lining appear to either excrete assimilated metals back into the gut lumen or bind them for storage in a metabolically unavailable form. Given the high availability of proteinaceous binding sites intracellularly, excretion of metals back into the lumen might involve energy-requiring mechanisms working against a concentration gradient. Thus, the intracellular storage of metals, in either a soluble or insoluble form, may be energetically less expensive. Hopkin (1989) proposed that metal granules should be classified into four types. Type A granules consist of concentric layers of calcium and magnesium phosphates which may contain other class A and borderline metals such as manganese and zinc. For example, granules in the basophil cells of the hepatopancreas of the snail *Helix aspersa* (Mason & Simkiss, 1982). Type B granules are structurally more heterogeneous than type A granules, ranging in appearance from membrane-bound accumulations, for example, the 'debris vesicles' in earth-worms (Morgan & Morris, 1982) to the 'copper granules' in the S cells of the hepatopancreas of isopods (Hopkin & Martin, 1982b) and always contain large amounts of sulphur in association with class B and borderline metals including Cd, Cu, Hg and Zn. Type C granules are composed almost entirely of iron and represent intracellular breakdown products of unwanted ferritin. For example, Moore & Rainbow (1984) described the production of iron-rich ferritin granules in the ventral caecal cells of the amphipod *Stegocephaloides christianiensis* as a physiological adaptation for dealing with an iron-rich food supply, cell breakdown resulting in the liberation of crystals into the caecal lumen and their expulsion into the gut. Finally, type D granules are generally much larger than the first three and are composed of concentric layers of calcium carbonate. Type D granules are produced extracellularly, and metals other than calcium have not been detected in them. In this sense, barnacles 'regulate' heavy metals even though they are often referred to as 'accumulators' or 'non-regulators' (Rainbow, 1988).

A final piece of evidence against the use of the 'regulator/accumulator' terminology is that it also tends to disguise biologically significant events. Many studies only consider the ways in which whole animal metal concentrations change with seawater metal concentrations. This neglects the internal tissue distributions of metals and metal species, which are likely to be the key determinants of metal toxicity and adverse effects. A situation can easily be envisaged in which increased exposure to metal might result in excess uptake with the need for enhanced excretion to maintain an approximately constant whole-body concentration. However, such 'regulation' is unlikely to be perfect, so there is the possibility that metal ions might accumulate in some target tissue in the body. If this occurred in nervous tissue, heart or even respiratory structures (gills and lungs), which constitute a relatively small proportion of the whole-body weight in many animals, metal concentrations might increase markedly without this being reflected in a statistically significant change in whole-body load. This could easily occur because usually the 'normal' range for whole-body trace metal concentrations is rather large. For example, the midgut gland of a typical brachyuran crab constitutes approximately 1–2% of the dry weight of the animal, and contains typical copper concentrations of c 80 µg Cu/g dry weight. This value could increase by an order of magnitude without a statistically significant departure of the whole-body copper concentration from the normal range given above for decapods. In other words, in a biomonitoring study, such a change would go undetected even though it may be of biological importance.

The key weakness of the 'regulator/accumulator' concept is that its mechanistic basis has not been established. The idea that an organism can 'regulate' its whole-body metal load implies that it is capable of detecting metals, ascertaining what its whole-body metal load is at any moment in time, then taking some action to ensure that metal is either taken up or excreted, to maintain the *status quo*. Furthermore, since there are several essential metals and metalloids, it also implies that this system is either very versatile or is replicated many times. At present there is no evidence to support such a hypothesis. Metal receptors and associated feedback mechanisms that might fulfil such roles have not been found.

The regulator/accumulator terminology was developed primarily on the basis of laboratory studies in which trace metal uptake via the seawater route alone has been considered. It has yet to be shown that similar biological responses pertain in organisms *in situ*. Indeed, Alliot & Frenet-Piron (1990) found that Cu, Zn, Cd and Pb concentrations in the shrimp *Palaemon serratus* from the Brittany coast fluctuated in accordance with changes in heavy metal concentrations in the sea. This indicates that although shrimps of this genus were designated 'zinc regulators' by Rainbow (1992) and yet 'accumulate' the non-essential metal cadmium, whole-body metal concentrations can still be used for monitoring fluctuating environmental bioavailability of metals. However, further verification of Alliot & Frenet-Piron's (1990) findings is required, as the rates of metal depuration from the shrimps that are implied by their study are very high compared with depuration rates actually measured in laboratory studies.

5.6 MECHANISMS OF METAL TOXICITY

The acute and chronic, lethal and sublethal toxicities of heavy metals to biota have been assessed in a wide range of species using conventional toxicity test procedures. The body of data available is enormous, but paradoxically, only concerns a minute proportion of species occurring in natural ecosystems. Also, the ecological relevance of such data is still questionable, as factors modulating heavy metal toxicity *in situ* are not taken into account by current test procedures. Furthermore, the data available have not been obtained in one set of standard conditions and often cannot be reproduced with precision.

The mechanisms by which metals exert toxicity are extremely diverse (Goyer, 1991) and cannot be fully reviewed here. A few generalizations are, however, possible.

Metals exert toxic effects on organisms if they enter into biochemical reactions in which they are normally not involved. Metal toxicity is influenced by their chemical form. These have been classified by Jørgensen & Jensen (1984) as:
1 simple hydrated metal ions;

2 metal ions complexed by inorganic anions;
3 metals complexed with organic ligands.

Toxic effects of metals on invertebrates include inhibition of growth, suppression of oxygen consumption and scope for growth, impaired reproduction, inhibition of byssus thread formation (in molluscs), damage to gills, impairment of tissue repair processes, melanogenesis and effects on ecdysis (in crustaceans) (see review by Viarengo, 1989).

Molecular mechanisms of heavy metal cytotoxicity include damage to plasma membranes following binding to proteins and phospholipids, inhibition of Na,K-dependent ATPases, inhibition of transmembrane amino acid transport, lipid peroxidation, enzyme inhibition (especially DNA polymerases, Ca-ATPases, Na,K-ATPases) and depletion of reduced glutathione (Viarengo, 1989).

Once inside the cell, metals may disrupt intermediary metabolism and even enter the nucleus where DNA synthesis and repair may be altered. Interestingly, the entry of certain metals into the nucleus may enhance the synthesis of RNA that codes for metallothionein (Hamer, 1986; Viarengo, 1989).

Heavy metals have been shown to inhibit a variety of important enzymes in plants, and disrupt several aspects of their biochemistry including photosynthesis, pigment synthesis and membrane structure and integrity (Fernandes & Henriques, 1991). In particular, damage is associated with metals blocking the electron transport chain in photosynthesis, leading to the production of free radicals which initiate lipid peroxidation in membranes (Fernandes & Henriques, 1991).

A guide to some useful references regarding the relative toxicities of heavy metals to a wide range of organisms is provided in Table 5.4.

5.7 INTERACTIONS AMONG HEAVY METALS

Most studies of heavy metal concentrations in biota ignore the influence that one metal may have on the uptake, accumulation and toxicity of other metals, or indeed, other pollutants. However, a number of interactions have been identified. Perhaps the most widely recognized example is that of the antagonistic interaction

Table 5.4 A guide to some useful references concerning heavy-metal toxicity

An extensive collection of toxicity data for plants and animals is to be found in Jørgensen *et al.* (1991) and for animals alone in Samiullah (1990)

The Water Research Centre (WRc), UK, has produced a series of reports dealing with environmental quality standards (EQS), which provide extensive information regarding toxicity and bioaccumulation of metals in aquatic organisms. The reports deal with the following metals and metalloids: chromium (TR207), inorganic lead (TR208), zinc (TR209), copper (TR210), nickel (TR211), arsenic (TR212), vanadium (TR253), inorganic tin (TR245), organotins (TR255), iron (TR258)

Acute and chronic toxicity in aquatic animals (Mance, 1987)

Developmental toxicity of metals and metalloids in fish (Weis & Weis, 1992)

Metal toxicity in soil microbes (Babich *et al.*, 1983)

Metal toxicity in terrestrial invertebrates (Hopkin, 1989; Van Straalen, 1993)

Metal toxicity in mammals (Goyer, 1991)

Lethal and sublethal effects of metals in plants (Lepp, 1981)

between mercury and selenium first demonstrated in the rat kidney (Parizek & Ostadalova, 1967). In many marine vertebrates, mercury and selenium concentrations are positively correlated in specific tissues (Koeman *et al.*, 1973, 1975; Mackay *et al.*, 1975; Norheim, 1987). The mechanism by which mercury toxicity is ameliorated by selenium in vertebrates remains obscure (see Pelletier, 1985). Cadmium–selenium interactions have also been reported (Magos & Webb, 1980).

Iron–mercury interactions have been described in the bivalve *Mercenaria mercenaria*. Mercury concentrations in the range $0.1-1.0$ mg/l resulted in reduced iron concentrations in the mantle fringe tissues of the clams (Fowler *et al.*, 1975). Also, low concentrations of selenium reduce mercury uptake in *Mytilus edulis* (Davis & Russell, 1988).

Copper–manganese interactions may have special significance with regard to phytoplankton growth. Thus, with rising manganese ion availability, the growth rate of *Thalassiosira pseudonana* increases. However, simultaneous addition of copper, resulted in competitive inhibition of manganese uptake and consequently slower growth rates (Sunda & Huntsman, 1983).

Another way in which one metal may influence tissue concentrations of another is via exchange. For example, Engel & Brouwer (1989) proposed that, during the moult cycle of the crab, as copper availability increases at ecdysis (perhaps resulting from haemocyanin catabolism and copper release), the metal is bound to metallothionein in the midgut gland, displacing zinc as it does so. Consequently, as copper concentration increases, so zinc concentration falls.

Bulk metals, such as calcium, have been shown to influence heavy-metal concentrations in tissues. For example, Bjerregaard & Depledge (1989) demonstrated that for the gastropod, *Littorina littorea*, calcium ion concentration exerts a greater effect on cadmium uptake than does alteration in salinity. As chloride ion complexation of cadmium ions in seawater falls with reduction in salinity, one might expect cadmium uptake to increase. However, if calcium concentration in seawater is increased as salinity is reduced, cadmium uptake also falls. This is thought to result from a competitive interaction of cadmium and calcium ions for active uptake via calcium pumps. In other species—for example, *Carcinus maenas* and *Mytilus edulis*—uptake of cadmium by active transport appears to be of less significance, and cadmium uptake increases as salinity is reduced, whether or not calcium concentration in the seawater is increased.

One final consideration when studying interactions among heavy metals in the environment is that observed correlations between, for example, the concentrations of two metals in an organism may be causative. However, the correlation may merely reflect a common dependency of the two metals on a third factor, such as the weight or age of the organism. Statistical techniques are available for dealing with this problem (Packard & Boardman, 1987).

5.8 INTERINDIVIDUAL VARIABILITY IN TRACE-METAL CONCENTRATIONS

Depledge (1990a) emphasized the importance of considering interindividual differences among the representatives of a population in ecotoxicological studies. To assess the ecological significance of, for example, heavy-metal concentrations in organisms, it is important to know what proportion of the individuals within a population have a metal burden which does not affect their Darwinian fitness, and what proportion of the population have diminished fitness or, indeed, have accumulated potentially lethal concentrations of pollutants (including heavy metals) in their tissues (Depledge, 1990a). Such assessments depend on knowledge of the abiotic and biotic factors contributing to interindividual variability in metal accumulation and effects.

The importance of determining differences in bioavailability has already been mentioned. However, intraorganismal factors may be just as important as external abiotic factors in determining responses to heavy metals. There have been numerous studies in which the physiological state of the organism has been shown to have a very marked influence on the uptake, distribution and effects of heavy metals (McLusky *et al.*, 1986; Depledge, 1989b, 1990b; Donker, 1992). For example, nutritional state may influence rates of metal uptake and subsequent tissue distribution. Alterations in environmental salinity may alter permeability and urinary excretion rates of metals in aquatic organisms.

5.8.1 Are heavy-metal concentrations in organisms normally distributed?

Depledge & Bjerregaard (1989) and Lobel *et al.* (1989) pointed out that the frequency distributions of particular trace metal concentrations are often not 'normal'. This fact is almost invariably ignored in biomonitoring surveys. Depledge & Bjerregaard (1989) found that, in some tissues sampled from individuals of a *Carcinus maenas* population, the concentrations of metals were not normally distributed. Cu, Zn and Cd concentrations showed a skewed distribution in midgut gland samples from crabs. A similar situation applied for cadmium in muscle tissue. However, copper and zinc concentrations were normally distributed in muscle tissue samples.

The problem that arises by assuming a 'normal' distribution when the data are skewed has been elegantly demonstrated by Lobel *et al.* (1982). Thus, for zinc in *Mytilus edulis* collected from the Tyne estuary, UK, the mean zinc concentration was only 75% of the mid-range value. Furthermore, in a comparison of three sites that were contaminated to different extents by zinc, lowest tissue concentrations recorded at each site were very similar (0.83, 1.5 and 1.11 µmol Zn/g), whereas highest concentrations were markedly different (3.32, 10.0 and 20.5 µmol Zn/g). The distributions of tissue concentrations from animals at each site were positively skewed. Statistical techniques are available for quantifying and comparing the residual variability of trace-metal concentrations in biological tissues (Lobel *et al.*, 1989).

An additional problem is that many authors have ignored the variability inherent in analytical techniques. A more detailed discussion of analytical errors was presented in the previous Chapter 4.

5.8.2 The influence of climate and geographic range on metal concentrations in biota

It is well known that temperature is a key factor influencing the kinetics of metal uptake and excretion in biota (see for example, Nugegoda & Rainbow, 1987). Other factors such as salinity, oxygen tension, particulate content of air and water, are known to affect metal bioavailability. Since these factors vary with climate, and among different geographic regions, they may give rise to considerable variability in tissue metal concentrations. This is important when trying to compare the extent of metal pollution at different localities. For example, there is limited evidence that trace metal concentrations in the tissues of some oceanic marine invertebrates and in selected crustaceans from subtropical and tropical inshore waters are lower than those found in temperate species from relatively clean areas (Hungspreugs, 1988). When compared with temperate scallop species (Eisler, 1981), the concentrations of Fe, Zn, Cd, Cu, Mn, Pb and Ni found in *Adamussium colbecki* collected from the Antarctic are much lower (Berkman & Nigro, 1992). Depledge *et al.* (1992) have pointed out that, in such circumstances, it is vital to know what the normal metal concentrations are in organisms before attempting to recognize abnormal concentrations that

might result from pollution. It is seldom appropriate to use normal values obtained for one species to compare with values obtained for another species (or even the same species) from a different climatic zone.

Alliot & Frenet-Piron (1990) found marked differences in Zn, Pb, Cu and Cd concentrations off the coast of Brittany at different times of the year. They attributed the changes in metal concentration solely to variations in boating activities and associated tourism, although it seems likely that other factors, such as seasonal differences in temperature, wave action, etc., were also involved. Interestingly, heavy-metal concentrations measured in the shrimp *Palaemon serratus* fluctuated with changes in trace-metal concentrations in the sea. This suggests that, following uptake, the shrimps are able to depurate excess metal when seawater concentrations decline. Furthermore, the concentrations of essential metals to which the shrimps were exposed were well within the range in which Rainbow (1988) claims shrimps regulate body loads independent of environmental concentrations. On the other hand, non-essential metals, which Rainbow (1988) found to accumulate, were in fact depurated when ambient seawater concentrations fell. This demonstrates an important principle; namely, that simplified laboratory experiments often do not mimic the responses of organisms in the natural environment, and, consequently, great care should be taken when predicting what will happen in the field from laboratory data alone. A further example of this is a recent study on arsenic uptake in crabs. Animals exposed to low concentrations of arsenic in the laboratory were found to have markedly different tissue distributions of arsenic to crabs collected from an estuary in which arsenic concentrations were elevated due to exposure to naturally high environmental arsenic concentrations (Andersen & Depledge, unpublished).

Other examples of seasonal/climatic changes in heavy-metal concentrations are provided by scallops (*Pecten maximus* and *Chlamis opercularis*) whose trace metal loads apparently vary in relation to heavy-metal concentrations in, and the availability of, phytoplankton (Bryan, 1973). Schulz-Baldes (1992) also compiled an interesting table of Cd, Cu and Pb concentrations in open-ocean organisms. Strikingly high cadmium concentrations have been reported for surface-inhabiting hyperiid amphipods (Hamanaka, 1981; Rainbow, 1989), but also for mesopelagic decapods (Ridout *et al.*, 1989).

5.9 ALTERNATIVE METHODS OF ASSESSING METAL POLLUTION

It is not always necessary to measure heavy metal concentrations in organisms to determine whether they have been exposed to metals. Alternative methods involve the use of bio-chemical biomarkers. For a detailed discussion of the design and application of biomarker studies the reader is referred to recent accounts by Peakall & Shugart (1992) and Depledge (1993).

In animals, metallothioneins offer the greatest potential as biomarkers of metal pollution (Engel, 1988). The properties of these small (c 10 kDa), cysteine-rich proteins were reviewed by Engel & Brouwer (1989), Benson *et al.* (1990) and Petering *et al.* (1990). Key features are that they are normal constituents of all cells investigated so far, and probably serve as intracellular storage sites for essential metals to fulfil metabolic requirements. They have been detected in Echinodermata, Annelida, Mollusca, Arthropoda, and a wide range of vertebrates. Metallothioneins are inducible by exposure to raised environmental concentrations of some heavy metals (e.g., Cd, Cu and Zn). Consequently, the detection of elevated metal-lothionein concentrations in selected tissues of organisms can be used to map metal exposure (Engel & Roesijadi, 1987; Benson *et al.*, 1990). However, before using these biomarkers of metal exposure, natural fluctuations in tissue metal-lothionein concentrations must first be well characterized.

Although the genes coding for metallothioneins are also present in higher plants, a group of pep-tides known as phytochelatins appear to be more important with regard to metal binding (Ernst *et al.*, 1992). These may play a part in metal transport, but there is little evidence that they confer increased tolerance on plants exposed to heavy metals (Schat & Kalff, 1992). Nonethe-less, they are inducible. In cell suspensions of *Rauvolfia serpentina* induction ability varies as

follows: Hg > Cd, As, Fe > Cu, Ni > Sb, Au > Sn, Se, Bi > Pb, Zn (see Ernst *et al.*, 1992). Conse-quently, phytochelatins are potential biomarkers of heavy metal exposure in plants, although, once again, natural variations in concentration must first be characterized.

Sanders (1990) has demonstrated the potential of suites of stress proteins (sometimes referred to as heat-shock proteins) as indicators of metal exposure. Most stress proteins are present at very low levels (or are undetectable) in the tissues of unstressed animals. Exposure to stress results in their rapid synthesis and diminished synthesis of normal proteins (Sanders, 1990). It may be possible to recognize specific sites of metal induced stress proteins and so distinguish responses to metals from those to other stressors. Sanders *et al.* (1991) also demonstrated that stress proteins might be used as biomarkers of the adverse effects of metals, and not only as biomarkers of exposure.

5.10 ECOLOGICAL EFFECTS OF HEAVY METAL POLLUTION

Biomonitoring studies are very valuable for determining fluctuations of heavy metal concen-trations in biota and for identifying polluted localities. Their limitation is, however, that they do not provide information regarding the ecologi-cal significance of a particular metal concen-tration. As the ultimate goal of studies on heavy metals in the environment is to ensure that anthropogenic sources of heavy metals do not give rise to significant adverse effects on natural communities of plants and animals, it is import-ant that techniques be developed for measuring pollution-induced ecological change. Contempor-ary reviews of such techniques have recently been carried out by Hopkin (1993) and Depledge & Hopkin (in press). Unfortunately, none of the techniques available is without drawbacks and limitations. For example, the pollution-induced community tolerance (PICT) approach is aimed at detecting ecologically significant metal effects (Blanck *et al.*, 1988; Blanck & Wangberg, 1988). The underlying concept here is that communities that are exposed to a particular pollutant will be more tolerant of exposure to that pollutant under controlled laboratory conditions than will a community that has never been exposed. Blanck & Wangberg (1988) demonstrated this phenomenon in a series of elegant experiments in which arsenate exposure was shown to induce 'community tolerance' in periphyton. A difficulty with this approach is that co-tolerance may occur, in which periphyton may show tolerance to one metal as a result of exposure to a quite different metal (or even another pollutant). Secondly, the natural variability in the tolerance of populations and communities to pollutant exposure in the laboratory appears to vary widely, for reasons not related to prior exposure to pollutants *in situ* (see review by Depledge & Hopkin, in press). Thus, PICT may be useful, but only in very restricted circumstances.

A point related to the PICT approach, that should be considered in both biomonitoring and effects studies alike, is that when natural populations of organisms have been chronically exposed to raised heavy metal concentrations for several generations, the animals in the population *may* show greater tolerance to metal exposure than individuals of the same species from clean sites. This might result from the selection of individuals that are less likely to take up metals (and which therefore may not signal an especially polluted environment in biomonitoring studies) or, more likely, that individuals have high body burdens of heavy metals, but are apparently able to grow and reproduce normally (Donker, 1992). For example, Bryan & Hummerstone (1971, 1973) found that a population of *Nereis diversicolor* exposed chronically to Cd, Pb, Cu and Zn in an estuary, showed greater tolerance to copper exposure than individuals collected from a clean site. However, enhanced resistance to cadmium and lead toxicity was not detectable.

5.11 REGULATION OF HEAVY-METAL CONCENTRATION IN THE ENVIRONMENT

After entry into the environment, metals pass along diverse pathways until captured by receptors in soil, sediment or water, or in target organisms, ranging from microbes to humans. It has been suggested that if a heavy metal does not reach target organisms in damaging quantities, then there has been no pollution as such. Thus, emissions of metals into the environment do not necessarily constitute pollution, and thus to eliminate metal pollution it is not necessary to restrict emissions completely.

Environmental standards or objectives can be viewed as legislative tools to permit control at any point along the pathway of a particular pollutant. The control points may be at the source, in various environmental compartments or in the target organism. The strategy involved has been discussed in great detail by Holdgate (1980). The case of lead will be discussed here to illustrate the principles involved. Lead can reach humans from a number of sources. For example, it occurs naturally in soils and may be taken up by crop plants. It may be washed from soils into river waters where it enters the water supply and so reaches household supplies. Similarly, discharges containing lead from factories and sewage may contaminate river water. Lead is a constituent of paints which, if applied to toys and other articles that small children may chew or suck, can constitute an important and potentially dangerous source of lead. Ore smelters may emit particulate lead into air, exposing humans via inhalation. Settling of particulate lead as a dust on food, or on soils in which edible crops are sown, represents another route of uptake. Finally, since the 1920s, lead has been used extensively as a petrol additive and is widely dispersed by combustion engine use. Centres of population where heavy traffic occurs receive particularly high exposures from this source (Nriagu, 1978).

Many of the above sources (and other natural sources) are extremely difficult to control. However, others are clearly within the realm of legislation. As part of its environmental policy the European Community agreed directives to control lead at a number of points along pathways to humans (and other targets).

In 1975, the European Commission proposed a Directive setting lead concentrations not to be exceeded in human blood. Although this no longer sets a biological standard in a legally binding way, it does set certain reference concentrations that serve to indicate when too much lead is present.

These 'biological standards' permit control over the combined release of lead from all sources where it matters, i.e. the target to be protected. However, biological standards suffer the disadvantage that they provide a signal only when exceeded, when the pollutant has already reached the target — possibly in damaging concentrations. When sampling indicates that reference concentrations have been exceeded, 'Member States shall take action to trace the exposure sources responsible ... and shall take all appropriate measures' (Haigh, 1991). When such situations arise, remedial measures must be taken further back along the pathway, as it is often not possible to take further control measures at the target point — other than physically removing targets from exposure.

5.11.1 Exposure standards

One of the most important points at which control can be exerted is at the point of entry of the metal into the target. The standard here is termed the 'exposure standard' or, sometimes, the 'primary protection standard'. As an example, in the EC Directive setting standards for the quality of drinking water, including the maximum concentration of lead (and other heavy metals) permitted, the Community has sought to ensure that the amount of lead swallowed with water is minimized. The domestic water supply can be tested, and the water treatment or the supply system adjusted to ensure that the standard is met. Other standards limit the quantity of lead (and other metals) in a multitude of other substances.

5.11.2 Environmental quality standards (EQS)

If the pathway of heavy-metal passage through the environment is retraced towards the point of entry, further standards can be applied at other points. For example, one Directive sets a quality standard for surface water from which drinking water is to be abstracted. If the constituents of river water (including lead and other heavy metals) exceed given concentrations, then river water from the vicinity must either not be used at all, or only for specified purposes. Many other

environmental quality standards for water have been addressed by Directives. With regard to biological and exposure standards, the breaching of an EQS does not provide an immediate indication of the action to be taken, but serves only as a signal that the pathway to the target contains too much of the pollutant.

5.11.3 Emission standards

Heavy metals may be emitted to the environment from point sources, such as an industrial outlet pipe into a river or from a chimney into the air. Diffuse metal releases occur from, for example, the use of lead in petrol. Only when the metal comes from a point source is it possible to set emission standards. These are often set for each individual discharge, or, alternatively, uniform standards for a particular class of discharge can be applied across a whole area or country, or even the entire EC. Limit values (upper limits) for emission standards are laid down in EC Directives for particularly dangerous substances, including mercury and cadmium (see list I, Table 5.5). For potentially less-dangerous substances set out in list II (Table 5.5), emission standards standards are the responsibility of Community Member States, and are to be set by reference to quality objectives. Since lead appears on list II and not on list I, the Community has no plans for setting emission standards for lead discharged to water. However, the Community has already laid down an environmental quality standard for abstracted surface water that is to be used for drinking, and any emission standards laid down in Member States must be such that those quality standards are met at the abstraction points. It is only for the list I substances that the Commission has proposed limit values which emission standards are not to exceed.

5.11.4 Product standards

The product of a manufacturing process may itself give rise to pollution when in use or upon disposal, in addition to any pollution that may have been caused during its manufacture. Accordingly, product standards may be set to control the

Table 5.5 Heavy metals placed on list I (the 'black list') or on list II (the 'grey list'). Those metals on list I have limit values and EQSs agreed at Community level (e.g. Directive 88/513 EEC limits cadmium discharges)

List I heavy metals (the 'black list')
Mercury and its compounds
Cadmium and its compounds

List II heavy metals ('the grey list')
The following metals and metalloids and their compounds:

Zinc	Arsenic	Beryllium	Tellurium
Copper	Antimony	Boron	Silver
Nickel	Molybdenum	Uranium	Thallium
Chromium	Titanium	Vanadium	Barium
Lead	Tin	Cobalt	Selenium

composition or construction of the product. This also has a bearing on the control of heavy metal in the environment. An example is the directive setting standards for the lead content of petrol (Haigh, 1991).

Rather than setting emission standards for each source of heavy metal (or other pollutant) from a plant, it is possible to set an upper limit for all emissions irrespective of origin. In the USA this is known as the 'bubble' concept: a notional bubble is drawn around a plant or area and an upper limit is put on the total amount of a heavy metal allowed to pass into the bubble. Thus, if manufacturers succeed in reducing diffuse discharges of metals they may emit more through a chimney, or vice-versa.

The approaches described above are mostly attempts to control heavy metal and other forms of pollution, rather than to anticipate and so prevent it. Some EC controls require that potentially toxic effects of chemicals are identified before they are marketed (Chapter 1). Another directive, known as the 'Seveso' Directive, requires manufacturers to identify and take steps to reduce the risks to the environment from a major accident.

The use of one legislative tool does not necessarily exclude the use of others, and they are usually used in combination with one another to provide a network of protection. Further details of combined approaches are provided by Haigh (1991).

The above account emphasizes strategies for the control of heavy metals in the European environment. In North America other legislation and regulatory approaches apply. The United States Environmental Protection Agency was created by a Presidential Order in 1970. With regard to the protection of water resources from heavy metal (and other contamination), the Safe Drinking Water Act was enacted in 1974. By 1976 the EPA was further empowered by the Toxic Substances Control Act (TSCA) to restrict the use of heavy metals and other chemicals in commercial processes that pose an unreasonable risk to health or environment. Control of waste materials (including heavy-metal contaminants) is included under the Resources Conservation and Recovery Act (1976). The potential carcinogenic, mutagenic and teratogenic effects of heavy metals are covered by controls in, for example, the Consumer Product Safety Act (1972) and the amended Water Pollution Control Act of 1987. A fuller review of United States regulatory procedures and sources of more detailed information concerning regulation of heavy metal concentrations are to be found in Merrill (1991).

5.12 SUMMARY AND CONCLUSIONS

This chapter identifies sources of metals in the environment and gives some indication of the concentrations that can accumulate in marine freshwater and terrestrial biota. The literature covering this field is extensive and therefore key sources of data have been referred to, to enable the reader to find more details. It is emphasized that the extent of accumulation of heavy metals and their ecotoxicological effects are determined not only by abiotic factors influencing bioavailability, but also by biotic factors such as physiological condition, endogenous rhythms, etc. Current management procedures are discussed. It is clear, however, that whilst much is known of the chemistry, environmental fate and toxicity of heavy metals, the long-term ecological consequences of heavy metal pollution are still poorly known, and the validity of current management procedures must constantly be improved.

REFERENCES

Alliot, A. & Frenet-Piron, M. (1990) Relationship between metals in sea-water and metal accumulation in shrimps. *Mar. Pollut. Bull.* **21**, 30–33.

Alloway, B.J. (Ed.) (1990) *Heavy Metals in Soils.* John Wiley & Sons, New York.

Avery, R.A., White, A.S., Martin, M.H. & Hopkin, S.P. (1983) Concentrations of heavy metals in common lizards (*Lacerta vivipara*) and their food and environment. *Amphibia–Reptilia* **4**, 205–213.

Baars, A.J. (1990) Heavy metals and fluoride pollution in the Scheldt estuary: levels in soil and vegetation, and state of health of locally grazing sheep. In: *Estuarine Ecotoxicology* (Eds P.L. Chambers & C.M. Chambers), Japaga, pp. 101–109. Wicklow, Ireland.

Babich, H., Bewley, R.J.F. & Stotzky, G. (1983) Application of the ecological dose concept to the impact of heavy metals on some microbe-mediated ecological processes in soil. *Arch. Environ. Contam. Toxicol.* **12**, 421–426.

Baker, A.J.M. & Brooks, R.R. (1989) Terrestrial higher plants which hyperaccumulate metallic elements – a review of the distribution, ecology and phytochemistry. *Biorecovery* **1**, 81–126.

Baker, A.J.M. & Walker, P.L. (1990) Ecophysiology of metal uptake by tolerant plants. In: *Heavy Metal Tolerance in Plants: Evolutionary Aspects* (Ed. A.J. Shaw), pp. 155–177. CRC Press, Boca Raton, FL.

Barak, N.A.E. & Mason, C.F. (1989) Heavy metals in water, sediment and invertebrates from rivers in eastern England. *Chemosphere* **19**, 1709–1714.

Bayne, B.L. (1989) Measuring the biological effects of pollution: the mussel watch approach. *Water Sci. Technol.* **21**, 1089–1100.

Beckett, P.H.T. (1989) The use of extractants in studies on trace metals in soils, sewage sludges, and sludge-treated soils. *Adv. Soil Sci.* **9**, 144–176.

Benson, W.H., Baer, K.N. & Wilson, C.F. (1990) Metallothionein as a biomarker of environmental metal contamination. In: *Biomarkers of Environmental Contamination* (Eds J.F. McCarthy & L.R. Shugart), pp. 255–266. Lewis Publishers, Boca Raton, FL.

Berkman, P.A. & Nigro, M. (1992) Trace metal concentrations in Scallops around Antarctica. *Mar. Pollut. Bull.* **24**, 322–323.

Bienvenue, E., Boudou, A., Desmazs, J.P., Gavach, C., Sandeaux, R. & Seta, P. (1984) Transport of mercury compounds across bimolecular lipid membranes: effect of lipid composition, pH and chloride concentration. *Chem. Biol. Interact.* **48**, 91–101.

Bjerregaard, P. & Depledge, M.H. (1989) Effect of salinity

and calcium concentration on cadmium uptake in *Littorina littorea* (L.). In: *Collected Abstracts of the 1st European Conference on Ecotoxicology*, Copenhagen, Denmark, p. 68.

Black, W.A.P. & Mitchell, R.L. (1952) Trace elements in the common brown algae and seawater. *J. Mar. Biol. Assoc., UK.* **30**, 575–584.

Blanck, H. & Wangberg, S.-Å. (1988) Induced community tolerance in marine periphyton established under arsenate stress. *Can. J. Fish. Aquat. Sci.* **45**, 1816–1819.

Blanck, H., Wangberg, S.-Å & Molander, S. (1988) Pollution-induced community tolerance – a new ecotoxicological tool. In: *Functional Testing of Aquatic Biota for Estimating Hazards of Chemicals* (Eds J. Cairns & J.R. Pratt). ASTM STP 988, pp. 219–230. American Society for Testing and Materials, Philadelphia, PA.

Bowen, V.T. & Sutton, D. (1951) Comparative studies of mineral constituents of marine sponges. *J. Mar. Res.* **10**, 153–167.

Bradshaw, A.D. & Hardwick, K. (1989) Evolution and stress – genotypic and phenotypic components. *Biol. J. Linnean Soc.* **37**, 137–155.

Brooks, R.R., Lee, J., Reeves, R.D. & Jaffre, T. (1977) Detection of nickeliferous rocks by analysis of herbarium specimens of indicator plants. *J. Geochem. Explor.* **7**, 49–57.

Bruenig, E.F. (1989) Ecosystem of the world. In: *Ecotoxicology and Climate*, (Eds P. Bourdeau, J.A. Haines, W. Klein & C.R.K. Murti), SCOPE 38, pp. 29–40. John Wiley & Sons, Chichester.

Bryan, G.W. (1973) The occurrence and seasonal variation of trace metals in the scallops *Pecten maximus* (L.) and *Chlamys opercularis* (L.). *J. Mar. Biol. Assoc., UK* **53**, 145–166.

Bryan, G.W. (1976) Heavy metal contamination in the sea. In: *Marine Pollution* (Ed. R. Johnson), pp. 185–302. Academic Press, London.

Bryan, G.W. (1984) Pollution due to heavy metals and their compounds. In: *Marine Ecology* (Ed. O. Kinne), pp. 1289–1431. John Wiley & Sons, Chichester.

Bryan, G.W. & Hummerstone, L.G. (1971) Adaptation of the polychaete *Nereis diversicolor* to estuarine sediments containing high concentrations of heavy metals. I. General observations and adaptation to copper. *J. Mar. Biol. Assoc., UK* **51**, 845–863.

Bryan, G.W. & Hummerstone, L.G. (1973) Adaptation of the estuarine polychaete *Nereis diversicolor* to estuarine sediments containing high concentrations of zinc and cadmium. *J. Mar. Biol. Assoc., UK* **53**, 839–857.

Bryan, G.W. & Langston, W.J. (1992) Bioavailability, accumulation and effects of heavy metals in sediments with special reference to United Kingdom estuaries: a review. *Environ. Pollut.* **76**, 89–131.

Burrows, I.G. & Whitton, B.A. (1983) Heavy metals

in water, sediment and invertebrates from a metal-contaminated river free of organic pollution. *Hydrobiologia* **106**, 263–273.

Cantillo, A.Y. & Segar, D.A. (1975) Metal species identification in the environment: a major challenge for the analyst. In: *International Conference. Heavy Metals in the Environment*, Toronto, 1975, pp. 183–204. CEP Consultants, Edinburgh.

Chao, T.T. (1984) Use of partial dissolution techniques in geochemical exploration. *J. Geochem. Explor.* **20**, 101–135.

Clark, D.R. Jr (1979) Lead concentrations: bats vs terrestrial small mammals collected near a major highway. *Environ. Sci. Technol.* **13**, 338–340.

Connell, D.W. & Miller, G.J. (1984) *The Chemistry and Ecotoxicology of Pollution*. John Wiley & Sons, New York.

Coombs, T.L. & George, S.G. (1978) Mechanisms of immobilisation and detoxification of metals in marine organisms. In: *Physiology and Behaviour of Marine Organisms* (Eds D.S. McLusky & A.J. Berry), pp. 179–187. Pergamon Press, Oxford.

Davis, I.M. & Russel, R. (1988) The influence of dissolved selenium compounds on the accumulation of inorganic and methylated mercury compounds from solution by the mussel *Mytilus edulis* and the plaice *Pleuronectes platessa*. *Sci. Total Environ.* **68**, 197–205.

Depledge, M.H. (1989a) Re-evaluation of copper and zinc requirements in decapod crustaceans. *Mar. Environ. Res.* **27**, 115–126.

Depledge, M.H. (1989b) Studies on copper and iron concentrations, distributions and uptake in the brachyuran, *Carcinus maenas*, following starvation. *Ophelia* **30**, 187–189.

Depledge, M.H. (1990a) Interactions between heavy metals and physiological processes in estuarine invertebrates. In: *Estuarine Ecotoxicology* (Eds P.L. Chambers & C.M. Chambers), pp. 89–100. Japaga, Wicklow, Ireland.

Depledge, M.H. (1990b) New approaches in ecotoxicology: can inter-individual physiological variability be used as a tool to investigate pollution effects? *Ambio* **19**, 251–252.

Depledge, M.H. (1993) The rational basis for the use of biomarkers as ecotoxicological tools. In: *Nondestructive Biomarkers in Vertebrates* (Eds M.C. Fossi & C. Leonzio), pp. 261–285. Lewis Publishers, Boca Raton, FL.

Depledge, M.H. & Bjerregaard, P. (1989) Explaining variation in trace metal concentrations in selected marine invertebrates: the importance of interactions between physiological state and environmental factors. In: *Phenotypic Response and Individuality in Aquatic Ectotherms* (Ed. J.C. Aldrich), pp. 121–126. Japaga, Wicklow, Ireland.

Depledge, M.H., Forbes, T.L. & Forbes, V.E. (1992) Evaluation of cadmium, copper, zinc and iron concentrations and tissue distributions in the benthic crab, *Dorippe granulata* (De Haan, 1841) from Tolo Harbour, Hong Kong. *Environ. Pollut.* **81**, 15–19.

Depledge, M.H. & Hopkin, S.P. (in press) Methods for the assessment of the effects of chemicals on brackish, estuarine and near coastal waters. *Scientific Group on Methodologies for the Safety Evaluation of Chemicals* **10**.

Depledge, M.H. & Rainbow, P.S. (1990) Models of regulation and accumulation of trace metals in marine invertebrates: a mini-review. *Comp. Biochem. Physiol.* **97C**, 1–7.

Dixit, S.S. & Witcomb, D. (1983) Heavy metal burden in water, substrate and macro-invertebrate body tissue of a polluted river Irwell (England). *Environ. Pollut.* **6**, 161–172.

Donker, M. (1992) Physiology of metal adaptation in the isopod *Porcellio scaber*. PhD thesis, Free University of Amsterdam, Netherlands.

Eisler, R. (1981) *Trace Metal Concentrations in Marine Organisms*. Pergamon Press, Oxford.

Elbay-Poulichet, F., Martin, J.M., Huang, W.W. & Zhu, J.X. (1987) Dissolved Cd behaviour in some selected French and Chinese estuaries. Consequences on Cd supply to the ocean. *Mar. Chem.* **322**, 125–136.

ENDS (1984) Drought exacerbates effects of water pollution incidents. *Environ. Data Serv. Rep.* **115**, 7.

Engel, D.W. (1988) The effect of biological variability on monitoring strategies: metallothioneins as an example. *Water Res. Bull.* **24**, 981–987.

Engel, D.W. & Brouwer, M. (1989) Metallothionein and metallothionein-like proteins: physiological importance. In: *Advances in Comparative and Environmental Physiology*, Vol. 5, pp. 53–75. Springer-Verlag, Berlin.

Engel, D.W. & Roesijadi, G. (1987) Metallothioneins: a monitoring tool. In: *Pollution and Physiology of Estuarine Organisms* (Eds F.J. Vernberg, F.P. Thurberg, A. Calabrese & W.B. Vernberg), pp. 421–437. University of South Carolina Press, Columbia, SC.

Ernst, W.H.O., Verkleij, J.A.C. & Schat, H. (1992) Metal tolerance in plants. *Acta Bot. Neerl.* **41**, 229–248.

Fernandes, J.C. & Henriques, F.S. (1991) Biochemical, physiological and structural effects of excess copper in plants. *Bot. Rev.* **57**, 246–273.

Fowler, S.W. (1990) Critical review of selected heavy metal and chlorinated hydrocarbon concentrations in the marine environment. *Mar. Environ. Res.* **29**, 1–64.

Fowler, B.A., Wolfe, D.A. & Hettler, W.F. (1975) Mercury and iron uptake by cytochromes in mantle epithelial cells of quahog clams (*Mercenaria mercenaria*) exposed to mercury. *J. Fish. Res. Board Can.* **32**, 1767–1775.

Furness, R.W. & Rainbow, P.S. (1990) *Heavy Metals in the Marine Environment*. CRC Press, Boca Raton, FL.

GESAMP (1990) *The State of the Marine Environment*. Blackwell Scientific Publications, Oxford.

Gibbs, P.E., Langston, W.J., Burt, G.R. & Pascoe, P.L. (1983) *Tharyx marioni* (Polychaeta): a remarkable accumulator of arsenic. *J. Mar. Biol. Assoc., UK* **63**, 313–325.

Godan, D. (1983) *Pest Slugs and Snails*. Springer-Verlag, Berlin.

Goldberg, E.D. (1989) Aquatic transport of chemicals. In: *Ecotoxicology and Climate* (Eds P. Bourdeau, J.A. Haines, W. Klein & C.R.K. Murti), SCOPE 38, pp. 51–64. John Wiley & Sons, Chichester.

Goyer, R.A. (1991) Toxic effects of metals. In: *Casarett and Doull's Toxicology*, 4th edn (Eds M.O. Amdur, J. Doull & C.D. Klaasen), pp. 623–680. Pergamon Press, New York.

Gray, J.S. (1979) Pollution-induced changes in populations. *Phil. Trans. R. Soc. B* **286**, 545–561.

Gutknecht, J. (1981) Inorganic mercury (Hg^{++}) transport through lipid bilayer membranes. *J. Membrane Biol.* **61**, 61–66.

Haigh, N. (1991) *EEC Environmental Policy & Britain*. Longman, London.

Hamanaka, T. (1981) Cd and Zn concentrations in zooplankton in the subarctic region of the North Pacific. *J. Oceanogr. Soc. Jpn* **37**, 160–172.

Hamer, D.H. (1986) Metallothionein. *Annu. Rev. Biochem.* **55**, 913–951.

Hellawell, J.M. (1986) *Biological Indicators of Freshwater Pollution and Environmental Management*. Elsevier Applied Science Publishers, London.

Hertling, Th. & Kuhnt, G. (1992) Chemical time bombs in urban soils. In: *Abstracts of an International Meeting on Bio-remediation, Toxicology, Environmental Fate and Ecology*, pp. B22–B28. Joint meeting of SETAC-Europe and Aquatic Ecosystem Health and Management Society, 21–24 June, Potsdam, Germany.

Hobden, D.J. (1967) Iron metabolism in *Mytilus edulis*, I. Variation in total content and distribution. *J. Mar. Biol. Assoc., UK* **47**, 597–606.

Holdgate, M.W. (1980) *A Perspective in Environmental Pollution*. Cambridge University Press, Cambridge.

Hopkin, S.P. (1989) *Ecophysiology of Metals in Terrestrial Invertebrates*. Elsevier Applied Science Publishers, London.

Hopkin, S.P. (1993) *In situ* biological monitoring of pollutants in ecosystems. In: *Handbook of Ecotoxicology*, Vol. 1 (Ed. P. Calow). Blackwell, Oxford.

Hopkin, S.P. & Martin, M.H. (1982a) The distribution of zinc, cadmium, lead and copper within the woodlouse *Oniscus asellus* (Crustacea, Isopoda). *Oecologia* **54**, 227–232.

Hopkin, S.P. & Martin, M.H. (1982b) The distribution of zinc, cadmium, lead and copper within the hepatopancreas of a woodlouse. *Tissue and Cell* **14**, 703–715.

Hungspreugs, M. (1988) Heavy metals and other non-oil pollutants in south-east Asia. *Ambio* **17**, 178–182.

Ireland, M.P. (1983) Heavy metal uptake and tissue distribution in earthworms. In: *Earthworm Ecology* (Ed. J.E. Satchell), pp. 247–265. Chapman & Hall, London.

Jackson, T.A. (1991) Effects of heavy metals and selenium on mercury methylation and other microbial activities in freshwater sediments. In: *Heavy Metals in the Environment* (Ed. J.P. Vernet), pp. 191–218. Elsevier, Amsterdam.

Jones, K.C. (1987) Honey as an indicator of heavy metal pollution. *Water, Air Soil Pollut.* **33**, 179–189.

Jørgensen, S.E. & Jensen, A. (1984) Processes of metal ions in the environment. In: *Metals in Biological Systems*, Vol. 18 (Ed. H. Sigel), pp. 61–103. Marcel Dekker, New York.

Jørgensen, S.E., Nielsen, S.E. & Jørgensen, L.A. (1991) *Handbook of Ecological Parameters and Ecotoxicology*. Elsevier, Amsterdam.

Kalk, M. (1963) Absorption of vanadium by tunicates. *Nature*, **198**, 1010–1011.

Kelly, M. (1988) *Mining and the Freshwater Environment*. Elsevier Applied Science Publishers, London.

Kheboian, C. & Bauer, C.F. (1987) Accuracy of selective extraction procedures for metal speciation in model aquatic sediments. *Anal. Chem.* **59**, 1417–1423.

Koeman, J.H., Peeters, W.H.M., Koudstaal-Hol, C.H.M., Tjioe, P.S. & de Goeij, J.J.M. (1973) Mercury–selenium correlations in marine mammals. *Nature*, **245**, 385–386.

Koeman, J.H., van de Ven, W.S.M., de Goeij, J.J.M., de Tjioe, P.S. & van Haaften, J.L. (1975) Mercury and selenium in marine mammals and birds. *Sci. Total Environ.* **3**, 279–287.

Kraak, M.H.S., Scholten, M.T.C., Peeters, W.H.M. & De Kock, W.C. (1991) Biomonitoring of heavy metals in the western European rivers Rhine and Meuse using the freshwater mussel *Dreissena polymorpha*. *Environ. Pollut.* **74**, 101–114.

Kullenberg, G. (1986) The IOC programme on marine pollution. *Mar. Pollut. Bull.* **17**, 341–352.

Lepp, N.W. (Ed.) (1981) *Effects of Heavy Metal Pollution on Plants*, Vol. 1. Applied Science Publishers, London.

Lobel, P.B., Mogie, P., Wright, D.A. & Wu, B.L. (1982) Metal accumulation in four molluscs. *Mar. Pollut. Bull.* **13**, 170–174.

Lobel, P.B., Belkhode, S.P., Jackson, S.E. & Longerich, H.P. (1989) A universal method for quantifying the residual variability of element concentrations in biological tissues using 25 elements in the mussel *Mytilus edulis* as a model. *Mar. Biol.* **102**, 513–518.

Long, D.T. & Angino, E.E. (1977) Chemical speciation

of Cd, Cu, Pb, and Zn in mixed freshwater, seawater, and brine solutions. *Geochim. Cosmochim. Acta* **41**, 1183–1191.

Luoma, S.M. (1989) Can we determine the biological availability of sediment-bound trace metals? *Hydrobiologia* **176/177**, 379–396.

Mackay, N.J., Kazacos, M.N., Williams. R.J. & Leedow, M.I. (1975) Selenium and heavy metals in black marlin. *Mar. Pollut. Bull.* **6**, 57–61.

Magos, L. & Webb, M. (1980) The interactions of selenium with cadmium and mercury. *CRC Crit. Rev. Toxicol.* **8**, 1–42.

Malaise, F., Gregoire, J., Morrison, R.S., Brooks, R.R. & Reeves, R.D. (1979) Copper and cobalt in vegetation of Fungurume, Shaba Province, Zaire. *Oikos* **33**, 472–478.

Mance, G. (1987) *Pollution Threat of Heavy Metals in Aquatic Environments*. Elsevier Applied Science Publishers, London.

Mantoura, R.F.C., Dickson, A. & Riley, J.P. (1978) The complexation of metals with humic materials in natural waters. *Estuar. Coast. Mar. Sci.* **6**, 387–408.

Martin, M.H. & Coughtrey, P.J. (1976) Comparisons between the levels of lead, zinc, and cadmium within a contaminated environment. *Chemosphere* **4**, 155–160.

Martin, M.H. & Coughtrey, P.J. (1982) *Biological Monitoring of Heavy Metal Pollution*. Applied Science Publishers, London.

Mason, A.Z. & Simkiss, K. (1982) Sites of mineral deposition in metal-accumulation cells. *Exp. Cell Res.* **139**, 383–391.

Mason, C.F. (1981) *Biology of Freshwater Pollution*. Longman, London.

Mathew, C. & Al-Doori, Z. (1976) The mutagenic effect of the mercury fungicide Ceresan M on *Drosophila melanogaster*. *Mutat. Res.* **40**, 31–36.

McKay, G.A. & Thomas, M.K. (1989) Climates of the World seen from an ecotoxicological perspective. In: *Ecotoxicology and Climate* (Eds P. Bourdeau, J.A. Haines, W. Klein & C.R. Krishna Murti), SCOPE 38, pp. 15–28. John Wiley & Sons, Chichester.

McLusky, D.S., Bryant, V. & Campbell, R. (1986) The effects of temperature and salinity on the toxicity of heavy metals to marine and estuarine invertebrates. *Oceanogr. Mar. Biol. Annu. Rev.* **24**, 481–520.

Merrill, R.A. (1991) Regulatory toxicology. In: *Casarett & Doull's Toxicology*, 4th edn (Eds M.O. Amdur, J. Doull & C.D. Klassen), pp. 970–984. Pergamon Press, New York.

Miura, T., Koyama, T. & Nakamura, I. (1978) Mercury content of museum and recent specimens of Chiropeter in Japan. *Bull. Environ. Contam. Toxicol.* **20**, 696–701.

Moore, P.G. & Rainbow, P.S. (1984) Ferritin crystals in the gut caeca of *Stegocephaloides christianiensis* Boeck and other Stegocephalidae (Amphipoda: Gammaridae): a functional interpretation. *Phil. Trans. R. Soc. Lond.* **306B**, 219–245.

Morgan, A.J. & Morris, B. (1982) The accumulation and intracellularly compartmentation of cadmium, lead, zinc and calcium in two earthworm species (*Dendrobaena rubida* and *Lumbricus rubellus*) living on highly contaminated soil. *Histochemistry* **73**, 589–598.

Munda, I.M. & Hudnik, V. (1991) Trace metal content in some seaweeds from the Northern Adriatic. *Bot. Mar.* **34**, 241–249.

Nieboer, E. & Richardson, D.H.S. (1980) The replacement of the nondescript term 'heavy metals' by a biologically and chemically significant classification of metal ions. *Environ. Pollut.* **B1**, 3–26.

Norheim, G. (1987) Levels and interactions of heavy metals in seabirds from Svalbard and the Antarctic. *Environ. Pollut.* **47**, 83–94.

Nriagu, J.O. (1978) *The Biogeochemistry of Lead in the Environment: Part A, Ecological Cycles. Part B, Biological Effects*. Elsevier/North Holland Biomedical Press, Amsterdam.

Nriagu, J.O. (1979) *Copper in the Environment*, Parts 1 and 2. Wiley Interscience, Chichester.

Nriagu, J.O. (1980a) *Zinc in the Environment*, Parts 1 and 2. Wiley Interscience, Chichester.

Nriagu, J.O. (1980b) *Cadmium in the Environment*, Part 1. Wiley Interscience, Chichester.

Nriagu, J.O. (1990) Trace metal pollution in lakes: a global perspective. *Proceedings of the Second International Conference on Trace Metals in the Aquatic Environment*. Sydney, Australia, July.

Nugegoda, D. & Rainbow, P.S. (1987) The effect of temperature on zinc regulation by the decapod crustacean, *Palaemon elegans*, Rathke. *Ophelia* **27**, 17–30.

Nugegoda, D. & Rainbow, P.S. (1989) Effects of salinity changes on zinc uptake and regulation by the decapod crustaceans, *Palaemon elegans* and *Palaemonetes varians*. *Mar. Ecol. Progr. Ser.* **51**, 57–75.

Orians, G.H. & Pfieffer, E.W. (1970) Ecological effects of war in Vietnam. *Science* **168**, 544–554.

Packard, G.C. & Boardman, T.J. (1987) The misuse of ratios to scale physiological data that vary allometrically with body size. In: *New Directions in Ecological Physiology* (Eds M.E. Feder, A.F. Bennett, W.W. Burggren & R.B. Huey), pp. 216–239. Cambridge University Press, Cambridge.

Pacyna, J.M., Munch, J. & Axanfeld, F. (1991) European inventory of trace metal emissions to the atmosphere. In: *Heavy Metals in the Environment* (Ed. J.P. Vernet), pp. 1–20. Elsevier, Amsterdam.

Parizek, J. & Ostadalova, I. (1967) The protective effect of small amounts of selenite on sublimate intoxication. *Experimentia* **23**, 142–143.

Peakall, D.B. (1992) *Animal Biomarkers as Pollution Indicators*. Chapman & Hall, London.

Peakall, D.B. & Shugart, L.R. (Eds) (1992) *Strategy for Biomarker Research and Application in the Assessment of Environmental Health*. Lewis Publishers, Boca Raton, FL.

Pelletier, E. (1985) Mercury—selenium interactions in aquatic organisms: a review. *Mar. Environ. Res.* **18**, 111–132.

Petering, D.H., Goodich, M., Hodgman, W., Krezooski, S., Weber, D., Shaw, C.F., Spieler, R. & Zettergren, L. (1990) Metal-binding proteins and peptides for the detection of heavy metals in aquatic organisms. In: *Biomarkers of Environmental Contamination* (Eds J.F. McCarthy & L.R. Shugart), pp. 239–254. Lewis Publishers, Boca Raton, FL.

Petersen, P.J. (1971) Unusual accumulations of elements by plants and animals. *Sci. Prog. Oxford* **59**, 505–526.

Peterson, P.J. (1990) Chemical etiology of endemic disease: a global perspective. In: *In Situ Evaluations of Biological Hazards of Environmental Pollutants* (Eds S.S. Sandhu, W.R. Lower, F.J. de Serres, W.A. Suk & R.R. Tice), pp. 195–201. Plenum Press, New York.

Phillips, D.J.H. (1977) The use of biological indicator organisms to monitor trace metal pollution in marine and estuarine environments—a review. *Environ. Pollut.* **13**, 281–317.

Phillips, D.J.H. (1980) *Quantitative Aquatic Biological Indicators*. Applied Science Publishers, London.

Phillips, D.J.H. (1990) Arsenic in aquatic organisms: a review, emphasizing chemical speciation. *Aquat. Toxicol.* **16**, 151–186.

Phillips, D.J.H. & Depledge, M.H. (1985) Metabolic pathways involving arsenic in marine organisms: a unifying hypothesis. *Mar. Environ. Res.* **17**, 1–12.

Phillips, D.J.H. & Depledge, M.H. (1986a) Chemical forms of arsenic in marine organisms with emphasis on *Hemifusus* spp. *Water Sci. Technol.* **18**, 213–222.

Phillips, D.J.H. & Depledge, M.H. (1986b) Distribution of inorganic and total arsenic in tissues of the marine gastropod, *Hemifusus ternatanus*. *Mar. Ecol. Progr. Ser.* **34**, 261–266.

Prosi, F. & Dallinger, R. (1988) Heavy metals in the terrestrial isopod *Porcellio scaber* Latreille. I. Histochemical and ultrastructural characterisation of metal containing lysosomes. *Cell Biol. Toxicol.* **4**, 81–96.

Pullen, J.H.S. & Rainbow, P.S. (1991) The composition of pyrophosphate heavy metal granules in barnacles. *J. Exp. Mar. Biol. Ecol.* **150**, 249–266.

QSRNS (Quality Status Report on the North Sea) (1987) Report of the scientific and technical working group. HMSO, Department of the Environment, UK.

Rahn, K.A. & McCaffrey, R.J. (1979) Long-range transport of pollution aerosols to the Arctic: a problem without borders. In: *Proceedings of WMO Symposium on Long-range Transport of Pollutants and its Relation to General Circulation including Stratospheric/Tropospheric Exchange Processes*. World Meteorological Office, Geneva.

Rainbow, P.S. (1985) The biology of heavy metals in the sea. *Int. J. Environ. Stud.* **25**, 195–211.

Rainbow, P.S. (1987) Heavy metals in barnacles. In: *Barnacle Biology* (Ed. A.J. Southward), pp. 405–417. A.A. Balkema, Rotterdam.

Rainbow, P.S. (1988) The significance of trace metal concentrations in decapods. *Symp. Zool. Soc. Lond.* **59**, 291–313.

Rainbow, P.S. (1989) Copper, cadmium and zinc concentrations in oceanic amphipod and eupausiid crustaceans, as a source of heavy metals to pelagic seabirds. *Mar. Biol.* **103**, 513–518.

Rainbow, P.S. (1992) The significance of trace metal concentrations in marine invertebrates. In: *Ecotoxicology of Metals in Invertebrates* (Eds R. Dallinger & P.S. Rainbow), pp. 3–23. Lewis Publishers, Boca Raton, FL.

Ramade, F. (1987) *Ecotoxicology*. John Wiley & Sons, New York.

Ridout, P.S., Rainbow, P.S., Roe, H.S.J. & Jones, H.R. (1989) Concentrations of V, Cr, Mn, Fe, Ni, Co, Cu, Zn, As and Cd in mesopelagic crustaceans from the North east Atlantic Ocean. *Mar. Biol.* **10**, 465–471.

Salomons, W. & Forstner, U. (1984) *Metals in the Hydrocycle*. Springer Verlag, Berlin.

Samiullah, Y. (1990) *Biological Monitoring of Environmental Contaminants: Animals*. MARC Report Number 37, Global Environmental Monitoring Programme.

Sanders, B. (1990) Potential as multi-tiered biomarkers. In: *Biomarkers of Environmental Contamination* (Eds J.F. McCarthy & L.R. Shugart), pp. 165–192. Lewis Publishers, Boca Raton, FL.

Sanders, B.M., Jenkins, W.G., Sunda, W.G. & Costlow, J.D. (1983) Free cupric ion activity in seawater: effects on metallothionein and growth in crab larvae. *Science* **222**, 53–55.

Sanders, B.M., Martin, L.S., Nelson, W.G., Phelps, D.K. & Welch, W. (1991) Relationships between accumulation of a 60 kDa stress protein and scope-for-growth in *Mytilus edulis* exposed to a range of copper concentrations. *Mar. Environ. Res.* **31**, 81–97.

Schat, H. & Kalff, M. (1992) Are phytochelatins involved in differential metal tolerance or do they merely reflect metal-imposed strain? *Plant Physiol.* **99**, 1475–1480.

Schindler, P.W. (1991) The regulation of heavy metals in natural aquatic systems. In: *Heavy Metals in the Environment* (Ed. J.P. Vernet), pp. 95–123. Elsevier, Amsterdam.

Schulz-Baldes, M. (1992) Baseline study on Cd, Cu and Pb concentrations in Atlantic neuston organisms. *Mar. Biol.* **112**, 211–222.

Severne, B.C. & Brooks, R.R. (1972) A nickel accumulating plant from Western Australia. *Planta* **103**, 91–94.

Sheffy, T.B. & St Amant, J.R. (1982) Mercury burdens in furbearers in Wisconsin. *J. Wildl. Manage.* **46**, 1117–1120.

Simkiss, K. (1976) Intracellular and extracellular routes in biomineralization. *Symp. Soc. Exp. Biol.* **30**, 423–444.

Simkiss, K. & Mason, A.Z. (1983) Metal ions: metabolic and toxic effects. In: *The Mollusca* (Ed. P.W. Hochachka), Vol. 2, pp. 101–164. Academic Press, New York.

Simkiss, K. & Taylor, M.G. (1989) Metal fluxes across membranes of aquatic organisms. *Rev. Aquat. Sci.* **1**, 173–188.

Simkiss, K. & Watkins, B. (1991) Differences in zinc uptake between snails (*Helix aspersa* (Muller)) from metal- and bacteria-polluted sites. *Funct. Ecol.* **5**, 787–794.

Sunda, W.G. & Guillard, R.R. (1976) The relationship between cupric ion activity and the toxicity of copper to phytoplankton. *J. Mar. Res.* **34**, 511–529.

Sunda, W.G. & Huntsman, S.A. (1983) Effect of competitive interactions between manganese and copper on cellular manganese and growth in estuarine and oceanic species of the diatom *Thalassiosira*. *Limnol. Oceanogr.* **28**, 924–934.

Sunda, W.G., Engel, D.W. & Thuotte, R.M. (1978) Effect of chemical speciation on toxicity of cadmium to grass shrimp *Palaemonetes pugio*: importance of free cadmium ion. *Environ. Sci. Technol.* **12**, 409–413.

Timmermanns, K.R. (1991) *Trace metal ecotoxicokinetics of chironomids*. PhD thesis, University of Amsterdam, Netherlands.

Tong, S.S.C., Morse, R.A., Bache, C.A. & Lisk, D.J. (1975) Elemental analysis of honey as an indicator of pollution. *Arch. Environ. Health* **30**, 329–332.

Turner, D.R. (1984) Relationships between biological availability and chemical measurements. In: *Metal Ions in Biological Systems*, Vol. 18: *Circulation of Metals in the Environment* (Ed. H. Sigel), pp. 137–164. Marcel Dekker, New York.

UNEP (1983) A review and the prospects for open ocean pollution monitoring. Internal report prepared for the Regional Seas Programme Activity Centre, UNEP by the Monitoring and Assessment Research Centre, University of London.

Van Straalen, N.M. (1993) Soil and sediment quality criteria derived from invertebrate toxicity data. In: *Ecotoxicology of Metals in Invertebrates* (Eds R. Dallinger & P.S. Rainbow), pp. 427–441. Lewis Publishers, Boca Raton, FL.

Vernet, J.P. (Ed.) (1991) *Heavy Metals in the Environment*. Elsevier, Amsterdam.

Viarengo, A. (1989) Heavy metals in marine invertebrates: Mechanisms of regulation and toxicity at the cellular level. *Rev. Aquat. Sci.* **1**, 295–317.

Weeks, J.M. & Rainbow, P.S. (1990) A dual-labelling technique to measure the relative assimilation efficiencies of invertebrates taking up trace metals from food. *Funct. Ecol.* **4**, 711–717.

Weeks, J.M. & Rainbow, P.S. (1991) The uptake and accumulation of zinc and copper from solution by two species of talitrid amphipods (Crustacea). *J. Mar. Biol. Assoc., UK* **71**, 811–826.

Weeks, J.M., Rainbow, P.S. & Depledge, M.H. (in press) Barnacles (*Chthamalus stellatus*) as biomonitors of trace metal bioavailability in the waters of Sao Miguel (Azores). *Proceedings of the Second International Workshop of Malacology and Marine Biology*, Sao Miguel, Azores.

Weis, P. & Weis, J.S. (1992) The developmental toxicity of metals and metalloids in fish. In: *Metal Ecotoxicology: Concepts and Applications* (Eds M.C. Newman & A.W. McIntosh), pp. 145–171. Lewis Publishers, Boca Raton, FL.

White, S.L. & Rainbow, P.S. (1984) Regulation of zinc concentration in *Palaemon elegans* (Crustacea: Decapoda): zinc flux and effects of temperature, zinc concentration and moulting. *Mar. Ecol. Progr. Ser.* **16**, 135–147.

Wieser, W. & Makart, H. (1961) Der sauerstoffverbrauch und der Gehalt an Ca, Cu und einigen anderen Spurenelementen bei terrestrischen Asseln. *Z. Naturforsch.* **16B**, 816–819.

Wieser, W., Busch, G. & Buchel, L. (1976) Isopods as indicators of copper content of soil and litter. *Oecologia* **1**, 38–48.

Wieser, W., Dallinger, R. & Busch, G. (1977) The flow of copper through a terrestrial food chain II. Factors influencing the copper content of isopods. *Oecologia* **30**, 265–272.

Williamson, P. & Evans, P.R. (1972) Lead: levels in roadside invertebrates and small mammals. *Bull. Environ. Contam. Toxicol.* **8**, 280–288.

Wood, C.M., Playle, R.C., Simons, B.P., Goss, G.G. & McDonald, D.G. (1988) Blood gases, acid–base status, ions and hematology in adult brook trout (*Salvelinus fontinalis*) under acid/aluminium exposure. *Can. J. Fish. Aquat. Sci.* **45**, 1575–1586.

Yamamoto, T., Otsuka, M. & Okamoto, K. (1980) A method of data analysis on the distribution of chemical elements in the biosphere. In: *Analytical Techniques in Environmental Chemistry* (Ed. J. Albaiges), pp. 401–408. Pergamon, Press, Oxford.

Zamuda, C.D. & Sunda, W.G. (1982) Bioavailability of dissolved copper to the American oyster *Crassostrea virginica*. I. Importance of chemical speciation. *Mar. Biol.* **66**, 77–82.

Zirino, A. & Yamamoto, S. (1972) A pH-dependent model for the chemical speciation of copper, zinc, cadmium and lead in seawater. *Limnol. Oceanogr.* **17**, 661–671.

Zook, B.C., Sauer, R.M. & Garner, F.M. (1970) Lead poisoning in Australian fruit bats *Pteropus poliocephalus*. *J. Am. Vet. Med. Assoc.* **157**, 691–694.

6: Organometallic Compounds in the Aquatic Environment

M.J. WALDOCK

6.1 INTRODUCTION

Only a narrow range of metals and metalloids are capable of forming compounds with metal to carbon bonds. There are some data for the environmental fate of Hg, Pb, Sn, Se, Ge, As and Sb compounds, but of these the Hg, Pb and Sn compounds have been shown to have the greatest impact on the aquatic environment, and they have been subject to the most detailed research. Although the metalloid arsenic has also been studied in some depth, organoarsenicals differ from the metals in that the organic forms are less toxic than the inorganic compounds, and arsenobetane, arsenocholine and arsenosugars are generally regarded as products of a detoxification process. Conversely, for the metals the alkyl or aryl conjugates are more toxic than the metal alone, and hence they merit particular attention.

It would be foolish to rule out other organometals as of little environmental significance; the production of synthetic organometal compounds is still at an early stage, and there is large scope for the formation of new molecules, e.g. asymmetric organometals where alkyl or aryl groups of different molecular size are grafted on to a metal template in order to alter persistency and toxicity. However, for the purposes of this chapter only some Hg, Pb and Sn compounds will be considered. This selection conveniently describes a range of chemical species with differing environmental fate, and hence potential for impact at a variety of trophic levels.

The chemistry and environmental fate of Hg, Pb and Sn compounds have been subject to a number of reviews (see for example Anon, 1971; D'Itri, 1972; Holden, 1973; Branica & Konrad,

1980; Craig, 1983; Chau, 1986; Seidel, *et al.*, 1980; Anon, 1986; Maguire, 1987, 1991; UNEP, 1989). This chapter deals with such aspects briefly, and will focus on the analytical requirements and biological effects programmes necessary to define the causal relationships, rather than mere positive correlations, between occurrence and environmental harm.

6.2 CHEMISTRY AND TOXICITY

6.2.1 Lead

The most common use of alkyl lead compounds is as an additive to petrol as a method of increasing octane rating; and as the use of leaded petrol is phased out, the risk of associated environmental problems should also diminish (see also Chapter 10). The compounds presently used in petrol are tetramethyl (TML) and tetraethyl leads (TEL). Both are only sparingly soluble in water (15 mg/l and 100 µg/l respectively), and break down sequentially to the tri-, di-, mono-alkyl and eventually inorganic lead compounds. The rate of decay from the tetra- to tri-species may be measured in days (0.5–2 days for TML and up to 7 days for TEL), but thereafter the breakdown rate is reduced, and only 25% of TEL and virtually no TML was found to be lost in a 60-day experiment (Grove, 1980). As the number of alkyl groups is reduced the compounds become more hydrophilic in nature, and there is a concomitant decrease in toxicity; in a comparative study, Maddock & Taylor (1980) demonstrated three orders of magnitude difference between the acutely toxic concentration of TML and trimethyl lead to plaice (*Pleuronectes platessa*)

(0.05 mg/l and 24.6 mg/l, respectively), the value for di-species being similar to those quoted for inorganic lead compounds (300 and 180 mg/l, respectively). There were similar but less marked differences for comparative studies on the ethylated leads. Clearly the lack of environmental persistence of the most toxic forms mitigates the possibility of widespread environmental damage. Based on fairly limited data sets, Maddock & Taylor (1980) suggest safe concentrations to be 10 μg/l for tetraalkyl species and 100 μg/l for the trialkyl species. All of the species are known to be bioaccumulative, however, and in 35-day exposure experiments, bioconcentration factors (BCF) for trimethyl and triethyl in mussels (*Mytilus edulis*) were demonstrated to be in the range 300−620-fold at exposure concentrations of 10−100 μg/l. When exposure was discontinued, tissue concentrations of lead rapidly decreased (half-life *c* 3 days, Maddock & Taylor, 1980).

Methylation of inorganic lead has been suggested as an alternative source of alkyl lead compounds in aquatic systems (Jarvie *et al.*, 1975; Wong *et al.*, 1975; Schmit & Huber, 1976), but the evidence for significant production of methyl species by this route remains equivocal (Thompson & Crerar, 1980; Reisinger *et al.*, 1981; Chau, 1986).

6.2.2 Mercury

Organic forms of mercury include a variety of mono- and dialkyl and aryl compounds which range in molecular size from the highly volatile dimethyl species (boiling point 92°C) to the large molecular weight C20 mercurochromes which are organomercury fluorescein derivatives (molecular weight *c* 750). Historically, the organic forms of mercury have had many medical uses including spermicides, antiseptics and diuretics, and agricultural uses as slimicides, fungicides and bactericides. Nowadays most such uses have been discontinued, and the contribution of industrially produced organomercurials to the overall mercury budget in the environment is low (e.g. D'Itri, 1972; Craig, 1983). However, biological methylation of inorganic mercury in marine sediments, fish and water has been demonstrated by a number of researchers (Jensen & Jernelov,

1969; Rudd *et al.*, 1980; Topping & Davies, 1981).

Factors controlling rates of methylation in sediments have been studied in some detail. Early work suggested that most methylation occurs when neither oxygen nor sulphide is present (Wollast *et al.*, 1975), but more recent studies indicate that methylation proceeds in oxic conditions (Compeau & Bartha, 1987), and sulphide concentrations are major controlling factors in methylmercury production (Craig & Moreton, 1986). The methylation process is thought to be bacterially mediated (Wood *et al.*, 1968), but chemical methylation is also possible (Jewett & Brinckman, 1974).

Conversely, biochemical decomposition of organomercury compounds may also be mediated by microbes, leading to complex equilibria between organic and inorganic forms in marine sediments, depending on the total mercury content of the sediment, silt and organic composition, pH, Eh, temperature, concentration of methanogenic bacteria, sulphide content and degree of complexation (see review by Craig, 1983).

Methyl mercuric compounds are therefore the most common forms of organic mercury in the aquatic environment. They are highly toxic, and comparative studies have shown that they are 4−31 times more toxic than the inorganic forms (Thain, unpublished data; EPA, 1986). Acutely toxic concentrations for a variety of fish and invertebrate species are in the low microgram to milligram per litre range, but some chronic exposure studies have shown toxicity at submicrogram per litre levels. Tests on methylmercury with *Daphnia magna* and brook trout produced chronic values of less than 0.07 μg/l (EPA, 1986).

More importantly in the context of the generally low concentrations of methylmercury in water and sediment (see later), they are highly bioaccumulative and more so than inorganic mercury species. Comparative studies by Hannerz (1968), for example, showed that for methyl- and methoxyethylmercury, bioaccumulation factors were 1000−2500 in salt water, whereas bioconcentration factors (BCFs) for mercuric nitrate were between 1 and 50. BCFs for mercuric chloride

beyond reasonable doubt that the bird kills were a direct result of the alkyllead release from the Manchester Ship Canal. Osborn *et al.* (1983) also demonstrated that a variety of sublethal responses found in low-dosed test animals were similar to those found in the field with no overt signs of toxicity. Based on the lowest lead concentrations in the birds with reduced nutritional reserves (2–6 mg/kg lead wet weight (liver), the researchers proposed a no-effect level of approximately 0.5 mg lead/kg in birds. This value is equivalent to 0.4 mg/kg of alkyllead in *Macoma* (Wilson *et al.*, 1986).

As a result of these studies, the controlling authority at the time (North West Water Authority) was able to produce a series of risk assessments for levels of discharge from the production plant, and introduce sufficiently restrictive consent conditions to safeguard the critical group of bird populations.

6.4.2 Mercury

Except for specific pollution incidents there is generally a large margin between the environmental levels of organomercury compounds and concentrations which would directly affect aquatic species. Certainly for fish and invertebrates there is an approximate 1000-fold safety margin between effect concentrations (in the microgram per litre range) and natural levels. Even at contaminated sites there is generally a margin of 100-fold. The main hazard appears to be the biomethylation of inorganic mercury compounds in sediments and accumulation of organic mercury in animals leading to effects at higher trophic levels.

In UK inland waters the European otter (*Lutra lutra*), as a largely fish-eating species, has been recognized as being particularly at risk from aquatic pollutants such as organochlorine compounds and heavy metals. One example where the concentrations of methylmercury in food species are close to those representing toxic thresholds for otters was identified in the mid-1980s in the River Yare and associated Broads in Norfolk. The Broads system of shallow lakes in Norfolk is one of the few areas of England where populations of otters remain. Although there is

little heavy industrial development in the area, a chemical company was found to be discharging mercuric halides into the sewerage system which, in turn, discharged into the Yare after processing at a local sewage treatment works (STW). A research programme was set in place by the then Anglian Water Authority to determine the extent of contamination of the River Yare, and the neighbouring Rockland and Surlingham Broads. Analysis of mercury in the surficial sediments of the Yare showed a plume of contamination extending from the sewage effluent discharge point, with a maximum concentration of 32.9 µg/g at 6 km downstream of the outfall. Mean concentrations for the various sampling transects undertaken were 0.81–11.9 µg/g. The methylmercury concentrations ranged between <0.7 and 13.2 ng/g, estimated to be between 0.01% and 0.2% of the total mercury burden (Bubb *et al.*, 1991). Bubb and colleagues suggest that Rockland and Surlingham Broads have acted as an absorptive sink for mercury since the discharge began in the early 1970s.

The bioavailability of this sedimentary mercury has been assessed by a survey of mercury concentrations in eels (*Anguilla anguilla*) taken upstream and downstream of the STW. As eels forage in detrital matter and possess a fatty body tissue, they provide a worst-case example for the uptake of xenobiotics in fish. Concentrations of mercury in muscle tissue in eels taken upstream of the STW in 1985/1986 were 0.18 (SD 0.15) µg/g wet weight in 25 individuals. The concentration in the bulked liver was 0.10 µg/g. Downstream of the STW the concentration in muscle was 0.58 (SD 0.27), and in bulked liver 0.89 µg/g (MAFF, unpublished). For many individuals the total body content would therefore have been well over 1 µg/g.

In a study of otter tolerance to mercury in the US, O'Connor & Neilsen (1980) dosed *Lutra canadensis* at concentrations of 2, 4 and 8 µg/g methylmercury hydroxide in the diet. Control otters remained healthy, whereas all of the treated otters developed mercury poisoning. The clinical development of intoxication was essentially the same in all treatments with progressive central nervous system deterioration of 10–14 days duration. O'Connor & Neilsen concluded that mercury poisoning would occur in areas where methylmercury levels in their food items were regularly 2 μg/g or greater. Since minimal tolerances were not determined, they also suggested that it is probable that prolonged exposure to lower levels would also be lethal, or produce subclinical effects adversely affecting survival or reproductive success.

Although otters in the vicinity of the Yare catchment would not feed exclusively on eels, there is clearly little margin of safety between measured and effect concentrations. Mason *et al.* (1986) have compared the tissue burdens of otters in this experiment, with those measured in dead otters from the UK. Tissue concentrations in the American study were 25–39 μg/g in liver, 20–57 μg/g in kidney and 13–22 μg/g in muscle. Mean concentrations in UK otters were 5.37 μg/g (range 1.2–20.5) in liver, 2.27 μg/g (range 1.35–6.79) in kidney and 1.66 μg/g (0.61–2.64) in muscle. Four otters found in the North Norfolk area were included in the study, one of which had elevated concentrations of mercury in hair. The authors concluded that heavy metals were not causing direct mortality of otters in Britain, but mercury levels in some individuals were approaching those known to produce sublethal effects. The specific example of the River Yare was not considered in the study by Mason and his colleagues.

Similarly in the marine environment animals at higher trophic levels could be most at risk, particularly fish-eating marine mammals such as seals, dolphins and some whales. There are, however, two major difficulties in assessment of the hazards to marine mammals. First, the size and life span of the animals prevent the type of experimentation possible with fish and invertebrates, and secondly the relationship between body burden and toxic effect is not a simple one, due to detoxification processes effected by selenium sequestration of mercury in these animals (see later).

A very few experiments have been carried out under controlled laboratory conditions. Ronald *et al.* (1977), for example, introduced methylmercuric chloride into the diet of four harp seals. At the low-dose exposure (0.25 mg/kg body weight) the animals showed a decline in appetite and body weight, but there was no sign of neurological dysfunction, and both animals survived. At the high-dose exposure (25 mg/kg body weight) the animals had to be force-fed after 4 days, showed a variety of behavioural disorders, and eventually died on day 20 and day 26 of the experiment. The methylmercury concentrations in the liver of control animals were 0.16 and 0.2 mg/kg, 18 and 76 mg/kg in low-dosed animals and 125 and 127 mg/kg in high-dosed animals.

In order to assess the environmental significance of such figures reviewers have compiled inventories of mercury values in food fish (see, for example, Law, in press) and have come to the conclusion that for the majority of areas the total dietary intake of mercury is of the order of 0.003–0.01 mg mercury/kg body weight per day. Therefore it is only in highly impacted areas such as parts of the Mediterranean, where concentrations in fish are up to 7 mg/kg, that the lower experimental exposure values could be approached.

The other problem with interpretation of this type of experiment is that it takes no account of immobilization and detoxification systems (see also Chapter 5) that govern the toxicity of organomercury compounds in biological systems. Detoxification systems exploit the affinity of methylmercury for thiols such as cysteine or for sulphydryl groups (which have been used as the basis of analytical methods and antidotes in the case of acute mercury poisoning) and the propensity of mercury compounds to form selenides. Selenide formation in sea mammals has been shown to be an important mechanism for mineralization, and hence detoxification, of mercury (Koeman *et al.*, 1973; Thibaud & Duguy, 1973; Joiris *et al.*, 1991); and, for example, in Cuvier's beaked whale and the bottlenose dolphin, particles of pure mercuric selenide have been found (Majorta & Viale, 1977).

As a result of the detoxification mechanism, extremely high levels of mercury can be tolerated in marine mammals with no signs of poisoning. Itano *et al.* (1984), for example, reported no signs of mercury poisoning in a striped dolphin when the body burden of mercury was 2.5 times as high as the lethal body burden for humans. However, there is still a question of whether or not old cetaceans die from mercury contamination. In attempting to provide an answer, Joiris *et al.* (1992) sampled some 30 stranded or by-caught common dolphins, 17 harbour porpoises, seven bottlenose dolphins, two striped dolphins, one sperm whale and six Minke whales. Concentrations of methylmercury showed an increase with increasing age of the animals, but the increase in total mercury was more marked. The ratio of methylmercury to the total was therefore decreasing as the animals grew older, confirming that the mineralization process was operating. The resulting concentrations of inorganic mercury in liver and kidney were about four times higher in old animals than in juveniles. In the harbour porpoises about half of the inorganic mercury in the liver was not bound to metallothioneins (Chapter 5) or to selenium, and hence was potentially toxic. The authors concluded that more speciation studies were needed on this toxic reservoir of inorganic mercury in older animals, and that inorganic mercury might influence the health status of older animals.

6.4.3 Tin

For TBT it has been clearly demonstrated that there is no margin of safety between environmental concentrations and effects on a variety of invertebrates, and particularly estuarine and marine molluscs. In this case demonstration of a causal relationship has been relatively easy to accomplish. Many mollusc species may be held in the laboratory for long periods, and therefore animals may be exposed for a significant proportion of the life cycle, allowing the possibility of testing a variety of endpoints, e.g. survival, growth and reproduction. Sensitive analytical techniques have also facilitated confirmation of accurate dosing of low levels of toxicant, and therefore biological effects are well documented. The case study of *Crassostrea gigas* is a good example of an impact assessment evaluation.

Early reports from France by Alzieu *et al.* (1982) suggested that the growth of *C. gigas* was reduced and the shells of the animals became abnormally thickened at submicrogram per litre concentrations of TBT. The species had been introduced to Britain in the mid-1970s, and although it was

shown to grow well at many sites, at others, such as the River Crouch in Essex, the animals had failed to grow normally and produced malformed thickened shells. At that time the malformation had been attributed to high silt loads in the water (Key *et al.*, 1976). In the light of the French findings, the reasons for the shell deformations in UK oysters were reinvestigated. The likelihood of environmental concentrations in the microgram per litre range resulting from the use of TBT as an antifouling paint on yachts was calculated using a simple model based on numbers of boats, release rate of toxicant from the hulls, and local hydrographic considerations. The first of these modelling exercises for the Crouch estuary suggested concentrations of TBT would be approximately 0.2 µg/l (Waldock, unpublished data).

Method development for the analysis of such concentrations of TBT followed, and in an initial survey in 1982 maximum concentrations in the open waters of the Crouch estuary in the vicinity of moored yachts were shown to be up to 0.43 µg/l and approximately 2 µg/l in marinas (Waldock & Miller, 1983).

In the next phase of the study, juvenile *C. gigas* were exposed to environmentally realistic concentrations of TBT in the laboratory. Since silt had been implicated in causing the effect, treatments included either exposure to TBT alone or in association with suspended solids (Waldock & Thain, 1983). The results of the experiment (shown in Table 6.4) proved that the then-present environmental concentrations of TBT would have induced stunted growth and the shell thickening response, whereas high concentrations of suspended solids enhanced oyster growth. The findings were further verified by field transplanting experiments where juvenile oysters were deployed at six sites around the UK which provided conditions of high, intermediate or low suspended solids regimes in association with many or no yachts present. Measurement of a variety of parameters including TBT concentrations in water, sediments and tissues, and the growth performance and thickness of the shells of the animals sampled each month showed that the presence of boats promoted the thickening response, whereas the animals grew normally away from the influence of boats despite a variety

Table 6.4 Biological data and tributyltin concentrations for *Crassostrea gigas* after a 56-day growth experiment

Treatment	Percentage wet weight increase	Percentage length increase	Length/thickness ratio of upper valve	Condition index[a]	Tributyltin content (μg/g) wet tissue	Bioconcentration value[b]	Remarks
Day 0	—	—	53.0	66	<0.08	—	
Control (filtered seawater)	53.7	4.7	26.9	62	<0.08	—	Growth and normal shell development
1.6 μg/l TBTO[c]	-1.1	-0.1	51.6	23	3.70	2300	No growth, no shell thickening
1.6 μg/l TBTO + 30 mg/l Crouch sediment	4.4	0.0	33.1	36	4.89	3100	No increase in length, small weight increase related to shell thickening
0.15 μg/l TBTO	16.1	-0.2	15.1	51	1.71	11 400	No increase in length, weight increase related to shell thickening
0.15 μg/l TBTO + 30 mg/l Crouch sediment	28.2	1.4	6.2	55	1.30	8700	Little increase in length, moderate weight increase related to considerable shell thickening
30 mg/l West Mersea sediment	72.5	7.8	10.2	64	0.59	—	Moderate length increase, large weight increase related to considerable shell thickening
30 mg/l Crouch sediment	99.4	22.4	23.0	63	0.09	—	Good growth and normal shell development
75 mg/l Crouch sediment	125.7	27.3	28.5	61	<0.08	—	Very good growth and normal shell development

[a] Condition index $= \dfrac{\text{wet metal weight}}{\text{internal shell volume}} \times 100$.

[b] Bioconcentration value $= \dfrac{\text{tissue concentration of TBT}}{\text{measured water concentration of TBT}}$

[c] TBTO, tributyltin oxide

of suspended solids regimes (Alzieu & Portmann, 1984).

The demonstration that environmental concentrations exceeded the threshold of the thickening response in both the laboratory and field provided compelling evidence of cause and effect, and subsequent experiments and field studies have confirmed these findings. However, the experiments only demonstrate that TBT may be one of many factors to induce the shell thickening in *C. gigas*, and perhaps the only way to prove that TBT was the growth-limiting factor in UK estuaries would be to remove TBT and monitor subsequent growth performance.

These experiments on *C. gigas* were only a small part of the accumulating evidence of environmental harm mediated by TBT, and studies were also carried out on a wide variety of species at several research institutes. In consideration of the scientific evidence, the British government took steps to control the use of TBT under the Control of Pollution Act (COPA, 1974) in 1985, and later under the Food and Environment Protection Act (1985), Control of Pesticide Regulations (1987). In effect, the implementation of this legislation has provided the final part of the study, by allowing field experiments designed to confirm that TBT was the sole cause of poor growth performance of the oysters and the shell thickening response.

After 1985 the amount of TBT allowed in paints for retail sale was reduced to 7.5% in copolymer paints and 2.5% in free association products, and by 1987 sale, supply or use of TBT on boats of less than 25 m was effectively banned. Water concentrations of TBT decreased in each year following these measures, and oysters transplanted each year at a number of sites in UK estuaries contained successively lower concentrations .of TBT, exhibited better meat yield and produced shells of decreasing thickness (Fig. 6.3) (Waite *et al.*, 1991; Waldock *et al.*, 1992). Climatic and nutritional variables can be largely discounted as factors in the recovery of the animals, because illegal sale of TBT close to one of the sites has produced a 'positive control', where TBT concentrations in both water and animals have reduced less rapidly, and consequently improvement in growth has been less marked.

6.5 CONCLUSIONS

The assessment of the environmental impact of organometallic compounds may be summarized in several stages: (i) an evaluation of the likely fate and persistence of the compounds based on chemical properties and laboratory experimentation; (ii) establishment of the toxicity of the compounds under controlled conditions; (iii) development of analytical methods to measure the concentrations in the environment; (iv) undertaking an environmental survey; and (v) establishment of a margin of safety between environmental concentration and biological effect. The first stage has not proven problematic for any of the organometals discussed here, and there is a plethora of data in the literature. Interestingly for TBT, an assessment based on this first stage alone would have underestimated environmental problems, because the physical and chemical properties of the molecule suggest it should be less bioaccumulative and toxic than it is. Perhaps also the significance of methylation of lead and tin, and disproportionation reactions between some of the species, have not yet been fully investigated, but in a general sense the fate of these compounds is well understood.

Certainly the effects of organolead, -mercury and -tin compounds at lower trophic levels in aquatic systems are well documented, and particularly for tributyltin an enormous amount of work has been carried out on lethal and sublethal exposures to aquatic life. The range of organometallic species discussed here incorporates the most toxic compounds found in aquatic systems, and while this is a well-established fact, the mechanisms of toxicity are poorly described. The effects of methylmercury at higher trophic levels have also yet to be established. In the absence of carefully controlled laboratory conditions a causal relationship between body burden and effect is difficult to prove, and, to date, determination of the effects of prolonged exposure to slightly elevated concentrations of methylated mercury found in coastal zones on the longevity and health of marine mammals is no more than educated guesswork. The debate on the causes of strandings of cetaceans, for example, has already prompted a vast amount of research on several fronts,

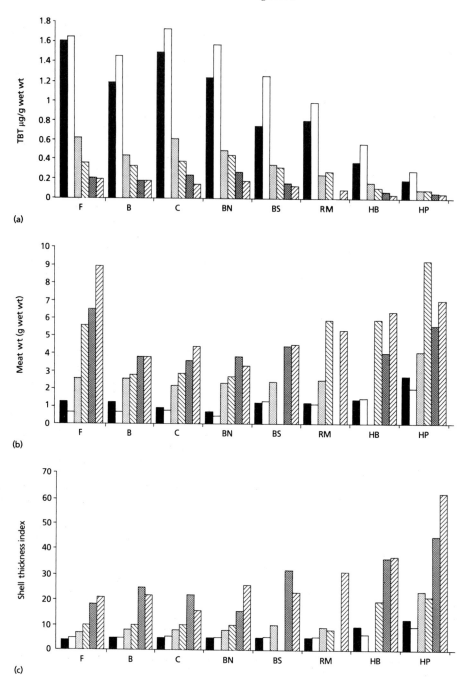

Fig. 6.3 Changes in (a) TBT content, (b) meat production and (c) shell thickness index of juvenile *C. gigas* oysters transplanted each year at sites on the River Crouch, UK, following the ban on the use of TBT on small vessels. F, Fambridge head of the estuary; B, Bridgemarh Isle; C, Creeksea; BN, Burnhan₁; BS, Bush Shore; RM, Roach Mouth; HB, Holliwell Buoy; HP, Holliwell Point mouth of the estuary. Sites are approx. 3 km apart. High shell thickness values are indicative of thin-shelled, normal-shaped animals. Sequential bars for each site show 1986–1991, respectively.

including work on mercury and polychlorinated biphenyls (see Chapter 9).

Fortunately, the increasing use of organometallic compounds has come at a time of rapid development of sensitive analytical instrumentation and sophisticated analytical methodology. It is an outstanding achievement to be able to measure these compounds accurately in the high picogram to low nanogram per litre range of concentration ($10^{-13}-10^{-12}$). The resulting data base for concentrations of organolead, and particularly -mercury and -tin is comprehensive. While some of the earliest data generated must be treated with some caution, the latest data appear to be derived from some very competent analyses.

The cumulative results of these efforts provide a reliable framework to establish the risks of many of these compounds to the aquatic environment, and enable the enforcement of tight regulatory control over the use of such chemicals when risk to the environment is shown to be unacceptably high.

REFERENCES

Alzieu, C. & Portmann, J.E. (1984) The effect of tributyltin on the culture of *C. gigas* and other species. *Proceedings of the Fifteenth Annual Shellfish Conference*, 15 and 16 May, Fishmongers Hall, London.

Alzieu, C., Héral, M., Thibaud, Y., Dardignac, M.J. & Feuillet, M. (1982) Influence des peintures antisalissures à base d'organostanniques sur la calcification de la coquille de l'huître *Crassostrea gigas*. *Rev. Trav. Inst. Pech. Marit.* **45**, 101–116.

Alzieu, C., Sanjuan, J., Michel, P., Borel, M. & Dreno, J.P. (1989) Monitoring and assessment of butyltins in Atlantic coastal waters. *Mar. Pollut. Bull.* **20**, 22–26.

Anon (1971) Methyl mercury in fish. A toxicologic–epidemiologic evaluation of risks. Report from an expert group. *Nordisk Hygienisk Tidskrift*, Supplementum 4. National Institute of Public Health, Stockholm 60, Sweden.

Anon (1986) Organotin in antifouling paints environmental considerations. Pollution Paper No. 25. Prepared by the Central Directorate of Environmental Protection, Department of the Environment. HMSO, London.

Bache, C.A. & Lisk, D.J. (1971) Gas chromatographic detection of organic mercury compounds by emission spectrometry in a helium plasma. Application to the analysis of methyl mercuric salts in fish. *Anal. Chem.* **41**, 950–952.

Beaumont, A.R. & Budd, M.D. (1984) High mortality of the larvae of the common mussel at low concentrations of tributyltin. *Mar. Pollut. Bull.* **15**, 402–405.

Birnie, S.E. & Hodges, D.J. (1981) Determination of ionic alkyl lead species in marine fauna. *Environ. Technol. Lett.* **2**, 433–442.

Bloom, N. (1989) Determination of picogram levels of methylmercury by aqueous phase ethylation, followed by cryogenic gas chromatography with cold vapour atomic fluorescence detection. *Can. J. Fish. Aquat. Sci.* **46**, 1131–1140.

Bloom, N.S. (1992) On the chemical form of mercury in edible fish and marine invertebrate tissue. *Can. J. Fish. Aquat. Sci.* **49**, 1010–1017.

Branica, M. & Konrad, Z. (Eds) (1980) *Lead in the Marine Environment*. Pergamon Press, Oxford.

Bryan, G.W., Gibbs, P.E., Hummerstone, L.G. & Burt, G.R. (1987) Copper, zinc and organotin as long-term factors governing the distribution of organisms in the Fal estuary in southwest England. *Estuaries* **10**, 208–219.

Bubb, J.M., Rudd, T. & Lester, J.N. (1991) Distribution of heavy metals in the river Yare and its associated Broads. I. Mercury and methylmercury. *Sci. Total Environ.* **102**, 247–168.

Chau, Y.K. (1986) The occurrence and speciation of organometallic compounds in freshwater systems. *Sci. Total Environ.* **49**, 305–323.

Chau, Y.K., Wong, P.T.S., Bengert, G.A. & Kramar, O. (1979) Determination of tetraalkyllead compounds in water, sediment and fish samples. *Anal. Chem.* **51**, 186–188.

Chau, Y.K., Wong, P.T.S., Bengert, G.A. & Dunn, J.L. (1984) Determination of dialkyllead, trialkyllead tetraalkyllead, and lead(II) compounds in sediment and biological samples. *Anal. Chem.* **56**, 271–274.

Cleary, J.J. & Stebbing, A.R.D. (1987) Organotin in the surface microlayer and subsurface waters of Southwest England. *Mar. Pollut. Bull.* **18**, 238–246.

Compeau, G.C. & Bartha, R. (1987) Effect of salinity on mercury-methylating activity of sulfate-reducing bacteria in estuarine sediments. *Appl. Environ. Microbiol.* **53**, 261–265.

COPA (1974) *The Control of Pollution Act* 1974. HMSO, London.

Control of Pesticide Regulations (1987) HMSO, London.

Craig, P.J. (1983) Organometallic compounds in the environment. In *Pollution: Causes, Effects and Control* (Ed. M. Harrison). Royal Society of Chemistry Special Publication No. 44, pp. 277–322.

Craig, P.J. & Moreton, P.A. (1986) Total mercury, methyl mercury and sulphide levels in British estuarine sediments—III. *Water Res.* **20**, 1111–1118.

Cruz, R.B., Lorouso, C., George, S., Thomassen, Y., Kinrade, J.D., Butler, L.R.P., Lye, L. & Van Loon, J.C. (1980) Determination of total, organic solvent extractable, volatile and tetraalkyllead in fish, vegetation, sediment and water samples. *Spectrochim. Acta* **35B**, 775–783.

Dalziel, J.A. & Yeats, P.A. (1985) Reactive mercury in the central north Atlantic Ocean. *Mar. Chem.* **15**, 357–361.

Decadt, G., Baeyens, W., Bradley, D. & Goeyens, L. (1985) Determination of methylmercury in biological samples by headspace analysis. *Anal. Chem.* **57**, 2788–2791.

D'Itri, E.M. (1972) *The Environmental Mercury Problem*. CRC Press, Cleveland, OH.

Donard, O.F.X. (1989) Tin and germanium. In: *Environmental Analysis Using Chromatography Interfaced with Atomic Spectroscopy* (Eds R.M. Harrison & S. Rapsomanikis), pp. 223–258. Ellis Horwood, Chichester.

EPA (1986) *Quality Criteria for Water*. US Environmental Protection Agency, Office of Water Regulations and Standards, Washington, DC. EPA/440/5-86/001.

Evans, C.J. & Smith, P.J. (1975) Organotin based antifouling systems. *J. Oil Col. Chem. Assoc.* **58**, 160–168.

Evans, D.W. & Laughlin, R.B. (1984) Accumulation of bis(tributyltin) oxide by the mud crab *Rhithropanopeus harrisi. Chemosphere* **13**, 213–219.

Fent, K. & Muller, M.D. (1991) Occurrence of organotins in municipal wastewater and sewage sludge and behavior in a treatment plant. *Environ. Sci. Technol.* **25**, 489–493.

Fileman, C.F., Althaus, M., Law, R.J. & Haslam, I. (1991) Dissolved and particulate trace metals in surface waters over the Dogger Bank, Central North Sea. *Mar. Pollut. Bull.* **22**, 241–244.

Food and Environment Protection Act (1985) HMSO, London.

Fowler, S.W. (1982) Biological transfer and transport processes. In: *Pollutants Transfer and Transport in the Sea* (Ed. G. Kullenburg), pp. 1–65. CRC Press, Cleveland, OH.

Gibbs, P.E., Pascoe, P.L. & Burt, G.R. (1988) Sex change in the female dogwhelk *Nucella lapillus*, induced by tributyltin from antifouling paints. *J. Mar. Biol. Assoc. UK* **68**, 715–731.

Gill, G.A. & Fitzgerald, W.F. (1987) Mercury in some surface waters of the open ocean. *Global Biogeochem. Cycles* **1**, 199–212.

Grove, J.R. (1980) Investigations into the formation and behaviour of aqueous solutions of lead alkyls. In: *Lead in the Marine Environment* (Eds M. Branica & Z. Konrad), pp. 45–53. Pergamon Press, Oxford.

Hannerz, L. (1968) Experimental investigations on the accumulation of mercury in water organisms. Fisheries Board of Sweden, Institute of Freshwater Research, Drottningholm, Report No. 48.

Harper, D.J., Fileman, C.F., May, P.V. & Portmann, J.E. (1989) Methods for analysis for trace metals in marine and other samples. *Aquat. Environ. Protect.: Analyt. Meth.* MAFF Direct. Fish. Res., Lowestoft, (3): 1–38.

Harrison, G.F. (1980) The Cavtat incident. In: *Lead in the Marine Environment* (Eds M. Branica & Z. Konrad), pp. 305–319. Pergamon Press, Oxford.

His, E. & Robert, R. (1985) Development des veligeres de *Crassostrea gigas* dans le Basin D'Arcachon, etudies sur les mortalities larvaires. *Rev. Trav. Peches. Marit.* **47**, 63–68.

Holden, A.V. (1973) Mercury in fish and shellfish: a review. *J. Food Technol.* **8**, 1–25.

Itano, K., Kawai, S., Miyazaki, N., Tatsukawa, R. & Fujiyama, T. (1984) Body burdens and distribution of mercury and selenium in striped dolphins caught off the Pacific coat of Japan. *Agric. Biol. Chem.* **48**, 1109–1116.

Iwata, H., Shinsuke, T. & Tatsukawa, R. (1993) Application on the analysis of organotin compounds to marine mammals. *Kankyo Kagaku* **3**, 402–403.

Jarvie, A.W.P., Markall, R.N. & Potter, H.R. (1975) Chemical alkylation of lead. *Nature* **255**, 217–218.

Jensen, S. & Jernelov, A. (1969) Biological methylation of mercury in aquatic organisms. *Nature* **223**, 753–754.

Jewett, K.L. & Brinckman, F.E. (1974) Trans-methylation of heavy metal ions in water. *Div. Environ. Chem.; Am. Chem. Soc.* **14**, 218–225.

Jian, W. & McLeod, C.W. (1991) Field sampling technique for mercury speciation. *Anal. Proc.* **28**, 293–294.

Joiris, C.R., Holsbeek, L., Bouquegneau, J.M. & Bossicart, M. (1991) Mercury contamination of the Harbour porpoise (*Phocoena phocoena*) and other cetaceans from the North Sea and Kattegaat. *Wat. Air Soil Pollut.* **56**, 283–293.

Joiris, C.R., Holsbeek, L., Antoine, N., Bouquegneau, J.M., Siebert, U. & Bossicart, M. (1992) Do older cetaceans die from mercury contamination? ICES, MMC.

Key, D., Nunny, R.J., Davidson, P.E. & Leonard, M.A. (1976) Abnormal shell growth in the Pacific oyster *Crassostrea gigas*. Some preliminary results from experiments undertaken in 1975. ICES CM 1976/K11 (mimeo).

Koeman, J.H., Peeters, W.H.M., Koudstaal-Hol, C.H.M., Tjioe, P.S. & de Goeij, J.J.M. (1973) Mercury–selenium correlations in marine mammals. *Nature* **245**, 385–386.

Kram, M.L., Stang, P.M. & Selegman, P.F. (1989) Adsorption and desorption of tributyltin in sediments of San Diego Bay and Pearl Harbour. *Appl. Organometal. Chem.* **3**, 523–536.

Laughlin, R.B. Jr, Pendoley, P. & Gustafson, R.G. (1987) Sublethal effects of tributyltin on the hard shell clam *Mercenaria mercenaria*. *Proceedings of Oceans 87 Conference and Exposition on Science and Engineering*. Halifax, NS, Canada, 28 Sept.–1 Oct. IEEE, Piscataway, NJ and Marine Technology Society, Washington DC, Vol. 4, pp. 1494–1498.

Law, R.J., Fileman, C.F., Hopkins, A.D., Baker, J.R., Harwood, J., Jackson, D.B., Kennedy, S., Martin, A.R. & Morris, R.J. (1991) Concentrations of trace metals in the livers of marine mammals (seals, porpoises, and dolphins) from waters around the British Isles. *Mar. Pollut. Bull.* **22**, 183–191.

Laws, R.J. (in press) Metals in marine mammals. In: *Interpreting Concentrations of Environmental Contaminants in Wildlife Tissues* (Eds N. Beyer & G. Heinz),

Lee, R.F., Valkirs, A.O. & Selegman, P.F. (1987) The fate of tributyltin in estuarine waters. In: *Proceedings of Oceans 87 Conference and Exposition on Science and Engineering*. Halifax, NS, Canada, 28 Sept.–1 Oct. IEEE, Piscataway, NJ and Marine Technology Society, Washington DC, Vol. 4, pp. 1411–1416.

Lee, Y. & Hultberg, H. (1990) Methylmercury in some Swedish surface waters. *Environ. Toxicol. Chem.* **9**, 833–841.

Maddock, B.G. & Taylor, D. (1980) The acute toxicity and bioaccumulation of some lead alkyl compounds in marine animals. In: *Lead in the Marine Environment* (Eds M. Branica & Z. Konrad), pp. 233–263. Pergamon Press, Oxford.

MAFF (1990) Monitoring and surveillance of non-radioactive contaminants on the aquatic environment, 1984–1987. *Aquat. Environ. Monit. Rep.*, MAFF Direct. Fish. Res., Lowestoft, (22): 1–60.

MAFF (1991) Monitoring and surveillance of non-radioactive contaminants on the aquatic environment and activities regulating the disposal of wastes at sea, 1988–1989. *Aquat. Environ. Monit. Rep.*, MAFF Direct. Fish. Res., Lowestoft, (26): 1–87.

MAFF (1992) Monitoring and surveillance of non-radioactive contaminants on the aquatic environment and activities regulating the disposal of wastes at sea, 1990. *Aquat. Environ. Monit. Rep.*, MAFF Direct. Fish. Res., Lowestoft, (30): 1–66.

Maguire, R.J. (1987) Environmental aspects of tributyltin. *Appl. Organometal. Chem.* **1**, 475–498.

Maguire, R.J. (1991) Aquatic environmental aspects of non-pesticidal organotin compounds. *Wat. Pollut. Res. J. Can.* **26**, 243–360.

Maguire, R.J. & Huneault, H. (1981) Determination of butyltin species in water by gas chromatography with flame photometric detection. *J. Chromatogr.* **209**, 458–462.

Maguire, R.J., Chau, Y.K., Bengert, G.A., Hale, E.J., Wong, P.T.S. & Kramar, O. (1982) Occurrence of organotin compounds in Ontario Lakes and rivers. *Environ. Sci. Technol.* **16**, 698–702.

Maguire, R.J. & Tkacz, R.J. (1985) Degradation of tri-n-butyltin species in water and sediment from Toronto harbor. *J. Agric. Food. Chem.* **33**, 947–953.

Maguire, R.J., Tkacz, R.J., Chau, Y.K., Bengert, G.A. & Wong, P.T.S. (1986) Occurrence of organotin compounds in water and sediment in Canada. *Chemosphere* **15**, 253–274.

Majorta, R. & Viale, D. (1977) Accumulation de granules de seleniure mercurique dans le foie d'Odontocetes (Mammiferes, Cetaces): un mecanisme possible de detoxication du methylmercure par le selenium. *C.R. Hebd. Acad Sci. Paris, D,* **285**, 109–112.

Mason, C.F., Last, N.I. & Macdonald, S.M. (1986) Mercury, cadmium and lead in British otters. *Bull. Environ. Contam. Toxicol.* **37**, 844–850.

Matthias, C.L., Bellama, J.M., Olsen, G.J. & Brinckman, F.E. (1986) Comprehensive method for determination of aquatic butylmethyltin species at ultratrace levels using simultaneous hydrization/extraction with gas chromatography–flame photometric detection. *Environ. Sci. Technol.* **20**, 609–615.

Matthias, C.L., Bushong, S.J., Hall, L.W. Jr, Bellama, J.M. & Brinckmann, F.E. (1988) Simultaneous butyltin determinations in the microlayer, water column and sediment of a northern Cheasapeake Bay marina and receiving system. *Appl. Organomet. Chem.* **2**, 547–552.

McCormack, A.J., Tong, S.C. & Cooke, D.W. (1965) Sensitive selective gas chromatography detector based on emission spectrometry. *Anal. Chem.* **37**, 1470–1475.

Neves, A.G., Allen, A.G. & Harrison, R.M. (1990) Determination of the concentration of alkyllead compounds in fish. *Environ. Technol.* **11**, 877–882.

O'Connor, D.J. & Nielsen, S.W. (1980) Environmental survey of methylmercury levels in wild mink (*Mustela vison*) and otter (*Lutra canadensis*) from the Northeastern United States and experimental pathology of methylmercurialism in the otter. In: *World Furbearer Conference Proceedings*, Vol. 3, pp. 1728–1745. University of Maryland, Frostburg, MD.

Olafsson, J. (1983) Mercury concentrations in the North Atlantic in relation to cadmium, aluminium, and oceanographic parameters. In: *Trace Metals in Sea Water* (Eds C.S. Wong *et al.*) pp. 475–487. Plenum Press, New York.

Osborn, D., Every, W.J. & Bull, K.R. (1983) The toxicity of trialkyl lead compounds to birds. *Environ. Pollut. A,* **31**, 261–275.

Osborn, D. & Leach, D.V. (1987) Organotin in birds: pilot study. Institute of Terrestrial Ecology: Final Report of the Department of the Environment, UK,

DOE/NERC contract F3CR/27/D4/01.

Page, D.S., Dassanayake, T., Gilfillan, E.S. & Kresja, C. (1990) Butyltin species in coastal Maine mussel samples. *Proceedings of the 3rd International Organotin Symposium, Monaco, 17–20 April,* pp. 170–175.

Quevauviller, Ph. & Donard, O.F.X. (1990) Variability of butyltin determination in water and sediment samples from European coastal environments. *Appl. Organometal. Chem.* **4**, 353–367.

Radojevic, M. (1989) Lead. In: *Environmental Analysis using Chromatography Interfaced with Atomic Spectroscopy* (Eds R.M. Harrison & S. Rapsomanikis), pp. 223–258. Ellis Horwood, Chichester.

Rapsomanikis, S. (1989) Mercury. In: *Environmental Analysis using Chromatography Interfaced with Atomic Spectroscopy* (Eds R.M. Harrison & S. Rapsomanikis), pp. 299–317. Ellis Horwood, Chichester.

Reisinger, K., Stoeppler, M. & Nurnberg, H.W. (1981) Evidence for the absence of biological methylation of lead in the environment. *Nature* **291**, 228–230.

Richie, L.S., Berrios-Duran, L.A., Frick, L.P. & Fox, I. (1964) Molluscicidal time–concentration relationships of organo-tin compounds. *Bull. WHO* **31**, 147–149.

Ronald, K., Tessaro, S.V., Uthe, J.F., Freeman, H.C. & Frank, R. (1977) Methylmercury poisoning in the Harp seal (*Pagophilus groenlandicus*). *Sci. Total Environ.* **38**, 153–166.

Rudd, J.W., Furutani, A. & Turner, M.A. (1980) Mercury methylation by fish intestinal contents. *Appl. Environ. Microbiol.* **40**, 777–782.

Scammel, M.S., Batley, G.E. & Brockbank, C.I. (1991) A field study of the impact on oysters of tributyltin introduction and removal in a pristine lake. *Arch. Environ. Contam. Toxicol.* **14**, 276–281.

Schmidt, U. & Huber, F. (1976) Methylation of organolead and lead(II) compounds to (CH$_3$)$_4$Pb by microorganisms. *Nature* **259**, 157–158.

Seidel, S.L., Hodge, V.F. & Goldberg, E.D. (1980) Tin as an environmental pollutant. *Thalassia Jugoslav.* **16**, 208–223.

Seligman, P.F., Grovhoug, A.O., Valkirs, A.O., Stang, P.M., Fransham, R., Stallard, M.O., Davidson, B. & Lee, R. (1989) Distribution and fate of tributyltin in the United States environment. *Appl. Organometal. Chem.* **3**, 31–47.

Seligman, P.F., Valkirs, A.O. & Lee, R.F. (1986) Degradation of tributyltin in marine and estuarine waters. In: *Proceedings of the Organotin Symposium of the Oceans '86 Conference,* Washington, DC, 23–25 Sept., Vol. 4, pp. 1189–1195. IEEE, Piscataway, NJ.

Shum, G.T.C., Freeman, R.C. & Uthe, J.F. (1979) Determination of organic (methyl) mercury in fish by graphite furnace atomic absorption spectro-

photometry. *Anal. Chem.* **51**, 414–416.

Soderquist, C.J. & Crosby, D.G. (1980) Degradation of triphenyltin hydroxide in water. *J. Agric. Food Chem.* **28**, 111–117.

Stewart, C. & de Mora, S.J. (1990) A review of the degradation of tri(*n*-butyl)tin in the marine environment. *Environ. Technol.* **11**, 565–570.

Thain, J.E. (1983) Acute toxicity of bis(tributyltin) oxide to the adults and larvae of some marine organisms. International Council for the Exploration of the Sea, CM1983, E13 (mimeo).

Thain, J.E. & Waldock, M.J. (1985) The growth of bivalve spat exposed to organotin leachates from antifouling paints. International Council for the Exploration of the Sea, CM1985, E28(mimeo).

Thain, J.E. & Waldock, M.J. (1986) The impact of tributyltin (TBT) antifouling paints on molluscan fisheries. *Water Sci. Technol.* **18**, 193–202.

Thain, J.E., Waldock, M.J. & Waite, M.E. (1987) Toxicity and degradation studies of tributyltin (TBT) and dibutyltin (DBT) in the aquatic environment. *Proceedings of Oceans 87 Conference and Exposition on Science and Engineering.* Halifax, NS, Canada, 28 Sept.–1 Oct. IEEE, Piscataway, NJ and Marine Technology Society, Washington DC, Vol. 4, pp. 1398–1404.

Thibaud, Y. & Duguy, R. (1973) Teneur en mercure chez les cetaces des Cotes de France. International Council for the Exploration of the Sea, CM1973, N2(mimeo).

Thompson, J.A.J. & Crerar, J.A. (1980) Methylation of lead in marine sediment. *Mar. Pollut. Bull.* **11**, 251–253.

Topping, G. & Davies, I.M. (1981) Methyl mercury production in the water column. *Nature* **290**, 243–244.

UNEP (1989) Assessment of organotin compounds as marine pollutants in the Mediterranean. MAP Technical Reports, No. 33. UNEP, Athens.

Unger, M.A., MacIntyre, W.G., Greaves, J. & Huggett, R.J. (1986) GC determinations of butyltins in natural waters by flame photometric detection of hexyl derivatives with mass spectrometric confirmation. *Chemosphere* **15**, 461–470.

Unger, M.A., MacIntyre, W.G. & Huggett, R.J. (1987) Equilibrium sorption of tributyltin chloride by Chesapeake Bay sediments. *Proceedings of Oceans 87 Conference and Exposition on Science and Engineering.* Halifax, NS, Canada, 28 Sept.–1 Oct. IEEE, Piscataway, NJ and Marine Technology Society, Washington DC, Vol. 4, pp. 1381–1386.

Waite, M.E., Evans, K.E., Thain, J.E. & Waldock, M.J. (1989) Organotin concentrations in the Rivers Bure and Yare, Norfolk Broads, England. *Appl. Organometal. Chem.* **3**, 383–391.

Waite, M.E., Waldock, M.J., Thain, J.E., Smith, D.J. & Milton, S.M. (1991) Reductions in TBT concentrations in UK estuaries following legislation in 1986 and 1987. *Mar. Environ. Res.* **32**, 89–111.

Waldock, M.J. & Miller, D. (1983) The determination of total and tributyl tin in seawater and oysters in areas of high pleasure craft activity. International Council for the Exploration of the Sea, Copenhagen, CM Papers and Reports/E:12 (mimeo).

Waldock, M.J. & Thain, J.E. (1983) Shell thickening in *Crassostrea gigas*: organotin antifouling or sediment induced? *Mar. Pollut. Bull.* **14**, 411–415.

Waldock, M.J., Thain, J.E. & Waite, M.E. (1987a) The distribution and potential toxic effects of TBT in UK estuaries during 1986. *Appl. Organometal. Chem.* **1**, 287–301.

Waldock, M.J., Waite, M.E. & Thain, J.E. (1987b) Changes in concentrations of organotins in UK rivers and estuaries following legislation in 1986. *Proceedings of Oceans 87 Conference and Exposition on Science and Engineering.* Halifax, NS, Canada, 28 Sept.–1 Oct. IEEE, Piscataway, NJ and Marine Technology Society, Washington DC, Vol. 4, pp. 1352–1356.

Waldock, M.J., Waite, M.E. & Thain, J.E. (1988) Inputs of TBT to the marine environment from shipping activity in the UK. *Environ. Technol. Lett.* **9**, 999–1010.

Waldock, M.J., Waite, M.E., Miller, D., Smith, D.J. & Law, R.J. (1989) The determination of total tin and organotin compounds in environmental samples. *Aquat. Environ. Protect.: Analyt. Meth.* MAFF Direct. Fish. Res., Lowestoft, (4): 1–25.

Waldock, M.J., Thain, J.E., Smith, D. & Milton, S. (1990) The degradation of TBT in estuarine sediments. *Proceedings of the 3rd International Organotin Con-ference*, Monaco 17–20 Apr. pp. 46–48.

Waldock, M.J., Thain, J.E., Waite, M.E. & Hart, V. (1992) Improvements in bioindicator performance in UK estuaries following the control of the use of antifouling paints. ICES CM1992/E:32.

Ward, G.S., Cramm, G.C., Parrish, P.R., Trachman, H. & Slessinger, A. (1981) Bioaccumulation and chronic toxicity of bis tributyltin oxide (TBTO): tests with a saltwater fish. *Aquatic Toxicology and Hazard Assessment:* Fourth Conference ASTM, Philadelphia STP 737 (Eds D.R. Bronson & K.L. Dickson), pp. 183–200. American Society for Testing and Materials.

Westoo, G. (1967) Determination of methylmercury in food stuffs. II. Determination of methylmercury in fish, eggs, meat, and liver. *Acta Chem. Scand.* **21**, 1790–1800.

Wilken, R. & Hintelmann, H. (1991) Mercury and methylmercury in sediments and suspended particles from the River Elbe, North Germany. *Water, Soil Air Pollut.* **56**, 427–437.

Wilson, K.W., Head, P.C. & Jones, P.D. (1986) Mersey estuary (U.K.) bird mortalities—causes, consequences and correctives. *Water Sci. Technol.* **18**, 171–181.

Wollast, R., Billen, G. & Mackenzie, F.T. (1975) Behaviour of mercury in natural systems and its global cycle. In: *Ecological Toxicology Research* (Eds A.D. McIntyre & C.F. Mills), pp. 145–166. Plenum Press, New York.

Wong, P.T.S., Chau, Y.K. & Luxon, L. (1975) Methylation of lead in the aquatic environment. *Nature* **253**, 263–264.

Wood, J.M., Kennedy, F.S. & Rosen, C.G. (1968) Synthesis of methylmercury compounds by extracts of a methanorganic bacterium. *Nature* **220**, 173–174.

7: Detergents

E.C. HENNES-MORGAN AND N.T. DE OUDE

7.1 INTRODUCTION

Detergents are complex mixtures of various ingredients such as surface-active agents (surfactants), builders and bleaches. The detergent formulation balances the properties and levels of the ingredients to achieve the required performance characteristics. The key ingredients only are discussed in this chapter. The names of many ingredients have been abbreviated, following generally accepted practice. They are defined in the text and in the Glossary at the end of the book.

In reading the following sections, some general points should be kept in mind. Surfactants are mixtures of homologues of a material, differing in chain length, degree of substitution, etc. The properties of these components are usually additive, i.e. the properties of the commercial materials can be calculated from those of the homologues. After use, detergents are discharged as domestic sewage. In most cases they reach the environment via sewers and sewage treatment plants. The homologues of the surfactants are biodegraded at different rates such that the mixtures reaching surface water have a composition different from those of the mixture used by the householder. Active substances of low solubility can be less bioavailable than highly soluble ones, and are therefore less toxic. This needs to be considered in environmental safety assessments.

Environmental safety assessments of household products, such as detergents, are based on a comparison of the predicted environmental concentrations and the concentrations at which no adverse effects are observed in the environment. Environmental concentrations can generally be predicted accurately for this group of products because consumption data and removal efficiencies are well known. Actual concentrations vary, depending on sewage treatment practices and on sewage dilution upon discharge. Broadly speaking, the primary biodegradation of surfactants, and hence the loss of surface activity, occurs rapidly in sewage treatment. No-adverse effect concentrations can be derived from laboratory tests or from (simulated) field tests. Taking into account the multiple processes (adsorption, precipitation, photodegradation, catalysis) that occur in nature, data derived from field tests should be considered more indicative than data obtained in laboratory tests.

This chapter draws on earlier reviews, in particular: Jakobi & Loehr, 1987; Schoeberl et al., 1988; Schoeberl & Huber, 1988; de Oude, 1992; Swisher, 1987 (analysis of surfactants). References to original literature are given only for publications that appeared after them, or provide additional valuable information. We concentrate on only the most important and/or useful data. The figures reported usually summarize the published data or are averages of them. In the case of biodegradation testing, there are numerous studies in different test systems, with different bacterial cell densities, test periods, etc. Emphasis was therefore put on standardized test methods and on results of monitoring or field studies if available.

7.2 ENVIRONMENTAL ANALYSES AND CONCENTRATIONS

The main active detergent ingredients reviewed are surfactants (anionic, nonionic, cationic), the

130

predominant builder, zeolite A, its builder additive, polycarboxylate, and the bleach component, perborate. The following gives the most widely applied procedures and some new developments for environmental analysis.

7.2.1 Anionic surfactants (AnS)

The standard method of ISO (International Standards Organization) and of many national standards organizations (e.g. DIN — Deutsche Industrie Norm) uses a cationic dye, methylene blue, to analyse the sum of methylene blue active substances (MBAS) (Swisher, 1987). Prior acid hydrolysis can distinguish between anionic sulphonates (LAS, SAS and AOS) and anionic sulphates (AS and AES). The latter group is readily hydrolysed and excluded from the MBAS response. The detection limit is about 0.02 mg/l MBAS. This method is a fast screening aid but unspecific and subject to substantial interference in environmental matrices; i.e. it overestimates anionic surfactants.

More specific analyses can be achieved by chromatographic techniques. Gas chromatography (GC) and high-pressure liquid chromatography (HPLC) can separate and measure individual homologues and isomers, e.g. phenyl isomers of LAS. Improved versions of GC can accurately determine microgram amounts of individual AnS (Waters & Garrigan, 1983). However, the method requires conversion to a volatile AnS derivative while preserving the original homologue and isomer pattern. This complex and tedious pretreatment methodology lowers its usefulness for routine analysis.

The application of HPLC for LAS analysis is simpler and yet efficient. It provides a lower resolution than GC, but allows separate analysis of main homologues without prior derivatization, thus combining selectivity and speed of analysis. High sensitivity can be achieved: 10 µg/l in aqueous samples and 100 µg/kg in solid samples for total LAS (Matthijs & Hennes, 1991). A HPLC method for alkyl sulfate (AS) analysis was recently described, which requires prior derivatization with phenylisocyanate in order to obtain an ultraviolet (UV)-detectable chromophoric group (Nitschke & Huber, 1993). Other recent sensitive

methods for the quantitative determination of AS are those of Fendinger *et al.*, and Popenoe *et al.*, 1994.

Environmental analyses have been carried out mainly for LAS or MBAS. Unless specified differently, the results in Table 7.1 are averages of reported values for European and US environments.

For primary AS, linear AES and SAS, estimates were made by Gilbert & Pettigrew (1984) for the UK (Table 7.2).

A few analyses are available for Japanese waters (Oba *et al.*, 1976). Average concentrations in sewage are: 1.4 mg/l AS, 1.3 mg/l AES, 0.162 mg/l AOS; concentrations in sewage effluent were below detection limit for AES and AOS.

7.2.2 Nonionic surfactants (NS)

The bismuth active substances (BiAS) procedure is the standard method in European and OECD guidelines for determining ethoxylated alcohols, the most commonly used nonionics in detergents (Swisher, 1987). It is based on the potentiometric or colorimetric measurement of the bismuth consumed by the sum of NS. The American SDA proposes the use of cobaltothiocyanate and a colorimetric determination of the nonionic complex (CTAS method) which is a simpler and faster procedure. Both methods give comparable results with a detection limit of *c* 0.05 mg/l. An optimized BiAS procedure corrects for underestimates in NS concentrations, particularly in sewage/sludge suspensions (Waters *et al.*, 1986). Optimized BiAS and CTAS methods provide a good estimate for the concentration of total NS (AE and APE) in raw sewage, but lead to a significant overestimate for sewage effluent and river water samples due to interferences.

Selective determinations using GC and HPLC are most promising. GC has been successfully applied to AE measurements in sewage and surface waters with a sensitivity of 10 µg/l, but not yet for APE analysis. A semi-quantitative reading of total nonionic concentrations is achieved by HPLC after derivatization with phenylisocyanate, and the method is being further improved for selective AE determination. APE can be selectively analysed by HPLC without derivatization

Table 7.1 Concentrations of LAS and MBAS in the environment. Data largely from Painter, 1992

Environmental sample	Concentrations
Sewage (raw or settled not always reported)	3–12 mg/l MBAS, 0.5–12 mg/l LAS, mostly 2–4 mg/l. Up to 10-fold higher in Spain and UK with 21 mg/l LAS and 24 mg/l MBAS, respectively
Sewage effluent	0.2–1.0 mg/l MBAS (biological filters), 0.01–0.13 mg/l LAS (activated sludge units), 0.3–1.2 mg/l LAS (trickling filters), 1.7–2.5 mg/l LAS (primary treatment) (Rapaport & Eckhoff, 1990)
Sewage sludge	2–12 g/kg LAS in primary and anaerobically digested sludge, mostly 4–10 g/kg LAS, 2.1–4.3 g/kg LAS in aerobically digested sludge, 0.09–0.86 g/kg LAS in activated sludge
Soil	0.9–1.3 mg/kg LAS in Germany and 0.5–40.3 mg/kg LAS in UK (contaminated site); 28 mg/kg LAS in top 7.5 cm of contaminated soil in US. Results from a more recent study in the UK: < 1 mg/kg LAS in a soil not recently spread with sludge and 0.2–20 mg/kg LAS in soils recently spread (Holt *et al.*, 1989). 13–47 mg/kg LAS (surface soil, US, 0–15 cm), 0.9–2.2 mg/kg LAS (surface soil, Europe), < 5 mg/kg LAS (15–90 cm) below surface, US) (Rapaport & Eckhoff, 1990)
River	0.005–6.9 mg/l MBAS, up to 1.6 mg/l LAS in contaminated sites: < 0.04 mg/l LAS and 0.005–0.2 mg/l MBAS in rivers not grossly polluted by sewage effluent. 0.01–0.3 mg/l LAS (below sewage outfall), 0.026–0.15 mg/l LAS (< 5 m downstream), < 0.005–0.12 mg/l LAS (> 5 m downstream), high concentrations measured at sites with 'worst-case', i.e. low flow, conditions (Rapaport & Eckhoff, 1990)
River sediment	1–10 mg/kg LAS for upstream of sewage effluents; up to 100–200 mg/kg LAS in vicinity of sewage discharge. 16–322 mg/kg LAS (below sewage outfall), 1.3–26.9 mg/kg LAS (< 5 m downstream), 1.3–14.6 mg/kg LAS (> 5 m downstream), high concentrations measured at sites with 'worst-case', i.e. low flow, conditions (Rapaport & Eckhoff, 1990)
Estuarine and coastal water; sediment	0.0008–0.09 mg/l LAS and 0.01–0.54 mg/l MBAS further or closer to sewage outfalls; 3–17 mg/kg LAS in sediments nearer to an outfall
Marine water	< 0.5–1.2 µg/l LAS in North Sea (salinity 28–35 g/l), < 0.5–9.4 µg/l LAS in Westerscheldt estuary up to open sea (salinity: 16–34 g/l) (Stalmans *et al.*, 1991)
Tap water	0.003–0.008 mg/l LAS; except some data for Osaka with 0.01–0.07 mg/l LAS.

Table 7.2 Estimated concentrations of AS, linear AES and SAS in the environment. From Gilbert & Pettigrew, 1984

Environmental sample	Concentrations of		
	Primary AS	Linear AES	SAS
Raw sewage (mg/l)	1	5	2
Settled sewage (mg/l)	0.7	3.5	1.4
Biologically treated sewage effluent (mg/l)	0.01	0.1	0.04
River water (worst-case 1:1 dilution) (mg/l)	0.005	0.05	0.02
Sewage sludge (g/kg)	0.8	4	1.6
Soil (mg/kg)	0.02	0.1	3.2

due to its UV-detectable chromophoric group, with a detection limit of 1 µg/l (Ahel & Giger, 1985). Recently, liquid chromatography/mass spectrometry (LC-MS) has been developed as the most suitable technique for AE determination in surface waters (Evans *et al.*, 1994).

Table 7.3 gives average values of reported environmental concentrations. These are values of total NS (BiAS or CTAS analysis), GC or HPLC data on AE, NPE, the main APE used, and major degradation products of the latter (NP, NP1EO, NP2EO), which may still occur on different sludges, although the use of APE in consumer detergents has been largely abandoned. This is due to the slow biodegradation of some of its metabolites in the environment.

7.2.3 Cationic surfactants (CS)

A colorimetric method is most commonly used for the analytical determination of CS (Swisher, 1987). After some labour-intensive extractions, needed due to the highly adsorptive characteristic of CS compounds, an ion-association complex with the anionic dye disulphine blue is formed. The sum of DBAS (disulphine blue active substances) includes all long-chain ammonium compounds. Applying a thin-layer chromatographic procedure in combination with the DBAS measurement allows a semi-quantitative analysis of ditallow dimethyl ammonium chloride

(DTDMAC), the major CS used in fabric softeners (Topping & Waters, 1982; Osburn, 1982).

HPLC with conductometric detection is a more sensitive technique to determine selectively CS in environmental samples. Detection limits as low as 2 µg/l for individual CS can be achieved. A further improved resolution of CS from interferences in complex matrices comes from HPLC with a continuous flow post-column ion-pair extraction detector in which the CS-containing column effluent is segmented by adding a UV/ visible absorbing or fluorescent counter-ion (Matthijs & Hennes, 1991). A recently developed powerful method is fast atom bombardment mass spectrometry (FAB/MS), which allows the simultaneous quantification of several components with a detection limit in the low microgram per litre range (Simms *et al.*, 1988). A recent method for environmental monitoring of DTDMAC is that of Gerike *et al.* (1994).

Data on environmental concentrations are reported for total CS (of which DTDMAC accounts for the greatest part), DTDMAC (HPLC) and monoalkyl quaternaries (HPLC) which are used in detergents and fabric softeners (Table 7.4).

7.2.4 Zeolite A

Zeolite A is a crystalline sodium aluminium silicate of similar structure to naturally occurring silicates. There is no fully developed analytical method available for environmental samples that would identify detergent-derived silicates. Since the very low amounts of zeolite A which are released to surface waters hydrolyse to amorphous silicates there is also no rational need for such a method.

In laboratory treatability tests, zeolite A was measured by a combination of photometrically analysed, molybdate-reactive silica following dissolution in dilute acid (intact zeolite) and atomic absorption determination of aluminium after dissolving all particulate material in strong acid (intact and altered zeolite) (King *et al.*, 1980). Both analyses require a careful correction for background concentrations of silica and aluminium. In field tests, zeolite A has often been labelled with indium, which can be characterized and quantified by neutron activation analysis.

Table 7.3 Concentration of nonionic surfactants in the environment. Data largely from Holt *et al.*, 1992

Environmental sample	Concentrations
Sewage (raw or settled not always reported)	0.03 – 10.0 mg/l BiAS/CTAS, mostly 2 – 4 mg/l; 0.2 – 0.9 mg/l AE; 0.01 – 2.5 mg/l NPE and metabolites
Sewage effluent	0.1 – 6.2 mg/l BiAS/CTAS, mostly < 0.5 mg/l; < 0.01 – 0.06 mg/l AE; 1 – 250 μg/l NPE and metabolites. These values are mainly for activated sludge treatment and very few for trickling filter plant effluents

Sludge
  anaerobically digested
  aerobically digested
  activated
  mixed primary and secondary
	
	< 0.001 – 2.5 g/kg NPE metabolites
	0.08 – 0.5 g/kg NPE metabolites
	< 0.001 – 0.15 g/kg NPE metabolites
	0.04 – 0.14 g/kg NPE metabolites

Sludge-amended soil	NPE metabolites analysed in top 5 cm: 0.1 – 4.7 mg/kg immediately after sludge application, 0.01 – 0.5 mg/kg after 320 days following sludge application
Groundwater	0.12 – 0.78 mg/l CTAS (Israel); 0.01 mg/l and < 0.005 mg/l CTAS 500 m and 3000 m from sewage infiltration pond, respectively (US). 2 – 4 μg/l NPE metabolites
River water	< 0.01 – 2.6 mg/l BiAS/CTAS, mostly 0.02 – 0.1 mg/l; < 0.005 – 0.2 mg/l AE; < 0.005 – 0.93 mg/l NPE; 0.002 – 18 μg/l NPE metabolites
River sediment	< 0.005 – 1.0 mg/kg AE; < 0.005 – 50 mg/kg NPE, mostly in the lower range
Marine water and sediment	1 – 5 μg/l NPE; 0.2 – 19.6 μg/l NPE metabolites; 0.1 – 6.6 mg/kg NPE; 8 – 10 mg/kg NPE metabolites in sediment nearer to sewage outfall

7.2.5 Polycarboxylates

Polycarboxylates can, in principle, be determined in environmental samples by polyelectrolyte titration with a polycation (e.g. polydiallyl dimethyl ammonium chloride) in the presence of an anionic, metachromic indicator. This reaction takes advantage of the high charge density of the polycarboxylates. The detection limit of the optimized procedure is 10 μg/l. However, the analysis is interfered with by natural anionic polyelectrolytes, such as humic acids, which react with the polycation. This procedure should therefore only be applied to drinking water samples where the interfering substances can be elimin-

ated by oxidative pretreatment of the samples (Schroeder *et al.*, 1991; Wassmer *et al.*, 1991).

The polyelectrolyte titration method has been applied to several drinking-water samples in which no polycarboxylates above the detection limit of 10 μg/l could be analysed. Environmental concentrations of polycarboxylates which derive mainly from detergents can be estimated from the consumption figures, knowledge about the elimination of polycarboxylates in sewage treatment, and sewage treatment practices. For example, in Germany the following concentrations of P(AA-MA)70 000 are predicted: 2.9 mg/l in untreated sewage, 0.26 mg/l in treated sewage effluent, 50 μg/l in surface water contami-

Table 7.4 Concentrations of cationic surfactants in the environment. Data largely from Boethling & Lynch, 1992; Versteeg *et al.*, 1992

Environmental sample	Concentrations
Raw sewage	30–1400 µg/l total CS on different weekdays; 180–330 µg/l DTDMAC; 25–50 µg/l A_{12-18} TMAC analysed at German, UK and US domestic sewage treatment plants.
Sewage effluent	3–700 µg/l total CS, the high value obtained at a poorly operating activated sludge plant; 22–56 µg/l DTDMAC; 1.0–2.2 µg/l A_{12-18} TMAC from activated sludge unit, c 6–7 µg/l A_{12-18} TMAC from trickling filter
Sewage sludge	3.8–5.1 g/kg in anaerobically digested sludge (Matthijs *et al.*, 1994)
River	5–50 µg/l total CS; < 2–c. 30 µg/l DTDMAC; < 1–4 µg/l A_{12-18} TMAC analysed in Germany, UK, France and USA
River sediment	An average of 23 mg/kg analysed at one US site

nated with sewage effluent (worst-case assumptions concerning sewage effluent dilution in receiving waters), 5 µg/l in drinking water, 14.2 g/kg in digested sewage sludge and 13–27 mg/kg in soil after application of sewage sludge according to German soil amendment practices (Fachgruppe Wasserchemie, 1990).

Radiolabelled (^{14}C at the COO⁻-group or at the backbone C-chain) substances have been used in studies on the fate of polycarboxylates in various environmental compartments, particularly when applying realistic concentrations.

7.2.6 Perborate

Several methods are available for the determination of boron in aqueous solutions (HMSO, 1980). The colorimetric azomethine-H method is currently the preferred analytical procedure, since it is simpler and quicker than earlier methods (curcumin and alkali titration). A detection limit of 0.04 mg/l can be achieved. In non-saline solutions the method based on a reaction with curcumin can provide a lower detection limit of 0.004 mg/l, but this is time-consuming. Titration of boric acid bound to mannitol with an alkali solution is useful for high boron concentrations (up to 1 g/l), but is also less preferred because of the long time needed in analysis.

Inductively-coupled plasma atomic emission spectrometry has been developed for boron analysis in surface and wastewater (Broeckaert & Leis, 1979). It measures boron accurately within a range of 0.05–20 mg/l and can also be used to analyse solutions with high salt concentrations.

Boron is released into the environment due to its use in detergents, but environmental concentrations derive also to a large extent from natural sources. Boron is continuously released by weathering and ultimately discharged into the marine environment. Levels in natural freshwaters are in the range of < 0.1–0.5 mg/l depending on the geochemistry of the drainage area; in areas with large masses of boron-bearing rocks concentrations up to 100 mg/l have been recorded (Butterwick *et al.*, 1989). Seawater has a consistent boron concentration of 4.6 mg/l. The concentrations in Table 7.5 were analysed in areas where boron derived from use in detergents was also present.

7.3 ENVIRONMENTAL FATE AND EFFECTS

7.3.1 Anionic surfactants

This section reviews the main AnS used in detergents, i.e. LAS, AS, AES, SAS and AOS. LAS is the most widely used and best studied of these.

Table 7.5 Boron concentrations in the environment. Data largely from Butterwick *et al.*, 1989 and Raymond & Butterwick, 1992

Environmental sample	Concentrations (expressed as Boron)
Sewage (raw and treated)	On average 2 mg/l, occasionally up to 3–5 mg/l
Sewage sludge	On average 70 mg/kg, range: 15–1000 mg/kg dried solids depending on the contribution of industrial effluents to domestic sewage
Surface water	0.001–2 mg/l in several European countries, e.g. 0.046–0.822 mg/l in the UK, 0.400–1.000 mg/l in Italy, 0.001–1.046 mg/l in Sweden, 0.040–0.330 mg/l in The Netherlands, and 0.078–2.000 mg/l in Germany. In more recent studies: < 1 mg/l in the German rivers Ruhr and Rhine
Soil	On average typically 10 mg/kg; 7–71 mg/kg found in the UK. Level depending on the mineral composition of the soil
Marine water	4.6 mg/l mainly due to natural sources; 4.66 mg/l recently found in the English Channel. Boron concentrations increase with the salinity
Rainwater	42 µg/l over Pacific and Indian oceans; 6.5–18 µg/l measured over an industrial zone, 8–12 µg/l over a city, and 8 µg/l over a rural area in Spain; 6–44 µg/l found in the UK

LAS

$CH_3(CH_2)_x\,CH\,(CH_2)_y\,CH_3$

$SO_3^-\,Na^+$

$x + y = n$, $n = 7$–11 carbon units

Environmental fate

Table 7.6 summarizes various screening and simulation test results regarding biodegradation and removal in biological sewage treatment. Removal and primary degradation of all AnS is high, mostly > 90%. Ultimate biodegradability, or mineralization, tests show a well-established ready biodegradation for all linear AnS. Moderately and highly branched AnS become increasingly less biodegradable. However, these 'hard' surfactants have been abandoned in favour of linear AnS, and are therefore not included in the following. Metabolic pathways are known in detail for LAS, AS, AES; they are less well established for SAS and only speculated on for AOS.

Several monitoring studies on LAS in full-scale activated sludge treatment plants have demonstrated that in properly operated plants the removal is > 95%, usually 98–99%, with 3–15% of that removed being adsorbed onto sludge in some tests (Painter, 1992). The removal in biological filters is 80–91%. LAS is not significantly degraded in anaerobic sludge treatment, although there are indications that a pre-aerobic wastewater treatment enhances a subsequent anaerobic degradation. Anaerobic sediment of a wastewater pond which was exposed to laundry effluents over several years did not show any mineralization of LAS (Federle & Schwab, 1992).

In rivers and sediments, LAS is also removed to a considerable extent (68–82% $^{14}CO_2$ evolution), with significantly increasing degradation rates in the presence of sewage effluents or sediment ($t_{1/2}$: 0.7–14 days). A recent study shows that LAS mineralization in river/sediment, seston, periphyton and leaf litter is positively correlated with ambient temperature (average $t_{1/2}$ at 22°C: 2.7 days, and at 4°C: 7.3 days) (Palmisano *et al.*, 1991). It was shown that a substantial mineralization is realized by the microbiota of submerged plant detritus (51% CO_2) (Federle & Ventullo, 1990). LAS is highly adsorbed onto suspended

Table 7.6 Surfactant biodegradation in laboratory activated sludge screening and STP simulation tests. Data largely from Jacobi & Loehr, 1987, Schoeberl et al., 1988, Boethling & Lynch, 1992, Holt et al., 1992, Painter, 1992

Surfactants	Screening tests				Simulation tests			
	Primary biodegradation	Ultimate biodegradation			Removal	Primary biodegradation	Ultimate biodegradation	
	OECD screening test (% MBAS/BiAS/DBAS removal)	CO$_2$ evolution test (% [14]ThCO$_2$)	Respirometric tests[a] (% ThOD)	Modified OECD screening test (% C removal)	OECD confirmatory test (% C removal)	OECD conf. test (% MBAS/BiAS/DBAS removal)	OECD conf. test (% [14]ThCO$_2$/COD)	Coupled units test (% C removal)
C10–14 LAS	95	45–76[b]	55–65	73–84	73–96	90–97	80–90 (ThCO$_2$)	73 ± 6
Commercially used: C < 13								
C12 AS								
C16–18 AS	99	64–96[c]	63–95[c]	88–96[c]	94–99	98–99[c]	107 ± 6 (COD)	97 ± 7
C12–15 oxo AS	99		91	88				
C12–14 AE2 S	98	65–83	86	96–100				
C12–15 oxo AE3 S			58–100[d]		79–89			67–68
C13–18 SAS	99	56–91	63–95	80–100		97–98		89–99
C14–18 AOS	99	65–80	85	85	88	98–100		83–96
C12–18 FA 5–50 EO	96–99	65–75	27–86	80–94	88–99	93–98		59–96
C9–15 oxo alcohol 3–20 EO	93		62	75	71	83–96		36–59
C8–10 AP 5–25 EO	6–85	40	5–29	8–17	82–97	85–97		48–91
C12–18 FA 2–6 EO2–8 PO	70–95	32	15–83	43–69		87–97		37–60
C12–18 alkylamino 2–20 EO	16–88	85 (2 EO)	33 (12–20 EO)		95	97–98	95 (2EO) (specific analysis)	6–70
C12–18 FA ester 5–29 EO	95–99		60–80	100	85	92–96		71–92
EO/PO block polymers	32		0–10	18	11–66	7–58		2 ± 4
DTDMAC (T = C10–18)	96[e]	0	5		94–97	62–94		108 ± 9
IQAMS	89[e]				92			111 ± 13
C12–18 TMAC	97–98[e,f]							
C16 TMAB		60–90[e,g,h]			91–98			104 ± 6

a Closed bottle, BOD$_5$, other die-away tests.
b half-life: 1.5–2.2 days.
c C12–18 AS and C12–15 oxo AS.
d Various CnAExS.
e Semi-continuous activated sludge test.
f Half-life: 2.5 h.
g Depending on position of radiolabel.
h Half-life: 28–40 h.

solids and sediments in river water. Distribution coefficients were described as increasing with decreasing LAS critical micelle concentration and with increasing organic carbon content or cation exchange capacity of the solid surface; K_d: $11->24$ l/kg for sediment/interstitial waters and $1000->5700$ l/kg for suspended solids/overlying waters (Hand *et al.*, 1990).

LAS is substantially biodegraded in acclimated marine and estuarine water (up to 60% $^{14}CO_2$ evolution) (Shimp, 1989). Biodegradation yielded up to 80% CO_2 in soil degradation studies; 33% mineralization was reached in an unexposed subsurface sediment, and up to c. 60% in subsurface sediments acclimated to LAS (Larson *et al.*, 1989; Knaebel *et al.*, 1990). Degradation rates vary depending on soil depth, moisture content, distance from the release point and acclimation ($t_{1/2}$: 1.1 to 27 days). Two field studies confirmed these laboratory data: in a porous sand underneath a laundry pond LAS concentrations declined from c. 220 µg/g to <2 µg/g over a vertical distance of less than 3 m; concentrations in an effluent/groundwater plume of a domestic septic tank discharging into a sand/gravel aquifer decreased from 10 mg/l to 30 µg/l over a horizontal distance of 10 m (Larson *et al.*, 1989).

The primary biodegradability of LAS generally increases with increasing chain length. For the isomers of a given homologue it has been observed that the greater the distance between the sulphonic group and the distant terminal methyl group on the alkyl chain the faster the degradation. Complete mineralization of the benzene ring takes place.

For *AS*, a field study in biological filters showed 96–98% MBAS removal (Mann & Reid, 1971). AS are well removed in anaerobic treatment (>88% MBAS removal); 88–95% ultimate anaerobic degradation is reported for stearyl sulphate. A high MBAS removal in river water (>90%) was reported for LPAS, C_{12}-AS and C_{16}-AS. AS show ready primary biodegradation in marine and estuarine water. In studies with and without sediment, 75% MBAS removal was observed for C_{12ave}-AS. AS are further well degraded in soil, with degradation qualitatively reported as progressing faster than LAS. There seems to be only little effect of chain length on the biodegradation of AS.

As a class, *AES* show a high primary aerobic and anaerobic degradation, with those used in detergents also highly mineralized under aerobic conditions. A Japanese activated sludge treatment field study found 100% removal of combined AES + AS (Oba *et al.*, 1976). Anaerobic degradation produced between 72% and 95% MBAS removal. In parallel anaerobic digester studies AES removal was somewhat lower than AS. Mineralization in marine and estuarine water is high. An unspecified AES was degraded up to 100% (measured as DOC) in uninoculated marine water. Radiolabelled AES materials were degraded in estuarine water to 70% when labelled on the ethoxy groups and up to 99% when labelled at 1-C. Biodegradation is not significantly influenced by chain length or the number of EO groups, as tested with C_{10-16}-AE_{2-6}S. However, linear secondary AES are less biodegradable than linear primary AES (39% and 80% DOC removal, respectively).

SAS are effectively removed in activated sludge treatment, ranging from 83% to 96% in simulation tests. The removal in small-scale trickling filters was 97–99% MBAS. In oxidation ditches and full-scale filters the removal is reported as being 'efficient'. SAS is qualitatively described as even easier to biodegrade than LAS. Like other sulphonates they are not anaerobically degradable. There is only a marginal effect of chain length on biodegradability, with higher chain length homologues having a longer lag phase.

AOS are mineralized with sewage inoculum at rates slightly lower than AS and slightly faster than LAS (Table 7.6). The various components of commercial AOS have a somewhat different biodegradation profile. Alkene sulphonates are equally or even more highly biodegradable than the corresponding hydroxyalkane sulphonates, and the monosulphonates have a slightly higher degradability than the disulphonates. Anaerobic degradation occurs probably only to a small extent. In river water and in seawater MBAS removal can be as high as 95–100%.

Soap showed 80–100% biodegradation in river water die-away and batch-activated sludge tests. In a batch anaerobic digester study, mineralization of a radiolabelled C_{16}-FA was >90% (Swisher, 1987).

Environmental effects

A summary of toxicity studies with aquatic and terrestrial organisms is shown in Table 7.7. Ranges are often wide due to studies with many organisms, differences in sample composition (e.g. surfactant chain-lengths) and interlaboratory variation. Fish species were rainbow trout (*Oncorhynchus mykiss*, formerly called *Salmo gairdneri*), fathead minnow (*Pimephales promelas*), bluegill sunfish (*Lepomis macrochirus*) and golden ide (*L. idus melanotus*). 'Other invertebrates' cover a particularly broad range of values. These species include sensitive ones like *Ceriodaphnia* and marine species, such as oyster larvae (*Crassostrea virginica*) and less sensitive ones like pink shrimps (*Penaeus duroorum*). The toxicity data for algae also cover a broad range of species with different sensitivities.

$L(E)C_{50}$ values and NOECs are mostly in the range of 0.3–50 mg/l. This is generally above concentrations in environments where these organisms can be expected to occur. In the case of LAS, in particular, many NOEC values derived from a variety of non-lethal end-points are published, and Table 7.7 shows only a few selected examples.

All AnS investigated have an increasing toxicity with increasing chain length, as long as the surfactant remains soluble. For example: LC_{50} of C_{13}-LAS for fathead minnow (*Pimephales promelas*) is 0.5 mg/l rising to 57.5 mg/l for C_{10}-LAS; EC_{50} of C_{13}-AS towards *Daphnia magna* is 42 mg/l rising to 8200 mg/l for C_4-AS. Such data allow calculation of the toxicity of homologue mixtures. In the case of LAS, the higher toxicity of longer C-chains has led to the commercial use of LAS with C < 13, average C11–12, in the most commonly used products.

Investigations on uptake by fish of intact LAS indicate BCF values in the range 10–100 for the whole body. These studies have shown that LAS is rapidly metabolized with breakdown intermediates found in the gall bladder. Indeed, fish may metabolize LAS in a similar way to natural bile salts which are also surfactants (Kimerle, 1989). AS is also readily metabolized, and has shown a BCF of 4 for the whole body, 50 for the hepatopancreas and 700 for the gall bladder in fish.

Field aquatic toxicity studies with LAS demonstrated the absence of any effect at realistic environmental levels on the structure and function of lake and stream microbial communities (Ventullo *et al.*, 1989). A recent experimental stream study showed no effects of 0.36 mg/l LAS after 45 days exposure to survival of fathead minnow (*Pimephales promelas*) or amphipods (*Hyalella azteca*); no sublethal effects were observed on bentic invertebrates or periphyton (Fairchild *et al.*, 1993).

Studies on sewage treatment processes (activated sludge) conducted for LAS showed no harmful effects on large-scale aerobic sewage treatment at realistic discharge levels. No harmful effect on laboratory systems up to 30 mg/l MBAS and even 70 mg/l MBAS was observed provided the increase did not occur in one step. Similarly, laboratory tests with other AnS, also at a high concentration of 20 mg/l, did not show any effect on activated sludge treatment (Gilbert & Pettigrew, 1984). Anaerobic sludge digestion is not inhibited by 15 g LAS/kg dry solids which is equivalent to 25 mg/l LAS in raw sewage.

7.3.2 Nonionic surfactants

The nonionic surfactants considered here are predominantly linear AE and APE (mainly NPE), fatty amine ethoxylates and block polymers, which are all used to some extent as detergent ingredients. Most used are fatty alcohol, Ziegler alcohol and oxoalcohol AE. The use of NPE is declining or discontinued in many products, for reasons described below.

Alcohol ethoxylates

$R(OCH_2CH_2)_nOH$

$R = C_{9-18}, n = 1$–40, usually averaging 7–12 for household detergents

Environmental fate

The results of screening and simulation tests for the fate of NS in biological sewage treatment are listed in Table 7.6. All broadly used NS show a primary degradability and removal of > 80%; AE, except for some oxo AE, are even removed to > 95%. AE are in general aerobically mineralized

Table 7.7 Toxicity of surfactants to various organisms (EC/LC$_{50}$: mg/l). Data largely from Jacobi & Loehr, 1987; Schoeberl et al., 1988; Boethling & Lynch, 1992; Holt et al., 1992; Painter, 1992.

Surfactants	Bacteria		Algae		*Daphnia magna* 24–96 h, mostly 48 h	Other invertebrates		Fish mostly 96 h	Terrestrial organisms
	Heterotrophic	Anaerobic	Freshwater 72–96 h	Marine 24–48 h		Freshwater 24–96 h	Marine 24–96 h		
C10–14 LAS	50[a]	20[b]	0.9–300	0.025–10	1–15	1.5–270	0.4–150	0.7–15, mostly 3–20	>500[i]
Commercially used: C<13		0.5[c]	30–300[d]		0.1–10[d,c]		0.04–0.9[d,c]	0.05–14[d,c], 1–2[d,g,h]	10–1000, mostly > 100[k]
C12–18 AS	>10	42[b]	4–30	1–2	5–70	0.6–>200		1–80	10[k]
C12–15 oxo AS	400[j]		60[j]					mostly 3–20	
C12–14 AE2S,	0.7–4[m] 2–18[m]		4–50	0.01–0.05	1–50	11–350		1–70, mostly 1.4–20	100[k]
C12–15 oxo AE3S	1.5–2.2[m]		65[d]					0.1–0.63[d,o,g]	
C13–18 SAS	20–>200[m]				1–250			1–144, mostly 1–24	
C14–18 AOS			10–100[d]	4–32	5–50			0.5–20, mostly 2–20	10[k]
Soaps:									
0 days								6.7	
3–23 days			10–50[d]					20–150	
C12–18 FA 2–4 EO				2–4	20–100			1.5–>100	
C12–18 FA 5–9 EO				4–50	0.3–200			0.7–30	
C12–18 FA 10–14 EO				0.3–10	1.4–60	1.0–6.8		1–3	
C12–18 FA 30 EO				30	700			100	
C9–15 oxoalcohol 2–10 EO				4–50[d]	2–10 0.43[d]			0.25–4	
C9–15 oxoalcohol >10 EO					4–20			1–40	
C9 AP 5–7 EO				20–50[d]				1.5–5	
C9 AP 9–10 EO					4–50			5–11	
C9 AP 11–30 EO								11–>1000	
C12–18 FA 2–6 EO 2–6 PO				0.5–0.7	2.0–6.0			0.7–5.7	
FA EO/PO adducts (>80% biodegradable)					0.3–1			0.5–1	
C12–18 alkylamino 2 EO					1.3			0.7	
C12–18 alkylamino 3,8–20 EO					3–10			1.0–2.5	
C12–18 alkylamino 12–25 EO					10–30			5–10	
C12–18 FA ester 5–29 EO EO/PO block polymers									
DTDMAC			0.05–32[p]	0.5–1.0	>100 0.1–3.0, mostly 0.1–1.0, 0.38[d,r]	0.32–10.7[d,r]	0.2–>50,	35–>100 >100 0.6–36[r] mostly 1–6 0.05–1.0[d,r]	
IQAMS	28		0.2–18[p]		4–100				
A12–18TMAC			0.1–6.1[p]		0.06–5.8,	0.75–>1000[g]		1.5–40	
A16TMAB	10–40		0.03–2.6[p]		0.11[q]		1.3–1.8	0.36–8.6	

[a] EC10.
[b] maximum acceptable concentration (g/kg).
[c] Nitrifying bacteria (soil) (g/kg).
[d] NOECs for different chain lengths.
[e] Kimerle, 1989.
[f] Mysid shrimp (*Mysidopsis bahia*).

[g] Survival of fathead minnow (*Pimephales promelas*) fry.
[h] Commercial mixtures.
[i] Earthworms (mg/kg).
[j] Terrestrial plants, concentration in watering solution.
[k] Threshold concentration.
[l] Sewage-derived mixed cultures.

[m] Marine bacteria.
[n] LOEC (g/l).
[o] Growth inhibition of fathead minnow (*Pimephales promelas*).
[p] Laboratory—in-situ studies.
[q] A12TMAC (Woltering et al., 1987).
[r] Versteeg et al., 1992.

to $> 60\%$, often to $> 80\%$. Higher ethoxylate (EO) numbers lower the biodegradability. The low removal or biodegradability in this compilation refer to compounds with high EO numbers. However, AE with more than 20 EO units are of marginal use in modern household detergents. Metabolism studies were conducted with a number of activated sludge, river water, and soil inocula and with plant detritus. AE are completely mineralized with degradation half-lives ranging from a few hours to a maximum of 8 days for plant detritus.

Factors influencing the ultimate degradability of straight chain primary or secondary alcohols are:

1 Structure of hydrophobic moiety (linear hydrophobes undergo most rapid biodegradation, secondary and highly branched hydrophobes retard ultimate biodegradation increasingly).
2 Chain length of the hydrophile (length of alkyl chain or number of EO units have little effect on rate and extent of ultimate biodegradation up to 11 EO. However, above 20 EO units a significant reduction of ultimate biodegradation occurs).
3 Incorporation of other glycols into the hydrophobe (insertion of oxypropylene groups into polyoxyethylene hydrophile of oxoAE decreases rate and extent of biodegradation).

Straight- or branched-chain APE are not biodegraded as rapidly, nor are they completely mineralized. APE undergo substantial primary biodegradation, which means fast mineralization of the polyglycol unit. The AP1EO and AP2EO formed are partly adsorbed to sludge solids, and during the subsequent anaerobic sludge treatment these compounds are converted to the poorly biodegradable AP.

High removal in activated sludge treatment plants and in trickling filters was confirmed in several monitoring studies showing $> 90\%$ BiAS/CTAS and $> 98\%$ AE removal, and 80–90% BiAS/CTAS and $c.$ 90% AE removal in these two processes, respectively (Holt *et al.*, 1992). The removal of APE is related to plant efficiencies and ambient temperature. A German study showed 89% removal in activated sludge treatment and 70 and 75% removal in a trickling filter during summer and winter, respectively (Brown *et al.*, 1986, 1987). An older UK trickling filter study

about two decades ago resulted in a stronger temperature dependence from $> 80\%$ down to 20% removal in cold weather conditions (Mann & Reid, 1971).

AE are also highly anaerobically degradable. An anaerobic sludge treatment study showed over 90% ultimate biodegradation of stearyl AE after 4 weeks. For APE only, a substantial degradation of the polyglycol unit is observed, whereas the AP part is not anaerobically degradable. A methane production of 70–80% was found with different AE compounds in anoxic sediments. In the same matrix only 45–50% methane production deriving from APE was reported. In anaerobic sediments of a wastewater pond receiving laundry effluents for several years a complete mineralization of AE to CO_2 and CH_4 was observed (Federle & Schwab, 1992).

Rapid and ultimate biodegradation of the AE hydrophobe and hydrophile unit occurs in river water and groundwater. Mineralization of LAE ($C_{12}FA9EO$ and $C_{16}FA3EO$) is $> 80\%$ in river water ($t_{1/2} = 1.2–1.5$ days) and $> 60\%$ in groundwater ($t_{1/2} = c.$ 6–14 days), at realistic environmental concentrations. Microbiota of submerged plant detritus yielded 64% CO_2 when exposed to LAE (Federle & Ventullo, 1990).

In estuarine water, LAE is mineralized to $> 75\%$ with half-lives for the alkyl chain of $c.$ 2 days and for the ethoxylate of $c.$ 6 days. Primary biodegradation of APE is $c.$ 35% in freshwater and $> 95\%$ in marine water. However, the latter drops to 15% at 3–4°C. In soil biodegradation studies a linear AE was biodegraded up to $c.$ 70% CO_2 production with degradation half-lives of 1–10 days depending on the soil type (Knaebel *et al.*, 1990).

Environmental effects

Aquatic and terrestrial toxicity data are given in Table 7.7. The acute aquatic toxicity to various species of algae, fish and to *Daphnia magna* is in the range of 0.3–> 100 mg/l; for surfactants mostly used in detergents the range is usually 1–10 mg/l. A comparison of acute toxicity data of several AE and APE indicates a somewhat higher sensitivity towards fish (non-specified) compared with *Daphnia magna* (Guhl & Gode,

1989). Other authors concluded that NS were most sensitive towards *Daphnia*. A very low toxicity of 500–10 000 mg/kg body weight was recorded for terrestrial organisms. Chronic responses measured in parallel tests are usually within one order of magnitude lower than the acute data.

Although APE generally show a lower toxicity of the parent molecule than the equivalent AE, the relative toxicity of biodegradation intermediates is reversed. The cleavage of AE into its alcohol and glycol components during biodegradation immediately removes surface activity and, thus, toxicity. However, the biodegradation pathway of APE first only shortens the EO chain, and surfactancy is retained. Toxicity decreases with increasing EO numbers of the hydrophobic unit of AE or APE given that the concentrations in the test medium are below water-solubility. Besides these observations a general relationship between structure and toxicity cannot be derived from the available data because many factors such as sample composition and various other study parameters influence the test results.

Exposure of aquatic microbial communities to long-chain linear AE over 21 days in the field had no long-term detrimental effect on bacterial heterotrophy. The bacteria showed an adaptive response upon chronic exposure (Ventullo *et al.*, 1989).

No negative effect on aerobic and anaerobic sewage treatment processes is reported from laboratory tests, which were usually conducted at higher than environmentally relevant concentrations.

7.3.3 Cationic surfactants

Three classes of CS are mostly used in detergents and detergent aids: DTDMAC ($T = C_{10-18}$, with the main one being C_{18}), ATMAC/B ($A = C_{12-18}$, with mainly C_{12} TMAC being used), and IQAMS, which are all quaternary ammonium compounds (QAC).

Environmental fate

Concerning removal and biodegradability of QAC in sewage treatment and in receiving waters some

DTDMAC

$R = C_{10-18}$, mainly C_{18}

ATMAC

$R = C_{12-18}$, mainly C_{12}; $X = Cl$ or Br, mainly Cl

common characteristics should be noted. QAC have a strong tendency to adsorb onto suspended solids in well-mixed systems. They form complexes with anionic compounds, e.g. with AnS, such as LAS in particular. Their biodegradation in sewage treatment, except for ATMAC, requires prior acclimation. In sewage treatment and receiving waters QAC should normally exist in the form of 1 : 1 complexes with AnS.

The fate of CS in biological sewage treatment as derived from laboratory screening and simulation tests is summarized in Table 7.6. Overall removal generally exceeds 90%. All CS studied show ready mineralization, mainly occurring on the sludge solids to which CS are rapidly sorbed. The two figures derived from ultimate biodegradation tests for DTDMAC in Table 7.6 are inconsistent with other data on this chemical. Although details on the tests are not available for clarification, it can be said that such screening tests can give low results, whereas simulation tests produce more reliable figures. Ultimate biodegradation of A_{18} TMAC was tested with a radiolabelled compound, and showed mineralization of both the methyl groups and the alkyl chain.

Monitoring studies conducted in German and UK sewage treatment plants confirmed high removal, mainly during biological treatment (Topping & Waters, 1982; Matthijs & De Henau, 1987). In activated sludge plants 88–93% DBAS, 94–96% DTDMAC and 96–97% ATMAC removals were found. Removal of DTDMAC in trickling filter plants is reported to be 84–100% in laboratory systems, and 75–85% removal was measured for ATMAC in a UK trickling filter plant. A recent study on the fate of DTDMAC in two German municipal sewage treatment plants confirmed an overall removal of 95–98% (Matthijs *et al.*, 1994). Elimination in activated sludge treatment was 91–95% DTDMAC with 61–64% due to primary biodegradation; the occurrence of complete mineralization was not analysed.

DTDMAC and ATMAC are rapidly biodegraded in river water, with half-lives for ultimate degradation in the range of several days or less. For different ATMAC/B homologues, $t_{1/2}$ is 2–3 days with no detectable or short lag phase of 2 days in unacclimated water. Again, microbial adaptation plays an important role. In pre-exposed river waters a half-life for A_{12}TMAC degradation of 90 h was analysed with a lag phase of 24 h. Re-application within a short time interval resulted in a $t_{1/2} = 20$ h without lag phase and continuous exposure gave a $t_{1/2} = <2$–15 h.

DTDMAC degradation was found to be much slower in sediment-free river water, but significantly enhanced when sorbed on suspended solids. Adsorption onto sediments of C_{12}TMAC (<0.01–100 mg/l), probably via an ion-exchange mechanism, was found to be inversely related to sediment concentration; K_d: 440 l/kg at 248 000 mg/l sediment concentration to 251 000 l/kg at 230 mg/l (Hand *et al.*, 1990b). The presence of sediment positively influences adaptation which persists on solid surfaces even after a temporary cessation of QAC discharge. Sediment-associated and periphytic microbial communities adapted to C_{12}TMAC within 5–10 days of exposure, reaching a $t_{1/2}$ of around 1 day (Shimp & Schwab, 1991). A_{18}TMAC were to some extent mineralized by microbiota of submerged plant detritus without a lag phase (45% and 16% CO_2) (Federle & Ventullo, 1990).

Some studies on A_{12}TMAC and DA_{18}DMAC upon exposure to sunlight suggest that QAC have a potential for photodegradation.

Biodegradation of A_{18}TMAC occurs in aerobic surface and subsurface soils (compost, organic, sandy and silt loam), with half-lives of 1.4–8.7 days, and in sludge-amended soils with a $t_{1/2}$ of *c.* 1 month. In an aerobic groundwater/subsurface soil slurry A_{12}TMAC was biodegraded with a $t_{1/2}$ of 1 week. Investigations in a laundry profile

(porous sand) showed extensive mineralization of A_{18}TMAC and D_{18}TMAC. However, anaerobic sediment in a wastewater pond with several years of exposure to laundry effluents did not show any mineralization of A_{18}TMAC (Federle & Schwab, 1992).

Environmental effects

QAC have an acute toxicity to aquatic organisms in the milligram per litre range and lower. Most data in Table 7.7 are between 0.1 and 50 mg/l. QAC show the highest acute toxicity to algae, although a remarkable difference exists between laboratory and *in-situ* lake and river studies. The laboratory-derived EC_{50} values are orders of magnitude lower than first-effect concentrations on algal community structure and functions measured in the field. A recent study on the effects of A_{12}TMAC on benthic microbial communities in experimental stream channels showed, generally transitory, first-effect concentrations on growth processes and trophic relationships at 250 μg/l. This is higher than most single-species responses (McCormick *et al.*, 1991).

The reduction of toxicity in natural matrices is mainly due to sorption to suspended solids and complex formation with anionics. This was also observed with daphnids and midge larvae (*Chironomus riparius*) populations chronically exposed to A_{12}TMAC and DTDMAC in a model stream. No effects were observed below the mixing zone at concentrations 35–75 times higher than those causing mortality in laboratory tests. Also some NOEC values for DTDMAC show the same tendency. In well-water, a NOEC of 0.053 mg/l was found for fathead minnow (*Pimephales promelas*), whereas in river water 0.23 mg/l for the same species and 0.38 mg/l for *Daphnia magna* were observed (Lewis & Wee, 1983).

Acclimation of microbial communities in surface waters increases the biodegradation of QAC and reduces mortality. The first-effect concentrations for inhibition of the heterotrophic activity of freshwater bacterial communities analysed in laboratory tests were 0.1 mg/l for ATMAC and 1.0 mg/l for DTDMAC. This is considerably higher than expected surface water concentrations.

Fish uptake and tissue distribution was traced with radiolabelled DTDMAC applied to carp (*Cyprinus carpio*). The radiolabel was concentrated in the intestinal tract and gall bladder. Some residual ^{14}C was found in the gills, probably due to filtration of contaminated faecal particles of leftover feed. BCFs of DTDMAC in bluegill sunfish (*Lepomis macrochirus*) were 256 and 94 for inedible tissue and 32 and 13 for the whole body tested after 49 days exposure in well-water and river water, respectively (Lewis & Wee, 1983). The data from these studies suggest that QAC bind preferentially to the negatively charged surfaces of gill tissues and/or incorporate into gill membranes but are not transported across membranes. This was also found in uptake and distribution studies of another QAC in clams (*Corbicula fluminea*), tadpoles (*Rana catesbeiana*) and fathead minnow (*Pimephales promelas*) (Knezovich et al., 1992).

The bioavailability of QAC in water is reduced by the presence of sediment onto which they adsorb. This is particularly effective for the bioavailability to fish gill tissue, but not necessarily for uptake into the intestinal tract. The gill tissue is the main acutely affected organ, though, and these observations are consistent with findings of reduced toxicity in the presence of suspended solids.

At realistic levels in sewage treatment (< 1.5 mg/l), detergent CS do not change/ influence aerobic and anaerobic microorganisms. However, in unacclimated test systems, only containing the CS investigated, germicidal activity at relatively low concentrations (< 1 mg/l) was found, in particular for ATMAC. In real municipal sewage matrices which are continuously exposed to QAC, inhibitory effects only occur at higher than realistic levels. Acclimation of microorganisms to low QAC concentrations and complex formation with AnS are the main reasons for this different behaviour. Only a sudden release of high CS dose (> 3 mg/l) can result in an inhibitory effect on biological treatment, in particular on the more sensitive nitrification process.

7.3.4 Zeolite A

Environmental fate

The fate of zeolite A after its use in the household was investigated in laboratory and monitoring studies for all stages of a state-of-the-art municipal sewage treatment (Christophliemk et al., 1992). Zeolite A used in laboratory experiments (mostly up to 10 mg/l zeolite in influent sewage) was either the sodium form present in laundry detergents or the calcium form which is the predominant species in wastewater. Field trials were conducted by supplying homes participating in the study with zeolite A-containing detergents.

Regarding transport properties of the zeolite in sewers, studies were carried out with copper, iron and polyvinyl chloride (PVC) piping charged for several weeks with laundry effluent (King et al., 1980). A minor visible deposit was found only in rough spots of the pipes (Fe > Cu > PVC) which did not affect the wastewater flow. Large-scale, long-term (several months to about 2.5 years) field trials demonstrated that zeolite A shows no excessive sedimentation in the pipes of household drains and communal sewage systems and, thus, does not cause any clogging (Roland et al., 1979). The maximum amount found in sewer sediments was 5% of intact zeolite in horizontal pipes in one series of tests.

In static and dynamic settling tests simulating primary wastewater treatment, zeolite A was removed similarly to other suspended solids. The removal of the smallest 10%-size fraction of the zeolite and of the bulk zeolite passed separately through a continuous-flow settling apparatus was comparable; 62% for the smallest 10%-size fraction and 77% for the bulk zeolite (King et al., 1980). Suspended solids removal was 33–41%. The previously mentioned field trial also demonstrated that approximately two-thirds of zeolite A is eliminated in the sand trap and the primary settler of STP.

Removal in activated sludge treatment was tested in semi-continuous and continuous activated sludge tests. The bulk of the zeolite was retained on sludge solids to 77–98%. The smallest 10%-size fraction was only investigated in the semi-continuous test with about 97% being eliminated from the effluent wastewater (King *et al.*, 1980). Addition of chemicals to wastewater for phosphorus removal (ferric chloride alum, or lime) further enhances zeolite A elimination. In large-scale field trials 96% of the zeolite was eliminated after the secondary treatment and sludge settling, i.e. only 4% of the original material was left in the final effluent. The removal in laboratory trickling filters averaged 89%. In a full-scale trickling filter plant, zeolite A was eliminated to about 81%; after passing a dual-media filter even 99% was removed (Baumann *et al.*, 1981).

The small amounts potentially released to surface waters undergo hydrolysis which transforms the zeolite to amorphous aluminosilicate. In most natural waters hydrolysis half-lives were typically 1–2 months. In waters with pH values below 7, hydrolysis was more rapid (half-life of a few days), and in the presence of high concentrations of calcium and silicate, hydrolysis was slower.

Laboratory septic tanks removed >97% of the zeolite, which was more than the corresponding suspended solids removal (57–77%) (Holman & Hopping, 1980). In the field septic tanks removal averaged 81%. The effluent concentration of zeolite correlated in these systems with the effluent concentration of the suspended solids and accounted for about 2% of the concentration of suspended solids in the septic tank effluent.

Environmental effects

Zeolite A shows a very low ecotoxicity. Acute toxicity data for *Daphnia magna*, marine macro-invertebrates (e.g. oysters, *Crassostrea virginica*, and pink shrimp, *Penaeus duororum*), freshwater fish (e.g. bluegill sunfish, *Lepomis macrochirus*, and rainbow trout, *Oncorhynchus mykiss*) and marine fish (pin fish, *Lagodon rhomboides*) (Table 7.8) are well above 500 mg/l. Subacute or chronic effects were not observed up to concentrations above any likely concentrations in aquatic environments (well below 1 mg/l) with bacteria, fungi and yeast, algae (green algae, *Selenastrum capricornutum*; blue-green algae, *Microcystis aeroginosa*; and a diatom, *Navicula seminulum*), freshwater and marine invertebrates, fish and benthic organisms (tubificids, mussels and

Table 7.8 Toxicity of zeolite A to various organisms (EC/LC$_{50}$; mg/l)

Organism	Concentration
Bacteria, fungi, yeast	25 (32 h-NOEC)[a]
Algae, 72–96 h	50–1000 (first-observed effect concentrations)[b]
Daphnia magna, 48 h	>500[b,d], >1000[c], 500 (21 days-NOEC)[a], 129–264 (21 days-NOEC)[b]
Marine macroinvertebrates, 96 h	>780[b]
Fish, 96 h	
freshwater	>500[b,d], >16 000[c], 250 (4 weeks-NOEC)[a], 87 (20 days-NOEC, egg and fry-test with fathead minnow, *Pimephales promelas*)[b]
marine	>780[b]
Benthic organisms	250 (4 weeks-NOEC)[a] 270–550, 100–200 (NOEC) (both 30 days full life cycle-test with dipteran midge, *Paratanytarsus parthenogenica*)[b]

[a] Christophliemk *et al.*, 1992.
[b] Maki & Macek, 1978.
[c] Schoeberl & Huber, 1988.
[d] Fischer & Gode, 1977.

dipteran midge, *Paratanytarsus parthenogenica*). No bioconcentration was observed in benthic organisms, which could be the most exposed to zeolite A settled on river sediments (Fischer & Gode, 1977).

With algae, no growth-inhibiting effects were observed in nutrient-rich media; whereas under oligotrophic conditions zeolite A (at concentrations up to a maximum of 10 mg/l in untreated effluents) can cause an effect on algal growth by depleting essential trace heavy metals. However, this is not of practical significance because wastewater will always contain sufficient other nutrients compensating for this trace depletion. In other natural waters only a marginal ion-exchange effect for trace elements, less than 10%

for Cd, Zn, Cu and Ni, and usually about 30% for lead was observed at a high zeolite A concentration of 1 mg/l, which will be even smaller at realistic concentrations (Allen *et al.*, 1983).

The discharge of zeolite A may result in enhanced silicon levels in surface waters. This does not cause or contribute to eutrophication in freshwater lakes, though. Silicon becomes a limiting factor for freshwater diatom growth below 0.1 mg/l, a concentration that is usually exceeded in surface waters. Input of silicon into lake waters with silicon > 0.1 mg/l does not cause any enhanced diatom growth which is independent of the silicon concentration above the limiting value. In eutrophic surface waters (caused by large phosphorus discharges), which can become silicon limited near the surface due to excessive growth of diatom populations, the undesired green and blue-green algae will eventually replace the diatoms. In such cases dominance of diatom species can be supported by dissolved silica levels.

Zeolite A does not adversely affect the performance of sewage treatment processes. In laboratory tests and field trials on activated sludge treatment, no differences were observed in general purification parameters such as removal of chemical oxygen demand and MBAS, sludge retention compared with control units and no effect on the chemical treatment for phosphorus removal from wastewater. The treatment efficiency of trickling filters is not adversely affected, either (Fischer *et al.*, 1978; Baumann *et al.*, 1981). The effluent concentrations of heavy metals were not significantly changed, but a depletion of calcium and magnesium levels was observed due to ion-exchange on the zeolite (Roland & Schmid, 1978; King *et al.*, 1980). In the presence of zeolite A no effects on the rate and extent of LAS biodegradation were observed. This is consistent with studies about adsorption of organic compounds, including LAS, onto zeolite A, showing that it is a relatively weak adsorptive surface for organics (Savitsky *et al.*, 1981). Some authors describe an enhanced nitrification in zeolite A containing activated sludge units; however, this was not observed in trickling filters and in large-scale activated sludge field trials.

The results of sludge thickening and dewatering processes showed that the efficiency of full-scale processes should not be adversely affected by zeolite A concentrations up to 20% of suspended solids in wastewater sludge, but may be in fact somewhat improved (Holman & Hopping, 1980). The concentration of zeolite A in sewage sludge is up to 2−3% of total suspended solids at broad-scale use of phosphorus-free detergents. Similarly, a zeolite A content of up to 20% of suspended solids in anaerobic digester sludge did not affect the process efficiency measured as gas production. The concentrations of heavy metals in the supernatant and mixed digester sludge were not altered. Dewatering processes of unconditioned and chemically conditioned activated sludges and anaerobic digester sludge were also not affected at higher than realistic concentrations of zeolite A.

No effects were observed on the process efficiency of septic tank systems, in the laboratory or as field trials, even at high influent concentrations of 55 mg/l zeolite A (Holman & Hopping, 1980). Neither were there increases in the effluent concentrations of heavy metals. At realistic concentrations of zeolite A the accumulation rate of the septic tank sludge volume and the percolation rate in the sand columns underneath the septic tank units were not impaired.

7.3.5 Polycarboxylates

Polycarboxylates used in detergents are homopolymers of acrylic acid (hereafter abbreviated as P(AA) followed by the molecular weight) and copolymers of acrylic acid and maleic acid (abbreviated as P(AA-MA) followed by the molecular weight). These are water-soluble, linear polymers with the basic building block of acrylic acid, and are commonly referred to as polycarboxylates. The mean molecular weights of these polymers are in general between 1000 and 100 000.

Homopolymers of acrylic acid

$$-\left[-CH_2-\underset{\underset{COOH\,(Na)}{|}}{CH}-\right]_n-$$

Copolymers of acrylic and maleic acid

Environmental fate

The environmental behaviour of polycarboxylates is to a large extent influenced by their adsorptive properties towards solid surfaces, due to electrostatic effects, and their tendency to form insoluble calcium complexes. The latter tendency increases with the molecular weight; P(AA)1000 forms virtually no calcium precipitate, whereas P(AA-MA)70 000 is completely precipitated in a solution with excess calcium — almost always the case in natural matrices with polycarboxylate concentrations in the lower milligram per litre range in sewage and in the lower microgram per litre range in surface waters.

Biodegradation of polycarboxylates has been tested in different environmental matrices: activated sludge suspensions, river water and sludge amended soil (Opgenorth, 1992). In general their biodegradation potential is rather low.

In respirometric tests (BOD5/ThOD, BOD30/ COD, modified MITI and closed bottle test) less than 20% mineralization was found with various polycarboxylates, the biodegradable fraction presumably being low molecular weight constituents or impurities. Also, in batch-activated sludge tests with higher bacterial concentrations, less than 20% mineralization was found, except for P(AA)1000 with 43% $^{14}CO_2$ production.

The partitioning of P(AA-MA)70 000 in a batch-activated sludge test system was found to be c. 4% evolved as $^{14}CO_2$ and >80% adsorbed on sludge solids (Yeoman *et al.*, 1990). In a sewage treatment plant model system the distribution of the same polymer was 2–5% $^{14}CO_2$ evolution (pulse and continuous dosing), >90% adsorption on sludge and 2–3% remaining in the supernatant. The biodegradable fraction is again assumed to be the low molecular weight content of the polycarboxylate (Schumann, 1990).

Mineralization of polycarboxylates in river water is <20%, except for P(AA)1000 in pre-adapted river water with 63% $^{14}CO_2$ production

and in river water containing sediments with 58% $^{14}CO_2$ production. Biodegradation by soil micro-organisms is also low (<20%), except again for P(AA)1000 with 35% mineralization. Polycarboxylates undergo only slight (<5%) anaerobic degradation. Studies with P(AA-MA)70 000 showed that the polymer remained to 95% on the digester sludge.

The removal of polycarboxylates has been tested in several relevant water treatment processes (Opgenorth, 1992). During primary clarification, removal depends on molecular weight. In a model dynamic settling tank of 2 h hydraulic residence time P(AA)4500 was removed to 13%, P(AA-MA)12 000 to 8% and P(AA-MA)70 000 to 29%, compared with 70–84% total suspended solids removal.

The removal in activated sludge treatment is also a function of molecular weight. Table 7.9 gives the percentage elimination in the semi-continuous activated sludge (SCAS) test, a screening study, and the continuous activated sludge (CAS) test. The relatively high elimination extent of P(AA)1000 in the SCAS test is mainly due to biodegradation when compared with the percentage mineralization in the batch-activated sludge test mentioned before. However, the removal of the remaining polycarboxylates is mainly caused by adsorption/precipitation of the calcium-polymer complex. Elimination in the CAS test is generally somewhat lower compared with the screening test, simulating more closely the activated sludge treatment process. For P(AA)4500 quite different levels of removal were found in separate studies, presumably due to different influent concentrations of the polymer; the higher removal was found with a lower influent concentration (Schaefer *et al.*, 1991). Adding 100 mg/l $FeCl_3$ to simulate simultaneous phosphorus precipitation in activated sludge treatment achieved a considerably higher elimination (>90%) of the tested P(AA)4500 and P(AA-MA)12 000. The same will be the case for other polycarboxylates, especially those of higher molecular weight.

In lysimeter tests P(AA)4500, P(AA-MA)12 000 and P(AA-MA)70 000 spiked into simulated groundwater were percolated through a medium-grained sandy soil; 84–93% of the polycarboxylates was non-mobile, and remained within the

Table 7.9 Removal of polycarboxylates in activated sludge simulation tests

Polymer	SCAS test (%)[a]	CAS test without $FeCl_3$ (%)	CAS test with $FeCl_3$ (%)
P(AA)1000	45	16[a]	n.d.
P(AA)2000	21	15[a]	n.d.
P(AA)4500	40	22[a], 76[b]	95[a]
P(AA)10 000	58	n.d.	n.d.
P(AA)60 000	93	n.d.	n.d.
P(AA-MA)12 000	83	74[a]	96[a]
P(AA-MA)70 000	95	87[a], 94[c], 97–98[d]	n.d.

n.d., Not determined.
[a] Average of several tests (Hennes, 1991).
[b] Schaefer *et al.*, 1991.
[c] Opgenorth, 1987.
[d] Continuous dosing and pulse loading (Schumann, 1990).

first 15 mm of the 100 mm-thick soil layer. The small mobile fraction in the lysimeter eluate consisted of the low molecular weight, biodegradable polymer fraction which would not reach the soil environment via sewage sludge. In a similar study, P(AA-MA)70 000 adsorbed onto sewage sludge was applied on a soil column and remained completely in the upper soil layer (Opgenorth, 1992).

In case small amounts (lower microgram per litre range) of polycarboxylates reach raw drinking water, flocculation studies with P(AA)4500, P(AA-MA)12 000 and P(AA-MA)70 000 have shown that more than 90% are eliminated with aluminium and iron salts.

Table 7.10 Aquatic toxicity of polycarboxylates (EC/LC_{50}; mg/l)[a]

Polymer	Bacteria[b]	*Daphnia magna* 48 h	Fish 96 h
P(AA)1000	> 100	> 200	> 200
P(AA)2000	> 100	> 200	> 200
P(AA)4500	> 100	> 200	> 200
P(AA)10 000	> 1000	> 770	> 1000
P(AA-MA)12 000	> 100	> 200	> 200
P(AA-MA)70 000	> 200	> 100	> 100

[a] Hennes, 1991.
[b] O_2 consumption and glucose uptake of activated sludge.

Environmental effects

The aquatic and terrestrial toxicities of polycarboxylates are very low. All effect concentrations found with various test organisms are well below any expected environmental concentration. Table 7.10 summarizes acute aquatic toxicity tests with activated sludge bacteria, *Daphnia magna* and fish (zebra fish, *Brachydanio rerio*; bluegill sunfish, *Lepomis macrochirus*; and golden ide, *L. idus melanotus*). The indicated concentrations are the maximum tested concentrations. They vary depending on the particular test or the test substance, but this need not

indicate a difference in toxicity. For P(AA)4500 also, the acute toxicity to chironomid larvae (*Chironomus riparius*) was tested with 96 h-LC_{50} of 4500 mg/kg dry matter (Hennes, 1991).

Subchronic and chronic toxicity to aquatic organisms has been tested with algae, fish, *Hydra littoralis*, and *Daphnia magna* (Opgenorth, 1987; Schumann, 1990; Hennes, 1991). The 96 h-EC_{10} with *Scenedesmus subspicatus* for P(AA)4500 is 180 mg/l, and for P(AA-MA)70 000 ranges between 32 mg/l and > 200 mg/l in different tests. The 4-week NOEC with zebra fish (*Brachydanio rerio*) for P(AA)4500 was 450 mg/l, and the 6 week NOEC for P(AA-MA)70 000 was 40 mg/l (highest concentrations tested in an early life stage test). The 14-day EC_{10} for inhibition of colony growth of *Hydra littoralis* was 40 mg/l for P(AA-MA)70 000. The 21-day reproduction tests with *Daphnia magna* yielded a NOEC of 450 mg/l for P(AA)4500 and some divergent results for P(AA-MA)70 000, ranging from 1.3 to 350 mg/l. This can be explained by the fact that at low concentrations the polycarboxylate is precipitated as a calcium salt with the excess water hardness, and physical rather than toxic effects are responsible for *Daphnia* (and algal) toxicity at lower concentrations. In reality this polycarboxylate is preferentially adsorbed to solids and therefore not present in the surface waters in the form it was tested. The lowest NOEC found is still considerably higher than estimated surface water concentrations (in the lower microgram per litre range).

For the terrestrial environment, toxicity to higher plants and the earthworm (*Eisenia fetida*) has been tested (Opgenorth, 1992). In growth tests with corn, soybeans, wheat and grass, P(AA)4500 had no adverse effect on germination, growth or yield up to 225 mg/kg soil (highest concentration tested); 400 mg/kg soil of P(AA-MA)70 000 had no effect on the growth or the appearance of oats. With earthworms, an LC_0 of 1600 mg/kg soil is reported for P(AA-MA)70 000.

Bioaccumulation of polycarboxylates is unlikely because of the high water solubilities of the parent compounds and their propensity to form insoluble calcium salts in natural waters. Besides, absorption through biological membranes is assumed to occur only for substances with a molecular weight less than 600.

Polycarboxylates do not interfere with the operation of sewage treatment plants at relevant concentrations in municipal sewage. First-effect concentrations of polycarboxylates with different molecular weights on the settling behaviour of suspended solids in primary clarification, on the simultaneous phosphorus precipitation in activated sludge treatment and on the dewatering of digested sewage sludge were 10 mg/l or higher in waste water (see Section 7.2.5). In experiments on the coagulation of raw sewage by $FeCl_3$, first-effect concentrations of 3 mg/l for P(AA)4500, 10 mg/l for other P(AA) and 30 mg/l for P(AA-MA) were observed (Hennes, 1991). A field trial of several weeks with P(AA-MA)70 000 (10 mg/l influent concentration) showed no adverse effects on the activated sludge treatment with respect to the relevant parameters DOC-, COD- and BOD_5-removal, effluent concentrations of settleable solids and the sludge characteristic. Anaerobic digestion of sewage sludge is also not disturbed as shown in qualitative tests with P(AA-MA)70 000 (Opgenorth, 1992).

The complex binding capacity of polycarboxylates towards heavy metals is low compared with typical complexing agents such as NTA and EDTA. Several batch tests showed that the mobility of heavy metals in activated sludge treatment, river sediments and soils is not affected at polymer concentrations of at least 10 mg/l in suspensions of these different solids, which is higher than realistic environmental concentrations (cf. Section 7.2.5) (Opgenorth, 1989; Hennes, 1991). In addition, the removal of heavy metals in the presence of P(AA-MA)70 000 (10 mg/l influent concentration) was tested in two model STPs operated in parallel. No differences in the elimination of Cr, Cu, Ni, Pb or Zn were observed; cadmium and mercury concentrations were at the detection limit in both plant effluents (Opgenorth, 1987).

7.3.6 Perborate

The two main forms of perborate used are sodium perborate tetrahydrate, $NaBO_3.4 H_2O$, and sodium perborate monohydrate, $NaBO_3.H_2O$. The basic structural form is:

$$\left[\begin{array}{c} HO \\ HO \end{array} B \begin{array}{c} O-O \\ O-O \end{array} B \begin{array}{c} OH \\ OH \end{array} \right]^{2-} 2Na^+ . x H_2O$$

Where $x = 1$ and 4

Environmental fate

After its use in the household, boron contained in detergents and cleaning products is discharged into domestic sewage. Since no elimination takes place from sewage during the treatment process, all the boron from domestic sources is released to surface waters receiving sewage effluent.

The dominant species in environmental aqueous solutions are boric acid and the monomeric borate anion. The equilibrium distribution in surface waters depends mainly on pH, i.e. at freshwater pH values the dominant species is boric acid. Polyborate species occur only at significant concentrations far above environmentally relevant levels. The presence of metal cations enhances the dissociation of boric acid; e.g. in marine water the predominant species are borate and its metal complexes. Removal of boron from the aqueous phase takes place via adsorption on suspended clay particles which undergo sedimentation. However, this seems not to be a very effective removal mechanism in surface waters where boron distribution favours the aqueous phase.

Boron is strongly held by the solid substances of soils where the ratio of mineral matter to porewater is orders of magnitude higher than in surface waters. The behaviour of boron in the soil is determined by factors such as boron concentration in the soil solution, soil pH, cation exchange capacity, type of clay, soil moisture, etc. Generally, only < 5% of the boron content of soils is available for plant uptake. Therefore, a boron deficiency often occurs in leaching environments. On the other hand, irrigation schemes in arid regions, which provide higher boron concentrations than occur naturally in porewater to soils of low clay content, can considerably increase the available boron.

Environmental effects

Levels of boron found to be toxic to aquatic organisms are generally above concentrations measured in surface waters. Also, significant differences in the concentrations causing toxicity in laboratory tests and in natural waters have been observed, e.g. as described below for rainbow trout (*Oncorhynchus mykiss*). Boron is an essential nutrient for the growth of vascular plants and some algae, and has been found to be beneficial at low levels to other freshwater organisms. Often, the levels necessary for a healthy development and those giving a toxic response are relatively close to each other; in particular in the case of plants. In soil, 10−100 mg/kg boron was found to be optimal for micro-organisms; only at very high concentrations (> 1000 mg/kg boron) did an

adverse effect on growth and the population's composition occur (Butterwick *et al.*, 1989; Eisler, 1990).

Acute, subchronic and chronic studies on different aquatic organisms (algae, invertebrates, amphibians and fish) are summarized in Table 7.11. As an essential nutrient for freshwater and marine diatoms boron is incorporated into the shell. Toxic effects on *Chlorella* and marine phytoplankton were found only at concentrations much higher than occur naturally in the environment. Adverse effects on *Daphnia magna* growth and development also occur only at rather high concentrations, e.g. LOEC is 13.6 mg/l. The low values on *Daphnia* toxicity reported in Table 7.11, representing a threshold concentration for immobilization, were measured in sodium perborate solutions, and it is not reported whether any pH adjustment to natural values had been made. The wide range of effect concentrations reported for the same kind of test suggests a very flat dose−response curve that makes it difficult to establish precisely an effect concentration.

Several freshwater fish, mainly channel catfish (*Ictalurus punctatus*), goldfish (*Carassius auratus*), rainbow trout (*Oncorhynchus mykiss*), and marine fish, in particular coho salmon (*Oncorhynchus kisutch*) and dab (*Limanda limanda*), have been tested in different life stages. The early life stages of the rainbow trout tested in reconstituted water gave the lowest response (smallest NOEC and LOEC values in Table 7.11). Flat concentration−response relationships often influence the definition of the NOEC and LOEC values; e.g. from a consistent set of studies with early life stages of rainbow trout (*Oncorhynchus mykiss*) LOEC values were 0.1 − > 18 mg/l boron (Black *et al.*, 1993). When tested in natural waters the embryo larval stages of trout were found to be much less sensitive to boron. Levels of 0.75 mg/l boron (the natural background level in the tested medium) and even 17 mg/l boron (highest concentration tested) did not cause any adverse effect in natural waters. Indeed, in the wild, healthy rainbow trout have been found in waters with up to 13 mg/l boron, and trout hatcheries with water supplies of 0.02−1 mg/l boron are known.

As mentioned above, boron is an essential element for the development of vascular plants,

Table 7.11 Boron toxicity to aquatic organisms. Data largely from Raymond & Butterwick, 1992

Organism	Concentration (mg/l boron)
Algae	
freshwater	50–100 (toxic concentration)[a]
marine	30–50 (50% and 26% growth reduction)[b]
	10 (NOEC-growth)[b]
Daphnia magna	133–226 (LC_{50})
	6.4 (21 days-NOEC)
	13.6 (21 days-LOEC)
	<0.38–<240[c]
Other invertebrates	1.0 (toxicity threshold)[d]
	25–125 (92% and 100% mortality)[e]
Amphibians	7.04–48.7 (7 days-NOEC)[f]
	9.60–96.0 (7 days-LOEC)[f]
Fish	
freshwater	4.6–31.45 (24–144 h medium tolerance limit = LC_{50})
	0.001–26.5 (9–32 days-NOEC)[g]
	0.01–49.7 (9–32 days-LOEC)[g]
	0.75–24 (30–60 days-NOEC)[h]
	1.0–88 (30–60 days-LOEC)[h]
marine	12.2–88.3 (24–283 h LC_{50})

[a] Several *Chlorella* species.
[b] 19 different species of marine phytoplankton.
[c] Threshold concentration for immobilization.
[d] Protozoan (*Entosiphon sulcatum*).
[e] Mosquito larvae (*Anopheles quadrimaculatus*).
[f] Embryo-larval stages of Fowler's toad (*Bufo fowleri*) and Leopard frog (*Rana pipiens*).
[g] Embryo-larval stages and freshly distilled eggs of different fish in reconstituted water.
[h] Early life stages of rainbow trout (*Oncorhynchus mykiss*) and fathead minnow (*Pimephales promelas*) in natural waters.

but it is also toxic at higher levels. There is no evidence for any boron requirements of other terrestrial organisms. Boron deficiency seems to occur more often than boron toxicity, and thus more studies have been undertaken on the boron requirements of crops. Table 7.12 summarizes studies on many crops resulting in specific threshold concentrations, i.e. the maximum concentration that a given plant species tolerates without showing any adverse effects. Here, wide ranges of threshold concentrations are given, each representing many crop species. Amongst the most boron-sensitive crops are fruit. Since fruit plants, especially citrus, are often subject to extensive irrigation schemes, particular attention was paid to the boron content of irrigation water. Attempts to develop guidelines for boron limits in irrigation water must take into account not only the tolerance of the crops but also the specific climatic and soil conditions of the considered area. Therefore, different countries or regions developed their own irrigation water quality guidelines for boron. Table 7.13 shows the UK irrigation guidelines as an example, which take into account total boron loadings to soils, i.e. 2.0, 3.0 and 4.0 kg/ha per year boron are considered to be safe for sensitive, semitolerant and tolerant crops (MAFF, 1981).

Boron does not affect aerobic microorganisms in sewage treatment plants below 20 mg/l boron, and no significant inhibition of anaerobic sludge

Table 7.12 Threshold concentration ranges for boron in soil water according to crop species. Derived from Raymond & Butterwick, 1992

Tolerance group	Threshold concentration range (mg/l boron)
Sensitive crops	0.30–1.01
Semitolerant crops	1.01–4.00
Tolerant crops	4.00–15.03

Table 7.13 Safe boron concentrations (mg/l boron) in irrigation water according to seasonal water need (MAFF, 1981)

Tolerance group	Seasonal irrigation need (mm)		
	50	100	200
Sensitive crops	4.0	2.0	1.0
Semitolerant crops	6.0	3.0	1.5
Tolerant crops	8.0	4.0	2.0

digestion occurs at levels below 200 mg/l boron. At concentrations above 100 mg/l, which do not occur in domestic sewage, though, boron may initially affect sludge settleability (Butterwick *et al.*, 1989).

ACKNOWLEDGEMENT

The authors thank Ms Didem Kiyidelen for her kind help in producing the tables for this chapter.

REFERENCES

Ahel, M. & Giger, W. (1985) Determination of non-ionic surfactants of the alkylphenol type with high-performance liquid chromatography. *Anal. Chem.* **57**, 1577–1583, 2584–2590.

Allen, H.E., Cho, S.H. & Neubecker, T.A. (1983) Ion exchange and hydrolysis of type A zeolite in natural waters. *Water Res.* **17**, 1871–1879.

Baumann, E.R., Hopping, W.D. & Warner, F.D. (1981) Field evaluation of the treatability of type A zeolite in a trickling filter plant. *Water Res.* **15**, 889–901.

Black, J.A., Barnum, J.B. & Birge, W.J. (1993) An integrated assessment of the biological effects of boron to the rainbow trout. *Chemosphere* **26**, 1383–1413.

Boethling, R.S. & Lynch, D.G. (1992) Quaternary ammonium surfactants. In: *Anthropogenic Compounds*, Vol. 3, Part F. *Handbook of Experimental Chemistry* (Ed.-in-Chief O. Hutzinger), pp. 145–177. Springer-Verlag, Berlin.

Broekaert, J.A.C. & Leis, F. (1979) An injection method for the sequential determination of boron and several metals in waste water samples by inductively-coupled plasma atomic emission spectrometry. *Analyt. Chim. Acta* **109**, 73–83.

Brown, D., De Henau, H., Garrigan, J., Gerike, P., Holt, M., Keck, E., Kunkel, E., Matthijs, E., Waters, J. & Watkinson, R.J. (1986) Removal of nonionics in a sewage treatment plant. *Tenside Surfactants Detergents* **23**, 190–195.

Brown, D., De Henau, H., Garrigan, J., Gerike, P., Holt, M., Kunkel, E., Matthijs, E., Waters, J. & Watkinson, R.J. (1987) Removal of nonionics in sewage treatment plants II. *Tenside Surfactants Detergents* **24**, 14–19.

Butterwick, L., de Oude, N.T. & Raymond, K. (1989) Safety assessment of boron in aquatic and terrestrial environments. *Ecotoxicol. Environ. Safety* **17**, 339–371.

Christophliemk, P., Gerike, P. & Potokar, M. Zeolites. In: *Anthropogenic Compounds*, Vol. 3, Part F. *Handbook of Experimental Chemistry* (Ed.-in-Chief O. Hutzinger), pp. 205–228. Springer-Verlag, Berlin.

de Oude, N.T. (Ed.) (1992) *Anthropogenic Compounds*, Vol. 3, Part F. *Handbook of Environmental Chemistry* (Ed.-in-Chief O. Hutzinger). Springer-Verlag, Berlin.

Eisler, R. (1990) Boron hazards to fish, wildlife and invertebrates: a synoptic review. *Contaminant Hazard Reviews*, Rep. 20. Biological Report 85 (1.20). Fish and Wildlife Service, US Department of the Interior.

Evans, K.A., Dubey, S.T., Kravetz, L., Dzidic, I., Gumulka, J., Mueller, R. & Stork, J.R. (1994) Quantitative analysis of linear primary alcohol ethoxylate surfactants in environmental samples by thermospray LC/MS (liquid chromatography/mass spectrometry). *Anal. Chem.* (accepted for publication).

Fachgruppe Wasserchemie (1990) Stellungnahme Umweltverträglichkeit von Polycarboxylaten aus Waschmitteln. Gesellschaft Deutscher Chemiker Muenchen.

Fairchild, J.F., Dwyer, F.J., La Point, T.W., Burch, S.A. & Ingersoll, C.G. (1993) Evaluation of a laboratory-generated NOEC for linear alkylbenzene sulphonate in outdoor experimental streams. *Environ. Toxicol. Chem.* **12**, 1763–1775.

Federle, T.W. & Ventullo, R.M. (1990) Mineralization of surfactants by the microbiota of submerged plant detritus. *Appl. Environ. Microbiol.* **56**(2), 333–339.

Federle, T.W. & Schwab, B.S. (1992) Mineralization of surfactants in anaerobic sediments of a laundromat wastewater pond. *Water Res.* **26**(1), 123–127.

Fendinger, N.J., Begley, W.M., McAvoy, D.C. & Eckhoff, W.S. (1992) Determination of alkyl sulfate in natural waters. *Environ. Sci. Technol.* **26**(12), 2493–2498.

Fischer, W.K., Gerike, P. & Kurzyca, G. (1978) Sodium aluminium silicate in detergents — the effect on the aerobic biological purification of sewage. *Tenside Surfactants Detergents* **15**, 60–64.

Fischer, W.K. & Gode, P. (1977) Testing of sodium aluminium silicates as detergent additives for their toxicity to aquatic organisms. *Vom Wasser* **49**, 11–26.

Gerike, P., Klotz, H., Kooijman, J.G., Matthijs, E. & Waters, J. (1994) The determination of dihardenedtal-lowdimethyl ammonium compounds (DHTDMAC) in environmental matrices using trace enrichment techniques and high performance liquid chromatography with conductivity detection. *Water Res.* (accepted for publication).

Gilbert, P.A. & Pettigrew, R. (1984) Surfactants and the environment. *Int. J. Cosmet. Sci.* **6**, 149–158.

Guhl, W. & Gode, P. (1989) Correlations between lethal and chronic/biocenotic effect concentration of surfactants. *Tenside Surfactants Detergents* **26**, 282–287.

Hand, V.C., Rapaport, R.A. & Pittinger, C.A. (1990a) First validation of a model for the adsorption of linear alkylbenzenesulfonate (LAS) to sediment and comparison to chronic effects data. *Chemosphere* **21**(6),

741–750.

Hand, V.C., Rapaport, R.A. & Wendt, R.H. (1990b) Adsorption of dodecyltrimethylammonium chloride (C_{12}TMAC) to river sediment. *Environ. Toxicol. Chem.* **9**, 467–471.

Hennes, E.C. (1991) Fate and effects of polycarboxylates in the environment. Procter & Gamble, *Strombeek-Bever*, **B**, 1–33.

HMSO (1980) *Boron in waters, effluents, sewage and some solids. Methods for the examination of waters and associated materials.* HMSO, London.

Holman, W.F. & Hopping, W.D. (1980) Treatability of type A zeolite in wastewater – part II. *J. Water Pollut. Contr. Fed.* **52**(12), 2887–2905.

Holt, M.S., Matthijs, E. & Waters, J. (1989) The concentrations and fate of linear alkylbenzene sulphonate in sludge amended soils. *Water Res.* **23**(6), 749–759.

Holt, M.S., Mitchell, G.C. & Watkinson, R.J. (1992) The environmental chemistry, fate and effects of nonionic surfactants. In: *Anthropogenic Compounds*, Vol. 3, Part F. *Handbook of Experimental Chemistry* (Ed.-in-Chief O. Hutzinger), pp. 89–144. Springer-Verlag, Berlin.

Jakobi, G. & Loehr, A. (1987) Ecology. In: *Detergents and Textile Washing*, pp. 167–187. VCH Verlagsgesellschaft, Weinheim.

Kimerle, R.A. (1989) Aquatic and terrestrial ecotoxicology of linear alkylbenzene sulfonate. *Tenside Surfactants Detergents* **26**(2), 169–176.

King, J.E., Hopping, W.D. & Holman, W.F. (1980) Treatability of type A zeolite in wastewater – part I. *J. Water Pollut. Contr. Fed.* **52**(12), 2875–2886.

Knaebel, D.B., Federle, T.W. & Vestal, J.R. (1990) Mineralization of linear alkylbenzene sulfonate (LAS) and linear alcohol ethoxylate (LAE) in 11 contrasting soils. *Environ. Toxicol. Chem.* **9**, 981–988.

Knezovich, J.P., Lawton, M.P. & Inouye, L.S. (1989) Bioaccumulation and tissue distribution of a quaternary ammonium surfactant in three aquatic species. *Bull. Environ. Contam. Toxicol.* **42**, 87–93.

Larson, R.J., Federle, T.W., Shrimp, R.J. & Ventullo, R.M. (1989) Behaviour of linear alkylbenzene sulfonate (LAS) in soil infiltration and groundwater. *Tenside Surfactants Detergents* **26**(2), 116–121.

Lewis, M.A. & Wee, V.T. (1983) Aquatic safety assessment of cationic surfactants. *Environ. Toxicol. Chem.* **2**, 105–118.

MAFF (1981) *Water quality for crop irrigation. Guidelines on chemical criteria.* Leaflet 776. Ministry of Agriculture, Fisheries and Food, UK.

Maki, A.W. & Macek, K.J. (1978) Aquatic environmental safety assessment for a nonphosphate detergent builder. *Environ. Safety Technol.* **12**(5), 573–580.

Mann, A.H. & Reid, V.W. (1971) Biodegradation of synthetic detergents. Evaluation by community trials. 2. Alcohol and alkylphenol ethoxylates. *J. Am. Oil Chem. Soc.* **48**, 794–797.

Matthijs, E. & De Henau, H. (1987) Analysis of mono-alkylquaternaries and assessment of their fate in domestic wastewater, river waters and sludges. *Vom Wasser* **69**, 73–83.

Matthijs, E. & Hennes, E.C. (1991) Determination of surfactants in environmental samples. *Tenside Surfactants Detergents* **28**(1), 22–27.

McCormick, P.V., Cairns Jr, J., Belanger, S.E. & Smith, E.P. (1991) Response of protistan assemblages to a model toxicant, the surfactant C12-TMAC (dodecyltrimethyl ammonium chloride), in laboratory streams. *Aquat. Toxicol.* **20**, 41–70.

Oba, K., Miura, K., Sekiguchi, H., Yagi, R. & Mori, A. (1976) Microanalysis of anionic surfactants in waste water by infrared spectroscopy. *Water Res.* **10**, 149–155.

Opgenorth, H.J. (1987) Umweltvertraeglichkeit von Polycarboxylaten. *Tenside Surfactants Detergents* **24**, 366–369.

Opgenorth, H.J. (1989) Polycarboxylate in Abwasser und Klaerschlamm. In: *Muenchener Beitraege zur Abwasser-, Fischerei- und Flussbiologie*, Vol. 44: *Umweltvertraeglichkeit von Wasch- und Reinigungsmitteln*, pp. 338–351. Oldenburg, Muenchen.

Opgenorth, H.J. (1992) Polymeric materials, polycarboxylates. In: *Anthropogenic Compounds*, Vol. 3, Part F. *Handbook of Experimental Chemistry* (Ed.-in-Chief O. Hutzinger), pp. 337–350. Springer-Verlag, Berlin.

Osburn, Q.W. (1982) Analytical method for a cationic fabric softener in waters and wastes. *J. Am. Oil Chem. Soc.* **59**, 453.

Painter, H.A. (1992) Anionic surfactants. In: *Anthropogenic Compounds*, Vol. 3, Part F. *Handbook of Experimental Chemistry* (Ed.-in-Chief O. Hutzinger), pp. 1–88. Springer-Verlag, Berlin.

Palmisano, A.C., Schwab, B.S. & Maruscik, D.A. (1991) Seasonal changes in mineralization of xenobiotics by stream microbial communities. *Can. J. Microbiol.* **37**, 939–948.

Popenoe, D.D., Morris III, S.J., Horn, P.S. & Norwood, K.T. (1994) Determination of alkyl sulfates and alkyl ethoxysulfates in waste water treatment plant influents and effluents and in river water using liquid chromatography/ion-spray mass spectrometry. *Anal. Chem.*, (accepted for publication).

Rapaport, R.A. & Eckhoff, W.S. (1990) Monitoring linear alkyl benzene sulfonate in the environment: 1973–1986. *Environ. Toxicol. Chem.* **9**, 1245–1257.

Raymond, K. & Butterwick, L. (1992). In: *Anthropogenic Compounds*, Vol. 3, Part F. *Handbook of Experimental Chemistry* (Ed.-in-Chief O. Hutzinger), pp. 287–318. Springer-Verlag, Berlin.

Roland, W.A. & Schmid, R.D. (1978) Sodium aluminum silicates in detergents – tests on the ion-exchange behaviour towards heavy metal ions in wastewater. *Tenside Surfactants Detergents* **15**, 281–285.

Roland, W.A., Graupner, W. & Hoffmann, W. (1979) Praxisversuche zum Ablagerungsverhalten von Zeolith A in der Kanalisation. In: *Die Pruefung des Umweltverhaltens von Natrium-Aluminium-Silikat Zeolith A als Phosphatersatzstoff in Wasch- und Reinigungsmitteln.* pp. 34–38. Materialien 4/79. Erich Schmidt Verlag, Berlin.

Savitsky, A.C., Wiers, B.H. & Wendt, R.H. (1981) Adsorption of organic compounds from dilute aqueous solutions onto the external surface of type A zeolite. *Environ. Sci. Technol.* **15**(10), 1191–1196.

Schaefer, E.C., Crapo, K.C., Orvos, D.R. & Williams, R.T. (1991) Assessing the removal of acrysol LMW-45N during secondary wastewater treatment. Weston study 90–059, pp. 2–3. Rohm and Haas. Spring House, PA.

Schoeberl, P. & Huber, L. (1988) Oekologisch relevante Daten von nichttensidischen Inhaltsstoffen in Wasch- und Reinigungsmitteln. *Tenside Surfactants Detergents* **25**, 99–107.

Schoeberl, P., Bock, K.J. & Huber, L. (1988) Oekologisch relevante Daten von Tensiden in Wasch- und Reinigungsmitteln. *Tenside Surfactants Detergents* **25**, 86–98.

Schroeder, U., Horn, D. & Wassmer, K.H. (1991) Bestimmung von Polycarboxylaten mit Hilfe der Poly-elektrolyt-Titration in Wasserproben. *Seifen, Oele, Fette, Wachse* **117**, 311–314.

Schumann, H. (1990) Elimination von ^{14}C-markierten Polyelektrolyten in biologischen Laborreaktoren. Fortschritt-VDI Berichte, Reihe 15. Umwelttechnik 81 pp. 1–190, VDI, Duesseldorf.

Shimp, R.J. (1989) LAS biodegradation in estuaries. *Tenside Surfactants Detergents* **26**(6), 390–393.

Shimp, R.J. & Schwab, B.S. (1991) Use of a flow-through in situ environmental chamber to study microbial adaptation processes in riverine sediments and periphyton. *Environ. Toxicol. Chem.* **10**, 159–167.

Simms, J.R., Keough, T., Ward, S.R., Moore, B.L. & Bandurraga, M.M. (1988) Quantitative determination of trace levels of cationic surfactants in environmental matrices using fast atom bombardment mass spectrometry. *Anal. Chem.* **60**(23), 2613–2620.

Stalmans, M., Matthijs, E. & de Oude, N.T. (1991) Fate and effect of detergent chemicals in the marine and estuarine environment. *Water Sci. Technol.* **24**(10), 115–126.

Swisher, R.D. (1987) *Surfactant Biodegradation*, 2nd edn. Marcel Dekker, New York.

Topping, B.W. & Waters, J. (1982) Monitoring of cationic surfactants in sewage treatment plants. *Tenside Surfactants Detergents* **19**, 164.

Ventullo, R.M., Lewis, M.A. & Larson, R.J. (1989) Response of aquatic microbial communities to surfactants. In: *Aquatic Toxicology and Environmental Fate*, Vol. 11 (Eds G.W. Suter & M.A. Lewis), pp. 41–58. ASTM STP 1007.

Versteeg, D.J., Feijtel, T.C.J., Cowan, C.E., Ward, T.E. & Rapaport, R.A. (1992) An environmental risk assessment for DTDMAC in the Netherlands. *Chemosphere* **24**(5), 641–662.

Wassmer, K.H., Schroeder, U. & Horn, D. (1991) Characterization and detection of polyanions by direct polyelectrolyte titration. *Makromol. Chem.* **192**, 553–565.

Waters, J. & Garrigan, J.T. (1983) An improved micro-desulfonation/gas liquid chromatography procedure for the determination of linear alkylbenzene sulfonates in U.K. rivers. *Water Res.* **17**, 1549–1562.

Waters, J., Garrigan, J.T. & Paulson, A.M. (1986) Investigations into the scope and limitations of the bismuth active substances (Wickbold) for the determination of nonionic surfactants in environmental samples. *Water Res.* **20**(2), 247–253.

Woltering, D.M., Larson, R.J., Hopping, W.D., Jamieson, R.A. & de Oude, N.T. (1987) The environmental fate and effects of detergents. *Tenside Surfactants Detergents* **24**(5), 286–296.

Yeoman, S., Lester, J.N. & Perry, R. (1990) The partitioning of polycarboxylic acids in activated sludge. *Chemosphere* **21**, 443–450.

8: Pesticides

D.R. NIMMO AND L.C. McEWEN

8.1 INTRODUCTION

8.1.1 General definitions of chemicals or substances used as pesticides

What is a pesticide? Perhaps the most rudimentary definition is given in *Since Silent Spring* (Graham, 1970) as 'an agent used to kill pests'. However, this definition is too simple considering current technology. For instance, Graham was only considering the *synthetic chemical* agents that *kill* pests, yet the definition should also take into account new developments in naturally occurring pest control, such as chitin inhibitors, pheromones or pathogenic bacteria, or non-chemical pesticides such as light attraction— electrocution devices and any wood preservatives that prevent microorganisms, moulds and bacteria from decomposing wood. Also, later in this chapter, we discuss the use of specialized chemical pesticides that, instead of *killing* target pests, cause other actions, e.g. plants to defoliate before the crops are harvested or birds to emit distress calls that send the flock elsewhere. Chemicals that target various taxa will also be mentioned, such as acaracides, algicides, bactericides, fungicides, grain preservatives (fumigants to control a variety of microbiological and insect pests during the storage of grain), growth regulators for insects, herbicides, insecticides, nematicides, rodenticides and wood preservatives. There is also a host of chemicals that are not considered or classed as pesticides but are utilized in formulations to either enhance the action of a chemical or to perform a function that is much like a pesticide. Adjuvants, for example, are materials that are added to a pesticide mixture to improve its mixing and application or to enhance its performance. Emulsifiers are included to allow chemicals to be either soluble or to stick to the surfaces of the intended plant pest. Various inert substances are used as carriers to deliver chemicals to the intended targets, and synergists are substances that increase the activity of a pesticide as it is placed into the environment.

And finally, as a special note, most pesticides are formulated and marketed under the manufacturers' label. For example, chlorpyrifos is a pesticide that is often used alone or formulated and sold under such labels as Dursban, Lorsban, and others. For simplicity, we have tried to limit our discussion to only the active ingredients such as with chlorpyrifos.

Past concerns linking pesticides with pollution have been justified, but with a better understanding of the proper use of pesticides, coupled with changing patterns of human use that lean toward more environmentally labile or natural materials, unintended pesticide pollution should diminish. The crux of the problem, however, about pesticides as pollutants is the distinction between the target species that pesticides are designed to kill, and the non-target species inhabiting the treated areas, that are not intended to be killed or affected in some undesirable way. Frequently it is not possible to kill just the target species; other coexisting species are also affected. Also, the more stable pesticides tend to move into parts of the environment where they were not intended to be, e.g. to groundwater, or in the tissues of birds and mammals, as is the case for the residues of most chlorinated pesticides. Therefore, it is the *unintentional* aspects of pesticide use that result in environmental pollution. An 'ideal' pesticide should affect only target species, such as an unwanted plant, microorganism or insect, and then degrade immediately to its non-toxic elemental constituents.

It is difficult to identify the first published scientific report that dealt with pesticides as pollutants in waters. The first meeting of the North American Wildlife Conference, held in Washington DC in 1936 (US Government Printing Office, 1936), did not, in the discussions, mention pesticides when addressing the issue of water pollution. Topics dealt only with pollution abatement, cooperative efforts among various states, mining impacts, oil pollution in coastal waters, effects of oil on waterfowl, effects of pollution (mostly effluents or wastewaters) on fish, and discharges from pulp and paper mills. According to Tarzwell (1978), the first research on effects of pesticides was conducted in the late 1930s and early 1940s when Tennessee Valley Authority personnel began extensive field investigations into non-target effects of oil and Paris Green used on mosquito vectors of malaria. The first published paper on effects of modern-day pesticides was written by Goodnight (1942). However, in this paper, wood preservatives were not mentioned as being pesticides, but rather as industrial chemicals. Apparently, the connection between pesticides and pollution had not been made at that time. Interestingly, however, the research conducted by Goodnight on pentachlorinated chemicals barely preceded the advent of the modern-day expansion of pesticide development beginning with the first environmentally disastrous insecticide, DDT.

As background to this chapter it is useful to refer to some earlier reviews of pesticides which focus on effects on aquatic species (Nimmo, 1985, 1987). The first is more comprehensive, and contrasts effects of pesticides on freshwater species with those on marine species, comparing data from laboratory and field studies, and acute vs chronic effects. The second reviews the history of assessing the effects of pesticide use 25 years after Rachel Carson wrote *Silent Spring*, in 1962.

A comprehensive review of all the various types of available pesticides is not possible in this chapter. The exact chemical structure of all pesticides may not be known because of their derivation

from biological or natural sources, or they may be mixtures of substances, as with pheromones. In contrast, some pesticide chemical structures are well known because they are simply salts of metals, such as copper sulphate which is used to control algae. But for the large number of pesticides that can be represented by chemical structure, the most common elements are C, N, S, Cl, O, Br, P and F, as well as metals such as Cu, Sn, Mg, As and I. Carbon rings with ethyl or methyl groups and straight or branched chains of carbon with side groups containing Cl, S, N or CN are often found. Some chemicals are made up almost entirely of ring compounds, whereas others are predominantly straight-chain carbon linkages. One pesticide, mirex, has a box-like structure made up of carbon and chlorine atoms. Therefore, instead of a comprehensive listing of all pesticides, we shall begin our discussion with representative features of a few pesticides listed in the *Farm Chemicals Handbook* (1990) such as: (i) chemical structure, (ii) mode of action (if known) that makes them effective against target species, (iii) some unique features about their use, and (iv) why each is representative of a pesticide group. We shall continue the chapter with summaries of pesticide concentrations occurring in environmental samples, studies using single and multiple species, acute and chronic endpoints, and direct and indirect effects of pesticides on various test species. The latter portion of the chapter emphasizes field studies, risk assessment approaches, derivation of criteria and the application of pesticide standards.

8.1.2 Examples of structure and modes of action of representative pesticides

Organophosphate pesticides

Acephate, *O,S*-dimethyl acetylphosphoramido-thioate, is representative of systemic insecticides that interfere with transmission of nerve impulses. The list of insects for which acephate is recommended includes more than 40 species. Its chemical structure is:

Methamidophos, *O,S*-dimethyl phosphoramidothioate, is a more toxic metabolite of acephate that lacks the acetyl group. This small difference in chemical structure limits its use and therefore allows it to target only five species. Its chemical structure is:

Acetanilides

Alachlor, 2-chloro-2'-6'-diethyl-*N*-(methoxymethyl)-acetanilide, a widely used herbicide, was chosen to represent both the extensive uses and methods of application of various herbicides. Because alachlor is classified as a pre-emergent herbicide it is commonly used to control annual grasses and certain broad-leaf weeds in soybeans, corn, peanuts, dry beans, sunflowers and milo. The formulations can be applied by either ground or aerial equipment, in water or sprayable fluid fertilizers, or impregnated on dry bulk fertilizers. Because of its stability and extensive use, alachlor has been detected in groundwater. Its chemical structure is:

Carbamates

Maneb, manganese ethylenebisdithiocarbamate, is a well-known fungicide. It controls blights on potatoes, tomatoes, fruits, vegetables, and field crops including onions, tobacco, groundnuts and sugar beets. An interesting feature of maneb is its use in formulations containing other chemicals such as insecticides (lindane), fungicides (captan), or low concentrations of zinc, used as a plant nutrient. One formulation lists 19.2% maneb, 18.4% copper hydroxide plus added nutritional zinc. Its chemical structure is:

Pyrethroids

Fenvalerate, cyano (3-phenoxyphenyl) methyl 4-chloro-(1-methylethyl) benzeneacetate, is an example of the increasing use of pyrethroids — insecticides that are environmentally stable analogues of natural extracts derived from plants that have long been known to have insecticidal properties. The original extracts, called pyrethrums or pyrethrins, are derived from the chrysanthemum family. The pyrethroids, however, are 'engineered' variants that, while possessing the toxicty of the natural extracts, provide additional advantages (reduced cost of production, more efficient application, longer active life in the environment). Unfortunately, disadvantages include an extreme toxicity to many aquatic species. Most importantly, though, they retain the ability to degrade in sufficient time to minimize serious threats to most non-target species. Fenvalerate is used for a wide range of mass plantings including Christmas tree plantations, pine seed orchards and forest tree nurseries. It targets Lepidoptera, Diptera, Orthoptera, Hemiptera and Coleoptera. Its chemical structure is:

Growth regulators

Methoprene, isopropyl (2E-4E)-11-methoxy-3, 7, 11-trimethyl-2,4-dodecadienoate, is an example of increasingly used natural or 'chemically engineered' substances that mimic the pests' physiology. Compounds in this group are also sex attractants that confuse the pest species at the improper time for mating; some prevent the development to succeeding larval stages or preclude their emergence from aquatic environments; and others interfere with the laying down of cuticle in the succeeding life stage. The primary use of methoprene is as a selective mosquito larvicide, but it is also active against Coleoptera, Diptera, Hemiptera and Siphonoptera. Its chemical structure is:

Rodenticides

Warfarin, 3(a-acetonylbenzyl)-4-hydroxycoumarin, is designed specifically for terrestrial species, although an approach for addressing the question of impacts of rodenticides on both aquatic and terrestrial environments has been published (Brown *et al.*, 1988). It is applied in bait form and, in theory, has no water-borne connection. Warfarin is an anticoagulant for the control of rats and mice. The onset of death with warfarin is slow, and intake of treated bait must continue over time. The rodents usually do not become bait-shy after tasting the chemical, but succeeding generations develop resistance. Its chemical structure is:

Avi-repellents

Avitrol, 4-aminopyridine, is effective on the behavioural response of the target species. When ingested, it causes birds to emit vocal distress calls and signal physical distress to the other birds in the flock. It is used to frighten away blackbirds, cowbirds, crows, gulls, pigeons and starlings from areas of human concern. Avitrol is a highly toxic pesticide that can be used only by certified applicators. The acute oral LD_{50} for mallards is 4.4–5.2 mg/kg (Hudson *et al.*, 1984). When used properly in a ratio of one treated bait particle to 100 untreated particles, Avitrol is effective in repelling large numbers of pest birds with

little direct mortality to them. Its structural formula is:

Biorationals

Insecticidal agents that are increasing in use are often grouped together in what are called biorational, biological, microbial or so-called 'natural' agents used to infect various insect pests. Probably the best known of this group is the bacterium *Bacillus thuringiensis*, often referred to as Bt. One variety is used to control wax moth larval infestations in stored grains or honey combs; another variety is used against mosquito and blackfly larvae; whereas still another is used to control Lepidoptera larvae such as armyworms, cabbage loopers, gypsy moths and spruce budworms.

8.2 PESTICIDES AND THE ENVIRONMENT

8.2.1 Classes and amounts of pesticides applied to the environment

There are no exact estimates of the quantity of pesticides used in the world, but the data below illustrate current usage for a few countries.

A report from the Centre for Aquatic Weeds, Institute of Food and Agricultural Sciences (Joyce & Ramey, 1986), provided fate, persistence and movement of the major herbicides used in Florida. They listed the most widely used chemicals as endothal, 2,4-D, floridone, diquat, copper and glyphosate. Data for pesticide use by the United States Forest Service were obtained for fiscal year 1991 (Table 8.1). During the tabulation of the data it was evident that scores of pesticide types are used with many formulations that combine two and sometimes three chemicals. Heavy, widespread use of the bacterium *Bacillus thuringiensis* indicated an attempt to control the destructive gypsy moth biologically in forests, and accounted for the majority (in kilograms) of the total use of pesticides.

According to Pimentel *et al.* (1991), of the

Table 8.1 United States Forest Report, Pesticide-Use for Forest Management, fiscal year 1991 (US Department of Agriculture, 1992)

Pesticide type	Quantity used (kg)[a]
Insecticides	
Bacillus thuringiensis	839 841
Carbaryl	8 841
Malathion	723
Chlorpyrifos	260
Herbicides	
2,4-D	28 379
Triclopr	25 422
Hexazinone	6 980
Dicloram	6 569
Glyphosate	4 354
Fungicides and fumigants	
Methyl bromide	10 170
Dazomet	8 074
Borax[b]	7 194
Chlorpicrin	5 414

[a] Estimates for the (major) quantities used were rounded to the nearest US pound and converted to kilograms.
[b] Used for the control of Annosus root disease.

estimated 434 million kg of pesticides applied each year in the USA, 69% are herbicides, 19% are insecticides and 12% are fungicides (Table 8.2). A report by Gianessi & Puffer (1991) estimated that for the years 1986–1988 approximately 209 million kg of herbicide active ingredients were used in US agricultural production. Corn and soybean crop applications accounted for most of the herbicide use. Atrazine was the herbicide used most. Since their introduction, use of synthetic pesticides in the USA, based on weight in kilograms, has increased 33-fold (Pimentel *et al.*, 1991). Pesticide usage peaked in the mid-1970s, but some decrease has occurred since then. According to Pimentel *et al.* (1991), the decline in total amount applied is primarily due to the 10–100-fold increase in toxicity and effectiveness of new pesticides.

Pesticides are also widely used in Egypt, Africa and Asia. The most commonly used insecticide in the rich agricultural area surrounding Egypt's Lake Quarun is bayluscide, used for the eradication of *Bilharzia* snails. Lannate and dimethoate, insecticides used for controlling Egypt's agricultural pests, are also widely used (Academy of Scientific Research and Technology, 1981). In the temperate climates of Africa, DDT, dieldrin and endosulfan continue to be the primary insecticides used for controlling the tsetse fly (Goldenman, 1987). Asian rice fields are receiving applications of endosulfan, BHC, endrin, dieldrin, carbofuran, parathion and azinphos-ethyl to control the brown planthopper (Bull, 1982; Isensee & Tayaputch, 1986). Because the brown planthopper has developed resistance to so many insecticides, larger volumes and more frequent applications of these pesticides are typical.

Of the pesticides used in the UK for soft fruit crops (i.e. strawberries, blackberries and others), fungicides accounted for 56%; herbicides accounted for 22%; insecticides, 11%; acaricides, 9%; molluscides, 1%; and 1% remaining unaccounted for (Davis *et al.*, 1990). The three active ingredients that accounted for half of the fungicide treatments were dichlofluanid, bupirimate and chlorothalonil. The active ingredients most extensively used in herbicides were simazine, paraquat, propyzamide and diquat.

Trends in pesticide usage in nine Asian countries were reported by Soerjani (1988). In 1985, 760 million kg of pesticide formulations were produced in those countries and an additional 325 million kg were imported. Insecticides accounted for 62.5% and herbicides accounted for 21.4% of those materials. The associated problems of improper pesticide usage, human poisoning and environmental contamination are discussed. The incidence of anticholinesterase poisoning in Asian countries related to heavy use of organophosphates and other pesticides is reviewed by Phoon (1992). The lack of education and training for pesticide use in Third World countries is discussed by Boardman (1986). The need for improved practices such as integrated pest management (IPM) is world-wide (Kraus, 1988).

8.2.2 Transport of pesticides after application

Although the scientific community has not yet studied all aspects of pesticide transport, the following information about the fate of pesticides after application is generally accepted. The entry

Table 8.2 US hectarage[a] treated with pesticides (in kilograms) (Pimentel *et al.*, 1991). Numbers are in millions

Land-use category	Total hectares	All pesticides		Herbicides		Insecticides		Fungicides	
		Treated hectares	Quantity	Treated hectares	Quantity	Treated hectares	Quantity	Treated hectares	Quantity
Agricultural	472	114	320	86	220	22	62	4	38
Government and industrial[b]	150	28	55	30	44	n.a.	11	n.a.	n.a.
Forest	290	2	4	2	3	<1	1	n.a.	n.a.
Household	4	4	55	3	26	3	25	1	4
Total	916	148	434	121	293	26	99	5	42

[a] Total for hectarage treated with herbicides, insecticides and fungicides exceeds total treated hectares because the same land area can be treated several times with several classes of chemicals.

[b] Government and industrial.

n.a., Not available.

of pesticides into wetlands and waterways is either by direct application or inadvertent drifting of droplets and vapours via storm runoff and seepage. In consequence, pesticides are being identified in treated effluents from municipal wastewater treatment plants (Amato *et al.*, 1992).

The likelihood of pesticide contamination is lowest if it is applied in a dry bait form, higher if applied as a liquid spray from a tractor-mounted boom sprayer and highest when applied aerially from a spray plane. With the latter, indirect contamination is dependent on meteorological events and the specific properties of the chemicals. After application, pesticides are lost through volatilization, degradation by chemical and biological processes, removal in runoff water when dissolved in solution, or attached to particles of soil or organic materials (Table 8.3). Pesticides are also transported to other parts of the environment in the tissues of animals that have come in direct contact with the chemicals, or have fed on contaminated plants or animals.

One of the most important factors in the movement of pesticides is precipitation. The amounts and frequency of rainfall, snow, or in some cases irrigation water, affect the volume of runoff and the leaching and transport of pesticides. In the cases of rainfall or snowmelt events the pesticides are carried into streams where they become a factor in aquatic and often eventually terrestrial environments, and biota as well. According to Reynolds (1989), high runoff events shortly after application generally cause the greatest loss from the terrestrial environments to the aquatic ecosystems. In some instances pesticides become incorporated in groundwater aquifers where they can eventually contaminate drinking-water sources. In one report (Williams *et al.*, 1988), 46 pesticides were detected in groundwater from 26 states. Atrazine was detected in 13 states and alachlor was found in 12 states. Both pesticides are extensively used herbicides. Perhaps in response to these detections, the Ciba-Geigy Corporation made a recent decision to place voluntary restrictions on the use of their herbicide atrazine (US EPA, 1992a). Some of the label restrictions included:

1 deletion of non-crop uses, including use on rights-of-ways, highways and railroads;

2 reduction in application rates for corn and sorghum from about 3.4 kg/ha to 2.5 kg/ha;

3 no ground or aerial application within 60 m around all natural or impounded waters, or within 20 m where field surface runoff water enters perennial or intermittent streams and rivers;

4 all mixing and loading operations must have 15-m setbacks from intermittent streams, rivers, reservoirs, impounded and natural lakes and wells, including drainage wells, abandoned wells and sink holes.

8.2.3 Analyses of pesticides in environmental samples

An excellent introduction to determining pesticides in environmental samples is given by Gilbert

Table 8.3 Maximum expected and typical residues of pesticides on differing categories of vegetation types in ppm after an application of about 0.92 kg/ha (1 US lb/acre) (US EPA, 1988a)

Plant category	Immediately after application		Six weeks after application	
	Upper limit	Typical limit	Upper limit	Typical limit
Range grass	240	125	30	5
Grass	110	92	20	1.5
Leaves and leafy crops	125	35	20	< 1
Forage crops (small insects)	58	33	1	< 1
Pods containing seeds (large insects)	12	3	1.5	< 1
Grain (large insects)	10	3	1.5	< 1
Fruit (large insects)	7	1.5	1.5	< 0.2

& Kakareka (1985). However, they focus only on the final step in the analyses, which involves using instruments to identify and measure pesticide concentrations. Although equipment and procedures are important, the proper collection and preservation of the pesticide sample may be even more so; therefore the discussion here begins with sampling.

Comprehensive guidance for sampling, preserving, transporting, holding and analysing pesticides in environmental samples can be found in both Plumb (1981) and the *Federal Register* (1984). All procedures require that samples for analysis only come in contact with glass or Teflon for water, or metals (such as in a coring device for sediments, or metallic instruments used for tissue preparation). Coring devices, dippers, water samplers and glass containers must be rinsed thoroughly with solvents used in the extraction. Water samples must be stored in amber-coloured glass containers to prevent degradation by sunlight. In water samples, with some classes of pesticides, the pH must be adjusted to circumneutral, and with other classes sodium thiosulphate must be added. Tissue samples must come in contact only with metallic instruments used to cut or separate the tissues. To store the tissue sample it must first be wrapped in aluminium foil, then sealed in plastic bags. All water and sediment (or soil) samples must be refrigerated at 4°C. Whole test organisms or tissues should be frozen immediately and kept frozen until preparation for analysis. Animal tissues held for organophosphate or carbamate analysis should be stored at −40°C or lower.

Analysis of samples usually involves extraction procedures where the pesticides are concentrated from the aqueous phase into an organic solvent phase. In some instances a series of extractions is used; depending on the efficiency of the procedure, combinations of both polar and non-polar solvents are used. Methylene chloride and hexane are common solvents for extraction of pesticides. As an example, Wylie *et al.*, 1990), working with pentachlorophenol in water samples, extracted with methylene chloride in the field and then extracted again, with the same solvent, in the laboratory. Extraction and clean-up (separation of the pesticide from solids and aqueous components)

were completed as described in Method 608 (*Federal Register*, 1984). This step involves using a florisil column capped with anhydrous sodium sulphate. After wetting and rinsing the column with hexane, the pesticide was eluted with several combinations of ethyl ether in hexane. The rinses were then concentrated, using a Kuderna-Danish (K-D) concentrator, into iso-octane and analysed by gas chromatography. The tissue sample must be homogenized prior to extraction. Specialized columns or separation devices are often used to elute and concentrate pesticides for both tissue and sediment analysis. In a recent example, extraction of organophosphorus and carbamate pesticides from sediment was accomplished by using a column in which the sediments were mixed with equal amounts of sand and sodium sulphate (Swineford & Belisle, 1989). The mixture was extracted with a combination of acetone, methylene chloride and a small amount of toluene. The extract was then concentrated on an evaporator at 45°C, transferred to a 50-ml tube, further concentrated and finally analysed by gas chromatography. Two instruments commonly used in the analysis of pesticides are the gas chromatograph (GC), equipped with various detectors dependent on different types of pesticides, and a mass spectrometer (MS), which is used with a GC. In some instances the thin-layer chromatography (TLC) technique is used (*Federal Register*, 1984).

Where a variety of pesticide classes is being studied, combinations of techniques and instruments can be, and often are, used in the analyses (McKim *et al.*, 1987). For instance, for analysis of benzaldehyde and carbaryl mixtures, isocratic, reversed-phase, direct aqueous injection liquid chromatography equipped with ultraviolet detection has been used. A high-performance liquid chromatography (HPLC) was used in which a sample of water (with minimal organic solvents) was injected directly into the column. For analysis of acrolein and malathion mixtures, McKim *et al.* (1987), used a GC equipped with a Ni 65 electron-capture detector. The analysis of acrolein followed a step referred to as derivatization by using pentafluorophenyl hydrazine and acidification with acetic acid. Swineford & Belisle (1989) also used multiple techniques to extract and analyse pesticides with differing chemical characteristics. Pond water from North Dakota was extracted with a preconditioned solid-phase C_{18} high-capacity column attached to a filtering flask. After the unfiltered water sample was eluted, the sample was air-dried to remove water. Acetone was then used to remove individual pesticides from the column. The sample was then subjected to the various procedures in Fig. 8.1. Percentages recovered from 500 g samples of fortified pond water ranged from 80% to 100%. Other examples of different combinations of procedures used to detect mixtures of pesticides are the research conducted by Beyers *et al.* (1991) and by Burmaster *et al.* (1991). The former used C_8 columns to extract carbaryl and malathion from field-fortified samples, and the latter used hexane-filled membrane bags *in situ* to concentrate chlorinated pesticides from water below a hazardous-waste site.

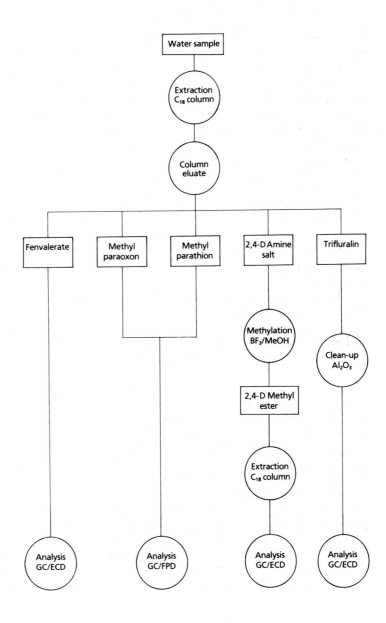

Fig. 8.1 Flow diagram of sample preparation. After Swineford & Belisle, 1989.

8.2.4 Pesticides in environmental samples

Pesticides are carried in streams (Butler & Arruda, 1985; US EPA, 1983) to eventually sequester into environmental niches, often many miles from where they were intentionally applied. In the USA, pesticides have been found in surface waters in rural Kansas at substantial concentrations and frequencies. Between 1977 and 1984, 21 chemicals were found that exceeded detection limits of the instruments used. Alachlor, atrazine, dual, sencor and 2,4-D (Table 8.4) accounted for 77% of the total. The ranges and means of pesticide concentrations detected in federally managed or private lakes are shown in Table 8.5. Atrazine, a herbicide used primarily on corn and sorghum, is one of the most widely used pesticides in the USA, and is also the most widely detected pesticide in water monitoring studies conducted in the Midwest Corn Belt of the United States (US EPA, 1992a).

Pesticides were also found in runoff water from cities (US EPA, 1983); in water, sediments and biota in the New York Bight (O'Connor et al., 1982); in benthos in the Great Lakes (Nalepa & Landrum, 1988); and in municipal wastewater (Amato et al., 1992). The National Urban Runoff Program analysed samples obtained over 18 months from 28 cities; nine were coastal urban centres (Table 8.6). In urban runoff, metals were the most frequently found pollutants, detected in 90% of the samples. Organic compounds were detected in 20% of the samples. The most common organic industrial chemical, bis(2-ethylhexyl)phthalate, was found in 22% of the samples. The most frequently detected pesticides were alpha-hexachlorocyclohexane found in 20% of the samples followed by alpha-endosulfan in 19%, chlordane in 17%, and gamma-hexachlorocyclohexane (lindane) in 15%. In the New York Bight, O'Connor et al. (1982) identified the potentially harmful pesticides as chlorinated hydrocarbons (Table 8.7) which were similar to those found in benthic species taken from the Great Lakes (Table 8.8). Wylie et al. (1990) found penta-

Table 8.4 Frequency of pesticide detections in surface water in Kansas from 1977 to 1984 from the Ambient Stream Water Quality Network (Butler & Arruda, 1985)

Substance	1977	1978	1979	1980	1981	1982	1983	1984	Total
Alachlor	2	6	4	2	3	7	15	11	50
Aldrin	0	0	0	0	0	1	1	0	2
Alpha-BHC	0	0	0	0	3	0	0	0	3
Atrazine	13	28	18	9	14	17	39	34	172
Chlordane	0	0	0	1	2	0	1	0	4
Dachthal	5	1	3	1	1	0	1	1	13
DDE	0	0	0	4	1	0	0	0	5
Diazinon	0	0	0	0	2	0	1	0	3
Dieldrin	0	0	0	0	0	1	1	0	2
Dual	0	0	0	5	5	9	12	12	43
Dursban	0	1	0	0	0	0	0	0	1
HCB	2	0	1	1	0	0	0	0	4
Lindane	0	0	1	1	4	2	2	5	15
Malathion	1	1	0	1	1	0	1	2	7
Sencor	0	4	1	1	1	5	3	9	24
Propazine	0	0	1	0	3	0	1	0	5
Ramrod	0	1	0	0	2	3	3	2	11
1-Hydroxychlordene	0	0	3	6	4	0	1	0	14
2,4-D	0	0	4	4	9	6	10	7	40
2,4,5-T	0	0	2	1	1	2	1	1	8
2,4,5-TP	0	0	0	0	0	0	1	0	1
Total detections	23	42	38	37	56	53	94	84	427
Total samples	209	161	119	103	106	100	120	117	1035

Table 8.5 Summary of pesticide data from Kansas lakes sampled in the Lake Monitoring Program from 1979 to 1984. Data are in micrograms per litre (Butler & Arruda, 1985)

Substance	Federal Lakes			Small Lakes		
	Range	Mean	n	Range	Mean	n
Alachlor	0.10–3.10	0.82	21	—	0.36	1
Atrazine	1.40–23.0	4.80	43	1.20–2.80	2.00	5
Dual	0.26–2.60	0.74	19	—	—	—
Sencor	0.05–0.31	0.18	8	—	0.21	1
Ramrod	0.25–2.90	1.00	5	0.27–1.30	0.79	2
2,4-D	0.69–2.40	1.37	4	0.42–0.48	0.45	2
Propazine	—	2.60	1	—	—	—

Table 8.6 Most frequently detected priority pollutants in samples from the National Urban Runoff Program (NURP)[a] (US EPA, 1983)

Inorganics	Organics
Priority pollutants detected in 75% or more of the NURP samples	
Lead (94%)	None
Zinc (94%)	
Copper (91%)	
Priority pollutants detected in 50–74% of the NURP samples	
Chromium (58%)	None
Arsenic (52%)	
Priority pollutants detected in 20–49% of the NURP samples	
Cadmium (48%)	Bis(2-ethylhexyl) phthalate (22%)
Nickel (43%)	Alpha-hexachlorocyclohexane (20%)
Cyanides (23%)	
Priority pollutants detected in 10–19% of the NURP samples	
Antimony (13%)	Alpha-endosulfan (19%)
Beryllium (12%)	Pentachlorophenol (19%)
Selenium (11%)	Chlordane (17%)
	Gamma-hexachlorocyclohexane (Lindane) (15%)
	Pyrene (15%)
	Phenol (14%)
	Phenanthrene (12%)
	Dichloromethane (methylene chloride) (11%)
	4-Nitrophenol (10%)
	Chrysene (10%)
	Fluoranthene (16%)

[a] Based on 121 sample results received as of 30 September 1983, adjusted for quality control review. Does not include special metals samples.

chlorophenol concentrations that ranged from 130 to 970 µg/l as well as other phenolic compounds in wastewaters in south-western Missouri. A survey of diazinon in municipal wastewaters from different regions of the USA was initiated as a result of an in-depth toxicity identification study of a single wastewater (Amato *et al.*, 1992). Preliminary results indicated that diazinon was a

Table 8.7 Pesticides detected in water, sediments and biota of the Hudson River, the Hudson estuary and the New York Bight region (O'Connor et al., 1982). Copyright 1982, Estuarine Research Federation, reprinted with permission. Permission limited to use in referenced volume. All other rights, including sublicensing, reserved

1,2-Dibromo-3-chloropropane	Heptachlor epoxide
Hexachlorobenzene	Alpha-chlordane
Alpha-benzenehexachloride	*Trans*-nonachlor
Lindane	Endrin
Heptachlor	Mirex[a]
Aldrin	DDT (and metabolites)
Dieldrin	

[a] Mirex detected in a single sample may represent a laboratory-induced artifact. Identity of the compound was verified by GC-MS analysis.

relatively common contaminant in water from municipal wastewater treatment systems.

Pesticide residues and effects on flora and fauna in several African, South American and European countries were examined by Müller (1988). He concluded that certain pesticide uses could be effective and environmentally compatible, but only after vigorous environmental research and validating studies.

8.2.5 Pesticides in tissues: results of monitoring programmes

Pesticides are found in the tissues of coastal molluscs in the USA. In a report prepared by the National Oceanic and Atmospheric Administration (NOAA, 1987) data indicated that, despite the fact that chlorinated hydrocarbon pesticides are used less today because of their harmful ecological affects, they are still found in molluscan tissues taken from US Pacific and Atlantic coastal waters. For instance, pesticide analyses of molluscs taken from 145 sites indicated that total DDT, including metabolites (tDDT), a pesticide banned over 20 years ago in the USA, was found in molluscs collected as recently as 1986. Non-DDT chlorinated hydrocarbon pesticides (tChlP) in molluscan tissues included aldrin, alpha-chlordane, dieldrin, heptachlor, heptachlor epoxide, hexachlorobenzene, lindane, mirex and

trans-nonachlor (a contaminant of chlordane). The 145 sites were then put into 'contamination groups', ranked and prioritized by degree of contamination. The 12 chemicals prioritized as toxic were: tChlP, tDDT, tPAH (polyaromatic hydrocarbons), tPCB (polychlorinated biphenyls), As, Cd, Cr, Pb, Hg, Ni, Ag and S. Forty-one sites selected on a national scale were ranked from highest to lowest. Of the 41 sites, 14 ranked in the top 10% with respect to the concentrations of tChlP; whereas nine of the 41 ranked as most contaminated due to the concentrations of total DDT. Of the 10 most contaminated sites, seven had high non-DDT chlorinated contamination and six had high DDT contamination.

Not explained, but duly noted from the report (NOAA, 1987), was that, based on the molluscan samples, 14 of the 16 most contaminated sites were in coastal areas in the upper northeastern United States, whereas, based on the fish-liver samples, 14 of the 16 most contaminated sites were in coastal areas along the Pacific Coast. Clearly, both Atlantic and Pacific coast species continue to harbour chlorinated pesticides that had become more sparingly used in the past 15–20 years.

Pesticides are also found in fish tissues in the USA. In fish taken from five national wildlife refuges in the southeastern USA, average pesticide residue concentrations of total organochlorine chemicals, mainly DDT and toxaphene, exceeded $2.0 \mu g/g$ (wet weight), concentrations that the authors believed were a direct threat to fish-eating wildlife (Winger et al., 1988). Contamination, while lower at five other refuges, where average concentrations in fish were $0.6–1.0 \mu g/g$, was still potentially toxic when fish were consumed by predators in the food chain (Winger et al., 1985). In all bluegill and common carp samples collected from San Joaquin Valley, California, p,p'-DDE was detected (Saiki & Schmitt, 1986). Also, six other pesticides, chlordane; p,p'-DDD; o,p'-DDT; p,p'-DDT; DCPA and dieldrin, were detected in both bluegill and carp at one or more of the collection sites. During a larger survey, fish livers were analysed for contamination from 43 sampling sites (NOAA, 1987). Similarly, sites with 'contamination groups', as with the molluscan data, were used to rank the sites in the upper 10% for immediate attention.

Table 8.8 Concentrations of pesticides in Great Lakes benthos (ng/g)[a]

Pesticide	Organism	Ontario	Erie	Michigan	Superior
DDT	Net plankton	60–80	41	—	tr.[d]
	Mysis relicta	59–280	—	—	tr.–53
p,p'-DDT	Net plankton	30–40	14.3	—	tr.
	Mysis relicta	48–140	—	—	—
DDE	Oligochaetes	—	—	5.9[c]	—
	Pontoporeia hoyi[b]	34	—	—	—
	Pontoporeia hoyi	—	—	154	—
	Pontoporeia hoyi	—	58	—	—
	Pontoporeia hoyi	292	—	—	—
	Pontoporeia hoyi	730	—	—	—
DDD	Oligochaetes	—	—	1.5[c]	—
	Pontoporeia hoyi[b]	1.8	—	—	—
	Pontoporeia hoyi	—	—	—	96
	Pontoporeia hoyi	—	62	—	—
Dieldrin	Net plankton	17–19	23	—	tr.
	Mysis relicta	—	—	—	<1
	Pontoporeia hoyi[b]	2.9	—	—	—
	Pontoporeia hoyi	—	—	141	—
	Pontoporeia hoyi	—	35	—	—
	Pontoporeia hoyi	226	—	—	—
	Pontoporeia hoyi	376	—	—	—
Chlordane	Net plankton	20	10	—	—
	Oligochaetes	—	—	4.5[c]	—
	Pontoporeia hoyi	—	—	252	—
Toxaphene	Net plankton	—	—	85–560	—
	Mysis relicta	—	—	190–432	—
HCB	Oligochaetes	—	—	3.6[c]	—
	Oligochaetes	105–735	—	—	—
	Pontoporeia hoyi	230–895	—	—	—

[a] Table compiled from Nalepa & Landrum (1988) and Day (1990). Net plankton and *Mysis relicta* are expressed as nanograms per gram as lipid weight of pesticide. All remaining data are expressed as nanograms per gram dry weight.
[b] Benthos, but mostly *Pontoporeia hoyi*.
[c] Detroit River.
[d] tr., Trace.

Of the uppermost 16 ranked sites, tChlP and tDDT were each responsible for five (total of 10) sites.

Other studies (US EPA, 1992b,c) confirm that pesticides persist in the tissues of fish in diverse areas around the USA. Although there was a total of 60 compounds studied in the 388 monitoring sites nationwide, pesticides were well represented in what were termed as the most frequently detected bioaccumulative substances. For example, DDE was found in 99%, biphenyl in 94% and mercury in 92% of the samles. *Trans*-nonachlor, a contaminant of chlordane, was found in 77% of the samples and *cis*- and *trans*-chlordane were found in 64% and 61% of the samples, respectively. Understandably, metabolically stable compounds such as dieldrin, alpha-BHC, hexachlorobenzene (found in over 50% of the samples) are still resident in the environment. However, it is surprising that chloropyrifos (dursban) was found in 26% of the samples, dicofol in 16% and trifluralin in 12%, since these chemicals were previously believed to be more ecologically labile.

With regard to terrestrial species, a survey of birds that are preyed upon by peregrine falcons (*Falco peregrinus*) during their breeding season in the western USA revealed the extent of organochlorine pesticide contamination (DeWeese *et al.*, 1986). More than 1100 specimens of 38 avian prey species were collected in the vicinity of 36 historical or active peregrine nest sites in eight states. More than 94% of the 201 samples (77 specimens were analysed individually and the rest grouped into 124 samples) contained DDE ranging from <1 to 33 mg/kg. The mean DDE residues of 2.8–12 mg/kg in three important prey species were high enough to prevent natural reproduction in the nesting peregrines. Twelve organochlorine pesticides and polychlorinated biphenyl compounds (PCBs) were recovered in the prey species. Insectivorous species had significantly higher concentrations than year-round resident species, indicating that most of their pesticide burden was acquired on their wintering grounds or during migration rather than in the nesting habitat (DeWeese *et al.*, 1986). In a follow-up study in 1990–1991, residue concentrations in the prey species had dropped by as much as 50%, but all organochlorine compounds were still detected (L.R. DeWeese, personal communication). Organochlorine contamination in a wetland bird species, the black-crowned night heron (*Nycticorax nycticorax*), was also investigated in the western mountain states. Previous studies of this widely distributed species in other regions of North America indicated pesticide contamination in several geographic areas (McEwen *et al.*, 1984). Eggs of black-crowned night herons in Colorado and Wyoming each contained one or more of 14 organochlorines. As in peregrines, DDE was the most prevalent compound, occurring in 100% of 147 eggs at a mean concentration of 3.1 mg/kg and ranging from <1 to 44 mg/kg (McEwen *et al.*, 1984). The contamination depressed nesting success at some sites. There were no more nesting attempts in succeeding years in the most severely contaminated night heron nesting colony. Organochlorine compounds also continue to harm bird life in Mexico and Latin America (Mora & Anderson, 1991).

In view of the following information, it is not surprising that pesticides are currently found in tissues of aquatic species and non-target birds. There are reports of pesticides being transported by the atmosphere (Welch *et al.*, 1991) and also partitioned in plant tissues (Calamari *et al.*, 1991). For instance, in Welch's report, semivolatile organic compounds detected in the Canadian Arctic's 'brown snow' included polycyclic aromatic hydrocarbons, PCBs, polychlorinated camphenes and DDT-related compounds. Also found were trifluralin and methoxychlor, endosulfan and hexachlorocyclohexane. Similarly, analyses of plants (i.e. lichens and mosses from high altitudes and mango leaves from tropical areas) revealed hexachlorobenzene (HCB), hexachlorocyclohexanes and DDT-related compounds (Calamari *et al.*, 1991).

Furthermore, organochlorine pesticides and PCBs are still major contaminants in the oceans of the world (Hutchinson & Simmonds, 1992) including the Arctic Ocean (Hargrave *et al.*, 1992). Levels of organochlorines in the seas continue to rise and are adversely affecting the health and survival of sea life such as marine turtles (Hutchinson & Simmonds, 1992).

8.3 RESULTS OF PESTICIDE TOXICOLOGICAL STUDIES

8.3.1 Acute tests using single aquatic species

The purpose of an acute toxicity test is to determine the effective concentration of a test material (in this case a pesticide) over a brief period of time, usually in either 48 or 96 h (see Volume 1). A traditional acute test consists of about 20 organisms of one species of the same age-class exposed to five or six concentrations of a pesticide and control for 96 h. Temperature, pH and dissolved oxygen in the test solutions are controlled and recorded. An idealized acute data set consists of a direct increase in test species lethality with increasing pesticide concentrations. An optimum data set has less than 5% lethality in both the control and the lowest pesticide concentrations, and nearly 100% lethality in the highest pesticide concentration. With aquatic species the pesticide is administered dissolved in water so that the species can absorb it through its gills, through ingestion and through its body surface. This procedure is considered to be a fair approximation of the means by which pesticides enter aquatic organisms. The endpoint often noted from an acute test is the LC_{50} or EC_{50} (Volume 1). An example of a successful EC_{50} is with algal assays where the endpoint is reduction of population density. Results of 187 tests with unicellular algae using 21 pesticides and 18 tin-containing substances were reported by Walsh *et al.* (1987). Five species were tested with pesticides, and two species were tested with tin compounds. While many of the compounds were toxic to the algae in the milligram per litre range, five of the EC_{50}s were in the microgram per litre range. Amdro, a chemical used to control fire ants, was toxic to the alga *Thalassiosira pseudonana*, at a concentration of about $0.5\,\mu g/l$.

Comprehensive data bases reporting acute effects of pesticides on aquatic species are those of Johnson & Finley (1980), Mayer & Ellersieck (1986) and Mayer (1987). The first reference is a compilation of results from 1587 acute tests with 271 chemicals using 28 species of freshwater fishes and 30 species of freshwater invertebrates. The second is a data set of 410 chemicals and 66 freshwater species. The third reference provides a listing of 214 chemicals tested using 52 salt water species. In the data within the three works, 99% of the chemicals used in the studies were pesticides. Of the three studies, only Mayer & Ellersieck (1986) presented a thorough statistical analysis and interpretation of the data. Below, we provide highlights of their findings.
- As a taxonomic group, insects were the most sensitive to the chemicals, followed by crustaceans, fishes and amphibians.
- Among the four most commonly tested forms, daphnids were the most sensitive 58% of the time, followed by rainbow trout (*Oncorhynchus mykiss*) 35%, bluegills (*Lepomis macrochirus*) 5%, and fathead minnows (*Pimephales promelas*) 2%.
- The lower of the two $L(E)C_{50}$s of either daphnids or rainbow trout was ≤ 15 times that of the most sensitive species 95% of the time, and ≤ 25 times that value 100% of the time.
- Testing of three groups (daphnids, gammarids and rainbow trout) provided the lowest toxicity value 88% of the time, and could not be improved

more than 2.5% by adding any other single species.
- Frequency distribution of $L(E)C_{50}$s tended to be bimodal for many species. Toxicities of insecticides were mainly in the $<100\,\mu g/l$ category; herbicides, fungicides and industrial chemicals were in the $>1000\,\mu g/l$ category.

Regarding a summary of toxicity for both chemicals and species, Mayer & Ellersieck (1986) stated:

> The lower mode [of toxicity] contained almost all of the insecticides tested with invertebrates, whereas among fishes, insecticides were spread over more of the toxicity range. Insecticides occurring in the lower modes [of toxicity] of fishes were mainly botanicals and organochlorines plus some carbamates and organophosphates. The distribution of chemical toxicities in larval frogs and toads was similar to that for fishes, within the confines of the few chemicals tested against amphibians.

An example of data collected from acute tests using fenvalerate, a pyrethroid, is shown in Table 8.9; LC_{50} and EC_{50} values from both freshwater and saltwater tests (Mayer & Ellersieck, 1986; Mayer, 1987) are listed. Fenvalerate is one of the engineered varieties of natural pesticides (p. 157). As expected, since fenvalerate is an insecticide, algae and molluscs were relatively insensitive to it. Because crustaceans are closely related to insects taxonomically, fenvalerate was acutely toxic to both the amphipod *Gammarus pseudolimnaeus*, with an LC_{50} value of $0.032\,\mu g/l$, and the saltwater mysid *Mysidopis bahia*, with the lowest LC_{50} value of $0.008\,\mu g/l$. Unexpectedly, fenvalerate was also acutely toxic to fish (*Salmo* sp., *Ictalurus puctatus*, *Lepomis macrochirus* and *Cyprinodon variegatus*). If saltwater algae and molluscs are excluded, the high-to-low LC_{50} and EC_{50} values for the freshwater species were $0.032-2.4\,\mu g/l$, or a factor of 75. Similarly, the high-to-low values from the saltwater tests ranged from 0.008 to $5.4\,\mu g/l$, a factor of 675.

Effects other than lethality have been observed in acute tests with fish. For example, Little *et al.* (1990) showed that spontaneous swimming activity, swimming capacity, feeding behaviour and vulnerability to predation were sensitive indicators of sublethal pesticide toxicity in rainbow

Table 8.9 Comparison of species sensitivities to fenvalerate, a representative insecticide, in acute tests (μg/l)

Freshwater species[a]	LC_{50}[b]	Saltwater species[c]	LC_{50}/EC_{50}[b]
Daphnia magna	2.1	*Isochrysis galbana*	> 1000
Gammarus pseudolimnaeus	0.032	*Skeletomema costatum*	> 1000
Chironomus plumosus	0.43	*Thalassiosira pseudonana*	> 1000
Salmo gairdneri (Oncorhynchus mykiss[f]*)*	0.32–1.7[d]	*Nitzschia angularus*	> 1000
Pimephales promelas	2.15–2.35[d]	*Mysidopis bahia*	0.008–0.021[d]
Ictalurus punctatus	2.35–2.4[d]	*Penaeus duorarum*	0.84
Lepomis macrochirus	0.42–1.35[d]	*Crassostrea virginica*	> 1000
		Opsanus beta	2.4–5.4[d]
		Cyprinodon variegatus	4.4–5.0[d,e]
		Leuresthes tenuis	0.29–0.60[d]
		Menidia beryllina	1.0
		Menidia menidia	0.31
		Menidia peninsulae	0.95
		Mugil cephalus	0.41

[a] Mayer & Ellersieck, 1986.
[b] LC_{50}/EC_{50}: most of the data are from lethal concentration (LC_{50}) tests, but for certain organisms the effective concentration (EC_{50}) is calculated based on immobilization, growth or some other identifiable endpoint rather than lethality.
[c] Mayer, 1987.
[d] Ranges of LC_{50}/EC_{50} values reported.
[e] An LC_{50} value of 120 μg/l for sheepshead minnow was excluded from the table because it appeared to be a typographical error.
[f] Species name has changed since this table was first published.

trout (*Oncorhynchus mykiss*) exposed for 96 h to various chemicals (Table 8.10). Following exposure, the investigators observed sublethal toxicity that varied with different pesticides and concentrations. Fish exposed to a nominal concentration of 0.2 μg/l chlordane had a significant reduction in their swimming activity, strike frequency (the number of times fish struck at and successfully ate daphnids) and the total number of daphnids consumed. According to Little *et al.* (1990), feeding behaviour was similarly affected at lower concentrations of DEF, 2,4-DMA and methyl parathion; whereas vulnerability of trout to predation by a larger aggressive fish (bass) was heightened most by exposure to carbaryl and pentachlorophenol. A report presented by Little & Finger (1990) showed a compilation of comparisons of swimming behaviour and lethal concentration after fish were exposed to pesticides. Data from 35 tests, half involving pesticides, indicated alterations in swimming behaviour detected during exposure to small concentrations of chemicals and/or mixtures, 0.7–5% of their LC_{50} values.

8.3.2 Acute tests using avian and mammalian species

Acute pesticide toxicity tests are conducted by administering the toxicant in a single oral dose to test populations of birds or mammals. The dose (quantity by weight) lethal to 50% of the test population is called the LD_{50}. The LD_{50} is expressed as milligrams of toxicant per kilogram of body weight (mg/kg) of the test animal. This enables comparison of the toxicity of a particular pesticide to small young vs full-grown adults, and among bird and mammal species varying in body mass from a few grams to several kilograms (see Volume 1). To illustrate, consider that a small amount of a particular pesticide lethal to a mouse (for example, 10 mg) might be ingested by a large mammal such as a deer without toxic effect. On a milligram per kilogram basis, a 40 g mouse ingesting 10 mg of a toxic chemical would receive a 250 mg/kg dosage, whereas a 50 kg deer ingesting 10 mg of toxicant would receive only 0.2 mg/kg. To compare toxicity between the two species, the deer would have to receive 12 500 mg of the chemical.

Acute oral LD_{50} values for several of the commonly used insecticides and herbicides are given in Table 8.11. Note the variations in toxicity of a given chemical among different animal species. Vertebrate species may vary widely in their susceptibility to toxic effects of a particular pesticide. Consequently, only limited extrapolation to other species may be made with toxicity data from LD_{50} tests. Nevertheless, toxicity is generally more similar among closely related species. For example, dieldrin and methyl parathion (Table 8.12) are considerably more toxic to pheasants and quail (family Phasianidae) than they are to mallards (family Anatidae). The frequent comparability of a given compound's toxicity within the same genera makes it possible to estimate hazard to endangered species by testing toxic effects on close relatives that are not endangered. The related species are called surrogate species when tested or studied in this manner.

The variation in toxicity among different pesticides to a particular animal species can be very great, reaching many orders of magnitude. The insecticide endrin, for example, is more than 2000 times as toxic to pheasants and quail as are many more commonly used insecticides such as carbaryl and permethrin (Table 8.11). Thus, when selecting pesticides for field use that present the least hazard to non-target wildlife, it is important to know how toxic they are to a variety of species. Toxicity data are not the only factor, however; the rate and frequency of pesticide applications are also important. A less toxic pesticide may have more harmful effects than a more toxic pesticide if it is applied at a much heavier rate and at several times in a season. Some pesticides, including methoxychlor and various herbicides, are not very toxic, in a single dose, to any species (Table 8.11).

Herbicides are generally environmentally less toxic on an acute basis than insecticides (Table 8.11). However, if an oil carrier is used in a herbicide field formulation, the oil itself may be toxic to birds' eggs (Hoffman, 1990) and to sensitive plant species. Environmental effects of herbicides are primarily expressed through changes in vegetative cover and habitat. Loss of critical

habitat can be as deadly to the survival of a wildlife population as toxic poisoning. Total elimination of a sagebrush stand also eliminates several obligate wild vertebrates (McEwen & DeWeese, 1987). On the other hand, use of herbicides to create properly spaced openings and ecotones in extensive sagebrush stands is beneficial, and improves wildlife habitat (McEwen & DeWeese, 1987).

Another method for determining the short-term toxicity of a pesticide is by adding it to the food of the test species in measured concentrations (Hill & Camardese, 1986; Volume 1). With birds and mammals the dietary concentration lethal to 50% of the test population is termed the LC_{50} (Table 8.12). The LC_{50}s in Table 8.12 were largely determined in 5-day feeding tests. The test animals therefore received the dosages spread over a 5-day period rather than all in one single dose, and were able to metabolize more of the toxicant before being affected. As with the LD_{50} data, note the great differences in toxicity among the insecticides to a particular species, and the variation among species to a given chemical. Generally, herbicides are less toxic than insecticides via any exposure route.

Toxic signs in birds associated with lethal exposure to several commonly used pesticides are listed in Table 8.13 (see also Volume 1; Chapter 16). The organophosphates and carbamate chemicals often cause death in a few minutes or hours.

8.3.3 Effects of cholinesterase inhibitors

Use of organophosphates and carbamate chemicals increased greatly following the elimination of DDT and most other organochlorine insecticides from legal environmental use in the USA. Representatives of these two insecticide groups are applied to cropland, forests, rangelands and wetlands, and account for the majority of use (Smith, 1987). Although organophosphate and carbamate insecticides degrade more rapidly after application than the organochlorines, they generally are more toxic before they break down. These compounds are effective insecticides but can have harmful effects on non-target wild vertebrates by inhibiting acetylcholinesterase (AChE) activity (Grue et al., 1991). AChE and other cholinesterases (ChE)

Table 8.10 Influence of a 96-h exposure to six pesticides on the behavioural responses of rainbow trout compared to 96-h LC$_{50}$ values (Little et al., 1990)

Chemical (mg/l)[a]	96-h LC$_{50}$ (mg/l)	Mean behavioural response[b]					
		Swimming capacity (cm/s)	Swimming activity (s)	Strike frequency (5 min)$^{-1}$	Daphnia consumed	Percentage consuming Daphnia	Percentage survival from predation
Carbaryl	1.95						
0		24.7x	116x	1.0x	1.0x	100x	85x
0.01		25.4x	119x	1.2x	0.7x	70x	35yz
0.1		23.6x	113x	1.1x	0.6x	60x	50y
1.0		19.9y	79y	0.6x	0.3y	30y	25z
SE[c]		0.8	4.0	0.3	0.1		0.3
n[d]		26	10	10	10	10	4
Chlordane	0.042						
0		17.2x	115x	6.8x	4.8x	100x	67x
0.0002		18.1x	98xy	4.6xy	3.5xy	90x	73x
0.002		17.4x	77y	3.8y	2.2y	50y	20y
0.02		16.1x	69y	2.5y	1.7y	50y	7y
SE		0.8	12.0	1.0	0.6		0.3
n		17	10	10	10	10	3
DEF[e]	0.660						
0		25.5x	118x	2.3x	1.9x	80x	62x
0.005		26.6y	90xy	1.6x	0.8y	60x	60x
0.05		23.0yz	63y	0.6y	0.4yz	20y	42y
0.10		20.8z	30z	0.1y	0.0z	0y	32y
SE		1.2	11.0	0.3	0.2		0.5
n		10	10	10	10	10	4

	Nominal concentration[a]						
2,4-DMA[f]	100.0						
0		18.2x	109x	8.1x	2.3x	100x	58x
0.5		19.4y	110x	9.6x	2.0x	90x	48x
5.0		18.1x	86xy	7.0x	1.6x	60y	52x
50.0		16.7z	81y	1.9y	0.0y	0y	0y
SE		0.5	9.0	1.4	0.5		0.8
n		50	10	10	10	10	4
Methyl parathion	3.7						
0		19.9x	115x	0.9x	0.9x	90x	83x
0.01		21.5x	100y	0.9x	0.5y	50x	57xy
0.1		21.0x	70z	0.2y	0.2y	20y	33y
SE		0.8	5.0	0.2	0.1		0.7
n		12	10	10	10	10	3
Pentachlorophenol	0.052						
0		19.1x	111x	6.0xy	4.1x	100x	72x
0.0002		17.9x	111x	6.7xy	2.8xy	100x	32y
0.002		17.7x	86y	7.2x	2.1y	86x	52xy
0.02		17.5x	85y	3.8y	2.2y	93x	32y
SE		0.8	5.0	1.1	0.5		0.5
n		15	14	14	14	14	5

[a] Nominal concentration used during exposure.
[b] Means within a given column accompanied by a different letter (x,y,z) are significantly different from each other at $P \leq 0.05$.
[c] Pooled standard error of the mean.
[d] Number of fish tested per treatment.
[e] S,S,S-tributyl phosphorotrithioate.
[f] Dimethylamine salt of 2,4-dichlorophenoxyacetic acid.

Table 8.11 Acute oral toxicity (LD_{50}, mg/kg) of some common pesticides to birds and mammals[a]

	Species		
	Gallinaceous birds	Mallard	Rodent[f]
Insecticide			
Acephate	140[b]	234	866
Bacillus thuringiensis (Berliner)	—	>2000	>10 000
Carbaryl (technical)	>2000[b]	>2564	500
Carbaryl (Sevin-4-Oil)	707[b]; >2000[c]	—	—
Carbofuran	4.2[b]	0.5	8
Chlorpyrifos	17.7[b]; 68.3[c]	75.6	135
DDT	1334[b]; 595[c]	>2240	113
Diazinon	4.3[b]	3.5	300
Dieldrin	79[b]; 8.8[c]	381	40
Dimethoate	20[b]	63.5	225
Endrin	1.8[b]; 1.2[c]	5.6	7
Fenthion	17.8[b]; ≤4.0[d]	5.9	250
Lindane	≤100[b]	>2000	88
Malathion	167[b]	1485	1375
Methamidophos	8[d]	8.5	29.9
Methoxychlor	>2000[c]	>2000	6000
Methyl parathion	8.2[b]; 7.6[d]	10	9
Monocrotophos	2.8[b]	4.8	8
Parathion	24[b]; 16.9[c]	0.9	3
Resmethrin	>2000[c]	—	2500
Toxaphene	40[b]	30.8	49
Nosema locustae	(>5 × 10^9)[g]	(>5 × 10^9)[g]	(>4.5 × 10^9)[g]
Herbicide			
Atrazine	>2000[b]	>2000	1869
2,4-D	472[b]	>2000	639
Dicamba	673[b]	—	1040
Diquat (dibromide)	—	564	231
Glyphosate	—	—	4900
Paraquat CL	—	199	150
Picloram	>2000[b]	>2000	4012
2,4,5-T	<1000[b]	>2000	650
Trifluralin (Treflan EC)	>2000[b]	>2000	3700

[a] Toxicity data are derived primarily from Hudson, Tucker & Haegele (1984) and Smith (1987).
[b] Ring-necked Pheasant (*Phasianus colchicus*).
[c] California Quail (*Calipepla californica*).
[d] Northern Bobwhite Quail (*Colinus virginianus*).
[e] Japanese Quail (*Coturnix japonica*).
[f] White Rat (*Rattus rattus*).
[g] Spores/animal.

are essential for normal function of the nervous system. Toxic effects of AChE inhibition on fish and wildlife vary with the degree of exposure to, and potency of, the particular organophosphate or carbamate. Toxic effects can range from minor behavioural changes to severe effects such as paralysis, convulsions and death (associated with >50% to >90% inhibition) (Grue *et al.*, 1991). Inhibition of >20% is considered evidence of exposure to AChE-inhibiting compounds (Hudson *et al.*, 1984; Grue *et al.*, 1991).

Measurement of AChE and ChE activity in

Table 8.12 Dietary toxicity (LC$_{50}$, ppm) of common pesticides to mallards and gallinaceous birds[a]

	Species		
	Mallard (*Anas platyrhynchos*)	Pheasant (*Phasianus colchicus*)	Quail[b,c]
Insecticide			
Acephate (technical)	—	—	3275[b]
Acephate (orthene)	—	—	718[b]
Carbaryl (technical)	> 5000	> 5000	> 5000[c]
Carbofuran	190	573	746[b]
Chlorpyrifos	940	553	293[b]
Chlorpyrifos (Dursban)[d]	—	—	492[b]
DDT	1869	311	611[c]
Diazinon	191	244	245[c]
Dieldrin	153	58	37[c]
Dimethoate	1011	332	341[b]
Dimethoate (Cygon 2E)	—	—	496[b]
Endrin	18	14	14[c]
Fenthion	231	202	30[c]
Lindane	> 5000	561	882[c]
Malathion	> 5000	2639	3497[c]
Methamidophos	—	—	42[c]
Methoxychlor	> 5000	> 5000	> 5000[c]
Methyl parathion	336	91	90[c]
Monocrotophos	9.6	3.1	2.4[b]
Parathion	76	336	194[c]
Resmethrin	—	—	> 5000[b]
Toxaphene	538	542	828[c]
Herbicide			
Atrazine	> 5000	> 5000	> 5000[c]
2,4-D	> 5000	> 5000	> 5000[b]
Diallate	> 5000	—	> 5000[c]
Diquat (dibromide)	—	—	1337[b]
Glyphosate	—	—	> 5000[b]
Paraquat CL	4048	1468	981[c]
Picloram	> 5000	> 5000	> 5000[c]
2,4,5-T	> 5000	3950	3400[c]
Trifluralin (Treflan EC)	—	—	> 5000[b]

[a] Toxicity data are derived primarily from Hill *et al.* (1975), Hill & Camardese (1986) and Smith (1987).
[b] Japanese Quail (*Coturnix japonica*).
[c] Northern Bobwhite Quail (*Colinus virginianus*).
[d] Dursban formulation is 40.7% Chlorpyrifas active ingredient.

brain and/or blood is an effective technique for diagnosing exposure and effects of organophosphates and carbamates on non-target wildlife (Fairbrother *et al.*, 1991). Assays to measure AChE and ChE activity are conducted utilizing standardized spectrophotometric techniques, and require comparisons with 'normal' unexposed animals of

the same species, sex and age class (Fairbrother *et al.*, 1991; Volume 1, Chapter 16).

While analysis of AChE and ChE activity is useful in diagnosing toxicity and exposure of fish and wildlife to commonly used ChE inhibitors, some questions remain unanswered. The degree of inhibition can vary from a small percentage to

Table 8.13 Examples of toxic signs observed in birds exposed to insecticides and herbicides[a]

Insecticide	Species	Acute oral LD$_{50}$ (mg/kg)	Common toxic signs[b]	Time to death
Acephate	Mallard (*Anas platyrhynchos*)	234	Ataxia, tremors, immobility	2–18 h
Bacillus thuringiensis[c]	Mallard	>2000	None	No mortality
Carbaryl (technical)	Sharp-tailed grouse (*Tympanuchus phasianellus*)	<1000	Ataxia, paralysis, convulsions, coma	½ h–3 days
Carbaryl (Sevin-4-Oil)	Pheasant (*Phasianus colchicus*) California quail (*Calipepla californica*)	707 >2000	Ataxia, paralysis, convulsions, coma	½ h–3 days
Carbofuran	Mallard	0.5	Lacrimation, phonation, wing-beat convulsions	5 min
Chlorpyrifos	Pheasant Mallard	18 75.6	Lacrimation, tremors, coma, convulsions	1 h–9 days
DDT	Pheasant California quail	1334 595	Continuous tremors, convulsions	1–2 days
Diazinon	Pheasant Mallard	4.3 3.5	Lacrimation, prostration, wing-beat convulsions	<24 h
Dieldrin	California quail Pheasant	8.8 79	Ataxia, immobility, wing-beat convulsions	1–9 days
Dimethoate	Pheasant Mallard	20 63	Ataxia, weakness, clonic convulsions	15 min–24 h
Endrin	Mallard Sharp-tailed grouse	5.6 1.1	Opisthotonos, prostration, wing-beat convulsions	1 h–5 days
Fenthion	Mallard Northern Bobwhite quail	5.9 ≤4.0	Lacrimation, salivation, clonic convulsions	½–2 h
Malathion	Horned lark (*Eremophila alpestris*)	403	Ataxia, weakness and falling, tremors	1–18 h
Methamidophos	Mallard	8.5	Ataxia, immobility wing-beat convulsions	1–1½ h
Monocrotophos	Mallard Golden eagle (*Aquila chrysaetos*)	4.8 0.2	Lacrimation, tetany, convulsions	1–60 h
Nosema locustae	Mallard	(>5 × 10^9)[d]	None	No mortality
Parathion	Pheasant Mallard	12 1.4	Lacrimation, prostration, tremors, convulsions	½ h–13 days
Resmethrin	California quail	>2000	None	No mortality

Continued

Table 8.13 *continued*

Atrazine	Pheasant	>2000	Ataxia, weakness,	No mortality
	Mallard	>2000	tremors	
2,4-D	Pheasant	472	Ataxia, prostration,	No mortality
			tremors	
Diquat dibromide	Mallard	564	Ataxia, immobility	1–2 days
Paraquat dichloride	Mallard	199	Ataxia, swallowing,	3–20 h
			wing-drop	
Picloram	Pheasant	>2000	Mild ataxia,	No mortality
	Mallard	>2000	regurgitation	
2,4,5-T	Pheasant	<1000	Ataxia, regurgitation,	3–24 h
			slowness	

[a] Toxicity data are derived primarily from Hudson *et al.* (1984).
[b] Other toxic signs seen less frequently included: diarrhoea, anorexia, myasthenia, polydipsia, dyspnoea, rapid flickering of nictitating membranes, tachypnoea, circling and fluffed feathers.
[c] Variety Berliner.
[d] Spores/bird.

nearly 100%. Lethal exposures can be determined with a high degree of confidence when brain AChE inhibition exceeds 50% and the toxicant is recovered qualitatively from the gastrointestinal tract. Effects of sublethal exposures, however, need much more investigation (Grue *et al.*, 1991). Potential adverse effects include slowed reactions and greater vulnerability to capture by predators, lowered resistance to weather and food stress, behavioural changes and reproductive failure. Variation among individuals, species, chemicals and environmental factors makes it difficult to develop generalizations about the biological significance of, for example, 20% vs 40% inhibition of brain AChE in non-target animals exposure to inhibitors. Neurobehavioural toxicology of antiChE pesticides is reviewed by D'Mello (1992) indicating several effects on animal behaviours.

AChE assay of blood plasma is a non-destructive method of measuring inhibition. Individual animals can be sampled prior to exposure and, if they survive, sampled after exposure, until AChE activity has returned to pre-exposure levels. Utility of plasma AChE activity is limited to a short time period after exposure because recovery is much more rapid than in brain tissue (Peakall, 1992). A promising method of measuring plasma AChE and ChE inhibition is the use of a reactivator, 2-PAM (pyridine-2-aldoxime methochloride). Plasma from field specimens suspected of having been exposed to organophosphate or carbamate insecticides can be incubated with and without 2-PAM. Differences in activity can then be attributed to insecticide exposure (Fairbrother *et al.*, 1991). Other new and improved methods for diagnosing insecticide effects on ChE activity are being developed (Grue *et al.*, 1991; Kennedy, 1991).

8.3.4 Immunotoxicity and pesticide–disease interactions

Many pesticides are immunosuppressors that reduce an exposed animal's defences against carcinogens and diseases such as bacterial and virus infections (Peakall, 1992). Immune system processes are varied and complex, making it difficult to clarify effects of xenobiotic chemicals on immune function in wildlife. One of the earliest experiments to examine this problem was a test of mallard susceptibility to viral hepatitis after the ducks had been exposed to two levels of DDT (Friend & Trainer, 1974). Results indicated increased susceptibility to the virus in ducklings up to 6 weeks of age. Other interactions, such as increased DDT residues in older mallards, were also observed. Few wildlife field studies have been conducted, but laboratory investigation of immunotoxic effects of xenobiotic chemicals in birds and mammals is extensive (Peakall, 1992). Evidence suggests that cumulative effects from exposure to multiple xenobiotics can overwhelm the immune system function. It is also becoming clear that the reverse phenomenon is true. Disease-weakened animals have reduced ability to detoxify xenobiotics, such as insecticides, and are more susceptible to their toxic effects (Peakall, 1992).

There are other 'biomarkers' (other than ChE activity) that are becoming useful in evaluating pesticide effects on non-target wildlife. A bio-marker is defined as 'a xenobiotically induced variation in cellular or biochemical components or processes, structures, or functions that is measurable in a biological system or sample' (National Research Council, 1987). The subject of biomarkers is thoroughly reviewed in a monograph by Peakall (1992).

8.3.5 Sublethal effects of pesticides after chronic exposures

The principal reason for conducting chronic tests is to determine whether sublethal effects other than death can result from long-term exposure to pesticides whereas, as indicated earlier, the main point of an acute test is to establish the degree of toxicity in either 48 or 96 h (see Volume 1). A chronic test for invertebrates is conducted over several weeks or months. An example of acute vs chronic effects of pesticides with the same invert-ebrate species is illustrated in a study conducted by Nimmo *et al.* (1981). Presented are data from 11 studies, representing different chemical classes of pesticides (i.e. herbicides, insecticides, organo-phosphates, organochlorides, pheromones) using the estuarine mysid, *Mysidopsis bahia*. Although all are well-known pesticides, acute effects were not found with two of the pesticides (sevin or carbaryl and trifluralin), and the chronic end-points varied from effects on mysid growth, to fecundity, to chronic lethality, or combinations

of the above. For fish, several months or years (whichever is necessary to include an entire or substantial portion of the fish's life cycle) are used and defined as a chronic exposure to a toxi-cant. Currently, a 30-day early life-stage test or a partial-life-cycle test (US EPA, 1985) as well as a 7-day larval test are being widely used to estimate the chronic effects of chemicals and effluents (US EPA, 1989a). With respect to chronic effects of pesticides the conclusions of a review of 176 life-cycle, partial-life-cycle, and early-life-stage tests were examined for concentration–response relationships by non-linear regression (Suter *et al.*, 1987). Of the 176 sets of data, one-third (51) were chronic tests conducted with pesticides. Effects were most frequently noticed on: fecundity (42%), weight of early juveniles per initial egg (35%), parental survival (20%) — the survival of the early-life-stages that grew to reproductive maturity, weight (20%), larval survival (19%) and hatching success (12%). Suter *et al.* (1987) further indicated that the grand average of the chronic responses corresponded to the EC_{25} values (the theoretical concentration that is most often calcu-lated as chronically toxic to 25% of the test organisms).

Not mentioned by Suter *et al.* (1987), however, is that when designing chronic tests with fish, one should examine as many endpoints as possible to get a holistic view of the full environmental impact. This observation is valid for other taxa as well, including invertebrates (Table 8.14). The most important aspect of these data is that effects noted were at pesticide concentrations in the low parts-per-trillion to low parts-per-billion range. Another important observation is the number of behavioural changes noted after chronic exposure to pesticides. The behavioural changes have physiological or pathological bases that typically go unnoticed until distress signs are unmistakable to an observer. Important components of aquatic ecosystems are also not included as part of the species or observations listed in Table 8.14 (chronic data for plants, insects and amphibians were not found). Despite this, it would be useful to devise a hazard or risk assessment testing scheme, dis-cussed later in this chapter, using similar pro-cedures to arrive at the environmental risks from a selected chemical. A comparison of results of

Table 8.14 Summary of chronic sublethal effects of pesticides on aquatic test species

Test species	Pesticide	Observed effect and time of exposure	Effective concentration	Reference
Arenicola cristata (lugworm)	Mirex	Reduced burrowing and feeding activities during 75 days	< 0.003−0.062 µg/l	Schoor & Newman, 1976
Acactia tonsa (copepod)	Tributyltin	Reduced survival in 6 days	0.023−0.024 µg/l	Bushong *et al.*, 1990
Hyalella azteca (amphipod)	Kelthene	30% survival after 28 days	196.37 µg/l	Spehar *et al.*, 1982
Gammarus pseudolimnaeus (amphipod)	Fenvalerate	Loss of coordination after 4−9 days	0.022 µg/l	Anderson, 1982
Brachycentrus americanus (caddis fly)	Permethrin	'Pawing' activity followed by cessation of feeding after 21 days	0.029−0.52 µg/l	Anderson, 1982
Pteronarcys dorsata (stone fly)	Permethrin	Loss of equilibrium followed by loss of feeding after 21 days	0.029−0.5 µg/l	Anderson, 1982
Helisoma trivolvis (snail)	Permethrin	Reduced ability to avoid probing the first 7 days	0.33 µg/l	Spehar *et al.*, 1983
Mysidopsis bahia (mysid)	Aldicarb	Diminished development of brood pouches at 16 days	2.1 µg/l	US EPA, 1981
Mysidopsis bahia (mysid)	Methyl parathion/ phorate separately	Reduced swimming stamina against a current after 4 days	0.31−0.5 µg/l (mp) 0.078−0.18 µg/l (p)	Cripe *et al.*, 1981
Palaemonetes pugio (grass shrimp)	Carbophenothion	Diminished development of brood pouches at 16 days	2.9 µg/l	US EPA, 1981
Palaemonetes pugio (grass shrimp)	Methyl parathion	Impaired ability to escape predation by the gulf killifish at 5 days	0.475 µg/l	Farr, 1978
Mysidopsis bahia (mysid)	Fenthion	Reduced growth rates of juveniles at approximately 14 days	0.079 µg/l	McKenney, 1986
Cyprinodon variegatus (sheepshead minnow)	Carbophenothion	Darkened areas along body, lethargy, abnormal lateral body flexure at 28 days	2.8 µg/l	US EPA, 1981
Cyprinodon variegatus (sheepshead minnow)	Trifluralin	Symmetrical hypertrophy of vertebrae at 28 days	5.5−31 µg/l	Couch *et al.*, 1979
Cyprinodon variegatus (sheepshead minnow)	EPN	Reduced swimming stamina at 265 days	2.2 and 4.1 µg/l	Cripe *et al.*, 1984

Continued on page 180

Table 8.14 *continued*

Test species	Pesticide	Observed effect and time of exposure	Effective concentration	Reference
Cyprinodon variegatus (sheepshead minnow)	Guthion	Reduced acetylcholinesterase activity at 219 days	0.06–0.50 µg/l	Cripe *et al.*, 1984
Cyprinodon variegatus (sheepshead minnow)	Kepone	Scoliosis and related pathology after 11 days	0.8 µg/l	Couch *et al.*, 1977
Salmo salar (Atlantic salmon)	Fenitrothion	Affect on foraging behaviour after 7 days	0.004–0.005 µg/l	Morgan & Kiceniuk, 1990

ecological significance from the observations above with the more traditionally used lethality, growth or reproductive endpoints would be useful.

Pesticides have harmful affects on reproduction of wild birds and mammals at sublethal exposure levels (Peterle, 1991). The well-known DDE-caused eggshell thinning problem in peregrine falcons (*Falco peregrinus*), bald eagles (*Haliaeetus leucocephalus*) and many other raptors and aquatic bird species is described in hundreds of scientific papers. However, the effect is much broader than simple mechanical shell breakage. Associated effects can include smaller clutches, embryonic death, death due to unsuccessful egg pipping, mortality of small chicks and aberrant parental behaviour such as destruction of eggs or failure to care for young (McEwen & Stephenson, 1979). Additional problems due to chlorinated compounds, including disruption of endocrine functions in wildlife and humans, have been reported (Hileman, 1993). Organophosphate and carbamate pesticides can also adversely affect reproduction in several ways, including fewer eggs produced by birds and behavioural changes in breeding birds and mammals (Somerville & Walker, 1990; Grue *et al.*, 1991).

8.3.6 Multiple species testing and pesticides, microcosms, mesocosms and field studies

Mesocosm studies of pesticides use experimental ponds and *in-situ* enclosures (Touart, 1988) to which pesticides are applied. Using this definition, small laboratory-sized chambers and micro-organisms (bacteria, unicellular algae and small invertebrates) used for studies are referred to as microcosms (see also Volume 1). Investigations of pesticides applied for experimental use to larger tracts of water or land are referred to as field studies. Clark (1989) addressed the advantages and disadvantages of these different field approaches (Table 8.15). While most of his discussion involves direct experimental applications to estuaries, he points out the value of combining both laboratory and field data for meaningful ecological risk assessments. Examples of information necessary for assessment of pesticides through use of microcosms, mesocosms and field studies are presented in Table 8.16. The use of microcosms (Table 8.16) appears to be excellent for controlling physical aspects of systems to study processes including pesticide degradation or transformation, acute effects, or interactions of two species. Mesocosms appear useful to confirm exposure regimes, chronic effects, interactions of predator and prey, and changes in populations and communities. A handbook on mesocosm testing procedures using pesticides on freshwater environments is available (Anon., 1991a). Mesocosms and field studies appear to account uniquely for the physical aspects of habitats, as well as some of the interactions studied in microcosms and mesocosms. In other words, a pesticide's transport efficiency to its target area due to various wind directions, wind speed, physical barriers, differences in dilution rates, effective mixing in the water column and flushing (during tides or intense rainfall) is fully understood only through large-scale applications in field settings (Clark, 1989).

Table 8.15 Advantages and disadvantages of various approaches used to investigate contaminant effects in estuaries (Clark, 1989)

Approach	Advantages	Disadvantages
Chemical surveys	Field effects focused on sample collection	Biotic responses to exposure is not quantified
	Allows integration of water, sediment and biota residues	Residue data are biased towards persistent chemicals
	Bioavailability and exposure can be used to predict effects	Exposure–response relationships are deduced or hypothetical
Fish kill investigations	Effects obvious and indisputable	Total numbers of dead animals may be underestimated
	Important species recognized	Impact on plankton and microbiota not characterized
	Acute, lethal effects are assumed	Exposure–response relationships are not quantifiable
Caged animal studies	Animals tested are of known toxicological sensitivity and physiological health	Results may not be the same as ecological response
	Field exposure–response data are comparable to laboratory toxicity test data	Problems with cage fouling, loss through vandalism, storms or other events
	Acute or chronic, lethal or sublethal effects can be examined	Accurate field exposures difficult to quantify
		Efforts must be coordinated with contaminant release
Population sampling	Long-term effects can be estimated from population dynamics approach	Sampling must be done over a considerable time frame
	Data are relevant to ecotoxicological interpretation	Data interpretation more difficult than caged animal or fish kill studies
	Biomonitoring for growth or gross pathological effects can be incorporated	Resident test population sometimes difficult to delineate due to species mobility
		Exposure regime is difficult to quantify
Community studies	Effects on multiple species can be assessed	Sample collection and analyses require considerable effort
	Data are relevant to ecotoxicological interpretation	Sampling must be done over a considerable time frame
	Biomonitoring approaches can be incorporated	Exposure regime is difficult to quantify
Ecosystem	Provides the ultimate interpretation of the significance of effects	Integrative methods for data analyses are not fully understood
		Some estuaries are too large for intensive studies
		Cost and scale of ecosystem studies are prohibitive

Table 8.16 Findings derived from the use of microcosms, mesocosms and field studies with pesticides

Experimental design and major species tested	Pesticides	Findings	References
Microcosms with algae-dominated aufwuchs	Methyl parathion, 2,4-D (ester)	Transformation rate coefficients were similar in microcosm and natural waters. In microcosms, transformations were suppressed with high concentrations of microbiota	Lewis *et al.*, 1985
Microcosms with field-microbial communities	Methyl parathion, azinphosmethyl, Kepone	Various rates of transformation of two dissimilar-chemical pesticides can be studied in this microcosm. Test system provided sensitive indications of parent compound degradation	Portier, 1985
Microcosms with 10 species of algae; five species of invertebrates	Atrazine	Community responses fell into two categories, those with $60-200\,\mu g/l$ treatment and those with $500-5000\,\mu g/l$. In the lower treatment the responses in the microcosms were: an autotrophic phase, a daphnid bloom and an equilibrium phase. The major effect was an inhibition of primary production, not acute mortality	Stay *et al.*, 1985
Microcosms with algal and invertebrate mixtures from natural ponds	Atrazine	Ranges of sensitivity in the microcosms were less than the range reported for single-species bioassays using common test organisms. The use of microcosms accurately reflected concentrations causing ecosystem effects in experimental ponds	Stay *et al.*, 1989
Microcosm-3 comparts: algae, daphnids, bacteria or sand filter	Chlorpyrifos	After the addition of pesticide, daphnids decreased but population recovered; pH decreased two units; reduction in alga biomass. After 90 days, pH returned to that of controls. These effects were not predicted by single-species testing	Kersting & van Wisjgaarden, 1992
Small pond mesocosms using caged rainbow trout plus natural zooplankton	Cyfluthrin	Small cylindrical ponds, with interconnecting locks, were colonized before the test material was added. The locks were closed and caged fish could be removed without disturbing the systems. The design provided an almost identical biocenosis for 10 weeks. The distribution patterns of the pesticide in the ponds were similar to natural systems	Heimbach *et al.*, 1992

Continued

Table 8.16 *continued*

Littoral enclosures in a pond with macrophytes, micro- and macroinvertebrates and fishes	Esfenvalerate	One to 5 µg/l of the pesticide severely reduced or eliminated most crustaceans, chironomids, young bluegills and larval cyprinids. Copepods and insects were reduced at 0.08–0.2 µg/l, even up to 53 days. Recoveries of some invertebrates occurred by day 25 at ≤ 0.2 µg/l	Lozano *et al.*, 1992
Limnocorrals with zooplankton	Tetrachlorophenol	Two experimental designs were used: blocked analysis of variance (ANOVA) and regression. The no-effect for major zooplankton and other dominant species ranged from 0.28 µg/l for immature copepods to 0.50 mg/l for several daphnids. The regression design looks promising vs ANOVA for estimating specific endpoints such as EC_{50}	Liber *et al.*, 1992
Mesocosms with bluegill sunfish, benthic species and zooplankton	Esfenvalerate	Survival of fish, and reproductive success inversely correlated with pesticide concentrations. Zooplankton decreased at 0.25 µg/l although the pesticide's half-life was 10 h	Fairchild *et al.*, 1992
Mesocosms with bluegill sunfish, micro- and macroinvertebrates	Esfenvalerate	Ecosystem effects noted at mid and high treatments with microinvertebrates reduced at 4.1 and 23.3 g/0.1 ha ponds, but not at control and 0.23 g/0.1 ha ponds. High-rate ponds had significantly increased metabolism and reduced macroinvertebrates as well as the absence of 2-cm-sized bluegills at harvest	Webber *et al.*, 1992
Outdoor experimental streams with periphyton	Pentachlorophenol	Reduced biomass as low as the (previously derived) criterion concentration of 48 µg/l. Also noted, suppressed community metabolism in the 48 µg/l treatment channel and above	Yount & Richter, 1986
Outdoor experimental streams with fish and invertebrates	Pentachlorophenol	Some ecosystem effects noted at all treatments. Microinvertebrates were reduced; fathead minnows and bluegill sunfish died in high-treatment channel. The study indicated that the criterion (previously derived) did not fully protect the animals	Zischke *et al.*, 1985

Continued on page 184

Table 8.16 *continued*

Experimental design and major species tested	Pesticides	Findings	References
Outdoor experimental streams (intermittent vs continuous) dosing with naturally colonizing plants, invertebrates and stocked fathead minnows and bluegill sunfish	Chlorpyrifos	Similar results from intermittently and continuously dosed streams; species diversity decreased by equal amounts. Notable signs of poisoning of fish only in pulse-dosed streams but they survived, reproduced and grew equally well. Results of fish and macroinvertebrates of stream exposures were similar to those in laboratory studies	Eaton *et al.*, 1985
Outdoor experimental streams with fish, invertebrates and periphyton	Pentachlorophenol	This study was to compare the results of experimentally dosed streams with a water quality criterion (48 µg/l) using single-species tests. However, effects on fish, periphyton and system metabolism were found at 48 µg/l in the streams. The acceptability of the 48 µg/l criterion depended on the definition (and endpoints) of the aquatic life used	Hedtke & Arthur, 1985
Littoral enclosures within a 2 ha pond with ambient species	Chlorpyrifos	Growth of larval fathead minnows significantly reduced — especially noticeable 15 days post-treatment. Also noted, significant reductions in cladocerans, copepods, rotifers and chironomids, all forage species for fish	Brazner & Kline, 1990
Littoral enclosures within a 2 ha pond with ambient species with deliberate introductions of fathead minnows and bluegill sunfish	Chlorpyrifos	Pesticide degradations within treatments were similar; rapid decline in concentrations with slower than expected vertical mixing; toxicity to invertebrates more severe than expected but toxicity to the fish similar to that predicted from laboratory studies. Reduction of forage species was associated with reduced growth	Siefert *et al.*, 1989
Field application in a Florida salt marsh with truck-mounted ultra-low volume (ULV) equipment (part I of II)	Fenthion	Wide variation in the efficiency of the application on target species (mosquitoes); topographical features such as sand dunes and stands of trees affected wind speed and directions of spray	Clark *et al.*, 1985

Continued

Table 8.16 *continued*

Field application in a Florida salt marsh with truck-mounted ULV equipment (part II of II)	Fenthion	For all applications, the highest concentration detected in the upper portion of the water column was 0.48 µg/l, and detectable concentration persisted for up to 24 h. The pesticide did not accumulate in the tissues of caged shrimp or fish	Moore *et al.*, 1985
Field application in a Florida salt marsh with truck-mounted ULV equipment (deployment of caged pink shrimp)	Fenthion	Associated laboratory tests and field applications successfully predicted the range of lethal and non-lethal acute field exposures to this pesticide for pink shrimp in similar exposure regimes	Borthwick *et al.*, 1985
Field application in a Florida salt marsh with truck-mounted ULV equipment (deployment of caged mysids)	Fenthion	Increased mortality, reduced growth and increased rate of oxygen consumption in mysids	McKenny *et al.*, 1985
Aerial application to an estuary with caged mysids and pink shrimp	Fenthion	Deployed cages of mysids and pink shrimp resulted in toxicity from 'no effect' to 100% mortality, but these responses were expected based on laboratory tests that established the 24-, 48- and 72-h LC_{50} values. Previously derived laboratory pulse−exposure tests were predictive of the 'no-effect' and 'effect' pulse exposures in the field	Clark *et al.*, 1986
Aerial application to an estuary with caged mysids, pink shrimp, and sheepshead minnows	Fenthion	Two sites with different dilution/mixing regimes gave distinctive results. At site 1, with substantial dilution and mixing, pesticide concentrations decreased quickly from 1.5 and 0.29 µg/l to ≤ 0.02 µg/l with no observed mortality in caged species. At site 2, maximum concentrations were 2.6 and 0.51 µg/l and measurable concentrations (> 0.038 µg/l) persisted for 4 days. Mortalities occurred in the cages with mysids and pink shrimp, with no losses among the cages with sheepshead minnows	Clark *et al.*, 1987
Freshwater mesocosms with bluegills, zooplankton and phytoplankton	Esenvalerate	Zooplankton, macroinvertebrates and invertebrates decreased at 0.25 µg/l. Changes in	Fairbrother *et al.*, 1991

Continued on page 186

Table 8.16 *continued*

Experimental design and major species tested	Pesticides	Findings	References
		invertebrates were partly obscured by indirect effects, i.e. rotifers might have increased due to diminished grazing by fewer cladocerans and copepods. Reductions in recruits of bluegills had variable, secondary effects on subsequent growth rates, i.e lowest growth rates of fish occurred in the control and low-treatment ponds	
North Dakota wetlands with amphipods, conchostracans, odonate nymphs and gastropods	2,4-D Ethyl parathion, methyl parathion (tested separately)	Survival of enclosed invertebrates was monitored after aerial application to wetlands and watersheds. 2,4-D had an acute effect on survival of invertebrates. After the application of ethyl parathion no amphipods or odonate nymphs survived after 1 day. Amphipod survival in treated wetlands continued to be significantly lower until 21 days after application. Methyl parathion was less acutely toxic but survival of amphipods and odonate nymphs was less than 30% a day after application. Only after 14 days were amphipods again able to survive in the previously treated wetland	Borthwick, 1988

8.4 ASSESSMENTS OF PESTICIDE USE AND CONTROL

8.4.1 Risk, hazard and ecological impact assessments involving pesticides

Risk assessment, hazard assessment and environmental impact assessment are often used interchangeably, so definitions are appropriate. Since each has its own meaning, we shall use the definitions in a glossary provided by Parkhurst *et al.* (1990). Risk assessment is 'a set of formal scientific methods for estimating the probabilities and magnitudes of undesired effects resulting from the release of chemicals, other human action, or natural catastrophes'. Risk assessment includes quantitative determination of both exposure and effects; the latter is estimated using mathematical models to predict exposure of populations in a field setting. Ecological risk assessments usually consider not only the effects of chemicals and mixtures of complex materials (wastewaters) but also perturbations such as habitat changes that affect all components of an ecosystem. Hazard assessment is defined as a 'component of risk assessment that consists of the review and evaluation of toxicological data to identify the nature of the hazards associated with a chemical, and to quantify the relationship between dose and response', the latter usually determined in a laboratory setting. An environmental impact assessment is an attempt to analyse and evaluate the effects of human actions on natural environments.

Most assessment schemes use tiered protocols (e.g. Table 8.17). In the first or initial tiers the required short-term acute toxicity tests use either indicator or surrogate species in laboratory exposures. Lethal concentrations (LC_{50}) or effective concentrations (EC_{50}) are usually used as endpoints. As investigators progress through subsequent testing to the second tier, both a greater variety and an increased complexity of tests are required. Finally, tiers III and IV require full life-cycle tests using a fish species followed by field or mesocosm studies.

In their review, Parkhurst *et al.* (1990) stated that the most common approach to determining risk involves using the quotient method. This involves calculating the ratio of the estimated environmental concentration (the exposure concentration) to an estimated safe concentration (determined by toxicity tests). If the calculated ratios are equal to or less than one, a relatively lower risk is indicated; a ratio greater than one indicates a higher risk. An example of a simplified risk analysis is given (Table 8.18) where the estimated environmental concentrations (e.e.c.) are compared to effects based on toxicity tests. In their review of risk assessment protocols, Parkhurst *et al.* (1990) found five risk assessment protocols currently designed for use in assessing pesticides (Table 8.19). The standard approach for regulating pesticides is reflected in the data requirements for registration (Code of Federal Regulations, 1991) in the USA (Table 8.20). Although the data required and outlined in the table are not organized into definitive categories or tiers, the complexity of the tests and amount of toxicological information obtained increases if the test requirements are followed from the table's top to bottom.

The tests for terrestrial (avian and mammalian) and aquatic species can be required or conditionally required depending on specific questions about the pesticide, or they may be required only when an experimental use permit is requested.

Table 8.17 Aquatic species toxicity testing required by the Office of Pesticide Programmes, US EPA draft document (Rodier *et al.*, 1990)

Tier I
96-h LC_{50} test using a cold-water fish
96-h LC_{50} test using a warm-water fish
48-h (or 96-h) LC_{50}/EC_{50} test using a freshwater aquatic invertebrate
96-h EC_{50} test using an alga

Tier II
96-h LC_{50} test using an estuarine/marine fish
96-h LC_{50} test using an estuarine/marine crustacean
48-h EC_{50} test using a bivalve embryo−larvae or
96-h EC_{50} test using bivalve shell deposition
Fish early life-stage maximum acceptable toxicant concentration
  (MATC) or effect/no-effect level
Aquatic invertebrate life-cycle MATC or effect/no-effect level
Fish bioaccumulation factor, e.g. 1000×
Special aquatic organism test data (e.g. fish AChE level)
Aquatic plant growth testing

Tier III
Full fish life-cycle MATC or effect/no-effect level

Tier IV
Field testing of aquatic organisms (e.g. mesocosm or pond studies)

For example, a particular test may be required depending on the intended use, such as on aquatic or terrestrial environments, on food or non-food crops, or indoors in a greenhouse or outdoors in a forest. An important question that would influence the testing requirements is whether or not the pesticide is applied directly to a dry or wet environment, the latter posing a greater risk for pesticide movement. Special tests might also be necessary depending on whether the product formulation is solid, liquid, gas, a highly volatile liquid or a highly reactive solid. Additional caveats that trigger further testing involve questions about whether the pesticide is used repeatedly or continuously, the degree of mobility, its fate in water, soil/sediment or tissues, formation of metabolites, degradation rates of the metabolites, and effects on reproduction, growth or other physiological indicators used with various appropriate test species.

In Canada a proposed guidance manual has been developed for registering chemical pesticides for non-target plant testing (Boutin *et al.*, 1993). In the manual, considerable effort is given to the development of stepwise procedures including unicellular algae, aquatic vascular and terrestrial plants, microcosm, mesocosm and field testing. Particularly useful is the section on tier II tests, with accompanying suggestions for experimental design and statistical analysis of data.

There is increasing awareness of the need for longer-term environmental monitoring both before and after the pesticide is applied. Certainly, it should become clear that much more research is needed in this extremely important area. For example, the length of time required to conduct risk or hazard assessment protocols properly, such as multiple-species tests (mesocosms, or field studies with pesticides, or entire-life-cycle tests), is often limited. Mesocosms or field studies last only a few days, a few weeks or perhaps a growing season, but never long enough to detect some of the subtle effects that become a problem after decades of use. While a life-cycle test may be the longest test required in the protocols, the focus is limited to extremely narrow endpoints of one test species. Table 8.21 illustrates DDT effects (in birds) that were not detected until 7–10 years after its widespread use (US EPA, 1989b).

Before new pesticides are registered for use in the USA the assessment emphasis is on the use of aquatic mesocosms (Table 8.16), and a reliance on information derived from aquatic laboratory tests (Tables 8.18, 8.20 and 8.21). In the UK, however, there is a greater emphasis on monitoring terrestrial species after the pesticides are used (Greig-Smith *et al.*, 1988).

Because 26% of the UK is arable, the impact of pesticides on the diversity of terrestrial communities is substantial (Aldridge & Carter, 1992).

Table 8.18 US EPA regulatory risk criteria for aquatic organisms. From Urban & Cook, 1986; modified by Jenkins *et al.*, 1989

Regulatory presumption	Acute toxicity	Chronic toxicity
Presumption of no risk	e.e.c.[a] $< 1/10$ LC_{50}	e.e.c. $<$ Chronic no effect level
Presumption of risk that may be mitigated by restricted use[b]	$1/10$ $LC_{50} \leq$ e.e.c. $< 1/2$ LC_{50} e.e.c. $\geq 1/10$ LC_{50}	n.a.[c]
Presumption of unacceptable risk Non-endangered species	e.e.c. $\geq 1/2$ LC_{50}	e.e.c. $\geq$ Chronic effect levels including reproductive effects
Endangered species	e.e.c. $> 1/20$ LC_{50} or e.e.c. $> 1/10$ LC_{50}	e.e.c. $\geq$ Chronic effect levels including reproductive effects: also any adverse habitat modification

[a] e.e.c., Estimated environmental concentration.
[b] Restricted use is a classification of a pesticide whereby its use is limited to applications which have been certified by US EPA through EPA-approved training programmes.
[c] n.a., Not available.

Table 8.19 Synopsis of aquatic ecological risk assessment pesticide protocols and their applications. Modified from Parkhurst *et al.*, 1990

Protocol	Regulation	Source or class of chemicals	Environment
1 Burns *et al.*, 1990	FIFRA, TSCA, CWA	Pesticides, new chemicals	All surface waters
2 US EPA, 1984a	TSCA, CWA	Pesticides	All surface waters
3 Urban & Cook, 1986	FIFRA, TSCA, CWA	Pesticides, new chemicals	All surface waters
4 Rodier & Zeeman, 1994	FIFRA, TSCA, CWA	Pesticides, new chemicals	All surface waters
5 Onishi & Wise, 1982	FIFRA, CWA	Pesticides	Inland waters

Table 8.20 Data requirements for wildlife and aquatic organisms (Code of Federal Regulations, 1991)

Avian and mammalian testing
Avian oral LC_{50} test, preferably using a mallard or bobwhite
Avian dietary LC_{50} test, preferably using a mallard or bobwhite
Wild mammal toxicity test
Avian reproduction test, preferably using a mallard or bobwhite
Simulated and actual field testing of mammals and birds

Aquatic organism testing
Freshwater fish LC_{50} test, preferably using rainbow and bluegill
Acute freshwater invertebrate LC_{50} test, preferably using *Daphnia*
Acute LC_{50} test, using estuarine and marine organisms
Fish early life-stage and aquatic invertebrate life-cycle test
Fish life-cycle test
Aquatic organism accumulation test

Simulated or actual field testing — aquatic organisms

Table 8.21 Maximum bioaccumulation of DDT in a forest food chain (US EPA, 1989b)

Receptor	Chemical	Years to maximum concentration
Foliage	DDT	0
Forest litter	DDT/DDE	1
Litter invertebrates	DDT/DDE	2
Ground-feeding birds	DDE	4–5
Canopy-feeding birds	DDE	5–7
Bird-eating hawks and owls	DDE	7–10

There are various panels of experts involved in 10 areas of ecological concern regarding pesticides, including soil, groundwater, surface water, aquatic organisms, microorganisms, honeybees, arthropod natural enemies, terrestrial vertebrates, earthworms and mesofauna (Fig. 8.2). For instance, recommendations from an international workshop on earthworm ecotoxicology (Anon., 1991b), placed equal emphasis on using earthworms as bioindicators in the field, as well as in laboratory assessments, for post-pesticide registration study in the UK. Because the majority of pesticides are applied to terrestrial environments (i.e. soils), earthworm tests are useful indicators of pesticide pollution (Callahan *et al.*, 1991).

In a review of risk assessment in the UK, Greig-Smith (1992) discussed long-range terrestrial monitoring programmes including the United

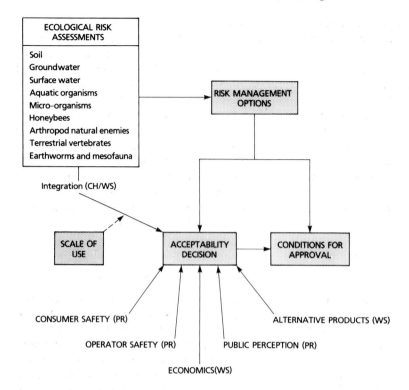

Fig. 8.2 An example of ecological risk integrated with other priorities affecting regulatory decisions about the acceptability of chemical uses. Appropriate methods of prioritization are indicated by: CH, checklist; PR, priority ranking; WS, weighted score. After Greig-Smith, 1992.

Kingdom Wildlife Incident Investigation Scheme that lasted 4 years, the gamebird chick-food (invertebrates) monitored for over 20 years, and information gained by national surveys of bird populations. The Boxworth Project is also an example of a long-term monitoring programme (Cooper, 1990; Greig-Smith, 1991). It started with 2 years of baseline data followed by 5 years of monitoring the density of animals and plants in the fields after pesticide application.

While the UK's long-range programmes are impressive, past widespread monitoring of pesticidal environmental effects have often proved inadequate by not alloting enough elapsed time for proper assessment. To illustrate this point Magnuson (1990) states:

> within this [our perception of a] time scale, ecosystems change during our lifetimes and the lifetimes of our children and our grandchildren. This is the time scale of acid deposition, the invasion of non-native plants and animals, the introduction of synthetic chemicals.

Furthermore, Schindler (1987), speaking of environmental impact assessment, agreed with others (Schindler, 1976; Rosenberg & Resh, 1981; Hecky *et al.*, 1984; and Larkin, 1984) 'who stressed the lack of long-term monitoring in the post-impact period to assess the accuracy of the predictions which had been made during pre-impact studies'. As mentioned in Section 8.2.4, pesticides that should have diminished years ago, such as banned DDT in the USA, are still present in organisms (e.g. molluscs and fish) today. Therefore, the persistence of toxic chemicals and their breakdown products (such as DDT and its metabolites) should determine the length of time allotted for post-impact assessment.

Parkhurst *et al.* (1990) reported on a review by a panel of investigators of 26 risk-assessment protocols. The review revealed that only 18 protocols contained potentially useful guidance. Protocol procedures were not comprehensive and flexible enough to be widely applicable, and most were designed for specific regulatory applications. While the protocol analyses of pesticides were specific, and would seem beneficial for regulation, the uncertainty factors could not be quantified. Further, the panel found that most of what was termed aquatic ecological risk assessments now in use, are actually hazard-assessment procedures that were not usable for determining estimates of risk for many substances. The result is that agencies faced with the serious responsibility of regulating pesticide use, manufacturing industrial chemicals or discharging even small amounts of pesticides in wastewater have little alternative but to set conservative limits regarding their application.

Some specific questions related to risk uncertainty have been posed by Greig-Smith (1990). He discussed the decisions involved in reaching a balance between the numbers of study sites to which pesticides are applied, and the scale of detail to be undertaken at each site. He also suggested that the considerations to be tested must include worst-case scenarios to deal with unacceptable hazards that may arise if the pesticide were applied under extreme conditions. Examples of extreme conditions include hazards to non-target species that vary with good agricultural practices, habitat or wildlife populations, spillage of pellets or granules, improper mixing or extreme weather conditions. Lastly, the point is made that comparable untreated control areas must also be examined where varying habitats, wildlife populations and weather conditions apply.

A review of aquatic risk assessment involving pesticides was recently published by the World Wildlife Fund (1992). Selected groups of investigators representing government, universities and industry compiled the conclusions and recommendations from a 16-month study which focused on four major areas: (i) the adequacy of current aquatic testing tiers, (ii) improvements made in modelling and estimating expected environmental pesticide concentrations for making risk-assessment decisions and for dosing in simulated field studies, (iii) practical challenges in mesocosm methodology and experimental design and (iv) interpretation of, and extrapolation from, mesocosm studies. The main conclusions drawn addressed both meso- and microcosms. The investigators concluded that while mesocosms can generate necessary and useful data for ecological risk assessment, they exhibit less variability than that observed between natural ponds. Use of small outdoor microcosms was suggested to provide a useful bridge between laboratory and field studies. They agreed that methods of modelling pesticide transport and fate to develop estimated environmental concentrations need to be improved. Additional conclusions were that further study of appropriate experimental designs and response evaluations of fish communities to pesticides in mesocosms were needed, as were more studies on direct or indirect effects, strength of exposure–response relationships and linkages among ecosystem components.

8.4.2 Approaches to risk assessments

Ecological risk assessment has recently been reviewed for the USA by Norton *et al.* (1992), and for the UK by Greig-Smith (1992). The first is a review of ecological risk assessment using non-specific examples of ecological effects that could have been caused by a variety of human activities. Examples ranged from draining a wetland to releasing chemicals into the environment. Also presented was a framework whereby the assessor was provided with a consistent approach to ecological risk assessment, assistance for identifying key issues and guidance for defining terminology. Factors of risk beyond a single species were emphasized, focusing on examining a population, a community or an ecosystem. The second review (Greig-Smith, 1992) involved ecological risk assessment in Europe including, as an example, pesticide registration procedures in the UK. Identification of 10 critical issues involved in designing risk-assessment procedures, and the advantages or disadvantages of predicting ecological effects of pesticide use, were presented. Five of the critical issues, often minimized in previous approaches to risk analysis or perhaps overlooked by the analyst, are as follows: (i) extrapolation between species and conditions, (ii) use of trigger (threshold) values in stepwise testing, (iii) setting margins of safety, (iv) use of uncertainty analysis, and (v) classification of risk, categorized as high, medium or low. The outline provided by Greig-Smith (1992) for predicting ecological risk

(Fig. 8.2) is designed to be used with the following prioritization techniques:

1. A checklist of possible ecological effects of concern, each of which should be examined for evidence of risks. Decisions about acceptability could then depend on whether a specified minimum percentage of cases showed adequate indications of ecological safety.
2. A priority ranking among species and effects to identify those for which assurance of safety is essential (for example, [recreationally or economically important] species . . . or the efficiency of important processes). Adverse results for those cases would override any information on the less important effects.
3. A weighted score, which involves computing a combined result across a number of species and effects, taking into account the higher priorities attached to a certain number of (recreationally and economically important species) (Greig-Smith, 1992).

8.4.3 Discussion of established criteria, advisories, standards or guidelines for pesticides

In the USA, water quality criteria have been developed and published for approximately 25 pollutants, with 15 of them being pesticides (US EPA, 1986b). This publication is usually referred to as either the 'Gold Book' or 'national criteria', because the abstracted versions of the guidelines for deriving the criteria referred to them as such. The first series of 64 documents (including some pesticides) was published pursuant to section 304(a) of the Clean Water Act (US EPA, 1980). Prior to the publishing of criteria, several versions of procedures for deriving water quality criteria were made available, beginning in the late 1970s and 1980s; these are summarized and discussed by Stephan (1985). The final guidance was published as 'Guidelines for deriving numerical national water quality criteria for the protection of aquatic organisms and their uses' (US EPA, 1984b). In the guidance section, before deriving a criterion for chemicals, including pesticides, the following tests were presented as necessary.

Data necessary for deriving a criterion for freshwater

Acute test using freshwater species should use:
1 a member of the family Salmonidae;
2 one other fish species (preferably a commercially or recreationally important warmwater species) in the class Osteichthyes, such as a bluegill sunfish or catfish;
3 one other family in the phylum Chordata such as a fish or amphibian;
4 a planktonic crustacean such as a daphnid;
5 a benthic crustacean such as an ostracod, isopod, amphipod, glass shrimp or crayfish;
6 an insect such as a mayfly, dragonfly, damselfly, stonefly, caddisfly, mosquito or midge;
7 a family in a phylum other than Arthropoda or Chordata such as a rotifer, annelid or mollusc;
8 a family in any insect order or phylum not already represented.

Chronic tests necessary for the derivation of acute or chronic ratios of aquatic species in at least three different families provided that:
1 at least one is a fish species;
2 at least one is an invertebrate species;
3 at least one is an acutely sensitive species.

Results of tests with a freshwater alga or vascular plant.

If plants are among the aquatic species most sensitive to the material (i.e. a herbicide), results of a test with a plant in another phylum (division) should also be available.

Results of one bioconcentration test, determined with a freshwater species.

Data set for the derivation of a criterion for saltwater

Acute tests using saltwater species should use:
1 two different families in the phylum Chordata;
2 a family in a phylum other than Arthropoda or Chordata;
3 either the Mysidacea or Penaeidae family;
4 three other families not in the phylum Chordata;
5 any other family.

Chronic tests required for the derivation of acute/chronic ratios with species of aquatic animals in at least three different families provided that:

1 at least one is a fish species;
2 at least one is an invertebrate species;
3 at least one is an acutely sensitive saltwater species.

Results of a test with a saltwater algal species or vascular plant species.

If plants are among the aquatic species most sensitive to the material (i.e. a herbicide), results of a test with a plant in another phylum (division) should also be available.

Results of one bioconcentration test determined with a saltwater species.

Derivation of acute and chronic limits

Acute limits are determined using the information from the acute tests above. The data are then reviewed, and if more than one test is available for each species, geometric means are calculated. After ranking the data from the least sensitive species (using LC_{50} or EC_{50}) to the most sensitive species, a cumulative probability of 0.05 is calculated based on the four lowest acute values using a linear extrapolation intercept along a 'Y' axis. If, on the other hand, the data set is larger than 59, then the lowest of the four values becomes the final acute value. However, if there is a commercially or recreationally important species in the data set, then the number for that species is considered the lowest acute value if it is lower than that computed from the cumulative probability. For the chronic value, emphasis is placed on the quality of the data collected from the chronic tests. For instance, a final chronic value is calculated provided that the data are from well-designed tests that employed flow-through methods with frequently measured concentrations of the toxicant; that the tests met the minimum requirements of acceptability of survival, growth and reproduction; that the test water was not unusually different from naturally occurring average values of hardness, organic carbon and particulates; and that the duration of the chronic tests was within acceptable guidelines. The guidelines suggest that the duration be linked to the life-cycle of the organisms; therefore,

acceptable data from chronic studies are limited to entire-lifecycle tests; partial-life-cycle tests, in which the exposures begin with immature juveniles before the development of gonads and continue through maturation and reproduction; or early-life-stage tests, in which the exposures begin after fertilization through embryonic, larval and early juvenile development. A chronic value is then obtained by calculating the geometric mean of upper and lower chronic limits of the test based on endpoints of survival, growth or reproduction, or, in some cases, a well-defined endpoint that has been accepted by the scientific community.

Other considerations for deriving a criterion

In the guidelines are provisions for deriving the final criterion based on toxicity data gathered from plants. If the test substance bioconcentrates in tissues of aquatic species, a criterion based on a final residue value may be calculated from a bioconcentration test.

Usually, a water quality criterion consists of two concentrations. First, a criterion maximum concentration (acute value) is the number equal to one-half the final acute value. Dividing the acute value in half provides greater species protection because the data from acute toxicity tests are from tests in which half of the test species were affected either by dying or with effects measured by other non-lethal endpoints. The second concentration, the criterion continuous concentration (chronic value), is equal to either the lowest final chronic value, the final plant value or the final residue value, unless other data indicate that a lower criterion be used.

For pesticides, as well as for other priority pollutants, the criterion usually consists of the two time-dependent concentrations, the acute and chronic values (Table 8.22). For example, the two-part criterion for chlorpyrifos (dursban) in freshwater is stated in the criteria document (US EPA, 1986a) as:

freshwater aquatic organisms and their uses should not be affected unacceptably if the four-day average concentration of chlorpyrifos does not exceed 0.041 µg/l more than once every three years on the average and if the one-hour average concentration does not exceed 0.083 µg/l more than once every three years on the average.

Interestingly, the two-part criterion for saltwater organisms was even lower. For chlorpyrifos, the 1-h average should not exceed 0.011 µg/l more than once every 3 years and the 4-day average should not exceed 0.0056 µg/l more than once every 3 years. For chlorpyrifos, the criterion was based primarily on toxicity, with major consideration given to chronic toxicity data for the fathead minnow (*Pimephales promelas*) where unacceptable effects occurred in second-generation larvae at 0.12 µg/l. However, growth of the saltwater mysid (*Mysidopis* sp.) was significantly reduced at the nominal concentration of 0.004 µg/l (US EPA, 1986a). Considerations that reduced the

chronic saltwater criterion to lower values were acute and chronic ratios that ranged from about 1.4 to 12.5 µg/l in five sensitive species. By calculating the geometric mean of these values the final acute–chronic ratio for chlorpyrifos resulted in a marine chronic criterion of 0.0056 µg/l (Table 8.22). When criteria are adopted as standards, they are listed as chronic values, as is the case in Florida with demeton, guthion, malathion, methoxychlor and mirex (Table 8.23). For malathion the single criterion is due to its rapid degradation in water, where occurrence is sporadic rather than continuous, and toxicity is exerted through the inhibition of the enzyme AChE. For mirex the criterion is based on chronic toxicity and the potential for bioaccumulation in aquatic food chains. It is unfortunate that, after nearly five decades of studying modern-day pesticides, only 15 (including some metals used as pesticides)

Table 8.22 Water quality criteria[a] for pesticides (µg/l) (US EPA, 1986b, 1991)

Pesticide	Freshwater acute	Freshwater chronic	Marine acute	Marine chronic	Number of states with aquatic life standards
Aldrin	3.0	—	1.3	—	16
Arsenic[+++b]	360.0	190.0	69.0	36.0	22
Chlordane	2.4	0.0043	0.09	0.004	12
Chlorine[b]	19.0	11.0	13.0	7.5	21
Chlorpyrifos[c]	0.083	0.041	0.011	0.0056	0
Chromium[+++b]	1700.0	210.0	10300.0	—	24
Copper[b]	18.0	12.0	2.9	—	20
DDT and metabolites	1.1	0.001	0.13	0.001	16
Dieldrin	2.5	0.0019	0.71	0.0019	16
Endosulfan	0.22	0.056	0.034	0.0087	10
Endrin	0.18	0.0023	0.037	0.0023	18
Heptachlor	0.52	0.0038	0.053	0.0036	12
Lindane[d]	2.0	0.08	0.16	—	12
Parathion	0.065	0.013	—	0.04	8
Toxaphene	0.73	0.0002	0.21	0.0002	17

[a] Water quality advisories are available for the following pesticides and metabolites: dimethyl tetrachloroterephthalate (DCPA), propachlor, metribuzin, metalochlor, alachlor, heptachlor epoxide, mirex, endosulphansulphate, chlorpyrifos (Criteria and Standard Division, Office of Water Regulations and Standards, US EPA, Washington, DC, 1986).
[b] Although copper and chlorine are not always considered as pesticides, copper salts are used widely for the control of algae, chlorine is used as a disinfectant and chromated copper arsenate is used widely as a fungicide, insecticide and wood preservative.
[c] Also known as Dursban or Lorsban.
[d] Hexachlorocyclohexane.

Table 8.23 Pesticides adopted 7 December, 1990; effective date 13 February, 1992, listed in Florida surface water classification (*Florida Surface Water Quality Standards*, 1992)

Pesticide (mg/l)[a]	Recreational freshwater, Class III		Recreational saltwater, Class III	
	Maximum	Average	Maximum	Average
Aldrin	3.0	0.00014[c]	1.3	0.00014[c]
Chlordane	0.0043	0.00059[c]	0.004	0.00059[c]
DDT	0.001	0.00059[c]	0.001	0.00059[c]
Demeton[b]	—	0.01	—	0.01
Dieldrin	0.0019	0.00014[c]	0.0019	0.00014[c]
Endosulfan	—	0.056	—	0.0087
Endrin	—	0.00023	—	0.0023
Guthion[b]	—	0.01	—	0.01
Heptachlor	0.0038	0.00021[c]	0.0036	0.00021[c]
Lindane	0.08	0.063[c]	0.16	0.063[c]
Malathion[b]	—	0.1	—	0.1
Methoxychlor[b]	—	0.03	—	0.03
Mirex[b]	—	0.001	—	0.001
Pentachlorophenol	30.0	pH-dependent[d]	—	7.9
Parathion	—	0.04	—	0.04
Toxaphene	—	0.0002	—	0.0002

[a] All pesticides expressed as $\leq$ values as µg/l (parts per billion).
[b] Only one number was published.
[c] Based on recalculated values for the protection of human health.
[d] $\leq e^{(1.005(pH)-5.29)}$, e.g. 5.7 µg/l when pH = 7.

have established use criteria (Table 8.22). This presents a serious dilemma when over 200 chemicals or natural substances are listed by the US EPA (1988b, 1990) as currently in use.

Criteria and standards for pesticides

After the Water Quality Criteria are reviewed and published, states have the option to promulgate, modify or disregard criteria published by the Environmental Protection Agency. If states adopt the criteria, as Colorado did, the numerical values become enforceable standards (Colorado Water Quality Control Commission, 1991). In Colorado there are basic standards that apply to all waters of the state, and limits for other pollutants such as metals, ammonia and nutrients are applied to specific stream segments. Standards are usually published with stream classifications that define the appropriate use, or uses, of the water resources for the segment. A good example of pesticide

limits are those set in Florida that, as mentioned before, uses a two-part standard for toxicants including pesticides which are presented as the maximum (acute) and average (chronic) limits (Table 8.23).

8.5 SUMMARY AND CONCLUSIONS

Much of what we have reported in this chapter about pesticides as environmental pollutants is alarming. There are, however, also positive trends that include termination or use restriction of the most harmful compounds. There is a noticeable movement towards more ecological and biological approaches to controlling unwanted pests. Use of integrated approaches to control pests (pheromones, pathogenic micro-organisms and environmentally labile insecticides) is also increasing. Hopefully, knowledge about better ecological approaches will result in less reliance on using an abundance of harmful chemicals to control pests, such as the myriad of chemicals that Asia uses in its rice fields to control the brown planthopper (Bull, 1982; Isensee & Tayaputch, 1986).

Although some trends are encouraging, we must not forget that there are sobering findings about pesticides in the environment. Particularly alarming are the chlorinated chemicals such as DDT, dieldrin and toxaphene – many of which were banned or restricted in the USA 10–20 years ago but are still being used in less developed countries today. Chlorinated chemicals became long-term environmental contaminants and, although the USA tried to correct a wrong by discontinuing their use, they are *still* being found in organisms living in both its coastal and inland waters. Similarly, DDE/DDT and 12 other chlorinated compounds are present at toxic levels in small migratory birds, and are affecting peregrine falcons and other raptors high in the terrestrial food chain. DDT, DDE and toxaphene have also been found in fish from national wildlife refuges in the south and southeastern USA and in fish exposed to water from return irrigation systems in the San Joaquin Valley, California. Also detected at other locations were numerous herbicides in surface water in Kansas, and chlorinated pesticides in water, sediments and biota in the New York Bight, and in benthic species taken from the Great Lakes. Evidence from monitoring organochlorine pesticides and PCBs indicates that they are also residing in remote areas such as Arctic samples and marine species. It is also not surprising that pesticides are identified in tissue samples from molluscs and fish, considering that pesticides are frequently found in runoff samples collected from many metropolitan centres. Very recent discoveries of pentachlorophenol and diazinon in municipal wastewaters signals the need for additional monitoring and research directed towards determining the sources of these chemicals in wastewater. Finally, before we detach ourselves from the environment, we too are at risk. Particularly disturbing from the standpoint of human health is the quantity of herbicides, such as atrazine and alachlor, that are currently being detected in groundwater supplies.

Pesticides will continue to present serious ecotoxicological problems, and additional immediate concerns continue to arise with some of the newer compounds. Published acute toxicity values for some of these pesticides are in the parts-per-trillon range. While it is good news that pesticides are being designed to be more toxic to pest species in the parts-per-billion range or even lower, some important non-target species in ecosystems are also affected at these concentrations. For example, data suggest that insects, very important in the functioning of our ecosystems, are typically, and understandably, the most sensitive taxa to pesticides. Next in sensitivity are the crustaceans that, like the insects, are very important to both aquatic and terrestrial food chains. The third most sensitive group to pesticides are the fishes. For proper ecological assessment, testing such organisms as daphnids, which are important as fish-food organisms in open-water lentic ecosystems; gammarids, important benthic fish-food organisms in both lentic and lotic environments; and rainbow trout, one of the most recreationally important species in the USA provides a low toxicity value for pesticides 88% of the time. A larger consideration, however, is that these organisms are representatives of important components of all aquatic ecosystems. Not only are they at the greatest risk from acute effects, they also appear to be subject to chronic effects at low concentrations as well (Table 8.14). To illustrate toxicity at minuscule concentrations, with regard to fenvalerate, the amphipod *Gammarus pseudolimnaeus* had an LC_{50} of 0.032 µg/l, and the saltwater mysid *Mysidopsis bahia* had an LC_{50} of 0.008 µg/l (Table 8.9).

Furthermore, although death is the most traditional endpoint considered in acute tests, research suggests that investigators also consider behavioural effect endpoints. Currently, there is a noticeable emphasis on toxicity endpoints in regulatory risk criteria for aquatic organisms (Tables 8.18 and 8.21). For fish, spontaneous swimming activity, swimming capacity, feeding behaviour and vulnerability to predation are sensitive indicators of exposures to sublethal pesticide concentrations. Typically, however, these are not currently included as endpoints in hazard and risk-assessment protocols. Sublethal affects on terrestrial wildlife include reproductive failure, reduced survival, critical behavioural changes and greater susceptibility to disease.

Additionally, it appears that more considerations must be given to endpoints other than lethality in chronic tests. Observations such as reduced burrowing and feeding activities of invertebrates, loss of coordination, diminished development of brood pouches, impaired ability to avoid predation and scoliosis in fish have been noted. Particularly important was the observation by Suter *et al.* (1987) that the most sensitive endpoint in chronic chemical tests was fecundity. Birds and small mammals may suffer the same adverse effects from chronic exposure as observed in response to sublethal acute exposure. Again, many of these endpoints are usually not currently considered in toxicological data bases for the assessment of pesticidal effects. Because chronic effects are important, however, the long-term monitoring programmes established in the UK for terrestrial species may hopefully become world-wide assessment models for pesticide research.

One of the greatest needs in pesticide research today is improvement of risk assessment procedures and their validation. As Parkhurst *et al.* (1990) summarized, 'Most protocols presently available are, in fact, hazard assessment protocols, or at best, qualitative risk assessment protocols. Few protocols quantify risks and uncertainties.' Desperately needed are longer-term monitoring programmes of mesocosms and field sites including multiple seasons and protocols determining fish fecundity endpoints through multiple generations. An important publication written by Greig-Smith (1992) should provide valuable data and guidance for risk assessments for pesticide use world-wide. Finally, as mentioned above, sublethal and behavioural, as well as lethal test endpoints and the implications of total ecosystem pesticide effects, appear to provide important research opportunities.

In the USA, despite four decades of research on non-target effects of synthetic pesticides coupled with the enactment of environmental laws and initiation of programmes to reduce the effects of pesticides on non-target species, fewer than 20 established criteria exist. Furthermore, of the 15 listed in Table 8.22, one is chlorine used for years as a disinfectant, and four are metals which have been extensively investigated as environmental pollutants because of the problems associated with their extraction from the earth and their use in industry. Generally, less than 20 states have adopted these criteria as state standards. By contrast, the list of pesticides currently registered for use in the USA today (US EPA 1988b, 1990) includes more than 200 chemicals or substances. We must consider also the hundreds of different formulations and combinations of chemical pesticides that are used in practice. Clearly, although we have come a long way in assessing pesticides, much work still needs to be done to control these compounds in our environment.

ACKNOWLEDGEMENTS

We thank Mary J. Willox for reviewing, formating and providing helpful suggestions while preparing this chapter.

REFERENCES

Academy of Scientific Research and Technology, Egypt. (1981) Investigation of level and effects of pollutants in saline lakes and littoral marine environments. *Inland Waters Lake Quarun Studies.* Report No. III.

Aldridge, C. & Carter, N. (1992) The principles of risk assessment for non-target arthropods: a UK registration perspective. *Aspects Appl. Biol.* **1**, 149–156.

Amato, J.R., Mount, D.I., Durhan, E.J., Lukasewycz, M.T., Ankley, G.T. & Robert, E.D. (1992) An example of the identification of diazinon as a primary toxicant in an effluent. *Environ. Toxicol. Chem.* **11**, 209–216.

Anderson, R.L. (1982) Toxicity of fenvalerate and permethrin to several nontarget aquatic invertebrates. *Environ. Entomol.* **11**, 1251–1257.

Anon. (1991a) Guidance document on testing procedures for pesticides in freshwater mesocosms. In: *Workshop at Monks Wood Experimental Station, Abbotts Ripton,* p. 46. Huntingdon, UK.

Anon. (1991b) *Recommendations from the International Workshop on Earthworm Ecotoxicology.* Intercept Ltd. Booklet available from B.G. Printers, 45 West Way, Walworth Estate, Andover, UK.

Beyers, D.W., Carlson, C.C. & Tessari, J.D. (1991) Solid-phase extraction of carbaryl and malathion from pond and well water. *Environ. Toxicol. Chem.* **10**, 1425–1429.

Boardman, R. (1986) *Pesticides in World Agriculture.* St Martin's Press, New York.

Borthwick, P.W., Clark, J.R., Montgomery, R.M., Patrick, Jr, J.M. & Lores, E.M. (1985) Field confirmation of a laboratory-derived hazard assessment of the acute toxicity of fenthion to pink shrimp, *Penaeus duorarum*. In: *Aquatic Toxicology and Hazard Assessment: 8th Symposium, ASTM STP 891* (Eds R.C. Bahner & D.J. Hansen), pp. 177–189. American Society for Testing and Materials, Philadelphia.

Borthwick, S.M. (1988) Impact of agricultural pesticides on aquatic invertebrates inhabiting prairie wetlands. MS thesis, Colorado State University. Fort Collins, CO.

Boutin, C., Freemark, K.E. & Keddy, C. (1993) *Proposed Guidelines for Registration of Chemical Pesticides: Nontarget Plant Testing and Evaluation*. Technical Report Series No. 145, Environment Canada. Canadian Wildlife Service.

Brazner, J.C. & Kline, E.R. (1990) Effects of chloropyrifos on the diet and growth of larval fathead minnows, *Pimephales promelas*, in littoral enclosures. *Can. J. Fish. Aquat. Sci.* **47**, 1157–1165.

Brown, R.A., Hardy, A.R., Greig-Smith, P.W. & Edwards, P.J. (1988) Assessing the impact of rodenticides on the environment. *OEPP/EPPO Bulletin* **18**, 283–292.

Bull, D. (1982) *A Growing Problem: Pesticides and the Third World Poor*. Oxfam, Oxford.

Burmaster, D.E., Menzie, C.A., Freshman, J.S., Burris, J.A., Maxwell, N.I. & Drew, S.R. (1991) Assessment of methods for estimating aquatic hazards at superfund-type sites: a cautionary tale. *Environ. Toxicol. Chem.* **10**, 827–842.

Burns, L.A., Barber, M.C., Bird, S.L., Mayer, Jr, F.L. & Suarez, L.A. (1990) *PIRANHA. Pesticide and Industrial Chemical Risk Analysis and Hazard Assessment*. Version 1.0. US EPA, Office of Research and Development, Athens, GA.

Bushong, S.J., Ziegenfuss, M.C., Unger, M.A. & Hall, Jr, L.W. (1990) Chronic tributyltin toxicity experiments with the Chesapeake Bay copepod, *Acartia tonsa. Environ. Contam. Toxicol.* **9**, 359–366.

Butler, M.K. & Arruda, J.A. (1985) Pesticide monitoring in Kansas surface waters: 1973–1984. In: *Perspectives on Nonpoint Source Pollution, Proceedings of a National Conference*, pp. 196–200. US EPA, Washington, DC. EPA440/5–85–001.

Calamari, D., Bacci, E., Focardi, G., Gaggi, C., Morosini, M. & Vighi, M. (1991) Role of plant biomass in the flobal environmental partitioning of chlorinated hydrocarbons. *Environ. Sci. Technol.* **25**, 1489–1495.

Callahan, C.A., Menzie, C.A., Burmaster, D.E., Wilborn, D.C. & Ernst, T. (1991) On-site methods for assessing chemical impact on the soil environment using earthworms: a case study at the Baird and McGuire Superfund site, Holbrook, Massachusetts. *Environ. Toxicol. Chem.* **10**, 817–826.

Carson, R. (1962) *Silent Spring*. Riverside Press, Cambridge, MA.

Clark, J.R. (1989) Field studies in estuarine ecosystems: a review of approaches for assessing contaminant effects. In: *Aquatic Toxicology and Hazard Assess-ment*, Vol. 12, ASTM STP 1027 (Eds U.M. Cowgill & L.R. Williams), pp. 120–133. American Society for Testing and Materials, Philadelphia.

Clark, J.R., Borthwick, P.W., Goodman, L.R., Patrick, Jr, J.M., Lores, E.M. & Moore, J.C. (1987) Effects of aerial thermal fog applications of fenthion on caged pink shrimp, mysids and sheepshead minnows. *J. Am. Mosq. Contr. Assoc.* **3**, 466–472.

Clark, J.R., Goodman, L.R., Borthwick, P.W., Patrick, Jr, J.M., Moore, J.C. & Lores, E.M. (1986) Field and laboratory toxicity tests with shrimp, mysids, and sheepshead minnows exposed to fenthion. In: *Aquatic Toxicology and Environmental Fate*, Vol. 9, ASTM STP 921 (Eds T.M. Poston & R. Purdy), pp. 161–176. American Society for Testing and Materials, Philadelphia.

Clark, J.R., Middaugh, D.P., Hemmer, J.J., Clements, Jr, B.W., Dukes, J.C. & Rathburn, Jr, C.B. (1985) Effects of ground ULV applications of fenthion on estuarine biota I. Study design and implementation. *J. Florida Anti-Mosq. Assoc.* **56**, 51–62.

Code of Federal Regulations (1991) Part 158 – Data requirements for registration. 40 CFR, Chapter 1 (7–1–91 edition). US Government Printing Office.

Colorado Water Quality Control Commission (1991) *The Basic Standards and Methodologies for Surface Waters*. 24–4–103(9). CRS, Denver, CO.

Cooper, D.A. (1990) Development of an experimental programme to pursue the results of the Boxworth Project. In: *Proceedings of the Brighton Crop Protection Conference – Pests and Diseases*, pp. 153–162.

Couch, J.C., Winstead, J.T. & Goodman, L.R. (1977) Kepone-induced scoliosis and its histological consequences in fish. *Science* **197**, 585–587.

Couch, J.C., Winstead, J.T., Hansen, D.J. & Goodman, L.R. (1979) Vertebral dysplasia in young fish exposed to the herbicide trifluralin. *J. Fish Dis.* **2**, 35–42.

Cripe, G.M., Goodman, L.R. & Hansen, D.J. (1984) Effect of chronic exposure to EPN and guthion on the critical swimming speed and brain acetylcholinesterase activity of *Cyprinodon variegatus. Aquat. Toxicol.* **5**, 255–266.

Cripe, G.M., Nimmo, D.R. & Hamaker, T.L. (1981) Effects of two organophosphate pesticides on swimming stamina of the mysid *Mysidopsis bahia*. In: *Biological Monitoring of Marine Pollutants* (Eds F.J. Vernberg, A. Calabrese, F.P. Thurberg & W.B. Vernberg), pp. 21–36. Academic Press, New York.

Davis, R.P., Garthwaite, D.G. & Thomas, M.R. (1990) *Pesticide Usage Survey Report 86*, pp. 1–3. Pesticide Usage Survey Group, Central Science Laboratory, Hatching Green, Harpenden, Herts AL5 2BD UK.

Day, K.E. (1990) Pesticide residues in freshwater and marine zooplankton: a review. *Environ. Pollut.* **67**, 205–222.

DeWeese, L.R., McEwen, L.C., Hensler, G.L. & Petersen, B.E. (1986) Organochlorine contaminants in passeriformes and other avian prey of the peregrine falcon in the Western United State's. *Environ. Toxicol. Chem.* **5**, 675–693.

D'Mello, G.D. (1992) Neurobehavioral toxicology of anticholinesterases. In: *Clinical and Experimental Toxicology of Organophosphates and Carbameses* (Eds B. Ballantyne, T.C. Marrs & W.N. Aldridge), pp. 61–74. Butterworth-Heinemann, Oxford.

Eaton, J., Arthur, J., Hermanutz, R., Kiefer, R., Mueller, L., Anderson, R., Erickson, R., Nordling, B., Rogers, R. & Pritchard, H. (1985) Biological effects of continuous and intermittent dosing of outdoor experimental streams with chlorpyrifos. In: *Aquatic Toxicology and Hazard Assessment. 8th Symposium.* ASTM STP 891 (Eds R.C. Bahner & D.J. Hansen), pp. 85–118. American Society of Testing and Materials, Philadelphia.

Fairbrother, A., Mardan, B.T., Bennett, J.K. & Hooper, M.J. (1991) Methods used in determination of cholinesterase activity. In: *Cholinesterase-inhibiting Insecticides, Chemicals in Agriculture*, Vol. 2 (Ed. P. Mineau), pp. 35–71. Elsevier, Amsterdam.

Fairchild, J.F., La Point, T.W., Zajicek, J.L., Nelson, M.K., Dwyer, F.J. & Lovely, P.A. (1992) Population-, community- and ecosystem-level responses of aquatic mesocosms to pulsed doses of a pyrethroid insecticide. *Environ. Toxicol. Chem.* **11**, 115–129.

Farm Chemicals Handbook (1990). Meister, Willoughby, OH.

Farr, J.A. (1978) The effect of methyl parathion on predator choice of two estuarine prey species. *Trans. Am. Fish. Soc.* **107**, 87–91.

Federal Register (1984) Guidelines establishing test procedures for the analysis of pollutants under the Clean Water Act; final rule and interim final rule and proposed rule. *Fed. Reg.* **49**(209), 43234–43442.

Florida Surface Water Quality Standards 17–302 (1992) Department of Environmental Regulation, Tallahassee, FL.

Friend, M. & Trainer, D.O. (1974) Experimental DDT–duck hepatitis virus interaction studies. *J. Wildlife Manage.* **38**, 887–895.

Gianessi, L.P. & Puffer, C. (1991) *Herbicide Use in the United States.* Quality of the Environment Division, Resources for the Future. National Summary Report, Washington, D.C.

Gilbert, T.R. & Kakareka, J.P. (1985) Analytical chemistry. In: *Fundamentals of Aquatic Toxicology* (Eds G.M. Rand & S.R. Petrocelli), pp. 475–494. Hemisphere, Washington, DC.

Goldenman, G. (1987) Pesticides and wetlands overseas: a framework for mitigating adverse impacts. In: *Report to the United States Environmental Protection Agency National Network for Water Policy Research and Analysis Pilot Project.* Washington, DC.

Goodnight, C.J. (1942) Toxicity of sodium pentachloro-phenate and penthachlorophenol to fish. *Ind. Eng. Chem.* **34**, 868–872.

Graham, F. Jr (1970) *Since Silent Spring.* Houghton Mifflin, Boston, MA.

Greig-Smith, P.W. (1990) Intensive study versus extensive monitoring in pesticide field trials. In: *Pesticide Effects on Terrestrial Wildlife* (Eds L. Somerville & C.H. Walker), pp. 217–239. Taylor & Francis, London.

Greig-Smith, P.W. (1991) The Boxworth experience: effects of pesticides on the fauna and flora of cereal fields. In: *Proceedings of the 32nd Symposium of The British Ecological Society with the Association of Applied Biologists* (Eds L.G. Firbank, N. Carter, J.F. Darbyshire & G.R. Potts), pp. 333–371. Blackwell Scientific Publications, Oxford.

Greig-Smith, P.W. (1992) A European perspective on ecological risk assessment, illustrated by pesticide registration procedures in the United Kingdom. *Environ. Toxicol. Chem.* **11**, 1673–1689.

Greig-Smith, P.W., Somerville, L., Walker, C.H., Hardy, A.R., Klein, W., Mogensen, B., Pfluger, W., Riley, D., Stanley, P.I. & Vighi, M. (1988) Pesticides and wildlife-field testing. *Recommendations of an International Workshop on Terrestrial Field Testing of Pesticides*, pp. 1–39. Selwyn College, Cambridge, UK.

Grue, C.E., Hart, A.D.M. & Mineau, P. (1991) Biological consequences of depressed brain cholinesterase activity in wildlife. In: *Cholinesterase-inhibiting Insecticides, Chemicals in Agriculture*, Vol. 2 (Ed. P. Mineau), pp. 152–209. Elsevier, Amsterdam.

Hargrave, B.T., Harding, G.C., Vass, W.P., Erickson, P.E., Fowler, B.R. & Scott, V. (1992) Organochlorine pesticides and polychlorinated biphenyls in the Arctic Ocean food web. *Arch. Environ. Contam. Toxicol.* **22**, 41–54.

Hecky, R.E., Newbury, R.W., Bodaly, R.A., Patalas, K. & Rosenberg, D.M. (1984) Environmental impact prediction and assessment: the Southern Indian Lake experience. *Can. J. Fish. Aquat. Sci.* **41**, 730–732.

Hedtke, S.F. & Arthur, J.W. (1985) Evaluation of a site-specific water quality criterion for pentachlorophenol using outdoor experimental streams. In: *Aquatic Toxicology and Hazard Assessment: 7th Symposium.* ASTM STP 854 (Eds R.D. Cardwell, R. Purdy & R.C. Bahner), pp. 551–564. American Society for Testing and Materials, Philadelphia.

Heimbach, F., Pflueger, W. & Ratte, H. (1992) Use of small artificial ponds for assessment of hazards to aquatic ecosystems. *Environ. Toxicol. Chem.* **11**, 27–34.

Hileman, B. (1993) Concerns broaden over chlorine and chlorinated compounds. *Chem. Eng. News* 19 April.

Hill, E.F. & Camardese, M.B. (1986) *Lethal Dietary Toxicities of Environmental Contaminants and Pesticides to coturnix.* US Fish and Wildlife Service. Technical Report 2. Washington, DC.

Hill, E.F., Heath, R.G., Spann, J.W. & Williams, J.D. (1975) *Lethal Dietary Toxicities of Environmental Pollutants in Birds.* US Fish and Wildlife Service Special Scientific Report. Wildlife No. 191. Washington, DC.

Hoffman, D.J. (1990) Embryotoxicity and teratogenicity of environmental contaminants to bird eggs. *Rev. Environ. Contam. Toxicol.* **115**, 39–90.

Hudson, R.H., Tucker, R.K. & Haegele, M.A. (1984) *Handbook of Toxicity of Pesticides to Wildlife.* US Fish and Wildlife Service Resource Publication 153. Washington, DC.

Hutchinson, J. & Simmonds, M. (1992) Escalation of threats to marine turtles. *Onyx* **26**, 95–102.

Isensee, A.R. & Tayaputch, N. (1986) Distribution of carbofuran in a rice-paddy–fish microecosystem. *Bull. Environ. Contam. Toxicol.* **36**, 763–769.

Jenkins, D.G., Layton, R.J. & Buikema, Jr. A.L. (1989) State of the art in aquatic ecological risk assessment. In: *Using Mesocosms to Assess the Aquatic Ecological Risk of Pesticides: Theory and Practice* (Ed. J. Reese Voshell), *J. Entomol. Soc. Am. MPPEAL.* **75**, 20–30.

Johnson, W.W. & Finley, M.T. (1980) *Handbook of Acute Toxicity of Chemicals to Fish and Aquatic Invertebrates.* Resource Publication 137. US Department of Interior, Fish and Wildlife Service, Washington, DC.

Joyce, J.C. & Ramey, V. (1986) *Aquatic Herbicide Residue Literature Review.* Center for Aquatic Weeds Institute of Food and Agriculture Sciences. University of Florida, Gainsville, FL.

Kennedy, S.W. (1991) The mechanism of organophosphate inhibition of cholinesterase-proposal for a new approach to measuring inhibition. In: *Cholinesterase-inhibiting Insecticides* (Ed. P. Mineau), pp. 73–87. Elsevier, Amsterdam.

Kersting, K. & van Wijsngaarden, R. (1992) Effects of chloropyrifos on a microecosystem. *Environ. Toxicol. Chem.* **11**, 365–372.

Kraus, P. (1988) Global pest management in the future. In: *Pesticides: Food and Environmental Implications,* pp. 1–9. Proceedings of International Atomic Energy Agency. November 1987, Nurenberg.

Larkin, P.A. (1984) A commentary on environmental impact assessment for large projects affecting lakes and streams. *Can. J. Fish. Aquat. Sci.* **41**, 1121–1127.

Lewis, D.L., Kellogg, R.B. & Holm, H.W. (1985) Comparison of microbial transformation rate coefficients of xenobiotic chemicals between field-collected and laboratory microcosm microbiota. *Validation and Predictability of Laboratory Methods for Assessing the Fate and Effects of Contaminants in Aquatic Ecosystems.* ASTM STP 865 (Ed. T.P. Boyle), pp. 3–13. American Society for Testing and Materials, Philadelphia.

Liber, K.N., Kaushik, N.K., Solomon, K.R. & Carey, J.H. (1992) Experimental Designs for Aquatic Mescocosm Studies: A Comparison of the "Anova" and "Regression" Design for Assessing the Impact of Tetrachlorophenolon Zooplankton Populations in Limnocorrals. *Environ. Toxicol. Chem.* **11**, 61–77.

Little, E.E. & Finger, S.E. (1990) Swimming behavior as an indicator of sublethal toxicity in fish. *Environ. Toxicol. Chem.* **9**, 13–19.

Little, E.E., Archeski, R.D., Flerov, B.A. & Kozlovskaya, V.I. (1990) Behavioral indicators of sublethal toxicity in rainbow trout. *Arch. Environ. Contam. Toxicol.* **19**, 380–386.

Lozano, S.J., O'Halloran, S.L., Sargent, K.W. & Brazner, J.C. (1992) Effects of esfenvalerate on aquatic organisms in littoral enclosures. *Environ. Toxicol. Chem.* **11**, 35–47.

Magnuson, J.J. (1990) Long-term ecological research and the invisible present. *Bioscience* **40**, 495–501.

Mayer, Jr, F.L. (1987) *Acute Toxicity Handbook of Chemicals to Estuarine Organisms.* FL. EPA/600/8–87/017. US Environmental Protection Agency, Gulf Breeze.

Mayer, Jr, F.L. & Ellersieck, M.R. (1986) *Manual of Acute Toxicity: Interpretation and Data Base for 410 Chemicals and 66 Species of Freshwater Animals.* Resource Publication 160. US Department of Interior, Fish and Wildlife Service, Washington, DC.

McEwen, F.L. & Stephenson, G.R. (1979) *The Use and Significance of Pesticides in the Environment.* John Wiley & Sons, New York.

McEwen, L.C., Stafford, C.J. & Hensler, G.L. (1984) Organochlorine residues in eggs of black-crowned night herons from Colorado and Wyoming. *Environ. Toxicol. Chem.* **3**, 367–376.

McEwen, L.C. & DeWeese, L.R. (1987) Wildlife and pest control in the sagebrush ecosystem: basic ecology and management considerations. In: *Integrated Pest Management on Rangeland. State of the Art in the Sagebrush Ecosystem* (Ed. J.A. Onsager), pp. 76–85. USDA Agricultural Research Service, ARS-50. Washington, DC.

McKenney, Jr, C.L. (1986) Influence of the organophosphate insecticide fenthion on *Mysidopsis bahia* exposed during a complete life cycle. I. Survival, reproduction, and age-specific growth. *Dis. Aquat. Org.* **1**, 131–139.

McKenney, Jr, C.L., Matthews, E., Lawrence, D.A. & Shirley, M.A. (1985) Effects of ground ULV applications of fenthion on estuarine biota. IV. Lethal and sublethal responses of an estuarine mysid. *J. Florida Anti-Mosq. Assoc.* **56**, 72–75.

McKim, J.M., Schmieder, P.K., Niemi, G.J., Carlson, R.W. & Henry, T.R. (1987) Use of respiratory–cardiovascular responses of rainbow trout (*Salmo gairdneri*) in identifying acute toxicity syndromes in fish: Part 2. malathion, carbaryl, acrolein and benzaldehyde. *Environ. Toxicol. Chem.* **6**, 313–328.

Moore, J.C., Lores, E.M., Clark, J.R., Mooddy, P., Knight, J. & Forester, J. (1985) Effects of ground ULV applications of fenthion on estuarine biota. II. Analytical methods and results. *J. Florida Anti-Mosq. Assoc.* **56**, 62–68.

Mora, M.A. & Anderson, D.W. (1991) Seasonal and geographical variation of organochlorine residues in birds from northwest Mexico. *Arch. Environ. Contam. Toxicol.* **21**, 541–548.

Morgan, M.J. & Kiceniuk, J.W. (1990) Effect of fenitrothion on the foraging behavior of juvenile Atlantic salmon. *Environ. Toxicol. Chem.* **9**, 489–495.

Müller, P. (1988) Effects of pesticides on fauna and flora. In: *Pesticides: Food and Environmental Implications*, pp. 1–27. Proceedings of International Symposium, November 1987, Nurenburg. International Atomic Energy Agency, Vienna.

Nalepa, T.F. & Landrum, P.F. (1988) Benthic invertebrates and contaminant levels in the Great Lakes: effects, fates, and role in cycling. In: *Toxic Contaminants and Ecosystem Health; A Great Lakes Focus* (Ed. M.S. Evans). *Advances in Environmental Sciences and Technology* **21**, 77–102.

National Oceanic and Atmospheric Administration. (1987) *National Status and Trends Program for Marine Quality, Progress Report* (A summary of selected data on chemical contaminants in tissues collected during 1984, 1985, and 1986). NOAA Technical Memorandum NOS OMA 38, Rockville, MD.

National Research Council (1987) Committee on Biological Markers. *Environ. Health Perspect.* **74**, 3–9.

Nimmo, D.R. (1985) Pesticides. In: *Fundamentals of Aquatic Toxicology* (Eds G.M. Rand & S.R. Petrocelli), pp. 49–65. Hemisphere, New York.

Nimmo, D.R. (1987) Assessing the toxicity of pesticides to aquatic organisms. In: *Silent Spring Revisited* (Eds G.J. Marco, R.M. Hollingworth & W. Durham), pp. 335–373. American Chemical Society, Washington, DC.

Nimmo, D.R., Hamaker, T.L., Matthews, E. & Moore, J.C. (1981) An overview of the acute and chronic effects of first and second generation pesticides on an estuarine mysid. In: *Biological Monitoring of Marine Pollutants* (Eds F.J. Vernberg, A. Calabrese, F.P., Thurberg & W.B. Vernberg), pp. 3–20. Academic Press, New York.

Norton, S.B., Rodier, D.J., Gentile, J.H., Van Der Schalie, W.H., Wood, W.P. & Slimak, M.W. (1992) A framework for ecological risk assessment at the EPA. *Environ. Toxicol. Chem.* **11**, 1663–1672.

O'Connor, J.M., Klotz, J.B. & Kneip, T.J. (1982) Sources, sinks, and distribution of organic contaminants in the New York Bight ecosystem. In: *Ecological Stress and the New York Bight: Science and Management* (Ed. G.F. Mayer), pp. 631–653. Estuarine Research Federation, Columbia, SC.

Onishi, Y. & Wise, S.E. (1982) *Mathematical Model, SERATRA, for Sediment-Contaminant Transport in Rivers and its Application to Pesticide Transport in Four Mile and Wolf Creeks in Iowa.* EPA-68–03–2613. Battelle Pacific Northwest Laboratory, Richland, WA.

Parkhurst, B.R., Bergman, H.L., Marcus, M.D., Creager, C.S., Warren-Hicks, W., Olem, H., Boelter, M. & Baker, J.P. (1990) *Evaluation of Protocols for Aquatic Ecological Risk Assessment and Risk Management* pp. xii–xvii and 227. Western Aquatics, Inc., Laramie, WY.

Peakall, D.B. (1992) *Animal Biomarkers as Pollution Indicators.* Chapman & Hall, London.

Peterle, T.J. (1991) *Wildlife Toxicology.* Van Nostrand Reinhold, New York.

Phoon, W.-O. (1992) Incidence, presentation and therapeutic attitudes to antiChE poisoning in Asia. In: *Clinical and Experimental Toxicology of Organophosphates and Carbamates* (Eds B. Ballantyne, T.C. Marrs & W.N. Aldridge), pp. 482–488. Butterworth-Heinemann, Oxford.

Pimentel, D., McLaughlin, L., Zepp, A., Lakitan, B., Kraus, T., Kleinman, P., Vancici, F., Roach, W.J., Graap, E., Keeton, W.S. & Selig, G. (1991) Environmental and Economic Effects of Reducing Pesticide Use. *Bioscience* **41**(6), 402–409.

Plumb, R.H. (1981) Procedures for handling and chemical analysis of sediment and water samples. Technical Report EPA.CE-81–1, prepared by Great Lakes Laboratory, State University College at Buffalo, Buffalo, NY.

Portier, R.J. (1985) Comparison of environmental effect and biotransformation of toxicants on laboratory microcosm and field microbial communities. In: *Validation and Predictability of Laboratory Methods for Assessing the Fate and Effects of Contaminants in Aquatic Ecosystems.* ASTM STP 865 (Ed. T.P. Boyle), pp. 14–30. American Society for Testing and Materials, Philadelphia.

Reynolds, P.E. (1989) *Proceedings of the Carnation Creek Herbicide Workshop.* Forest Pest Management Institute, Forestry Canada, Sault Ste. Marie, Ontario.

Rodier, D.J. & Zeeman, M.G. (1994) Ecological Risk Assessment. In: *Basic Environmental Toxicology* (Eds L.G. Cockerham & B.S. Shane), pp. 581–604. US EPA, Office of Pesticides and Toxic Substances, Washington, DC.

Rosenberg, D.M. & Resh, V.H. (1981) Recent trends in environmental impact assessment. *Can. J. Fish. Aquat. Sci.* **38**, 591–624.

Saiki, M.K. & Schmitt, C.J. (1986) Organochlorine chemical residues in bluegill and common carp from the irrigated San Joaquin Valley floor, California. *Arch. Environ. Contam. Toxicol.* **15**, 357–366.

Schindler, D.W. (1976) The impact statement boondoggle. *Science* **192**, 509.

Schindler, D.W. (1987) Detecting ecosystem responses to anthropogenic stress. *Can. J. Fish. Aquat. Sci.* **44** (Suppl), 6–25.

Schoor, W.P. & Newman, S.M. (1976) The effect of mirex on the burrowing activity of the lugworm (*Arenicola cristata*). *Trans. Am. Fish. Soc.* **105**, 700–703.

Siefert, R.E., Lozano, S.J., Brazner, J.C. & Knuth, M.L. (1989) Littoral enclosures for aquatic testing of pesticides: effects of chlorpyrifos on a natural system. In: *Using Mesocosms to Assess the Aquatic Ecological Risk of Pesticides: Theory and Practice* (Ed. J. Reese Voshell). Entomology Society of America. MPPEAL. **75**, 57–73.

Smith, G.J. (1987) Pesticide use and toxicology in relation to wildlife: organophosphate and carbamate compounds. US Fish and Wildlife Service Resource Publication 170. Washington, DC.

Soerjani, M. (1988) Current trends in pesticide usage in some Asian countries: environmental implications and research needs. In: *Pesticides: Food and Environmental Implications*, pp. 219–234. Proceedings of International Symposium, November 1987, Nurenberg. International Atomic Energy Agency, Vienna.

Somerville, L. & Walker, C.H. (1990) *Pesticide Effects on Terrestrial Wildlife*. Taylor & Francis, London.

Spehar, R.L., Tanner, D.K. & Gibson, J.H. (1982) Effects of kelthane and pydrin on early life stages of fathead minnows (*Pimephales promelas*) and amphipods (*Hyalella azteca*). In: *Aquatic Toxicology and Hazard Assessment: Fifth Conference*. ASTM STP 766. (Eds J.G. Pearson, R.B. Foster & W.E. Bishop), pp. 234–244. Elsevier Biomedical Press.

Spehar, R.L., Tanner, D.K. & Nordling, B.R. (1983) Toxicity of the synthetic pyrethroids, permethrin and AC 222, 705 and their accumulation in early life stages of fathead minnows and snails. *Aquat. Toxicol.* **3**, 171–182.

Stay, F.S., Katko, A., Rohm, C.M., Fix, M.A. & Larsen, D.P. (1989) The effects of atrazine on microcosms developed from four natural plankton communities. *Arch. Environ. Contam. Toxicol.* **18**, 866–875.

Stay, F.S., Larsen, D.P., Katko, A. & Rohm, C.M. (1985) Effects of atrazine on community level responses in Taub microcosms. In: *Validation and Predictability of Laboratory Methods for Assessing the Fate and Effects of Contaminants in Aquatic Ecosystems*, ASTM STP 865 (Ed. T.P. Boyle), pp. 75–90. American Society for Testing and Materials, Philadelphia.

Stephan, C.E. (1985) Are the 'guidelines for deriving numerical national water quality criteria for the protection of aquatic life and its uses' based on sound judgements? *Aquatic Toxicology and Hazard Assess-*

ment, 7th Symposium. ASTM STP 854. (Eds R.D. Cardwell, R. Purdy & R.C. Bahner), pp. 515–526. American Society for Testing and Materials, Philadelphia.

Suter, II, G.W., Rosen, A.E., Linder, E. & Parkhurst, D.F. (1987) Endpoints for responses of fish to chronic toxic exposures. *Environ. Toxicol. Chem.* **6**, 793–809.

Swineford, D.M. & Belisle, A.A. (1989) Analysis of trifluralin, methyl paraoxon, methyl parathion, fenvalerate and 2,4-D dimethyamine in pond water using solid-phase extraction. *Environ. Toxicol. Chem.* **8**, 465–468.

Tarzwell, C.M. (1978) A brief history of water pollution research in United States. In: *Proceedings of the First and Second USA-USSR Symposia on the Effects of Pollutants Upon Aquatic Ecosystems*, Vol. 1, pp. 11–31. EPA-600/3–78–076. US EPA, Minnesota.

Touart, L.W. (1988) *Aquatic Mesocosm Tests to Support Pesticide Registrations*. EPA 540/09–88–035. US EPA Office of Pesticide Programs, Washington, DC. US EPA, Minnesota.

Urban, D.J. & Cook, N.J. (1986) *Hazard Evaluation Division Standard Evaluation Procedure Ecological Risk Assessment*. EPA-540/9–85–001. US EPA Office of Pesticide Programs. Washington, DC. US EPA, Minnesota.

US Department of Agriculture (1992) Pesticide-Use Report. In: *Report of the Forest Service*, Fiscal Year 1991. US Department of Agriculture Forest Service, Washington, D.C.

US EPA (1980) Water quality criteria documents; availability. *Fed. Reg.* **45**, 79318–79379.

US EPA (1981) *Acephate, Aldicarb Carbophenothion, DEF, EPN, Ethoprop, Methyl Parathion, and Phorate: Their Acute and Chronic Toxicity, Bioconcentration Potential and Persistence as Related to Marine Environments*. EPA-600/4–81–023. Environmental Research Laboratory. Duluth, MN.

US EPA (1983) *Results of the Nationwide Urban Runoff Program, Volume 1 — Final Report*. Water Planning Division WH-554, Washington, DC.

US EPA (1984a) Estimating 'concern levels' for concentrations of chemical substances in the environment (unpublished). Available from the Office of Toxic Substances, Health and Environmental Review Division, Environmental Effects Branch, Washington, DC.

US EPA (1984b) Guidelines for deriving numerical national water quality criteria for the protection of aquatic life and its uses. *Fed. Reg.* **49**, 4551–4554.

US EPA (1985) Environmental effects testing guidelines. *Fed. Reg.* **50**, 39321–39397.

US EPA (1986a) *Ambient Water Quality Criteria for Chlorpyrifos*. Office of Water Regulations and Standards, Criteria and Standards Division,

Washington, DC.

US EPA (1986b) *Quality Criteria for Water* (not paginated) EPA 440/5–86–001. Office of Water Regulations and Standards. Washington, DC.

US EPA (1988a) *Guidance Document for Conducting Terrestrial Field Studies.* EPA 540/09–88–109. Office of Pesticide Programs. Washington, DC.

US EPA (1988b) *Pesticide Fact Handbook*, Vol. 1. Noyes Data Corporation, Park Ridge, NJ.

US EPA (1989a) Short-term methods for estimating the chronic toxicity of effluents and receiving waters to freshwater organisms. EPA/600/4–89/001. Washington, DC.

US EPA (1989b) Risk assessment guidance for superfund. Volume II. *Environmental Evaluation Manual* (Interim Final). EPA/540/1/89/001. Office of Emergency and Remedial Response, Washington, DC.

US EPA (1990) *Pesticide Fact Handbook*, Vol. 2. Noyes Data Corporation, Park Ridge, NJ.

US EPA (1991) Water quality criteria summary. Bulletin available from the Office of Science and Technology, Health and Ecological Criteria Division. Washington, DC.

US EPA (1992a) EPA accepts voluntary label changes for atrazine to reduce water contamination. *EPA News–Notes.* The condition of the environmental and the control of nonpoint sources of water pollution 21. Office of Water, Washington, DC.

US EPA (1992b) *National Study of Chemical Residues in Fish*, Vol. 1. EPA 823-R-92-008a. Office of Science and Technology (WH-551). US EPA, Minnesota.

US EPA (1992c) *National Study of Chemical Residues in Fish*, Vol. 2. EPA 823-R-92-008b. Office of Science and Technology (WH-551). US EPA, Minnesota.

US Government Printing Office (1936) *Proceedings of the First North American Wildlife Conference*, 3–7 February. Washington, DC.

Walsh, G.E., Deans, C.H. & McLaughlin, L.L. (1987) Comparison of the EC_{50}s of algal toxicity tests calculated by four methods. *Environ. Toxicol. Chem.* **6**, 767–770.

Webber, E.C., Deutsch, W.G., Bayne, D.R. & Seesock,

W.C. (1992) Ecosystem-level testing of a synthetic pyrethroid insecticide in aquatic mesocosms. *Environ. Toxicol. Chem.* **11**, 87–105.

Welch, H.W., Derek, C., Muir, G., Billeck, B.N., Lockart, W.L., Brunskill, G.J., Kling, H.J., Olson, M.P. & Lemoine, R.M. (1991) Brown snow: a long-range transport event in the Canadian Arctic. *Environ. Sci. Technol.* **25**, 280–286.

Williams, W.M., Holden, P.W., Parsons, D.W. & Lorber, M.N. (1988) Pesticides in ground water data base, interim report. US EPA, Office of Pesticide Programs, Washington, DC.

Winger, P.V., Schultz, D.P. & Johnson, W.W. (1985) Organochlorine residues in fish from the Yazoo National Wildlife Refuge. *Proceedings of the Annual Conference of the Southeast Association of Fish and Wildlife Agencies*, Vol. 39, pp. 125–131. US EPA, Minnesota.

Winger, P.V., Schultz, D.P. & Johnson, W.W. (1988) Contaminant residues in fish from national wildlife refuges in the southeast. In: *Proceedings of the National Symposium on Protection of Wetlands from Agricultural Impacts* (Ed. P.J. Stuber, Symposium Coordinator), Vol. 16, pp. 38–46. Biological Report 88. US Department of Interior, Fish and Wildlife Service, Research and Development, Washington, DC.

World Wildlife Fund (1992) Improving aquatic risk assessment under FIFRA. Report available from RESOLVE, Suite 500, 1250 24th NW, Washington, DC.

Wylie, G.D., Finger, S.E. & Crawford, R.W. (1990) Toxicity of municipal wastewater effluents contaminated by pentachlorophenol in Southwest Missouri. *Environ. Pollut.* **64**, 43–53.

Yount, D.J. & Richter, J.E. (1986) Effects of pentachlorophenol on periphyton communities in outdoor experimental streams. *Arch. Environ. Contam. Toxicol.* **15**, 51–60.

Zischke, J.A., Arthur, J.W., Hermanutz, R.O., Hedtke, S.F. & Helgen, J.C. (1985) Effects of pentachlorophenol on invertebrates and fish in outdoor experimental channels. *Aquat. Toxicol.* **7**, 37–58.

9: PCBs, PCDDs and PCDFs

A.J. NIIMI

9.1 INTRODUCTION

Polychlorinated biphenyls (PCBs), dibenzo-*para*-dioxins (PCDDs) and dibenzofurans (PCDFs) are presently considered among the more important environmental contaminants because of their toxicity, presence in all environmental matrices and persistence. There are 419 congeners included in these groups, among which 2,3,7,8-tetrachlorodibenzo-p-dioxin (2,3,7,8-TCDD) has received the most attention. Incidents during the last 30 years have raised public awareness of these chemicals. The 2,3,7,8-TCDD contaminated herbicide Agent Orange was widely used in Vietnam in the 1960s (Gross *et al.*, 1984). An accidental consumption of rice oil contaminated with PCBs and PCDFs occurred in Japan in 1968, and in Taiwan in 1979 (Masuda *et al.*, 1985). A chemical plant explosion at Seveso in 1976 resulted in widespread dioxin ground contamination (Reggiani, 1989). Runoff from a waste disposal site at Love Canal received wide attention during the late 1970s (Smith *et al.*, 1983). These and other events, and studies that indicate 2,3,7,8-TCDD is the most toxic anthropogenic chemical known, have continued to focus attention on these chemicals (Poland & Glover, 1977; Safe *et al.*, 1990; Schulz *et al.*, 1990). Studies have reported the lethal effects of these chemicals on laboratory and domesticated animals, and wildlife at Seveso, but there appear to be no documented cases where a human has died from acute PCB, PCDD, or PCDF poisoning.

An evaluation of the toxicological impact of PCBs, PCDDs and PCDFs on the biosphere is not a simple task because of the large number of compounds that are represented with a large range in toxicity. These chemicals are often present in the same sample matrices, although there are large differences in concentrations. PCBs can attain milligram per kilogram concentrations because of large quantities that were produced and released to the natural environment, while PCDDs and PCDFs are industrial by-products that serve no known useful purpose and are often present at nanogram per kilogram levels or lower (Tanabe, 1988; Fiedler *et al.*, 1990). There is increasing evidence that indicates that PCBs and PCDD/DFs elicit similar toxicological effects that may be initiated by similar mechanisms-of-actions (Safe, 1990). The toxicity of 2,3,7,8-TCDD may only represent less than 25% of total toxicity equivalence when contributions from the other PCDD/DF congeners are considered (LeBel *et al.*, 1990; Liem *et al.*, 1990). Studies that have used this approach to examine chemical concentrations in fish and humans indicated that the impact from PCBs could be greater than that from PCDDs and PCDFs (Kannan *et al.*, 1988; Niimi & Oliver, 1989b). Hence, an assessment of the toxicological concerns of each chemical group, or a specific congener, may not be appropriate for these chemicals.

This chapter examines the impact of PCBs, PCDDs and PCDFs in the atmospheric, aquatic and terrestrial environments. It examines their sources, the physical and chemical properties that could influence their behaviour and analytical methods appropriate for them. The latter often require detection limits below milligram per kilogram and sometimes below picogram per kilogram concentrations. The toxicological impact of these chemicals on animals and humans is also examined. Information is available on the chemical distribution patterns in different matrices, and experimental studies have been conducted on many different species, but comparatively little is known about their deleterious effects at ambient concentrations. This aspect will be examined by reviewing existing information.

9.2 PHYSICOCHEMICAL PROPERTIES

9.2.1 PCBs

There are 209 possible PCB congeners. This is due to the number of chlorine atoms and their substitution pattern at the *ortho, ortho', meta, meta', para* and *para'* positions on the biphenyl nucleus (Fig. 9.1). The 10 monochloro- to decachlorobiphenyl groups can include 1–46 isomers (Table 9.1). Each congener is often identified by a number from 1 to 209 that is based on

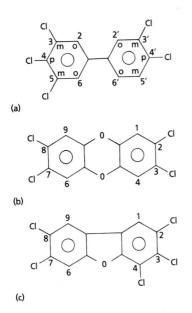

(a)

(b)

(c)

Fig. 9.1 Examples of (a) polychlorinated biphenyls (PCBs): 3,3',4,4',5-pentachlorobiphenyl, (b) dibenzo-*p*-dioxins (PCDDs): 2,3,7,8-tetrachlorodibenzo-*p*-dioxin and (c) dibenzofurans (PCDFs): 2,3,4,7,8-pentachloro-dibenzofuran showing their numerical chlorine substitution positions. The *ortho, meta* and *para* positions are indicated for PCBs. The examples shown represent the most potent cytochrome P-450 enzyme-inducing congener for each group.

the International Union of Pure and Applied Chemistry (IUPAC) rule for biphenyl substituent characterization (Ballschmiter & Zell, 1980).

PCBs have been produced under various trade names including Aroclor (Monsanto, US and UK), Clophen (Bayer, Germany), Kanechlor (Kanegafuchi, Japan), Phenoclor (Prodelec, France) and Fenclor (Caffaro, Italy). The more common products are Aroclors 1242–1260, Clophens A30–A60 and Kanechlors 300–600. The numerical designation of the Aroclors generally indicates the chlorine content of the product by molecular weight. PCBs can be synthesized in the laboratory by diazo coupling of chlorinated anilines and benzenes (Hutzinger, Safe & Zitko, 1974; Mullin *et al.*, 1984). Commercial PCBs are produced by the chlorination of biphenyl with a ferric catalyst, which yields a mixture of PCB isomers and congeners (Hutzinger *et al.*, 1974). For example, 50 of the 77 congeners reported in Aroclor 1242 represent only 0.05–1% each of the total concentration, while 16 congeners represent 1–3% each, and 11 congeners 3–7% each. Aroclor 1260 contains 75 congeners and has a composition pattern similar to Aroclor 1242, but only 29 congeners are common to both Aroclors (Schulz *et al.*, 1989).

Most congeners are solid at room temperature, with melting points generally increasing from 30°C to 300°C with increasing chlorine content (Hutzinger *et al.*, 1974). Congeners 2, 6, 7, 13, 149 and 154 occur as a liquid at room temperature. Commercial mixtures are liquids because of melting point depression, and their viscosity increases with increasing chlorine content. The specific gravity of Aroclors 1242 and 1260 is 1.392 and 1.566 at 15.5°C, respectively (Hutzinger *et al.*, 1974).

Aqueous solubilities of PCB congeners have been reported to range from 2 mg/l to 3 pg/l, or about log −5.1 to −10.2 mol, based on several methods. Patil (1991) tabulated the observed ·solubilities of 136 monochloro- to nonachloro-biphenyls which range from 31 pg/l to 1 mg/l. Solubilities of 26 dichloro- to hexachloro-biphenyls measured by the generator column method range from 3 pg/l to 2 mg/l (Dunnivant & Elzerman, 1988). Solubilities of 4–1124 µg/l were calculated for 80 monochloro- to heptachloro-

Table 9.1 Number of PCBs, PCDDs and PCDFs congeners and isomers. PCB congeners with chlorine substitution patterns at the *ortho* positions, and PCDD and PCDF congeners with 2,3,7,8-substituted positions that are of toxicological interest are also identified

	PCBs				PCDDs		PCDFs	
		Congeners[a]						
Group	No. of isomers	Non-*ortho*	Mono-*ortho*	Di-*ortho*	No. of isomers	2,3,7,8-substituted[b]	No. of isomers	2,3,7,8-substituted[c]
Mono-	3	0	0	0	2	0	4	0
Di-	12	1	0	0	10	0	16	0
Tri-	24	1	1	0	14	0	28	0
Tetra-	42	2	3	2	22	1	38	1
Penta-	46	1	4	3	14	1	28	2
Hexa-	42	1	3	6	10	3	16	4
Hepta-	24	0	1	4	2	1	4	2
Octa-	12	0	0	2	1	1	1	1
Nona-	3	0	0	0				
Deca-	1	0	0	0				
Total	209	6	12	17	75	7	135	10

[a]IUPAC numbers for non-*ortho*-substituted PCB congeners are 15, 37, 77, 81, 126, 169; mono-*ortho*-substituted congeners 28, 60, 66, 74, 105, 114, 118, 123, 156, 157, 167, 189; and di-*ortho*-substituted congeners 47, 75, 99, 115, 119, 137, 138, 153, 158, 166, 168, 170, 180, 190, 191, 194, 205.
[b]2,3,7,8-Tetrachloro-; 1,2,3,7,8-pentachloro-; 1,2,3,4,7,8-, 1,2,3,7,8,9-, 1,2,3,6,7,8-hexachloro-; 1,2,3,4,6,7,8-heptachloro-, and octachlorodibenzo-*p*-dioxin.
[c]2,3,7,8-Tetrachloro-; 2,3,4,7,8-, 1,2,3,7,8-pentachloro-; 1,2,3,4,7,8-, 1,2,3,7,8,9-, 1,2,3,6,7,8-, 2,3,4,6,7,8-hexachloro-; 1,2,3,4,6,7,8-, 1,2,3,4,7,8,9-heptachloro-, and octachlorodibenzofuran.

biphenyls based on their air–water equilibrium (Murphy *et al.*, 1987). The solubilities of Aroclors 1242 and 1254 have been estimated at 277 and 43 µg/l at 20°C, respectively (Murphy *et al.*, 1987). Other physical properties of PCBs – such as vapour pressure, boiling point, viscosity, colour, pour point, moisture content, flash point and fire point – have been compiled by Metcalfe *et al.* (1986).

Octanol–water partition coefficients (K_{ow}; see Chapter 13) reported for PCB congeners generally range between log 4.5 and 8.2. K_{ow} values for 58 monochloro- to heptachlorobiphenyl congeners measured by reverse-phase high-performance liquid chromatography (HPLC) range from log 4.40 to 8.13 (Rapaport & Eisenreich, 1984). Tabulated K_{ow} values for 140 monochloro- to nonachlorobiphenyl congeners range from log 4.71 to 7.80 (Patil, 1991). K_{ow} values for the 209 congeners estimated from total surface area range from log 4.46 to 8.18 (Hawker & Connell, 1988).

The Henry's law constants for many congeners have been measured, and values for Aroclors 1242 to 1260 range from 1.7 to 2.8 atm m^3/mol × 10^4 at 20°C (Burkhard *et al.*, 1985; Shiu & Mackay, 1986; Murphy *et al.*, 1987).

9.2.2 PCDDs and PCDFs

There are fewer PCDD and PCDF congeners than PCB congeners because PCBs have 10 potential substitution positions compared with eight positions for PCDDs and PCDFs. The same bonds on the phenyl rings for PCDDs result in a smaller number of isomers than for PCDFs with its different bonds. There are eight PCDD groups with 75 congeners containing 1–22 isomers, and eight PCDF groups with 135 congeners with 1–38 isomers (Table 9.1). The identity of each PCDD and PCDF congener is defined by its chlorine substitution position on the phenyl rings, because no IUPAC-based numbering system has been proposed. PCDDs and PCDFs are not manufactured commercially but can be synthesized for laboratory use. PCDDs have been synthesized by pyrolysis and alkali metal facile condensation from chlorophenols, chlorocatechols, chlorobenzenes and chloronitrobenzenes (Gray *et al.*, 1976a; Nestrick *et al.*, 1979; Patterson *et al.*, 1989). 2,3,7,8-TCDD can be synthesized by the dimerization of 2,4,5-trichlorophenol (Langer *et al.*, 1973). Various methods have been used to synthesize PCDFs from chlorinated *o*-phenoxyanilines, chlorodiphenyl ethers and PCBs (Gray *et al.*, 1976b; Norström *et al.*, 1979, Safe & Safe, 1984).

Melting points of PCDDs range from 80°C to 330°C (Pohland & Yang, 1972). There are few estimates of aqueous solubility and K_{ow} values for PCDDs and PCDFs. Solubility estimates for 2,3,7,8-TCDD range from 8 to 317 ng/l (Crummett & Stehl, 1973; Marple *et al.*, 1986; Shiu *et al.*, 1988). PCDD solubilities of 74 pg/l–417 μg/l at 25°C for 14 monochloro to octachlorocongeners were estimated by the generator column method (Shiu *et al.*, 1988). Solubilities for 2,3,7,8-substituted tetrachloro- to heptachlorodibenzofurans of 0.17–419 ng/l were determined by the same method (Friesen *et al.*, 1990). K_{ow} values from log 6.10 to 7.59 for tetrachloro- to hexachloro-PCDD congeners, and log 6.06 to 7.92 for tetrachloro- to heptachloro-PCDF congeners have been estimated by the stirring method (Sijm *et al.*, 1989). K_{ow} values of log 4.75–8.20 for 15 monochloro- to octachloro-PCDD congeners were estimated by HPLC retention time (Shiu *et al.*, 1988). Subcooled liquid vapour pressures of log −2 to −10 Pascals have been reported for PCDD and PCDF monochloro to octachlorocongeners (Eitzer & Hites, 1988; Rordorf *et al.*, 1990).

9.3 CHEMICAL ANALYSES

The large number of congeners, relatively low environmental concentrations, presence of other chemicals with similar physical properties, and different sample matrices can present special problems in sample analyses. Chemical analysis at the congener level is often based on the relative retention time of each chemical on a gas chromatograph (GC) that can use several types of detectors. Each type of detection system has its limitations, although several combinations have been used to characterize the chromatographic properties of each isomer and confirm its molecular structure. An electron-capture detector (ECD) is commonly used for PCB analyses, but its response is based on relative retention time of a chemical, which may not differentiate between co-eluting congeners or chemicals with different molecular weights with the same retention time. Mass spectrometers (MS) can resolve chemicals with different molecular weights, but may not always separate the different isomers. The latter can be resolved using nuclear magnetic resonance (NMR) and GC Fourier transform infrared spectroscopy, but their detection limit is not sensitive enough for most environmental samples.

9.3.1 PCB analyses

Most routine PCB analyses use a gas chromatograph with a capillary column equipped with an ECD, and some systems use a mass spectrometer to enhance identification and quantitation. The availability of instruments and accessories with different capabilities has contributed to the lack of a general protocol for routine PCB analyses. Use of two capillary columns with different polarities in parallel, and run with the same temperature programme, is one of the methods being used because it provides a low degree of confirmation for GC-ECD analyses (Oliver & Niimi, 1988; Durell & Sauer, 1990). No general method is available that would allow the 209 possible congeners in a sample to be examined by a single analysis because of co-elution, even though their chromatographic properties are known (Mullin *et al.*, 1984). Some of the closely or co-eluting congeners are 77 and 110; 105, 132 and 153; 118 and 149; 138 and 163; and 156 and 171 are important because of their toxicological significance, and additional steps may be required for their analyses (Duinker *et al.*, 1988; Duebelbeis *et al.*, 1989b; Larsen & Riego, 1990). Non-routine methods have been developed that could enhance the resolution of most congeners. Two-dimensional chromatographs have been used where part of the eluent is diverted to the second column that is connected in series and is operated using a different temperature programme. This is achieved by interfacing two instruments with a cold trap, or using an instrument with two ovens (Duinker *et al.*, 1988; Duebelbeis *et al.*, 1989a). The need to monitor all 209 PCB congeners is generally not required, because only about 140 congeners are present in commercial PCB mixtures (Schulz *et al.*, 1989; Kannan *et al.*, 1992). About 100 congeners could be present in environmental samples, and perhaps 50% of those are of environmental and toxicological significance (Hansen, 1987). Some congeners such as 30, 116 and 204 have been used as internal standards because they are not found in environmental samples (Luotamo, 1991).

There is no standardized method for reporting PCB concentrations because of the current use of capillary columns for routine analyses where over 100 peaks may be present; not all are identified because of the lack of standards. There is no single chemical standard from which total PCB quantification can be made, because the congener distribution frequencies of environmental samples are usually different from commercial products. The method of Webb & McCall (1973), that uses a glass packed column, and estimates Aroclors 1242 to 1260 concentrations based on the sum of 12−17 peaks, continues to receive attention because of its wide use for total PCB analyses in the past. Some studies have estimated PCB content using computer programs such as SIMCA, which examines congener patterns by principal-components analyses, and COMSTAR, which estimates the apparent contribution from the various PCB mixtures (Schwartz, Stalling & Rice, 1987; Swackhamer & Armstrong, 1988).

PCB profiles of some environmental samples can be similar to those of a commercial mixture from which quantitative estimates can be made (Bache *et al.*, 1972). Other chromatograms often suggest a more complex composition where an analytical standard can be prepared by mixing several Aroclors to estimate total PCB concentration (Christensen & Lo, 1986; Schwartz *et al.*, 1987). The International Council for the Exploration of the Sea (ICES) recommends that congeners 28, 52, 101, 118, 138, 153 and 180 be used for monitoring (Anon., 1986). Estimates based on

Aroclor mixtures and congener totals have been reported (Giesy *et al.*, 1986). Other studies have reported PCB concentrations as the sum of the 50−100 congeners monitored (Niimi & Oliver, 1989a). These methods will result in some different PCB estimates among studies; nevertheless, use of a capillary column is the method of choice for most analyses.

The total method detection limit for most routine analyses on biological tissues using a GC-ECD is about 1−10 µg/kg for most congeners. PCB analyses in other matrices such as air and water can generally be done without difficulties because similar methods are used for PCDDs, where environmental concentrations are much lower. Analysis for coplanar PCBs is becoming more common because of their greater toxicity relative to the other congeners. Additional clean-up methods such as carbon chromatography are required because of their concentrations in the low microgram per kilogram range and lower, and co-elution with more common congeners (Huckins, Stalling & Petty, 1980; Smith, 1981). Isotope dilution and GC-MS analyses can enhance detection of these congeners at nanogram per kilogram concentrations and lower (Tanabe *et al.*, 1987b; Kuehl *et al.*, 1991).

9.3.2 PCDD and PCDF analyses

All PCDD and PCDF analyses are done using a GC coupled with a MS. In the early 1960s 2,3,7,8-TCDD analyses were done by gas chromatography which gave detection limits in the low milligram per kilogram range but did not allow confirmation. Isomer-specific analysis became possible when a MS was added to the chromatograph as a detector (Ryhage, 1974). The spectrometric systems can include a mass selective detector (MSD), low resolution single quadrupole resolution mass spectrometer (LRMS), double focusing sector high resolution mass spectrometer (HRMS), and triple-stage quadrupole tandem or hybrid mass spectrometer (MS/MS). The instrument detection limit for 2,3,7,8-TCDD is 6 pg using an MSD system, 3 pg for LRMS, 0.3 pg for MS/MS and 0.2 pg for HRMS (McCurvin *et al.*, 1989). The MS/MS is less sensitive than a HRMS system but can resolve PCDDs and PCDFs in samples containing PCBs. Nevertheless, HRMS is the system of choice for most routine analyses (Charles & Tondeur, 1990; Huang *et al.*, 1991).

Total method detection limit for PCDDs and PCDFs will vary among matrices because of the sample weight or volume. Detection limits of 0.1–1 ng/kg for biological tissues can be attained with 5–20 g samples (Lindstrom & Rappe, 1990). Large-volume air samplers that can filter over $4000 \, m^3$ through a polyurethane foam absorbant can detect PCDDs and PCDFs in the $0.01 \, pg/m^3$ range (Christmann *et al.*, 1989c). The detection limit in water can vary widely where techniques that use traditional methods of extracting a bulk water sample can attain picogram per litre levels, whereas more recent methods that use a flow-through system containing an absorbant material and filter over $2 \, m^3$ water can attain detection levels in the low femtogram per litre range for 2,3,7,8-TCDD (Rappe *et al.*, 1989a; Broman *et al.*, 1991a). 2,3,7,8-TCDD is the most common PCDD monitored, and was often the only congener monitored in the past. Many programmes now routinely monitor all 2,3,7,8-substituted tetrachloro- to octachloro-PCDD/DF congeners, and report the minimum detection limits (Rappe *et al.*, 1987; Fürst *et al.*, 1990). There are many sample clean-up methods used for PCDD and PCDF analyses, and most include $^{13}C_{12}$ isotope analogues as internal sensitive standards (Ambidge *et al.*, 1990).

9.4 SOURCES

PCBs have wide industrial applications as plasticizers, hydraulic and dielectric fluids, and fire retardants because of their chemical and thermal stability (Hutzinger *et al.*, 1974). PCBs have been manufactured in large quantities since 1929, although production was curtailed after 1970 due to sale restrictions after they were identified as environmental contaminants in birds and fish (Risebrough *et al.*, 1968; Jensen, 1972). Estimates of total PCB production and use patterns are poorly known. Total production in Japan in the 1980s was about 59 000 tonnes, and about 155 000 tonnes in Germany, of which about one-third was used in closed systems (Tatsukawa, 1976, de Voogt & Brinkman, 1989). Domestic sales in the

USA amounted to about 360 600 tonnes between 1957 and 1974 (US EPA, 1976). Worldwide production is estimated at 1–1.5 million tonnes (de Voogt & Brinkman, 1989; World Health Organization (WHO), 1989).

The presence of PCBs in all environmental matrices is attributable to the substantial quantities of PCBs that were produced, the chemical stability of this substance and its use in various industrial applications (Tanabe, 1988). The most important use of PCBs has been in closed electrical systems, and this presently accounts for the largest contained source (Hansen, 1987; de Voogt & Brinkman, 1989). Leakages from the latter source and their use in many domestic products have contributed to the ubiquitous distribution of PCBs. About 23% of the 44 800 tonnes of PCBs used in Japan between 1962 and 1971 were included in dissipative products whose content could not be recovered (WHO, 1993). About 300 000 tonnes of PCBs have been placed in dump and landfill sites in the USA since 1930 (WHO, 1976). Chemical stability in rubberized paints was increased by adding 5–8% PCBs, and carbonless reproducing paper contained over 3% PCBs. Use of a protective coating material inside dairy silos that contained 18% PCBs was identified as a major source of contamination in dairy products (Fries, 1972; Anderson, 1989). Other uses for PCBs included additives to adhesives, lubricants and waxes. PCBs have also been dispersed through the use of contaminated oil for dust control purposes.

Methods to destroy PCBs have received considerable attention because of the large quantities that are currently use in capacitors or in storage. Thermal destruction using temperatures $> 900°C$ with a residence time over 2 s in excess oxygen has an efficiency exceeding 99.9999% (Ackerman *et al.*, 1981). Other techniques such as a chemical method of using an alkali metal to remove the chlorine to form an inorganic salt, and ultraviolet irradiation and biological degradation have been considered (Guertin, 1986).

PCDD/DFs have no industrial uses, and are not manufactured commercially, but are being inadvertently produced from various sources (Fiedler *et al.*, 1990). Their presence in the environment is largely attributed to anthropogenic activities. Natural sources of PCDD/DFs may include forest fires and volcanoes, although their contribution is minimal. Most studies examining sediment profiles indicate that concentrations have increased significantly this century. Semi-rural soil samples in the UK indicate PCDD/DFs increased after 1900 (Kjeller *et al.*, 1991). Sediments from Japanese coastal waters, the North American Great Lakes and Swiss lakes, indicate their concentrations sharply increased during the 1940s–1950s (Czuczwa & Hites, 1984, 1986; Czuczwa *et al.*, 1985; Hashimoto *et al.*, 1990). This relatively recent increase in PCDD/DF concentrations is supported by analyses of mummified humans; concentrations in Eskimo and Chilean remains dating back 400–2800 years were near or below analytical detection limits (Schecter *et al.*, 1988; Ligon *et al.*, 1989). Deep core sampling dating back 8000 years showed low concentrations of PCDFs, but no PCDDs (Hashimoto *et al.*, 1990). Historical samples generally show low concentrations of the octa-chlorocongeners relative to more recent samples (Czuczwa & Hites, 1984).

PCDD/DFs are often formed as by-products during chemical and industrial manufacturing processes. Major sources of PCDDs are products that contain chlorophenols and their derivatives because of the large quantities that are produced annually (WHO, 1989). Alkaline conditions, precursors with an aromatic moeity and temperatures in excess of 150°C can enhance the formation of PCDD/DFs from phenolic derivatives (Safe & Hutzinger, 1990). Trichlorophenols, which have been used extensively in the production of phenoxy herbicides — notably 2,4,5-trichlorophenoxyacetic acid (2,4,5-T) — have been an important source of PCDDs. 2,3,7,8-TCDD contamination of 2,4,5-T was an early indicator of PCDDs being an important environmental contaminant (Hutzinger *et al.*, 1985). Contaminated 2,4,5-T was used in a 1:1 ratio with 2,4-dichlorophenoxyacetic acid in the formulation of the herbicide Agent Orange that had an average 2,3,7,8-TCDD concentration of 1 mg/kg (Young *et al.*, 1983). Pentachlorophenol (PCP) can contain 500–990 mg/kg PCDDs, of which about 70% is present as its octachlorocongener, and is another source (National Research Council of Canada

(NRCC), 1981; Miles *et al.*, 1985). Chlorophenols are also widely used as wood preservatives where PCDD/DFs can average 56 mg/kg in various formulations, and 19 mg/kg in treated wood (Christmann *et al.*, 1989b). Elevated concentrations of heptachloro and octachlorocongeners have been monitored in the air of buildings where wood preservatives were used, and in harbour sediments near a wood-preserving plant (Päpke *et al.*, 1989; McKee *et al.*, 1990). More recently, pulp-and-paper mills that use a chlorine bleaching process have been identified as a major source (Rappe *et al.*, 1990a). Total annual discharges from these mills have been estimated at 100–150 g of 2,3,7,8-TCDD and 2–3 kg of 2,3,7,8-tetrachlorodibenzofuran (2,3,7,8-TCDF) in Canada, and 640 g and 5.1 kg in the USA, respectively (Government of Canada, 1990; Whittemore *et al.*, 1990).

PCDD/PCDFs are also present in fly ash and flue gas from municipal and industrial waste incinerators and residential sources (Olie *et al.*, 1977). Total annual PCDD/DF emissions from municipal and hospital solid waste incinerators have been estimated at 3 kg, or 50 g 2,3,7,8-TCDD equivalents in Sweden (Manscher *et al.*, 1990). Annual emission in flue gas from municipal incinerators was estimated at 21.6 kg PCDDs, which include 24 g 2,3,7,8-TCDD, and 18.6 kg PCDFs in the Netherlands (Hutzinger *et al.*, 1985). Emissions in fly ash from domestic coal use in the UK have been estimated at 180 g and 2.9 kg of the tetrachloro-PCDD/DF congeners respectively (Harrad *et al.*, 1991). Total PCDD/DF emissions in flue gas can range from 140 to 730 ng/m^3, and 10–28 mg per tonne of refuse burned (de Fré, 1986; Nottrodt & Ballschmiter, 1986). Polyvinyl chloride (PVC) could be an important source of organic carbon in the formation of PCDD/DF because it is present at about 3–15 kg/tonne of municipal waste (de Fré, 1986; Tysklind *et al.*, 1989). Laboratory studies that examined the combustion of PVC reported milligram per kilogram PCDD/DF concentrations (Christmann *et al.*, 1989a). Some observations at municipal incinerators containing PVC in waste materials reported no appreciable increase in PCDD/DF emissions, while operation of a PVC-coated metal incinerator at the same time as a waste incinerator increased total PCDD/DFs from 3 to 16 pg/m^3 in the stack emission (Giugliano *et al.*, 1989; Tiernan *et al.*, 1989).

Other sources of PCDD/DFs include metallurgical manufacturing and reclamation industries (Antonsson *et al.*, 1989). The quantities emitted from these sources can depend on the combustion processes used *et al.*, where an excess of oxygen, high carbon monoxide content and low furnace temperature can enhance their formation (Akimoto, 1989; Öberg & Allhammar, 1989; Nottrodt *et al.*, 1990). Industrial production of halogenated aromatic, aliphatic and inorganic compounds has been identified as a possible source, although their contributions have not been thoroughly evaluated (Hutzinger & Fiedler, 1989). Automative emissions could be a significant source because of the large quantities of fuel used annually. Combustion of 1 litre of gasoline that contains less than 2 pg of 2,3,7,8-TCDD equivalents can produce emissions with equivalents of 3.5 and 23 pg from unleaded and leaded gasoline, respectively (Marklund *et al.*, 1990). Studies that have monitored PCDD/DFs in air samples taken in traffic tunnels have indicated automotive exhaust could be a source (Zebühr *et al.*, 1989; Oehme *et al.*, 1991). PCBs are another source where commercial mixtures can contain 0.6–2.6 mg/kg PCDFs, but PCDDs are below 2 µg/kg (Wakimoto *et al.*, 1988). When PCBs are burned under uncontrolled conditions, PCDF concentrations in soot have been reported in the milligram per kilogram range and higher, although PCDD concentrations are considerably lower (des Rosiers, 1987). Methods to dispose of 2,3,7,8-TCDD are similar to those used for PCBs (Crosby, 1985).

9.5 DISTRIBUTION AND KINETICS

9.5.1 Atmospheric environment

Ambient air concentration of PCBs and PCDD/DFs at any location will depend on proximity to a point source (Table 9.2). Outdoor PCB concentrations range from 1 to 10 ng/m^3 in urban areas, with lower concentrations in more rural ones (Eisenreich *et al.*, 1981; Chevreuil *et al.*, 1989). Indoor levels in offices and homes are in the 10–100 ng/m^3 range (Wade, 1986). PCB concentrations close to point sources such as landfill sites and disposal facilities can attain 1–20 µg/m^3 levels (Bryant *et al.*, 1989; Hermanson & Hites, 1990). PCB concentrations at remote locations can also be relatively high, 1–25 pg/m^3 was reported in Arctic air and < 10–360 pg/m^3 was reported in the southwest Atlantic and Antarctic Oceans (Bidleman *et al.*, 1990; Weber & Montone, 1990). Airborne concentrations of PCDD/DFs show similar distribution patterns to PCBs, although concentrations are in the picogram per cubic metre range (Table 9.2). Airborne particulate concentrations also show a similar increase in levels in urban areas. Isomer-specific analyses indicate that PCDDs are generally more common than PCDFs, and the octachlorocongeners can represent nearly half of the total concentration.

The atmospheric behaviour of a chemical depends on its presence in the vapour or particle phase (Bidleman, 1988). Chemicals in a gaseous state can undergo photolysis or reaction with OH radicals, while those in the particle phase may be photolysed or removed by dry and wet deposition (Atkinson, 1991). Photodegradation of particle-bound PCDD/DFs is not an important process in reducing chemical levels in the atmosphere (Koester & Hites, 1992). Experimental data are not available on the photolytic rate of PCDD/DFs in the gas phase to determine its atmospheric lifetime. Wet deposition, as rain and snow, will vary seasonally, with the highest loadings occurring during spring and autumn (Brun *et al.*, 1991). Wet deposition of PCBs across Canada has been estimated as 1–4 µg/m^2 annually (Strachan, 1990). Annual dry and wet deposition rates for PCBs can vary from 2 µg/m^2 in Sweden, 7 µg/m^2 in Michigan and 40 µg/m^2 in Paris (Swackhamer & Armstrong, 1988; Chevreuil *et al.*, 1989). Annual PCDD/DF dry and wet fluxes of 3 ng/m^2 and 0.6 ng/m^2 have been estimated to occur along the Swedish Baltic coast (Broman *et al.*, 1991b).

Chemical transport through the atmosphere is a major pathway for the global distribution of PCBs, and at least the regional distribution of PCDD/DFs. Aerial deposition would be a plausible explanation for the picogram per litre concentrations of PCBs in the Arctic and Antarctic

Table 9.2 Total 2,3,7,8-substituted PCDD/DF concentrations, and their toxicity equivalents (see Section 9.6.2), reported in air and airborne particulate samples taken at various locations. Percentage PCDDs and octachlorocongeners of total PCDD/DF concentration are shown

Location	Concentration (pg/m³)	Percentage PCDDs	Percentage Octachloro-	I-TEF	References
Air					
Rural areas					
Sweden, coastal	0.08	69	51	0.003	1
Sweden, countryside	0.15	67	47	0.005	1
Urban areas					
Stockholm, suburb	0.45	75	53	0.01	1
Stockholm	0.56	62	44	0.02	2
Stockholm, centre	0.58	62	43	0.02	1
Hamburg, Germany	~1.5	~75	~33	~0.09	3
Dayton, Ohio	2.17	71	50	0.02	4
Niagara Falls, NY	~2.43	84	74	~0.14	5
Northeastern USA	3.61	76	64	0.13	6
Los Angeles	3.84	64	51	0.21	7
In traffic tunnel					
Stockholm	3.63	83	24	0.12	8
Norway, outlet	9.62	33	50	0.51	9
Hamburg	~12.8	~73	~58	~0.64	3
Near municipal incinerator					
Hamburg, Germany	~15.4	~76	~53	~0.65	3
Dayton, OH	34.7	51	35	1.43	4
Near industrial plant					
Brixlegg, Austria	~13.4	23	17	~1.70	10
Hamburg	72.1	~78	~65	~2.40	3
Air particulates					
Sweden, sea coast	~0.35	~22	~47	~0.01	3
Gothenburg, Sweden	~0.95	~55	~46	~0.08	3
Gothenburg, inversion	~5.56	~64	~43	~0.25	3
Rorvik, Sweden	~2.13	~53	~46	~0.09	3

References: 1, Broman *et al.*, 1991b; 2, Näf *et al.*, 1990; 3, Rappe & Kjeller, 1987; 4, Tiernan *et al.*, 1989; 5, R. Smith *et al.*, 1990; 6, Hunt & Maisel, 1990; 7, Maisel & Hunt, 1990; 8, Zebühr *et al.*, 1989; 9, Oehme *et al.*, 1991; 10, Christmann *et al.*, 1989c.

snowpack (Tanabe *et al.*, 1983; Gregor & Gummer, 1989). The relative importance of this pathway could be less in the aquatic than terrestrial environment because of input from other sources. Atmospheric input can account for 55–100% of annual PCB deposition to some areas of Lake Michigan, but only 3–11% in other areas due to loadings from terrestrial sources (Hermanson *et al.*, 1991). Nevertheless, significant quantities of PCBs are deposited from the atmosphere from which 2–10 tonnes were estimated to be deposited annually in each of the

Great Lakes in the early 1980s (Eisenreich *et al.*, 1981).

9.5.2 Terrestrial environment

Soil

Chemical concentration in soil is influenced by atmospheric deposition in most locations. A survey in the UK indicated PCB concentrations averaged 3 µg/kg, and ranged from 0.2 to 12 µg/kg among the 49 remote, rural and urban samples analysed (Jones, 1989). Soil concentrations in highly contaminated areas can exceed milligram per kilogram concentrations. One of the more intensive surveys in North America indicated mean PCDD/DF concentrations of 70 ng/kg in rural, 2070 ng/kg in urban and 8440 ng/kg in industrial areas (Table 9.3). PCDDs represented over 90% of total concentration, and the octachlorocongener was the most common in that survey. 2,3,7,8-TCDD concentrations up to 20 000 µg/m^2 were reported in the most contaminated zone in Seveso (Reggiani, 1989). Samples taken in Europe indicate similar differences between rural and industrial areas. Municipal and industrial waste incinerators are probably the largest contributors of PCDD/DFs to soil through wet and dry deposition. PCDFs can occur at lower or higher frequencies than PCDDs in different locations, depending on the emission source (Czuczwa & Hites, 1986; Vogg *et al.*, 1987).

Photodegradation of PCBs and PCDD/DFs in soil may not be an important process because of limited sunlight penetration and poor chemical mobility due to low vapour pressure and solubility. Laboratory studies indicate PCDDs can be photodegraded when a hydrogen source is present (Nestrick *et al.*, 1980; Watts *et al.*, 1991). There is a preferential loss of chlorines at the 2,3,7 and 8 substitution positions (Buser & Rappe, 1978). These conditions may have been evident when an olive oil emulsion was sprayed on a 2,3,7,8-TCDD-contaminated field in Seveso and concentrations decreased from 26 to 5 µg/m^2 after 9 days (Wipf *et al.*, 1978). Other studies indicate that photolysis can reduce PCDD levels by 10%, but suggest that the role of a solvent is to transport the chemical to the surface, where photolysis can occur (Kieatiwong *et al.*, 1990; Overcash *et al.*, 1991). Little is known about the microbial degradation of these chemicals in soil. It is considered that biotic degradation is not an important factor in soil because of the physical properties of these chemicals.

Plants

Less information is available on chemical levels in plants than animals because of their limited use for environmental monitoring programmes. The highest chemical concentrations are found in

Table 9.3 Total 2,3,7,8-substituted PCDD/DF concentrations, and their toxicity equivalents (see Section 9.6.2), reported in soil and settling particulate matter reported in various studies

Product	Concentration (ng/kg)	Percentage PCDDs	Percentage Octachloro-	I-TEF	References
Soil					
Rural Ontario, US mid-west (n = 30)	73	100	97	0.7	1
Urban Ontario, US mid-west (n = 47)	2 070	93	77	11.4	1
Industrial Ontario, US mid-west (n = 20)	8 440	88	70	40.8	1
Europe, rural (n = 3)	32	43	61	<4.3	2
Europe, industrial	~830	~46	~25	~51	2
	~4 800	~23	~5	94	2
Taiwan, industrial (n = 2)	510 000	~65	~18	9 000	3
Settling particulate matter					
Stockholm	580	86	64	9.9	4

References: 1, Birmingham, 1990; 2, Rappe & Kjeller, 1987; 3, Huang *et al.*, 1992; 4, Zebühr *et al.*, 1989.

samples collected downwind from a point source. PCB concentrations in different plants on a wet-weight basis generally indicate that levels in mosses and lichens are lower than in the leaves of deciduous plants, higher among conifer needles and highest in tree bark (Thomas *et al.*, 1984). Mosses and lichens with $10-110\,\mu g/kg$ dry weight PCBs were collected from remote areas in Scandinavia (Carlberg *et al.*, 1983). Background PCB levels in leaves of deciduous plants tend to be below microgram per kilogram concentrations, while tree bark can range from 10 to $400\,\mu g/kg$ (Buckley, 1982; Hermanson & Hites, 1990). Chemical concentrations can be higher in evergreen needles than deciduous leaves because of their waxy content and longer life span (Herrmann & Baumgartner, 1987; Reischl *et al.*, 1989). It is often not clear if the concentrations reported for leaves represent primarily surface adsorption or cellular absorption. The former probably contributed to the 12 mg/kg PCB concentration reported in leaves collected near a PCB dumpsite (Buckley, 1982).

Most studies indicate that the translocation of hydrophobic chemicals from soil to plants is poor because of poor chemical mobility (Strek & Weber, 1982). Studies on rooted vegetables such as carrots show high PCB content, although 97% was found adsorbed to skin (Iwata & Gunther, 1976). Soybeans had PCB levels of 0.2%, or less than the 100 mg/kg concentration in soil (Suzuki *et al.*, 1977; Fries & Marrow, 1981). Other studies on corn and goldenrod reported <1% of the PCBs was translocated after the growing season (Buckley, 1982). Similar observations were reported for 2,3,7,8-TCDD in soybean and corn where residues were found adsorbed to roots, but levels were non-detectable in shoots (McCrady *et al.*, 1990). The general property of poor chemical translocation and surface adsorption was evident in fruits harvested near the chemical plant in Seveso 1 year after the accident. Here 2,3,7,8-TCDD concentration in soil was about $10\,\mu g/kg$ and the skin of fruit contained 100 ng/kg, but in the edible portion the amount was below the 1 ng/kg detection limit (Wipf & Schmid, 1983). PCDD/DF concentrations in fruit and vegetables in Germany and North America are near or below the detection limit of 0.01 ng/kg, except for

octachlorocongener which can be present in the $0.6-8$ ng/kg range (Beck *et al.*, 1989a; Birmingham *et al.*, 1989). PCDD/DF concentrations of $0.002-1\,\mu g/kg$ in fruits and vegetables were reported in Japan (Ono *et al.*, 1987b; Takizawa & Muto, 1987).

Animals

PCB concentrations have been reported for a variety of animals in their natural environment. Bees, wasps and flies in the northeastern USA had $0.1-0.4$ mg/kg dry weight (Morse *et al.*, 1987). Snakes in the southern USA had 0.2–0.6 mg/kg (Sabourin *et al.*, 1984). Avian species associated with a terrestrial habitat had 0.03–0.4 mg/kg, while those associated with an aquatic habitat had 0.2–2.3 mg/kg in western Canada (Baril *et al.*, 1990). Game birds and small animals in Germany had 0.02–0.04 mg/kg (Brunn *et al.*, 1985). Raccoons in the eastern USA had 0.5 mg/kg in fat (Valentine *et al.*, 1988). Martens and fishers from central Ontario had 0.3–0.6 mg/kg in their livers (Steeves *et al.*, 1991). Up to 2 mg/kg PCBs was found in the adipose tissue of otters from western Canada, and 2–29 mg/kg in the lipid of mink from the Great Lakes region (Proulx *et al.*, 1987; Somers *et al.*, 1987). A mean concentration of 118 mg/kg PCBs was reported in otter from a contaminated area in Ireland (Mason & O'Sullivan, 1992). The higher PCB levels in marten, fisher, mink and otter can be attributed to their carnivorous and piscivorous feeding habits. Animals from remote areas have lower PCB levels; $1-20\,\mu g/kg$ was reported in muscle and liver of ptarmigan, hare and caribou from the Canadian Arctic (Thomas *et al.*, 1992). Little information is available on PCDD/DF concentrations in terrestrial animals from non-contaminated areas. Survey levels in wolf and moose tissues indicated most of the PCDD/DF congeners were below detection limits (de Wit *et al.*, 1990). 2,3,7,8-TCDD concentrations between 0.1 and $16\,\mu g/kg$ have been reported in invertebrates and vertebrates from highly contaminated sites (Lower *et al.*, 1989).

Humans

Measurements of PCBs in human adipose tissue, milk, blood serum and plasma show slightly higher levels in persons living in urban and industrialized areas than in rural areas (Jensen, 1989). Adipose PCB concentrations are often in the 1–3 mg/kg range or lower (Focadi *et al.*, 1986; Kashimoto *et al.*, 1989). PCB concentrations of 10–50 µg/l in mother's whole milk, and 1–5 µg/l in plasma or serum, are representative of the levels found in most countries (Jensen, 1989). Concentrations among Yusho and Yu-Cheng patients are higher, with 5–10 mg/kg PCBs in adipose and 6–60 µg/l in blood (Kuratsune, 1989).

Surveys of PCDD/DF levels in human adipose generally indicate concentrations in the low nanogram per kilogram range for 2,3,7,8-TCDD, and low microgram per kilogram levels of total 2,3,7,8-substituted PCDD/DFs (Andrews *et al.*, 1989). A survey of 2,3,7,8-TCDD levels among 500 persons in various occupational groups over a wide geographic range indicated a low-exposure group with a mean concentration of 7 ng/kg, a moderate-exposure group that included US military personnel and Vietnamese with a mean of 17 ng/kg, a high occupational exposure group with 193 ng/kg, and 1840 ng/kg in a woman exposed at Seveso (Lower *et al.*, 1989). PCDD/DF adipose levels of 0.4–29 µg/kg were reported among 45 workers exposed occupationally for up to 37 years (Beck *et al.*, 1989b).

Terrestrial trophodynamics

The mode of chemical exposure and accumulation for PCBs and PCDD/DFs is not as well defined in terrestrial as in aquatic ecosystems, because of shorter food chains and the ability of higher vertebrates to metabolize these chemicals. PCB congener 153 was found at 1 µg/kg in leaf matter, 10 µg/kg in caterpillar and 170 µg/kg in bird eggs on a dry-weight basis (Winter & Streit, 1992). Data from highly contaminated areas indicate 2,3,7,8-TCDD that may be biomagnified with concentrations of 109 ng/kg in mouse and 700 ng/kg in snake (Lower *et al.*, 1989). High chemical concentrations in soil could represent a

potentially significant pathway for chemical uptake, but few studies that have examined this aspect indicate that absorption is poor, and follow a first-order relationship (Brewster *et al.*, 1989; Banks & Birnbaum, 1991). Dermal application studies on mammals with PCBs and PCDD/DFs indicate that dermal absorption is not an important pathway, although some studies reported skin disorders at high exposure rates (Poland & Knutson, 1982; Puhvel *et al.*, 1982; Schmid *et al.*, 1992). Burrowing organisms such as grubs and earthworms can attain 2,3,7,8-TCDD concentrations similar to the 0.3–2 µg/kg levels in soil from contaminated areas (Lower *et al.*, 1989).

Dietary uptake is the most important pathway of these chemicals in animals and humans exposed to ambient concentrations. Products with high lipid content are the most important sources because of the lipophilic properties of these chemicals. Food accounts for 98.8%, air 1.1%, water 0.1% and soil 0.05% of the total 2,3,7,8-TCDD intake in humans (Travis & Hattemer-Frey, 1991). Among the food items, meat products represent about 28%, dairy products about 27%, fish about 30%, with the remainder from other items in a representative diet (Beck *et al.*, 1989a; Fürst *et al.*, 1990). A similar uptake pattern among the food groups was indicated for PCBs (Matsumoto *et al.*, 1987).

9.5.3 Aquatic environment

Water and sediments

PCBs are present in measurable concentrations in nearly all water bodies. Concentrations are lowest in the oceans where 0.04–0.07 ng/l were reported in the Antarctic, 0.007 ng/l in the Arctic, 0.002–0.02 ng/l in the north Atlantic and 0.04–0.6 ng/l in the north Pacific (Tanabe *et al.*, 1983, 1984; Schulz *et al.*, 1988; Hargrave *et al.*, 1988). Coastal marine waters near urban areas have higher PCB concentrations, 0.6 ng/l in the Dutch Wadden Sea; 0.7–3 ng/l in Puget Sound, Washington; and <2–11 ng/l in the Mediterranean Sea (Clayton, Paviou & Breitner, 1977; Duinker & Hillebrand, 1983; Marchand, Caprais & Pignet, 1988). Concentrations of 0.04 ng/l and 10–143 ng/l were found in a lake on Antarctica, in five rivers in northern

Ontario, respectively, where there were no known sources of local contamination (Tanabe *et al.*, 1983; McCrea & Fischer, 1986). PCB concentrations in riverine systems with local or regional sources of contamination were, for example, 50–300 ng/l in the River Seine, France; 100–200 ng/l in the Yodo River, Japan; and 650–1100 ng/l in the Shiawasse River, Michigan (Chevreuil *et al.*, 1987; Rice & White, 1987; Tanabe *et al.*, 1989). Concentration in large freshwater bodies such as the Great Lakes can range from 0.5 to 1 ng/l (Niimi & Oliver, 1988; Baker & Eisenreich, 1990).

PCDD/DF concentrations in aquatic systems and domestic water are often near or below the detection limits for most congeners except octachloro-PCDD, which is present in the 1–100 pg/l range (Meyer *et al.*, 1989; Rappe *et al.*, 1989a). Several studies that used large-volume samplers have reported detectable levels of several congeners where measurable concentrations were below picogram per litre levels (Tables 9.4 and 9.5). Concentrations of 0.5–60 pg/l of 2,3,7,8-TCDD have been reported in waters close to a point source (Näf *et al.*, 1990; Whittemore *et al.*, 1990). Concentrations in natural water systems are well below the detection limits of most sampling and analytical methods, although 2,3,7,8-TCDD was presented in Baltic Sea waters up to 2–3 fg/l (Broman *et al.*, 1991a).

Distribution patterns in inland waters could be different from oceanic waters because of the greater sediment surface-area/water-volume ratio and higher chemical concentrations. Sediments could contain over 90% of the hydrophobic chemicals such as PCBs that are present in contaminated ecosystems. High sediment concentrations in excess of 1 g/kg have been reported near point sources (Brown *et al.*, 1985; Sonzogni *et al.*, 1991). There are many factors, such as sediment size and organic carbon composition, that will influence chemical adsorption on and desorption from sediments (Karickhoff, 1980; Dickson *et al.*, 1984). PCDD concentrations tend to be higher than PCDFs in most systems, except near areas such as pulp-and-paper mills where the proportion of PCDFs is much higher (Table 9.4).

Some information is available on the degra-

dation of these chemicals in sediment and water. Microbial degradation of PCBs has been reported mainly under anaerobic conditions through dechlorination of the higher chlorinated congeners, primarily at the *meta*- and *para*-substituted positions (Nies & Vogel, 1990). The fastest rates were observed at the highest PCB concentrations, and the dechlorinated products were less toxic and more readily degradable by aerobic bacteria (Quensen *et al.*, 1988). A strain of *Alcaligenes eutrophus* was identified that predominantly metabolizes the *ortho*-substituted congeners (Bedard *et al.*, 1987). Microbial mineralization rates of PCBs in sediments with moderate to highly aerobic conditions of +250–500 mV are higher at pH 5.5 and 6.5 than at pH 8.0 (Pardue *et al.*, 1988). 2,3,7,8-TCDD appears to be resistant to microbial degradation in the aquatic environment. Only five of 100 microbial strains that can degrade persistent pesticides were capable of degrading 2,3,7,8-TCDD (Matsumura & Benezet, 1973). Degradation is by hydroxylation with 1-hydroxy-2,3,7,8-TCDD as a possible metabolite (Philippi *et al.*, 1982). Sediment-water studies using ^{14}C-labelled 2,3,7,8-TCDD indicated a half-life of 550–590 days in water (Ward & Matsumura, 1978). Photolysis of PCDDs in surface waters to 40°N latitude was calculated as being 1–225 days in winter and 0.4–68 days in summer (Atkinson, 1991).

Plants

PCBs, and probably PCDD/DFs, are accumulated by phytoplankton from water by partitioning or adsorption (Urey *et al.*, 1976). Accumulation is rapid following a first-order kinetic response with no differences between dead or live algae (Rhee, 1982). Most field studies in freshwater ecosystems have reported PCB concentrations in phytoplankton up to 100 µg/kg on a dry-weight basis (Mowrer *et al.*, 1982; Oliver & Niimi, 1988). Macrophytes can contain concentrations in the 100–600 µg/kg PCBs (Yap *et al.*, 1979; Anderson *et al.*, 1982; Mouvet *et al.*, 1985). PCB levels in whelks from the North and Irish Seas averaged 3 mg/kg on a lipid basis (Knickmeyer & Steinhart, 1989).

Table 9.4 Total 2,3,7,8-substituted PCDD/DF concentrations, and their toxicity equivalents, reported in different matrices in the aquatic environment. Percentage PCDDs and octachlorocongeners of total PCDD/DF are shown. All values expressed on a wet-weight basis, except sediment and sludge which is dry-weight

Matrix	Concentration (ng/kg)	Percentage PCDDs	Percentage Octachloro-	I-TEF	References
Water					
Baltic Sea	0.00004	86	75	0.000002	1
Eman River, Sweden	0.0023	77	71	0.00006	2
Marine sediment					
Baltic Sea	640	61	50	8.3	3
Swedish archipelago	~3200	~71	~41	~34	4
Water treatment plant					
Water, Uppsala	0.02	89	94	0.002	2
Sludge, Uppsala	10 230	98	90	22.8	2
Water, Stockholm	0.04	100	100	0.002	2
Sludge, Stockholm	8140	98	91	23.7	2
Water, Lockport, NY	560	8	54	10.3	5
Sludge, Lockport, NY	19 540	8	45	405	5
Sewage treatment plant					
Sludge, Stockholm	7640	99	79	32.9	6
Pulp-and-paper mill area, Sweden					
Sediment	95 000	4	86	770	7
Pike	8.5	26	0.4	2.4	7
Softwood effluent	0.8	95	45	0.02	8
Hardwood effluent	3.1	71	65	0.04	8
Sediment	660	70	52	31.2	8
Electrode sludge	304 000	2	27	41 000	8
Fish					
Marine species, UK	6.1	56	46	0.7	9
Marine species, Germany	20.7	26	8	1.8	10
Herring, Baltic Sea	23.1	36	< 1	~3.7	11
Salmon, Baltic Sea	78.2	15	< 1	11.9	11
Eel, Rhine River	31.5	70	41	4.3	12
Trout, Lake Superior	30.0	27	5	5.5	13
Trout, Lake Michigan	76.5	29	< 5	15.3	13
Trout, Lake Ontario	118	48	2	52.2	13
Marine mammals, blubber					
Seal, North Sea	34.1	64	19	3.1	14
Dolphin, New Zealand	64.4	47	20	12.6	15
Killer whale, Japan	88.0	0	< 1	7.42	16

References: 1, Broman *et al.*, 1991a; 2, Rappe *et al.*, 1989a; 3, de Wit *et al.*, 1990; 4, Rappe & Kjeller, 1987; 5, Meyer *et al.*, 1989; 6, Zebühr *et al.*, 1989; 7, Kjeller *et al.*, 1990; 8, Rappe *et al.*, 1990a; 9, Startin *et al.*, 1990; 10, Fürst *et al.*, 1990; 11, Rappe *et al.*, 1987; 12, Frommberger, 1991; 13, Zacharewski *et al.*, 1989; 14, Beck *et al.*, 1990a; 15, Buckland *et al.*, 1990; 16, Ono *et al.*, 1987a.

Table 9.5 Total 2,3,7,8-substituted PCDD/DF concentrations, and their toxicity equivalents (see Section 9.6.2), reported in food products. Percentage PCDDs and octachlorocongeners of total PCDD/DF concentrations are shown. All values are expressed on a wet-weight basis

Product	Concentration (ng/kg)	Percentage PCDDs	Percentage Octachloro-	I-TEF	References
Domestic drinking water					
Sweden	0.0002	100	75	0.00003	1
New York	<0.007	71	83	<0.0027	2
Cow's milk					
Sweden	0.25	73	60	0.018	3
UK, rural	0.52	65	52	0.045	4
Netherlands, rural	0.56	70	42	0.046	5
UK, industrial area	1.05	53	38	0.200	4
UK, near incinerator	1.06	57	40	0.142	4
Germany	1.34	64	32	0.097	6
Mother's milk					
CIS	3.75	71	49	0.45	7
Quebec	6.61	91	56	0.39	8
New York	7.18	89	61	0.43	7
Sweden	10.2	79	56	0.39	9
Germany	12.2	83	56	0.94	7
Los Angeles	17.1	94	70	0.65	7
Vietnam	~35	94	54	~6	10
Vietnam	38.7	91	73	1.1	11
Dairy products					
CIS, butter	14.4	69	63	~1.00	7
Germany, butter	20.4	78	58	0.54	6
Germany, cheese	3.5	70	53	0.23	6
CIS, cheese	6.9	93	78	~0.1	7
CIS, cream	6.6	56	48	~0.70	7
Meat products					
CIS, beef	1.2	67	52	~0.1	7
Germany, beef	5.5	71	22	0.46	6
Germany, pork	4.1	83	62	0.15	6
CIS, pork	14.7	88	78	~0.4	7
Germany, chicken	3.0	75	52	0.13	6
Food basket					
Sweden	1.7	88	88	0.48	12

References: 1, Rappe *et al.*, 1989a; 2, Meyer *et al.*, 1989; 3, Rappe *et al.*, 1990b; 4, Startin *et al.*, 1990; 5, Liem *et al.*, 1990; 6, Fürst *et al.*, 1990; 7, Schecter *et al.*, 1990a; 8, Dewailly *et al.*, 1991; 9, Rappe *et al.*, 1989b; 10, Schecter *et al.*, 1987; 11, de Jong *et al.*, 1989; 12, de Wit *et al.*, 1990.

Animals

PCBs and PCDD/DF concentrations in aquatic animals have been monitored in many water bodies for human health considerations and survey purposes (NOAA, 1987; Andersson *et al.*, 1988; D'Itri, 1988; O'Conner, 1991). In general, concentrations are lowest among organisms that represent the lower trophic levels, in open ocean or remote water bodies. PCB levels of 2–27 µg/kg dry weight in zooplankton, and 4 µg/kg in cod, were reported from Arctic waters (Muir *et al.*, 1988; Bidleman *et al.*, 1989). Fish from remote Arctic freshwater systems contain low microgram per kilogram levels (Lockhart *et al.*, 1992). Organisms from large contaminated ecosystems such as Lake Ontario can contain 50 µg/kg in plankton and 5 mg/kg in fish (Oliver & Niimi, 1988). A mean concentration of 145 mg/kg in fish muscle was reported from the highly contaminated Hudson River, New York (Brown *et al.*, 1985).

PCB levels in aquatic birds are influenced by the birds' habitat and feeding habits. Carcasses of two waterfowl species from San Francisco Bay had 0.9–3 mg/kg PCBs, while 8–11 mg/kg was reported in three species from the lower Great Lakes (Smith *et al.*, 1985; Ohlendorf *et al.*, 1991). Piscivorous birds can attain 3–40 mg/kg levels (Norstrom *et al.*, 1986). The highest concentrations of PCBs in animals may be found in marine mammals (Gilbertson, 1989). PCB concentrations in the 100 mg/kg range were reported in seals from the Atlantic (Law *et al.*, 1989). Surveys from the Pacific report about 400 mg/kg PCB in killer-whale blubber and 320 mg/kg in porpoise lipid (Ono *et al.*, 1987b; Kannan *et al.*, 1989).

Many of the earlier studies that reported PCDD concentrations were limited to only the 2,3,7,8-TCDD congener. More recent studies report values for some or all of the 2,3,7,8-substituted tetrachloro- to octachloro-PCDD and/or PCDF congeners (Broman *et al.*, 1992). Some studies also report some of the non-2,3,7,8-substituted congeners (Ono *et al.*, 1987b). Total 2,3,7,8-substituted PCDD/DF concentrations in the 1–100 ng/kg range for plankton, invertebrates and fish on a wet-weight basis were reported in two large ecosystems (Broman *et al.*, 1992; Whittle *et al.*, 1992). Concentrations in piscivorous birds can be higher; values in the 100–1000 ng/kg range were reported for several species (Van den Berg *et al.*, 1987; Elliott *et al.*, 1989; de Wit *et al.*, 1992). Concentrations in marine mammals can range from 100 ng/kg for 2,3,7,8-substituted PCDD/DF congeners, up to 400 ng/kg for total PCDD/DFs (Ono *et al.*, 1987b; Kannan *et al.*, 1989).

Aquatic trophodynamics

Laboratory studies have unequivocally demonstrated that aquatic organisms can take up many organic and inorganic chemicals from waterborne and dietary sources efficiently. The importance of each pathway is influenced by the physical properties, biological behaviour and environmental concentrations where chemicals accumulated through biomagnification are hydrophobic, persistent and present at low waterborne concentrations. PCB concentration in organisms at the higher trophic levels is accumulated more readily through biomagnification than through bioconcentration (Volume 1; Chapter 19). This view is supported by contaminant dynamic models, evaluation of the respiratory and feeding limitations of fish, and studies of some major ecosystems (Thomann, 1981; Niimi, 1985; Oliver & Niimi, 1988). Trophodynamic analyses of PCB concentrations in organisms that represent different trophic levels generally indicate smaller increments of food-chain magnification with increasing trophic levels in large ecosystems (Table 9.6). There is a well-defined increase in the bioaccumulation factor (BAF) with increasing trophic levels because of the increasing importance of dietary intake at higher trophic levels (Niimi, 1983). The trophodynamics of PCBs are similar to other persistent chemicals such as DDT and methylmercury, even when concentrations are examined on a congener-specific or total PCB basis. Benthic organisms can assume an important role in the transfer of chemicals in sediments into the food chain in a system such as Lake Ontario where the sediments appear to be a more important source of PCBs than atmospheric deposition (Oliver & Niimi, 1988). Concentration factors of 1–60 have been reported among different organisms, with the higher values associated with sediments with lower organic carbon content (McLeese *et al.*, 1980; Oliver, 1984).

The trophodynamics of PCDD/DFs in aquatic ecosystems are not as well defined as those for PCBs; this could be attributable to the differences in pharmacokinetics between PCBs and many of the PCDD/DFs. Waterborne concentrations of 2,3,7,8-TCDD are generally not available to estimate bioaccumulation factors. Limited information on 2,3,7,8-TCDD concentrations at three trophic levels in the Lake Ontario ecosystem suggest its food chain magnification potential may be lower than some PCBs, even though its

Table 9.6 Trophodynamics of total PCBs and 2,3,7,8-TCDD in aquatic ecosystems. Mean chemical concentration is shown for each group, from which the bioaccumulation factor (organisms/water), and food-chain magnification (predator/prey) were calculated

Lake Ontario, PCBs[a]	Lake Geneva, PCBs[b]	Western North Pacific, PCBs[c]	Lake Ontario, 2,3,7,8-TCDD[d]
Trout, 9.97 mg/kg BAF = log 6.96 FCM = log 0.84	Trout, 1.0 mg/kg FCM = log 0.79	Dolphin, 3.7 mg/kg BAF = log 7.12 FCM = log 1.73	Trout, 44 ng/kg FCM = log 0.64
Smelt, 1.4 mg/kg BAF = log 6.10 FCM = log 0.62	Gudgeon, 160 μg/kg FCM = log 0.97	Squid, 68 μg/kg BAF = log 5.38 FCM = log 0.15	Alewife, 10 ng/kg FCM = log 0.52
Mysid, 330 μg/kg BAF = log 5.48 FCM = log 0.82		Mictophid, 48 μg/kg BAF = log 5.23 FCM = log 1.43	
Plankton, 50 μg/kg BAF = log 4.65	Plankton, 17 μg/kg	Zooplankton, 1.8 μg/kg BAF = log 3.81	Zooplankton, 3 ng/kg
Water, 1.1 ng/l	Water, not reported	Water, 0.28 ng/l	Water, not reported

[a]Oliver & Niimi, 1988; Niimi & Oliver, 1989a.
[b]Mowrer *et al.*, 1982.
[c]Tanabe *et al.*, 1984.
[d]Whittle *et al.*, 1992.

physicochemical properties are similar to tetra-chlorobiphenyls (Table 9.6). 2,3,7,8-TCDD concentrations reported in phytoplankton, zooplankton, herring and cod from the Baltic Sea indicated a magnification factor of 5 (Broman *et al.*, 1992). The trophodynamic response for other PCDD/DF congeners could also differ from PCBs because of lower uptake efficiencies and faster elimination rates. Octachloro-PCDD/DFs are one of the most common congeners present in water, yet they occur at low concentrations in fish (Table 9.4). Concentrations of 930 and 140 ng/kg dry weight for other 2,3,7,8-substituted tetrachloro- to octachloro-PCDD/DFs in phytoplankton and cod would indicate a decreasing trophodynamic relationship (Broman *et al.*, 1992).

Food chain magnification of PCBs and PCDD/DFs from fish to piscivorous birds and marine mammals has also been reported by a number of studies. These relationships tended to be less clear than those shown at the lower trophic levels because of the biochemical differences between poikilotherms and homeotherms that may enhance elimination. Magnification factors reported for fish-to-piscivorous bird eggs include 21 for 2,3,7,8-TCDD in herons, 3 for PCDDs and 0.3 for PCDFs in gulls, and 11–32 for PCDD/DFs in eagles (Elliott *et al.*, 1989; Macdonald *et al.*, 1992; de Wit *et al.*, 1992). No magnification was suggested for PCDD/DFs for fish-to-seal in the Baltic Sea, while factors of 4–9 for fish-to-seal, and 7–14 for seal-to-bear were reported for PCBs in the Canadian Arctic (Muir *et al.*, 1988; Bignert *et al.*, 1989; de Wit *et al.*, 1992). Other studies have reported high PCDF levels in blubber of killer whale but no PCDDs, low 2,3,7,8-substituted PCDD/DFs in beluga whale blubber with high PCB concentrations, and polar bear with only six major PCB congeners (Ono *et al.*, 1987a; Norstrom *et al.*, 1990; 1991). These observations would suggest that the biomagnification factors observed among these animals may not accurately indicate chemical uptake patterns, and could be attributed to differences in metabolic capabilities (Muir *et al.*, 1992b). Residue levels for these animal groups are also often reported in eggs for birds, and specific tissues such as liver, muscle and blubber for mammals because of their large body masses, so chemical concentrations and congener profiles between trophic levels may not always be directly comparable.

9.5.4 Chemical kinetics among animals

The pharmacokinetic response of PCB uptake and elimination among lower vertebrates is more predictable than among mammals. Dietary absorption efficiencies of 70−>90% have generally been observed for PCBs among fish, birds and mammals exposed to congeners and commercial mixtures. (Albro & Fishbein, 1972; Niimi & Oliver, 1983). Results for fish fed with a Kanechlor mixture suggested a marginal decrease in efficiency with increasing chlorine content (Tanabe *et al.*, 1982a). Observations among birds and mammals fed with PCB mixtures indicated that the congeners were absorbed at similar efficiencies (Saschenbrecker *et al.*, 1972; deFreitas & Norstrom, 1974). Elimination rates for PCBs in most animals follow a first-order relationship. Half-lives among tetrachloro to decachloro-congeners that are usually in excess of 100 days in fish with increasing chlorine content may be attributed to their limited metabolic capabilities (Tanabe *et al.*, 1982a; Niimi & Oliver, 1983).

Many studies have shown that mammals can metabolize PCBs, although there are large differences between species in their capacity to metabolize different congeners. Most studies suggest PCBs are usually metabolized through an arene oxide intermediate and conjugated with glutathione to yield a hydroxylated polar product that is excreted in faeces more than in urine (Jerina & Daly, 1974; Sipes & Schnellmann, 1987). This metabolite has been reported in fish and various mammalian species (Melancon & Lech, 1976; Lutz & Dedrick, 1987). Hydroxylation is generally enhanced by decreasing chlorine content and the presence of vicinal hydrogens, although there are differences among animals (Jensen & Sundstrom, 1974; Wolff *et al.*, 1982; Hansen, 1987). The latter is exemplified for PCB congeners 136 and 153 where the metabolic rate differed by two orders of magnitude in rat, and by one order for dog and monkey (Lutz & Dedrick, 1987). The relative persistence in humans is suggested by PCB concentrations in Yusho

patients where adipose levels decreased three-fold over 15 years, and liver by three-fold over 8 years (Kuratsune, 1989). Half-lives of 124−338 days were reported for congeners 138, 153 and 180 in humans (Bühler *et al.*, 1988). The capability of different animals to metabolize PCBs is shown by examining congener profiles in food-chain relationships. Similar profiles among invertebrates and fish representing several trophic levels indicate their limited capability to metabolize and eliminate PCBs (Oliver & Niimi, 1988). A different conclusion is indicated from the 30 major congeners present in herring, that are reduced to 15 in seal, and the six congeners in polar bear that represent over 95% of its PCB content (Norstrom, 1986).

The kinetics of PCDD/DFs, excluding 2,3,7,8-TCDD, among most animal groups are less predictable than for PCBs. Dietary absorption efficiencies of 50−70% calculated for 2,3,7,8-TCDD in fish are considerably higher than efficiencies of 1−>30% reported for 10 other monochloro- to octachloro-PCDD/DF congeners (Kleeman *et al.*, 1986; Niimi & Oliver, 1986; Muir & Yarechewski, 1988; Opperhuizen & Sijm, 1990; Muir *et al.*, 1992a). Absorption efficiencies between 70% and 90% for 2,3,7,8-TCDD have been reported for mammalian species including humans (Piper *et al.*, 1973; Olson *et al.*, 1980; Poiger & Schlatter, 1986). In contrast, efficiencies of 1−10% were reported for the octachloro-congener (Williams *et al.*, 1972; Norback *et al.*, 1975). Even when differences in their toxicity are considered, poor absorption among other PCDD congeners is suggested by the oral LD_{50} values compiled for mice where five of the 12 congeners exceeded 1 g/kg (Kociba & Cabey, 1985). The molecular size of the higher chlorinated PCDD/DFs does not appear to be a limiting factor influencing their poor absorption in fish, although a low rate of membrane permeation could be a contributing factor (Niimi & Oliver, 1988; Opperhuizen & Sijm, 1990). There are also differences in kinetics among animal groups where the octachlorocongener represents a low percentage of the total PCDD/DF in fish (Table 9.4) compared with about 50% of the PCDD/DFs found in mammalian tissues (Table 9.5).

2,3,7,8-TCDD has a half-life of the order of

50–100 days in fish (Branson *et al.*, 1985; Kleeman *et al.*, 1986). Half-lives of other PCDD/DFs tend to be of the order of several weeks or less, with no consistent trend suggested with increasing chlorine content (Niimi & Oliver, 1986). The half-life of 2,3,7,8-TCDD in animals varies from 12 to 15 days for hamster, 24 to 31 days for rat, 22 to 43 days for guinea pig and about 5 years for human (Rose *et al.*, 1976; Olson *et al.*, 1980; Olsen, 1986; Poiger & Schlatter, 1986). 2,3,7,8-TCDF appears less persistent with half-lives of 2–4 days in rat and about 20 days in guinea pig (Decad *et al.*, 1981a,b). Half-lives of several tetrachloro- to octachloro-PCDD/DFs in human were estimated at 2–4 years (Gorski *et al.*, 1984). A number of studies using radiolabelled 2,3,7,8-TCDD on different animals generally reported that radioactivity in the faeces was about twice as high as that in urine, suggesting that most of the chemical is excreted as the parent compound but some is metabolized (WHO, 1989). Metabolic transformation studies on 2,3,7,8-TCDD in fish and mammals suggest glucuronide conjugation and hydroxylation as possible pathways (Ramsey *et al.*, 1982; Kleeman *et al.*, 1988). Even though there is limited information on the persistence of PCDD/DFs in animals, the tetrachloro to octachlorocongeners with the 2,3,7,8-chlorine substitution patterns are the most important because of their selective retention by animals (Rappe *et al.*, 1979; Van den Berg *et al.*, 1987; Zitko, 1992).

9.5.5 Maternal transfer

The transfer of chemicals from mother to offspring can occur during embryonic development and nurturing. The importance of this process varies with species; e.g. chemical concentrations in progeny can be high among oviparous species such as fish and birds. PCB transfer among five species of fish to their eggs ranged from 5% to 29% of maternal body burden (Niimi, 1983). This was influenced by the percentage of total lipid in fish deposited in eggs, but not by the percentage egg of body weight. Transfer in birds can also be high; e.g. PCB in Adelie penguin eggs was 4% of maternal body burden with no differences in congener patterns (Tanabe *et al.*, 1986). Body burden transfer rates of 24% for Herring gull and 45% for Arctic tern were also reported for PCBs

(Lemmetyinen *et al.*, 1982). PCDDs are also transferred from birds to eggs, based on observations among natural populations (Kubiak *et al.*, 1989).

Placental transfer among viviparous species, including humans, during embryonic development is less conclusive. PCB transfer rates in rats can be high; newborn pup livers contained 12 mg/kg from dams that were fed a 30 mg/kg Aroclor 1254-dosed diet (Shain *et al.*, 1986). Congener patterns in the pups were generally similar to the dam in that study, but preferential transfer of lower chlorinated PCB congeners across the placenta was reported in dolphins (Tanabe *et al.*, 1982b). PCB levels of 59 mg/kg lipid in a harbour porpoise fetus, and of 0.31 mg/kg lipid in a fetus from a bighorn sheep ewe with 0.4 mg/kg also suggests high transfer rates (Duinker & Hillebrand, 1979; Turner, 1979). Studies on humans indicate PCBs can cross the placenta with no preferential transfer among most of the congeners (Masuda *et al.*, 1978; Jacobson *et al.*, 1983; Bush *et al.*, 1984). Fetal PCB syndrome, characterized by brown pigmentation of the skin, was observed in infants from Yusho mothers that would indicate placental transfer (Yamashita & Hayashi, 1985). PCB levels of 0.1–0.3 mg/kg in human amniotic fluid were reported in 92% of the samples analysed, although the implications are unknown (Rao & Banerji, 1988). Transfer rates for PCDD/DFs across the placenta in rats are non-detectable to several percent of the dose administered (Nagayama *et al.*, 1980; Vodicnik & Lech, 1980; Weber & Birnbaum, 1985; Nau *et al.*, 1986). Limited PCDD/DF analyses of human stillborn fetuses and infants indicate transfer rates of up to several percent may be possible (Beck *et al.*, 1990b; Schecter *et al.*, 1990b; van Wijnen *et al.*, 1990).

Contaminant transfer during the nurturing period can be high when the mother's milk is utilized. Transfer of maternal PCBs and PCDD/DFs to progeny through milk has been shown in various animals, including humans, and most studies indicate that this source is more important than placental transfer (Nagayama *et al.*, 1980; Bowman *et al.*, 1989; Dewailly *et al.*, 1989). Human maternal milk contains higher PCDD/DF levels than cow's milk (Table 9.5). Mobilization of maternal lipid for milk production would suggest little or no preferential selection of congeners.

9.6 TOXICOLOGICAL EVALUATION

Two approaches can be used to assess the eco-toxicological effects of PCBs, PCDDs and PCDFs. The first is based on comparing the guidelines proposed by various regulatory agencies on chemical content in different matrices for the protection of human health. This approach is simple to apply because the concentration for a specific chemical such as 2,3,7,8-TCDD or total PCBs is defined, although the significance of concentrations below the guideline cannot be evaluated. The second approach focuses on the toxicological impact of a chemical or group of chemicals where each one is considered for its potency, and total chemical concentration becomes less important. This multi-chemical approach does have limitations because of its assumptions, such as additivity of effects, but does address the issue of chemical mixtures, and provides a relative evaluation for all chemical concentrations (NATO, 1988; Neubert *et al.*, 1992; see Chapter 12).

9.6.1 Assessment based on chemical concentrations

Regulatory guidelines or levels of concern have been proposed by agencies which define the concentrations of PCBs and 2,3,7,8-TCDD that are permissible in certain food products or environmental matrices. These action levels can be based on a concentration per unit weight or total daily intake basis for both PCBs and PCDD/DFs (Table 9.7). There is relatively good agreement among the different matrices except for 2,3,7,8-TCDD daily intake levels. Guidelines for PCDD/DFs are currently under review by some agencies because of the interest in 2,3,7,8-TCDD equivalents rather than its concentration. There is a range of concentrations for 2,3,7,8-TCDD, or its equivalents, of 0.006–10 pg/kg per day for humans where assessment is based on carcinogenic risk. These differences are due to the methods used and assumptions made, and this is demonstrated by comparing the US agencies. The

Dutch and Canadian values, which were all based on the study by Kociba *et al.* (1978). The EPA and CDC estimates were derived from a linearized multi-stage model, while the FDA, the Dutch and Canadian estimates were based on the 'no observed adverse effect level' (NOAEL) 'safety factor' approach. Most assessments apply a safety or uncertainty factor of 100–1000 to estimate levels of concern (Barnes, 1989). These differences indicate the degree of uncertainty about the effects of 2,3,7,8-TCDD, and these concentrations should be considered more as a 'level of concern' rather than viewed as a 'safe level of intake'.

9.6.2 Multi-chemical assessment based on toxic equivalency factors (TEFs)

Equivalency factors are an interim approach which is used to assess the toxicological risks of PCBs and PCDD/DFs. The concept of 'approximate activity relative to TCDD' or toxic equivalency factors (TEFs) was proposed by Grant (1977), who observed different induction levels of aryl hydrocarbon hydroxylase (AHH) activity in a chick embryo assay among different PCDF congeners, and expressed their potencies relative to 2,3,7,8-TCDD. This is based on the affinity of these chemicals to a specific cytosolic protein complex, called the Ah receptor (Poland *et al.*, 1976). Many studies, particularly those of Safe and his co-workers, have shown good correlations between the *in vitro* ED_{50} response enzyme induction and *in vivo* ED_{50} values for indices such as body weight change and thymic atrophy among different animals exposed to PCBs and PCDD/DFs (Safe, 1984, 1986, 1987). There are at least 10 TEF tables for PCDD/DFs proposed by various groups whose coefficients among different congeners differ slightly (World Health Organization, 1989; Kutz *et al.*, 1990). An international TEF (I-TEF) that was developed under the auspices of NATO is now widely used, and provides coefficients for seven PCDD and 10 PCDF congeners with the 2,3,7,8-chlorine substitution pattern (NATO, 1988). These coefficients range from 1.0 for 2,3,7,8-TCDD to 0.001 for the octachloro-congeners of PCDDs and PCDFs. There is good agreement on the use of I-TEF coefficients because many studies have shown congeners with this substitution pattern are more potent than those

Table 9.7 Regulatory and advisory guidelines of 2,3,7,8-TCDD, or its equivalents, and PCBs in different matrices for the protection of human health

Matrix/product	Chemical	Guideline	Agency
Air	2,3,7,8-TCDD equivalent	$1.0\,pg/m^3$ $5\,pg/m^3$	Connecticut, USA Canada
Air emissions	2,3,7,8-TCDD equivalent	$1\,ng/m^3$ $2\,ng/m^3$	Denmark Norway
Ambient water	2,3,7,8-TCDD	$0.013\,pg/l$	US EPA
Domestic water	2,3,7,8-TCDD equivalent	$15\,pg/l$	Ontario, Canada
Surface contamination	2,3,7,8-TCDD equivalent	$1\,ng/m^2$	Advisory panel
Fish	2,3,7,8-TCDD	$20\,ng/kg$	Canada
Human daily intake	2,3,7,8-TCDD equivalent	$0.006\,pg/kg$ per day $0.028\,pg/kg$ per day $0.06\,pg/kg$ per day $1-10\,pg/kg$ per day $4\,pg/kg$ per day $10\,pg/kg$ per day	US EPA US CDC US FDA Germany, UBA/BGA Netherlands Canada
Air	PCBs	$1.0\,\mu g/m^3$	USA, NIOSH
Mother's milk	PCBs	$5\,\mu g/kg$ per day	WHO
Fish	PCBs	$2\,mg/kg$ $2\,mg/kg$ $2\,mg/kg$	Canada Sweden US FDA

without it, and these congeners are selectively accumulated by animals and humans (Rappe *et al.*, 1979; Hutzinger & Fiedler, 1989; Kutz *et al.*, 1990).

The TEF concept for PCBs has not been widely applied because over 100 congeners may be present in some samples. Relatively few studies monitor the more potent congeners because of low concentrations and a lack of agreement on the coefficients for specific congeners. The TEFs proposed by Safe (1990) for the more potent non-*ortho*-substituted congeners have been used by some recent studies (Dewailly *et al.*, 1991). A non-*ortho*-substituted configuration is one with no chlorine substitution at the *ortho* positions, both *para* positions substituted, and two to four substitutions at the *meta* positions. There are six non-*ortho*-substituted congeners among which only four, with four or more chlorine substitutions, are of toxicological importance because of their 'coplanar' structural configuration that

are approximate stereoisomers of 2,3,7,8-TCDD (Table 9.1). Mono-*ortho*-substituted coplanar congeners also show enhanced potency, while some di-*ortho*-substituted congeners have shown enzyme induction capabilities, although at lower levels (Safe *et al.*, 1985). Based on a coefficient of 1.0 for 2,3,7,8-TCDD, the values proposed by Safe (1990) include a coefficient of 0.1 for congener 126, 0.05 for congener 169, 0.01 for congener 77, and 0.001 for congeners 105, 114, 118, 123, 156, 157, 167 and 189. A coefficient of 0.00002 is assigned for all di-*ortho*-substituted congeners.

TEF is applied by multiplying the coefficient of each congener by its concentration in the sample, and the sum is reported as 2,3,7,8-TCDD equivalents, or toxic equivalent quantity (TEQ). TEQ values compiled for PCDD/DFs that are reported in the tables of this chapter are based on I-TEF coefficients, and non-detectable concentrations were assigned a value of one-half of the reported detection limit. It was necessary to recalculate many of the values because the ones reported were often based on other TEF tables. Hence, the TEQ values reported here may not be the same as those reported in the studies cited. Furthermore, the studies that are cited were limited to those that reported many or all of the 2,3,7,8-substituted PCDD and PCDF congeners. Few TEQ values were compiled for PCB results because they were limited to those studies that reported concentrations of the three most potent congeners. The units of measurement should be noted in comparing TEQ values among different matrices, because PCDD concentrations in food products are reported as nanogram per kilogram, air as picogram per cubic metre and water as nanogram per litre.

9.6.3 Human health concerns

Laboratory studies have demonstrated that these chemicals can cause mortality, induce cancer, adversely affect reproduction, alter immune systems and induce chronic effects in different animals (Morrissey & Schwetz, 1989; Vos & Luster, 1989; Silberhorn et al., 1990). Their adverse effects on humans are less clear because evaluations are based on retrospective analyses from accidental exposures rather than experimental studies that could demonstrate cause-and-effect relationships. Most studies have focused on 2,3,7,8-TCDD with comparative studies on other congeners to develop structure–activity relationships. 2,3,7,8-TCDD is not an important acute toxicant to humans. Lipid concentrations of $1-2\,\mu g/kg$ 2,3,7,8-TCDD and $7-21\,\mu g/kg$ total PCDD/DFs were reported in two occupationally exposed workers; this 2,3,7,8-TCDD would be comparable to $0.4\,\mu g/kg$ wet weight assuming a 15% body lipid content (Beck et al., 1989b). LD_{50} concentrations compiled for five mammalian species ranged from 1 to $>3000\,\mu g/kg$ for guinea pigs and rats (WHO, 1989). 2,3,7,8-TCDD is classified as a 'probable human carcinogen', and its role as a cancer initiator or a cancer promoter has been the focus of intensive investigations (US EPA, 1985; IARC, 1987; McConnell, 1989). There have been epidemiological studies on humans exposed to 2,3,7,8-TCDD accidentally in Italy,

Japan, Taiwan, the USA and Vietnam, and occupationally in Denmark, Germany, New Zealand, Sweden and the USA. The general conclusions among the more comprehensive studies indicate that there is no clear evidence that 2,3,7,8-TCDD and PCBs have caused an increased incidence of reproductive impairment, cancer and immune deficiencies in humans when other factors that also induce the same effects are taken into consideration (Hoffman & Stehr-Green, 1989; Kuratsune, 1989; Reggiani, 1989; WHO, 1989; Schulz et al., 1990; A. Smith et al., 1990.) The antagonistic effect of PCBs against PCDD/DF toxicity must also be taken into consideration because of their common presence in many animals (Biegel et al., 1989; Silberhorn et al., 1990). The role of PCDD/DFs as cancer initiators is not resolved because of a long latency period, although more recent analyses may suggest a more potent role for 2,3,7,8-TCDD (Huff, 1992; Johnson, 1992).

A comparative examination of the biological effects of 2,3,7,8-TCDD and related halogenated aromatic hydrocarbons by Schulz et al. (1990) concluded that there is sufficient evidence to demonstrate serum lipid alterations, cancer, gastrointestinal, immunotoxicity, liver toxicity and reproductive effects in animals, but only limited or insufficient evidence in humans. Some of these effects have been observed in humans following chemical exposure, although the severity of the symptoms generally declines with time (Kuratsune, 1989; A. Smith et al., 1990). Other symptoms observed in animals but not in humans are wasting syndrome and thymic atrophy (Safe et al., 1990). Chloracne is one of the few clinical symptoms that has been observed in humans following exposure to high concentrations of PCBs and PCDD/DFs (WHO, 1989). A strong dose–response relationship has been observed in humans, and chloracne has been induced in monkey, hairless mouse and rabbit following chemical exposure. Induction of aryl hydrocarbon hydroxylase (AHH) and other enzymes is another clinical response observed in humans following exposure (Schulz et al., 1990).

The concentrations of these chemicals in ambient air, water and food indicate that the levels are generally within the advisory guidelines. Some exceptions are PCB concentrations in fish, and PCDD/DF levels in fish and shellfish from contaminated waters. Agencies have restricted the sale of consumable products such as fish that exceed the guidelines, but not the consumption of these items from non-commercial sources.

Persons consuming contaminated sportfish, and infants nurtured on breast milk, are two groups that have been studied for possibly toxicological effects. Consumption of contaminated fish was of particular concern in the Great Lakes region in the 1970s because PCB levels in salmonids exceeded 10 mg/kg. Persons consuming the largest amount of fish had PCB serum levels that were comparable with Yusho and Yu-Cheng patients. There were significant differences in PCB blood serum levels and some differences among other symptoms across the different studies (Anderson, 1989). Health measures against the consumption of contaminated fish, and the decline of PCB concentrations to near-guideline levels, have partially reduced this concern. The impact of PCBs and PCDD/DFs on infants is another issue that requires further examination. Based on an intake of 120 ml/kg/d of milk by an infant, the 1–6 µg/kg PCBs and 6–24 pg/kg PCDD/DFs (I-TEQ) ingested could exceed the guidelines proposed by various agencies (Table 9.7).

9.6.4 Terrestrial ecosystem concerns

Very little information is available to assess the adverse effects of PCBs and PCDD/DFs on plants and animals in the terrestrial environment. Chemical concentrations of animals raised for food purposes are routinely monitored by regulatory agencies, and adverse effects are not likely. Natural mink populations in the Great Lakes basin could be affected by PCBs where 1–34 mg/kg levels in lipid were reported (Proulx et al., 1987). Maternal lipid levels of 10–43 mg/kg can significantly affect reproduction in mink (Hornshaw et al., 1983). Fur-bearing animals in Michigan that were fed a diet containing contaminated fish had reproductive failure, and laboratory studies indicated PCBs as the likely cause (Aulerich et al., 1973; Byrne et al., 1975). Severe behavourial abnormalities were reported in otters

with over 100 mg/kg PCBs that also contained over 100 mg/kg ΣDDT (Mason & O'Sullivan, 1992).

9.6.5 Aquatic ecosystem concerns

Current evidence suggests that PCBs and PCDD/DFs could have relatively little direct impact on aquatic plants and invertebrates, some impact on fish, and probably a larger impact on piscivorous birds and marine mammals. There are localized habitats near urban centres where the sediment is heavily contaminated and is nearly devoid of aquatic life, but this could be attributed more to physical conditions such as hypoxia than to the presence of chemicals. Even if chemicals are a likely cause, the less hydrophobic chemicals could be more important than PCBs and PCDD/DFs. Waterborne PCB concentrations in some contaminated systems can exceed regulatory guideline levels in the 1–14 ng/l range, but most are well below the 0.1–2 µg/l concentrations of the 'lowest observed adverse effect level' (LOAEL) reported for larval fish and invertebrates (Nebeker & Puglisi, 1974; Schimmel et al., 1974). Laboratory studies on various aquatic organisms that attained mg/kg PCB concentrations reported changes in cellular and ultrastructural features and reproductive impairment (Hacking et al., 1977; Nebeker and Puglisi, 1977; DeFoe et al., 1978). Concentrations in the low picogram per litre range (or below) of 2,3,7,8-TCDD, and its equivalents, in natural and domestic waters, are also well below guideline levels and the LOAEL of 5–55 ng/l reported for larval fish (Wisk & Cooper, 1990; Walker et al., 1991). Fish exposed to high 2,3,7,8-TCDD concentrations at the larval stage can develop pericardial oedema and haemorrhages, while larger fish develop fin necrosis (Helder, 1980; Kleeman et al., 1988; Walker & Peterson, 1991).

Elevated frequencies of skin papillomas, hepatic carcinomas and similar anomalies have been reported in fish from many contaminated sites (Harshbarger & Clark, 1990; Black & Baumann, 1991). There is strong circumstantial information to suggest a chemical basis for these anomalies, although the specific role of PCBs and PCDD/DFs is less clear. DNA adducts have received considerable attention in carcinogenic studies because of their role in cancer initiation (Randerath *et al.*, 1982; Varanasi *et al.*, 1989). The formation of DNA adducts has been reported in fish collected from contaminated sites, and in laboratory exposure studies (Bailey *et al.*, 1987; Dunn *et al.*, 1987). It could be difficult to link the presence of these carcinogenic-like anomalies and the formation of DNA adducts causally to PCBs and PCDD/DFs because the role of these chemicals as carcinogenic initiators has not been resolved. Fish with these anomalies are often benthic and are collected from areas with high polycyclic aromatic hydrocarbon (PAH) content in the sediments. PAHs are carcinogenic promotors, and are good AHH inducers which could suggest a similar mechanism-of-action with PCBs and PCDD/DFs (Piskorska-Pliszczynska *et al.*, 1986). Studies that have induced carcinomas in aquatic organisms often use naturally contaminated sediments or sediment extracts that contain PAHs and other carcinogenic inducers (Metcalfe *et al.*, 1990; Gardner *et al.*, 1991). Aroclor 1254 itself does not induce hepatocellular carcinoma, and may not even be a carcinogenic promotor in fish (Shelton *et al.*, 1984; Bailey *et al.*, 1987). Some fish with tumours do not have alterations in cellular glutathione-*S*-transferase, suggesting that some tumours are not chemically promoted (Hayes *et al.*, 1990).

The effects of these chemicals on aquatic birds and marine mammals are inconclusive, but mortality and reproductive impairment have been related to increased chemical concentrations. Reproductive and teratogenic effects among piscivorous birds in the Great Lakes basin have been well documented, and correlated with PCBs and 2,3,7,8-TCDD tissue levels (Gilbertson, 1989; Kubiak *et al.*, 1989). While the effects of other chemicals that are present cannot be excluded, the presence of porphyrins would suggest that these chemicals could be at least a contributing factor (Fox *et al.*, 1988).

There have been increasing reports of marine mammal deaths from various locations. The cause of death is unknown in some cases, while viral infection have been identified in others (Osterhaus & Vedder, 1988). PCBs and other organochlorine measurements among marine mammals are among the highest concentrations reported (Kannan *et al.*, 1989). PCB levels in California sea-lion blubber averaged 820 mg/kg in the early 1970s (DeLong *et al.*, 1973). More recent studies on North Sea seals indicate PCB levels of 7–210 mg/kg (Law *et al.*, 1989). Reproductive impairment among some seal populations in European waters has been reported, and a chemical cause is suspected (Helle *et al.*, 1976; Reijnders, 1986). Decreased testosterone levels in porpoises were also observed in animals with elevated PCBs, although statistical significance was shown only with DDE concentrations. (Subramanian *et al.*, 1987). While PCBs and PCDD/DFs can adversely affect the reproductive and immune systems, these effects cannot be entirely attributed to these chemicals at this time.

9.7 COROLLARY

It is apparent from this survey that there are very few instances like Seveso where the effects of a single chemical can be assessed in the natural environment. Lack of conclusive evidence linking PCBs and PCDD/DFs to deleterious effects observed among some animals does not imply their toxicological concerns are diminished. On the contrary, it is becoming more evident that ecotoxicological assessments should not be made on a single chemical basis when other chemicals are present in the same matrix. This is the case for PCBs and PCDD/DFs where there are some similarities in their mechanisms of action. For example, it has been questioned if the clinical symptoms observed among Yusho patients were due to PCBs, PCDFs or polychlorinated quaterphenyls (Masuda *et al.*, 1986). It is probable that all of these chemicals were contributing factors of varying degrees. There have been recent changes in hazard assessment where the guidelines for 2,3,7,8-TCDD have been superseded by 2,3,7,8-TCDD equivalents, recognizing the contribution not only of the other PCDD congeners but also of the PCDFs from their toxicity (Table 9.7). The merits of using this approach have been shown in risk assessments of fish products containing PCBs, PCDDs and PCDFs (Niimi & Oliver, 1989b; Williams *et al.*, 1992). Further

Table 9.8 Relative toxicity between PCBs and total PCDDs and PCDFs in different tissues or animals based on TEQs and I-TEQs. Comparisons are based on chemical measurements reported in the same study or from two comparable studies. PCB estimates are based on non-*ortho*-substituted congeners 77, 126 and 169, and possibly other mono-*ortho*-substituted congeners, and their concentrations are reported. TEF coefficients for PCBs were proposed by Safe (1990). Total PCDD and PCDF concentrations are based on 2,3,7,8-substituted tetra- to octachlorocongeners.

	PCBs			PCDDs and PCDFs		
Sample matrix	No. congeners	Concentration	TEQ	Concentration (ng/kg)	I-TEQ	References
Human						
Milk, Canada	6	844 ng/kg	1.1	6.6	0.40	1
Milk, Sweden	3	7 ng/kg	0.5	10.2	0.4	2, 3
Adipose, Canada	3	454 ng/kg	37.5	1600	51.8	4
Adipose, Japan	3	771 ng/kg	41.0	552	37.0	5, 6
Adipose, Japan	3	1530 ng/kg	~72	1550	~30	7
Aquatic animals						
Trout, Lake Ontario	3	37 µg/kg	2080	143	22.04	8, 9
Marine mammals, blubber						
Baird's beak whale	6	0.77 mg/kg	137	~5	0.6	10
Killer whale	6	16.0 mg/kg	21 450	29	3.0	10
Forster's tern eggs	7	2.0 mg/kg	2330	37.3	37.7	11

References: 1, Dewailly *et al.*, 1991; 2, Nóren *et al.*, 1990; 3, Rappe *et al.*, 1989b; 4, Williams & LeBel, 1991; 5, Tanabe *et al.*, 1987a; 6, Kannan *et al.*, 1989; 7, Kashimoto *et al.*, 1989; 8, Niimi & Oliver, 1989a; 9, Niimi & Oliver, 1989b; 10, Kannan *et al.*, 1989; 11, Kubiak *et al.*, 1989.

examination of human and animal tissue residue concentrations also suggest that PCBs could have a greater toxicological significance than PCDDs and PCDFs, based on their TEQ contributions (Table 9.8).

ACKNOWLEDGEMENTS

The comments of Dr D.C.G. Muir, Freshwater Institute, Winnipeg, Manitoba and Dr J.P. Giesy, Michigan State University, East Lansing, on this manuscript were greatly appreciated.

REFERENCES

Ackerman, D.G., Scinto, L.L., Bakshi, P.S., Dulumyea, R.G., Johnson, R.J., Richard, G. & Takata, A.M. (1981) *Guidelines for the Disposal of PCBs and PCB Items by Thermal Destruction.* EPA-600-2-81-022. US EPA, Research Triangle Park, Washington.

Akimoto, Y. (1989) Formation of polychlorinated dibenzo-*p*-dioxins (PCDD) and dibenzofurans (PCDF) in rotary combustor. *Chemosphere* **19**, 393–400.

Albro, P.W. & Fishbein, L. (1972) Intestinal absorption of polychlorinated biphenyls in rats. *Bull. Environ. Contam. Toxicol.* **8**, 26–31.

Ambidge, P.F., Cox, E.A., Creaser, C.S., Greenberg, M., de M. Gem, M.D., Gilbert, J., Jones, P.W., Kibblewhite, M.G., Levy, J., Lisseter, S.G., Meredith, T.J., Smith, L., Smith, P., Startin, J.R., Stenhouse, I. & Whitworth, M. (1990) Acceptance criteria for analytical data on polychlorinated dibenzo-*p*-dioxins and polychlorinated dibenzofurans. *Chemosphere* **21**, 999–1006.

Anderson, H.A. (1989) General population exposure to environmental concentrations of halogenated biphenyls. In: *Halogenated Biphenyls, Terphenyls, Naphthalenes, Dibenzofarans and Related Products*, (Ed. R.D. Kimbrough & A.A. Jensen), pp. 325–344. Elsevier, Amsterdam.

Anderson, M.L., Rice, C.R. & Carl, C.C. (1982) Residues of PCB in a *Cladophora* community along the Lake Huron shoreline. *J. Great Lakes Res.* **8**, 196–200.

Andersson, Ö., Linder, C.-E., Olson, M., Reutergårdh, L., Uvemo, U.-B. & Wildeqvist U. (1988) Spatial differences and temporal trends of organochlorine compounds in biota from the northwestern hemisphere. *Arch. Environ. Contam. Toxicol.* **17**, 755–765.

Andrews, Jr, J.S., Garrett, Jr, W.A., Patterson, Jr, D.G.,

Needham, L.L., Roberts, D.W., Bagby, J.R., Anderson, J.E., Hoffman, R.E. & Schramm, W. (1989) 2,3,7,8-Tetrachlorodibenzo-*p*-dioxin levels in adipose tissue of persons with no known exposure and in exposed persons. *Chemosphere* **18**, 499–506.

Anon. (1985) Report of the ICES advisory committee on marine pollution, 1985. *ICES Cop. Res. Rep.* 135.

Antonsson, A-B., Runmark, S., Mowrer, J. & Kjeller, L-O. (1989) Dioxins in the work environment in steel mills. *Chemosphere* **19**, 699–704.

Atkinson, R. (1991) Atmospheric lifetimes of dibenzo-*p*-dioxins and dibenzofurans. *Sci. Total Environ.* **104**, 17–33.

Aulerich, R., Ringer, R.K. & Iwamoto, S. (1973) Reproductive failure and mortality in mink fed on Great Lakes fish. *J. Reprod. Fertil.*, Suppl. **19**, 365–376.

Bache, C.A., Serum, J.W., Youngs, W.D. & Lisk, D.J. (1972) Polychlorinated biphenyl residues: accumulation in Cayuga Lake trout with age. *Science* **177**, 1191–1192.

Bailey, G., Selivonchick, D. & Hendricks, J. (1987) Initiation, promotion, and inhibition of carcinogenesis in rainbow trout. Environ. *Health Perspect.* **71**, 147–153.

Baker, J.E. & Eisenreich, S.J. (1990) Concentrations and fluxes of polycyclic aromatic hydrocarbons and polychlorinated biphenyls across the air-water interface of Lake Superior. *Environ. Sci. Technol.* **24**, 342–352.

Ballschmiter, K. & Zell, M. (1980) Analysis of polychlorinated biphenyls (PCB) by glass capillary chromatography. Composition of technical Aroclor and Clophen-PCB mixtures. *Fres. Z. Anal. Chem.* **302**, 20–31.

Banks, Y.B. & Birnbaum, L.S. (1991) Absorption of 2,3,7,8-tetrachlorodibenzo-*p*-dioxin (TCDD) after low dose dermal exposure. *Toxicol. Appl. Pharmacol.* **107**, 302–310.

Baril, A., Elliott, J.E., Somers, J.D. & Erickson, G. (1990) Residue levels of environmental contaminants in prey species of the Peregrine Falcon, *Falco peregrinus*, in Canada. *Can. Field-Nat.* **104**, 273–284.

Barnes, D.G. (1989) Characterization of the risks posed by CDDs and CDFs. *Chemosphere* **18**, 33–39.

Beck, H., Breuer, E.M., Droß, A. & Mathar, W. (1990a) Residues of PCDDs, PCBs and other organochlorine compounds in harbour seals and harbour porpoise. *Chemosphere* **20**, 1027–1034.

Beck, H., Droß, A., Kleemann, W.J. & Mathar, W. (1990b) PCDD and PCDF concentration in different organs from infants. *Chemosphere* **20**, 903–910.

Beck, H., Eckart, K., Mathar, W. & Wittkowski, R. (1989a) PCDD and PCDF body burden from food intake in the Federal Republic of Germany. *Chemosphere* **18**, 417–424.

Beck, H., Eckart, K., Mathar, W. & Wittkowski, R. (1989b) Levels of PCDDs in adipose tissue of occu-pationally exposed workers. *Chemosphere* **18**, 507–516.

Bedard, D.L., Wagner, R.E., Brennan, M.J., Haberl, M.L. & Brown, Jr, J.F. (1987) Extensive degradation of Aroclors and environmentally transformed polychlorinated biphenyls by *Alcaligenes eutrophus* H850. *Appl. Environ. Microbiol.* **53**, 1094–1102.

Bidleman, T.F. (1988) Atmospheric processes. *Environ. Sci. Technol.* **22**, 361–367.

Bidleman, T.F., Patton, G.W., Hinckley, D.A., Walla, M.D., Cotham, W.E. & Hargrave, B.T. (1990) Chlorinated pesticides and polychlorinated biphenyls in the atmosphere of the Canadian Arctic. In: *Long Range Transport of Pesticides* (Ed. D.A. Kurtz), pp. 347–372. Lewis, Chelsea, MI.

Bidleman, T.F., Patton, G.W., Walla, M.D., Hargrave, B.T., Vass, W.P., Erickson, P., Fowler, B., Scott, V. & Gregor, D.J. (1989) Toxaphene and other organochlorines in Arctic Ocean fauna: evidence for atmosphere delivery. *Arctic* **42**, 307–313.

Biegel, L., Howie, L. & Safe, S. (1989) Polychlorinated biphenyl (PCB) congeners as 2,3,7,8-TCDD antagonists: teratogenicity studies. *Chemosphere* **19**, 955–958.

Bignert, A., Olsson, M., Bergqvist, P.A., Bergek, S., Rappe, C., de Wit, C. & Jansson, B. (1989) Polychlorinated dibenzo-*p*-dioxins (PCDD) and dibenzofurans (PCDF) in seal blubber. *Chemosphere* **19**, 551–556.

Birmingham, B. (1990) Analysis of PCDD and PCDF patterns in soil samples: use in the estimation of risk exposure. *Chemosphere* **20**, 807–814.

Birmingham, B., Thorpe, B., Frank, R., Clement, R., Tosine, H., Fleming, G., Ashman, J., Wheeler, J., Ripley, B.D. & Ryan, J.J. (1989) dietary intake of PCDD and PCDF from food in Ontario, Canada. *Chemosphere* **19**, 507–512.

Black, J.J. & Baumann, P.C. (1991) Carcinogens and cancers in freshwater fishes. *Environ. Health Perspect.* **90**, 27–33.

Bowman, R.E., Schantz, S.L., Weerasinghe, N.C.A., Gross, M.L. & Barsotti, D.A. (1989) Chronic dietary intake of 2,3,7,8-tetrachlorodibenzo-*p*-dioxin (TCDD) at 5 or 25 parts per trillion in the monkey: TCDD kinetics and dose–effect estimate of reproductive toxicity. *Chemosphere* **18**, 243–252.

Branson, D.R., Takahashi, I.T., Parker, W.M. & Blau, G.E. (1985) Bioconcentration kinetics of 2,3,7,8-tetrachlorodibenzo-*p*-dioxin in rainbow trout. *Environ. Toxicol. Chem.* **4**, 779–788.

Brewster, D.W., Banks, Y.B., Clark, A-M. & Birnbaum, L.S. (1989) Comparative dermal absorption of 2,3,7,8-tetrachlorodibenzo-*p*-dioxin and three polychlorinated dibenzofurans. *Toxicol. Appl. Pharmacol.* **97**, 156–166.

Broman, D., Näf, C., Rolff, C. & Zebühr, Y. (1991a) Occurrence and dynamics of polychlorinated dibenzo-*p*-dioxins and dibenzofurans and polycyclic aromatic hydrocarbons in the mixed surface layer of remote coastal and offshore waters of the Baltic. *Environ. Sci. Technol.* **25**, 1850–1864.

Broman, D., Näf, C., Rolff, C., Zebühr, Y., Fry, B. & Hobbie, J. (1992) Using ratios of stable nitrogen isotopes to estimate bioaccumulation and flux of polychlorinated dibenzo-*p*-dioxins (PCDDs) and dibenzofurans (PCDFs) in two food chains from the northern Baltic. *Environ. Toxicol. Chem.* **11**, 331–345.

Broman, D., Näf, C. & Zebühr, Y. (1991b) Long-term high- and low-volume air sampling of polychlorinated dibenzo-*p*-dioxins and dibenzofurans and polychlorinated aromatic hydrocarbons along a transect from urban to remote areas on the Swedish coast. *Environ. Sci. Technol.* **25**, 1841–1850.

Brown, M.P., Werner, M.B., Sloan, R.J. & Simpson, K.W. (1985) Polychlorinated biphenyls in the Hudson River. *Environ. Sci. Technol.* **19**, 656–661.

Brun, G.L., Howell, G.D. & O'Neill, H.J. (1991) Spatial and temporal patterns of organic contaminants in wet precipitation in Atlantic Canada. *Environ. Sci. Technol.* **25**, 1249–1261.

Brunn, H., Berlich, H.D. & Müller, F.J. (1985) Residue of pesticides and polychlorinated biphenyls in game animals. *Bull. Environ. Contam. Toxicol.* **34**, 527–532.

Bryant, C.J., Hartle, R.W., Crandall, M.S. & Roper, P. (1989) Polychlorinated biphenyl, polychlorinated dibenzo-*p*-dioxin and polychlorinated dibenzofuran contamination in PCB disposal facilities. *Chemosphere* **18**, 569–576.

Buckland, S.J., Hannah, D.J., Taucher, J.A., Slooten, E. & Dawson, S. (1990) Polychlorinated dibenzo-*p*-dioxin and dibenzofurans in New Zealand's Hector's dolphin. *Chemosphere* **20**, 1035–1042.

Buckley, E.H. (1982) Airborne polychlorinated biphenyls in foliage. *Science* **216**, 520–522.

Bühler, F., Schmid, P. & Schlatter, C. (1988) Kinetics of PCB elimination in man. *Chemosphere* **17**, 1717–1726.

Burkhard, L.P., Andrew, A.W. & Armstrong, D.E. (1985) Estimation of vapor pressures for polychlorinated biphenyls: a comparison of eleven predictive methods. *Eviron. Sci. Technol.* **19**, 500–507.

Buser, H.R. & Rappe, C. (1978) Identification of substitution patterns in polychlorinated dibenzo-*p*-dioxins by mass spectrometry. *Chemosphere* **7**, 199–211.

Bush, B., Snow, J. & Koblintz, R. (1984) Polychlorinated (PCB) congeners, *p,p'*-DDE, and hexachlorobenzene in maternal and fetal blood from mothers in upstate New York. *Arch. Environ. Contam. Toxicol.* **13**, 517–527.

Byrne, J.J., Reinke, E.P., Ringer, R.K. & Aulerich, R.K. (1975) Influences of polychlorinated biphenyl mixture (Aroclor 1254) administration on reproduction and thyroid function in mink (*Mustela vison*). *Fed. Proc. Fed. Soc. Exp. Biol.* **34**, 321.

Carlberg, G.E., Ofstad, E.B., Drangsholt, H. & Steinnes, E. (1983) Atmospheric deposition of organic micropollutants in Norway studied by means of moss and lichen analysis. *Chemosphere* **12**, 341–356.

Charles, M.J. & Tondeur, Y. (1990) Choosing between high-resolution mass spectrometry and mass spectrometry/mass spectrometry: environmental applications. *Environ. Sci. Technol.* **24**, 1856–1860.

Chevreuil, M., Chesterikoff, A. & Létolle, R. (1987) PCB pollution behaviour in the River Seine. *Water Res.* **21**, 427–434.

Chevreuil, M., Chesterikoff, A., Létolle R. & Granier, L. (1989) Atmospheric pollution and fallout by PCBs and organochlorine pesticides (Ile-de-France). *Water, Air, Soil Pollut.* **43**, 73–83.

Christensen, E.R. & Lo, C-K. (1986) Polychlorinated biphenyls in dated sediments of Milwaukee Harbour, Wisconsin, USA. *Environ. Pollut.* **12**, 217–232.

Christmann, W., Kasiske, D., Klöppel, K.D., Partscht, H. & Rotard, W. (1989a) Combustion of polyvinylchloride — an important source for the formation of PCDD/PCDF. *Chemosphere* **19**, 387–392.

Christmann, W., Klöppel, K.D., Partscht, H. & Rotard, W. (1989b) PCDD/PCDF and chlorinated phenols in wood preserving formulations for household use. *Chemosphere* **18**, 861–865.

Christmann, W., Klöppel, K.D., Partscht, H. & Rotard, W. (1989c) Determination of PCDD/PCDF in ambient air. *Chemosphere* **19**, 521–526.

Clayton, Jr, J.R., Paviou, S.P. & Breitner, N.F. (1977) Polychlorinated biphenyls in coastal marine zooplankton: bioaccumulation by equilibrium partitioning. *Environ. Sci. Technol.* **11**, 676–682.

Crosby, D.G. (1985) The degradation and disposal of chlorinated dioxins. In: *Dioxins in the Environment*, (Ed. M.A. Kamrin & R.W. Rodgers), pp. 195–204. Hemisphere, Washington, DC.

Crummett, W.B. & Stehl, R.H. (1973) Determination of chlorinated dibenzo-*p*-dioxins and dibenzofurans in various materials. *Environ. Health Perspect.* **5**, 15–25.

Czuczwa, J.M. & Hites, R.A. (1984) Environmental fate of combustion-generated polychlorinated dioxins and furans. *Environ. Sci. Technol.* **18**, 444–450.

Czuczwa, J.M. & Hites, R.A. (1986) Airborne dioxins and dibenzofurans: sources and fates. *Environ. Sci. Technol.* **20**, 195–200.

Czuczwa, J.M., Niessen F. & Hites, R.A. (1985) Historical record of polychlorinated dibenzo-*p*-dioxins and dibenzofurans in Swiss lake sediments. *Chemosphere* **14**, 1175–1179.

Decad, G.M., Birnbaum, L.S. & Matthews, H.B. (1981a) 2,3,7,8-tetrachlorodibenzofuran tissue distribution and excretion in guinea pigs. *Toxicol. Appl. Pharmacol.* **57**, 231–240.

Decad, G.M., Birnbaum, L.S. & Matthews, H.B. (1981b) Distribution and excretion of 2,3,7,8-tetrachlorodibenzofuran in C57/BL/6J and DBA/2J mice. *Toxicol. Appl. Pharmacol.* **59**, 564–573.

DeFoe, D.L., Veith, G.D. & Carlson, R.W. (1978) Effects of Aroclor® 1242 and 1260 on the fathead minnow (*Pimephales promelas*). *J. Fish. Res. Bd. Can.* **35**, 997–1002.

de Fré, R. (1986) Dioxin levels in the emissions of Belgian municipal incinerators. *Chemosphere* **15**, 1255–1260.

deFreitas, A.S.W. & Norstrom, R.J. (1974) Turnover and metabolism of polychlorinated biphenyls in relation to their chemical structure and the movements in pigeon. *Can. J. Physiol. Pharmacol.* **52**, 1080–1094.

de Jong, A.P.J.M., Liem, A.K.D., Den Boer, A.C., Van Der Heeft, E., Marsman, J.A., Van De Werken, G. & Wegman, R.C.C. (1989) Analysis of polychlorinated dibenzofurans and dibenzo-*p*-dioxins in human milk by tandem hybrid mass spectrometry. *Chemosphere* **19**, 59–66.

DeLong, R.L., Gilmartin, W.G. & Simpson, J.G. (1973) Premature births in California sea lions: association with high organochlorine pollutant residue levels. *Science* **181**, 1168–1170.

des Rosiers, P.E. (1987) Chlorinated combustion products from fires involving PCB transformers and capacitors. *Chemosphere* **16**, 1881–1888.

de Voogt, P. & Brinkman, U.A.T. (1989) Production, properties and usage of polychlorinated biphenyls. In: *Halogenated Biphenyls, Terphenyls, Naphthalenes, Dibenzofurans and Related Products.* (Eds R.D. Kimbrough & A.A. Jensen), pp. 3–45. Elsevier, Amsterdam.

Dewailly, E., Nantel, A., Weber, J.-P. & Meyer, F. (1989) High levels of PCBs in breast milk of Inuit women from Arctic Quebec. *Bull. Environ. Contam. Toxicol.* **43**, 641–646.

Dewailly, E., Weber, J.-P., Gingras, S. & Laliberté, C. (1991) Coplanar PCBs in human milk in the Province of Quebec, Canada: are they more toxic than dioxin for breast fed infants? *Bull. Environ. Contam. Toxicol.* **47**, 491–498.

Dickson, K.L., Maki, A.W. & Brungs, W.A. (Eds) (1984) *Fate and Effects of Sediment-bound Chemicals in Aquatic Ecosystems.* Pergamon Press, New York.

D'Itri, F.M. (1988) Contaminants in selected fishes from the Great Lakes. In: *Toxic Contamination in Large Lakes*, Vol. II (Ed. N.K. Schmidke), pp. 51–84. Lewis, Chelsea, MI.

Duebelbeis, D.O., Kapila, S., Clevenger, T., Yanders, A.F. & Manhan, S.E. (1989a) A two-dimensional reaction gas chromatography system for isomer-specific determination of polychlorinated biphenyls.

Chemosphere **18**, 101–108.

Duebelbeis, D.O., Pieczonka, G., Kapila, S., Clevenger, T.E. & Yanders, A.F. (1989b) Application of a dual column reaction chromatography system for confirmatory analysis of polychlorinated biphenyl congeners. *Chemosphere* **19**, 143–148.

Duinker, J.C. & Hillebrand, M.T.J. (1979) Mobilization of organochlorines from female lipid tissue and transfer to fetus in a harbour porpoise (*Phocoena phocoena*) in a contaminated area. *Bull. Environ. Contam. Toxicol.* **23**, 728–732.

Duinker, J.C. & Hillebrand, M.T.J. (1983) Composition of PCB mixtures in biotic and abiotic marine compartments (Dutch Wadden Sea). *Bull. Environ. Contam. Toxicol.* **31**, 25–32.

Duinker, J.C., Schultz, D.E. & Petrick., G. (1988) Multidimensional gas chromatography with electron capture detection for the determination of toxic congeners in polychlorinated biphenyl mixtures. *Anal. Chem.* **60**, 478–482.

Dunn, B.P., Black, J.J. & Maccubbin, A. (1987) ^{32}P-postlabeling aromatic DNA adducts in fish from polluted areas. *Cancer Res.* **47**, 6543–6548.

Dunnivant, F.M. & Elzerman, A.W. (1988) Aqueous solubility and Henry's law constant data for PCB congeners for evaluation of quantitative structure-property relationships (QSPRs). *Chemosphere* **17**, 525–541.

Durell, G.S. & Sauer, T.C. (1990) Simultaneous dual-column, dual detector gas chromatographic determination of chlorinated pesticides and polychlorinated biphenyls in environmental samples. *Anal. Chem.* **62**, 1867–1871.

Eisenreich, S.J., Looney, B.B. & Thornton, J.D. (1981) Airborne organic contaminants in the Great Lakes ecosystem. *Environ. Sci. Technol.* **15**, 30–38.

Eitzer, B.D. & Hites, R.A. (1988) Vapor pressures of chlorinated dioxins and dibenzofurans. *Environ. Sci. Technol.* **22**, 1362–1364.

Elliott, J.E., Butler, R.W., Norstrom, R.J. & Whitehead, P.E. (1989) Environmental contaminants and reproductive success of great blue herons *Ardea herodias* in British Columbia, 1986–87. *Environ. Pollut.* **59**, 91–114.

Fiedler, H., Hutzinger, O. & Timms, C.W. (1990) Dioxins: sources of environmental load and human exposure. *Toxicol. Environ. Chem.* **29**, 157–234.

Focardi, S., Fossi, C., Leonzio, C. & Romei, R. (1986) PCB congeners, hexachlorobenzene, and organochlorine insecticides in human fat from Italy. *Bull. Environ. Contam. Toxicol.* **36**, 644–650.

Fox, G.A., Kennedy, S.W., Norstrom, R.J. & Wigfield, D.C. (1988) Porphyria in herring gulls: a biochemical response to chemical contamination of Great Lakes food chains. *Environ. Toxicol. Chem.* **7**, 831–839.

Fries, G.F. (1972) Polychlorinated biphenyl residues in

milk of environmentally and experimentally contaminated cows. *Environ. Health Perspect.* **1**, 55–59.

Fries, G.F. & Marrow, G.S. (1981) Chlorobiphenyl movement from soil to soybean plants. *J. Agric. Food Chem.* **29**, 757–759.

Friesen, K.J., Vilk, J. & Muir, D.C.G. (1990) Aqueous solubilities of selected 2,3,7,8-substituted polychlorinated dibenzofurans (PCDFs). *Chemosphere* **20**, 27–32.

Frommberger, R. (1991) Polychlorinated dibenzo-*p*-dioxins and polychlorinated dibenzofurans in fish from south-west Germany: River Rhine and Neckar. *Chemosphere* **22**, 29–38.

Fürst, P., Fürst, C. & Groebel, W. (1990) Levels of PCDDs and PCDFs in food-stuffs from the Federal Republic of Germany. *Chemosphere* **20**, 787–792.

Gardner, G.R., Yevich, P.P., Harshbarger, J.C. & Malcolm, A.R. (1991) Carcinogenicity of Black Rock Harbor sediment to the eastern oyster and trophic transfer of Black Rock Harbor carcinogens from the blue mussel to the winter flounder. *Environ. Health Perspect.* **90**, 53–66.

Giesy, J.P., Newsted, J. & Garling, D.L. (1986) Relationships between chlorinated hydrocarbon concentrations and rearing mortality of chinook salmon (*Oncorhynchus tshawytscha*) eggs from Lake Michigan. *J. Great Lakes Res.* **12**, 82–98.

Gilbertson, M. (1989) Effects on fish and wildlife populations. In: *Halogenated Biphenyls, Terphenyls, Naphthalenes, Dibenzofurans and Related Products* (Eds R.D. Kimbrough & A.A. Jensen), pp. 103–127. Elsevier, Amsterdam.

Giugliano, M., Cernuschi, S. & Ghezzi, U. (1989) The emission of dioxins and related compounds from the incineration of municipal solid wastes with high contents of organic chlorine (PVC). *Chemosphere* **19**, 407–411.

Gorski, T., Konopka, L. & Brodzki, M. (1984) Persistence of some polychlorinated dibenzo-*p*-dioxins and polychlorinated dibenzofurans of pentachlorophenol in human adipose tissue. *Rocz. Pzh.* **35**, 297–301. (Cited in WHO, 1989)

Government of Canada (1990) Canadian Environmental Protection Act. Priority substances list assessment report no. 1: polychlorinated dibenzodioxins and polychlorinated dibenzofurans. Minister of Supply and Service Canada, Ottawa.

Grant, D.L. (1977) Dioxins — a toxicological review. In: *Proceedings of the 12th Annual Workshop on Pesticide Residue Analysis and the 5th Annual Workshop on the Chemistry & Biochemistry of Pesticides (Western Canada)*, pp. 251–269. Northstar Inn, Winnipeg. Manitoba, 10–13 May.

Gray, A.P., Cepa, S.P., Solomon, I.J. & Aniline, O. (1976a) Synthesis of specific polychlorinated dibenzo-*p*-dioxins. *J. Org. Chem.* **41**, 2435–2437.

Gray, A.P., Dipinto, V.M. & Solomon, I.J. (1976b) Synthesis of specific polychlorinated dibenzofurans. *J. Org. Chem.* **41**, 2428–2434.

Gregor, D.J. & Gummer, W.D. (1989) Evidence of atmospheric transport and deposition of organochlorine pesticides and polychlorinated biphenyls in Canadian Arctic snow. *Environ. Sci. Technol.* **23**, 561–565.

Gross, M.L., Lay, J.O., Lyon, P.A., Lippstreu, D., Kangas, N., Harless, R.L., Taylor, S.E. & Dumpuy Jr, A.E. (1984) 2,3,7,8-tetrachlorodibenzo-*p*-dioxin levels in adipose tissue in Vietnam veterans. *Environ. Res.* **33**, 261–268.

Guertin, J. (1986) PCB destruction. In: *Hazards, Decontamination and Replacement of PCB* (Ed. J.-P. Crine), pp. 175–184. Plenum Press, New York.

Hacking, M.A., Budd, J. & Hodson, K. (1977) The ultrastructure of the river of rainbow trout: normal structure and modifications after chronic administration of a polychlorinated biphenyl Aroclor 1254. *Can. J. Zool.* **56**, 477–491.

Hansen, L.G. (1987) Environmental toxicology of polychlorinated biphenyls. In: *Environmental Toxin Series.* (Eds S. Safe & O. Hutzinger), pp. 16–48. Springer-Verlag, Berlin.

Hargrave, B.T., Vass, W.P., Erickson, P.E. & Fowler, B.R. (1988) Atmospheric transport of organochlorines to the Arctic Ocean. *Tellus* **40**, 480–493.

Harrad, S.J., Fernandes, A.R., Creaser, C.S. & Cox, E.A. (1991) Domestic coal combustion as a source of PCDDs and PCDFs in the British environment. *Chemosphere* **23**, 255–261.

Harshbarger, J.C. & Clark, J.B. (1990) Epizootiology of neoplasms in bony fish of North America. *Sci. Total Environ.* **94**, 1–32.

Hashimoto, S., Wakimoto, T. & Tatsukawa, R. (1990) PCDDs in the sediments accumulated about 8120 years ago from Japanese coastal areas. *Chemosphere* **21**, 825–835.

Hawker, D.W. & Connell, D.W. (1988) Octanol–water partition coefficients of polychlorinated biphenyl congeners. *Environ. Sci. Technol.* **22**, 382–387.

Hayes, M.A., Smith, I.R., Rushmore, T.H., Crane, T.L., Thorn, C., Kocal, T.E. & Ferguson, H.W. (1990) Pathogenesis of skin and liver neoplasms in white suckers from industrially polluted areas in Lake Ontario. *Sci. Total Environ.* **94**, 105–123.

Helder, T. (1980) Effects of 2,3,7,8-tetrachlorobidenzo-*p*-dioxin (TCDD) on early life stages of pike (*Esox lucius* L.). *Sci. Total Environ.* **14**, 255–264.

Helle, E., Olsson, M. & Jensen, S. (1976) PCB levels correlated with pathological changes in seal uteri. *Ambio* **5**, 261–263.

Hermanson, M.H., Christensen, E.R., Buser, D.J. & Chen, L-M. (1991) Polychlorinated biphenyls in dated sediment cores from Green Bay and Lake Michigan. *J. Great Lakes Res.* **17**, 94–108.

Hermanson, M.H. & Hites, R.A. (1990) Polychlorinated biphenyls in tree bark. *Environ. Sci. Technol.* **24**, 666–671.

Herrmann, R. & Baumgartner, I. (1987) Regional variation of selected polyaromatic and chlorinated hydrocarbons over the South Island of New Zealand, as indicated by their content in *Pinus radiata* needles. *Environ. Pollut.* **46**, 63–72.

Hoffman, R.E. & Stehr-Green, P.A. (1989) Localized contamination with 2,3,7,8-tetrachlorodibenzo-*p*-dioxin: the Missouri episode. In: *Halogenated Biphenyls, Terphenyls, Naphthalenes, Dibenzofurans and Related Products* (Eds R.D. Kimbrough & A.A. Jensen), p. 471–484. Elsevier, Amsterdam.

Hornshaw, T.C., Aulerich R.J. & Johnson, H.E. (1983) Feeding Great Lakes fish to mink: effects on mink and accumulation and elimination of PCBs by mink. *J. Toxicol. Environ. Health* **11**, 933–946.

Huang, C.-W., Miyata, H., Lu, J.-R., Ohta, S., Chang, T. & Kashimoto, T. (1992) Levels of PCBs, PCDDs and PCDFs in soil samples from incineration sites for metal reclamation in Taiwan. *Chemosphere* **24**, 1669–1676.

Huang, L.Q., Eitzer, B., Moore, C., McGown, S. & Tomer, K.B. (1991) The application of hybrid mass spectrometry/mass spectrometry and high-resolution mass spectrometry to the analysis of fish samples for polychlorinated dibenzo-*p*-dioxins and dibenzofurans. *Biol. Mass Spectrom.* **20**, 161–168.

Huckins, J.N., Stalling, D.L. & Petty, J.D. (1980) Carbon-foam chromatographic separation of non-*o,o'*-chlorine substituted PCBs from Aroclor mixtures. *J. Assoc. Off. Anal. Chem.* **63**, 750–755.

Huff, J. (1992) 2,3,7,8-TCDD: a potent and complete carcinogen in experimental animals. *Chemosphere* **25**, 173–176.

Hunt, G.T. & Maisel, B.E. (1990) Atmospheric PCDDs/ PCDFs in wintertime in a northeastern U.S. urban coastal environment. *Chemosphere* **20**, 1455–1462.

Hutzinger, O., Berg, M.V.D., Olie, K., Opperhuizen, A. & Safe, S. (1985) Dioxins and furans in the environment: evaluating toxicological risk from different sources by multi-criteria analysis. In: *Dioxins in the Environment* (Eds M.A. Kamrin & P.W. Rodgers), pp. 9–32. Hemisphere, Washington, DC.

Hutzinger, O. & Fiedler, H. (1989) Sources and emissions of PCDD/PCDF. *Chemosphere* **18**, 23–32.

Hutzinger, O., Safe, S. & Zitko, V. (1974) *The Chemistry of PCBs.* CRC Press, Cleveland, OH.

International Agency for Research on Cancer (1987) Overall evaluations of carcinogenicity: an updating of IARC monographs volumes 1 to 42. In: *IARC Monographs on the Evaluation of Carcinogenic Risks to Humans,* Suppl. 7, pp. 350–351. IARC, Lyon.

World Health Organization (1993) *Environmental Health Criteria 140. Polychlorinated Biphenyls and Terphenyls* 2nd Edn. WHO, Geneva.

Iwata, Y. & Gunther, F.A. (1976) Translocation of the polychlorinated biphenyl Aroclor 1254 from soil into carrots under field conditions. *Arch. Environ. Contam. Toxicol.* **4**, 44–59.

Jacobson, S.W., Jacobson, J.L., Schwartz, P.M. & Fein, G.G. (1983) Intrauterine exposure of human newborns to PCBs: measures of exposure. In: *PCBs: Human and Environmental Hazards.* (Eds F.M. D'Itri & M.A. Kamrin), pp. 311–343. Butterworth, Boston, MA.

Jensen, A.A. (1989) Background levels in humans. In: *Halogenated Biphenyls, Terphenyls, Naphthalenes, Dibenxofurans and Related Products,* (Eds R.D. Kimbrough & A.A. Jensen), pp. 345–380. Elsevier, Amsterdam.

Jensen, S. (1972) The PCB story. *Ambio* **1**, 123–131.

Jensen, S. & Sundstrom, G. (1974) Structures and levels of most chlorobiphenyls in two technical PCB products and in human adipose tissue. *Ambio* **3**, 70–76.

Jerina, D.M. & Daly, J.W. (1974) Arene oxides: a new aspect of drug metabolism. *Science* **185**, 573–582.

Johnson, E.S. (1992) Human exposure to 2,3,7,8-TCDD and risk of cancer. *Crit. Rev. Toxicol.* **21**, 451–463.

Jones, K.C. (1989) Polychlorinated biphenyls in Welsh soils: a survey of typical levels. *Chemosphere* **18**, 1665–1672.

Kannan, N., Schulz-Bull, D.E., Petrick, G. & Duinker, J.C. (1992) High resolution PCB analysis of Kenechlor, Phenoclor and Sovol mixtures using multidimensional gas chromatography. *Int. J. Environ. Chem.* **47**, 201–215.

Kannan, N., Tanabe, S., Ono, M. & Tatsukawa, R. (1989) Critical evaluation of polychlorinated biphenyl toxicity in terrestrial and marine mammals: increasing impact of non-*ortho* and mono-*ortho* coplanar polychlorinated biphenyls from land to ocean. *Arch. Environ. Contam. Toxicol.* **18**, 850–857.

Kannan, N., Tanabe, S. & Tatsukawa, R. (1988) Toxic potential of non-*ortho* and mono-*ortho* coplanar PCBs in commercial PCB preparations: '2,3,7,8-T(4) CDD Toxicity Equivalence Factors approach'. *Bull. Environ. Contam. Toxicol.* **41**, 267–276.

Karickhoff, S.W. (1980) Sorption kinetics of hydrophobic pollutants in natural sediments. In: *Contaminants and Sediments,* Vol. 2 (Ed. R.A. Baker), pp. 193–205. Ann Arbor Science Publishers, Ann Arbor, MI.

Kashimoto, T., Takayama, K., Miyata, M., Murakami, Y. & Matsumoto, H. (1989) PCDDs, PCDFs, PCBs, coplanar PCBs and organochlorinated pesticides in human adipose tissue in Japan. *Chemosphere* **19**, 921–926.

Kieatiwong, L.V. Nguyen, Hebert, V.R., Hackett, M.,

Miller, G.C., Mille, M.J. & Mitzel, R. (1990) Photolysis of chlorinated dioxins in organic solvents and on soils. *Environ. Sci. Technol.* **24**, 1575–1580.

Kjeller, L-O., Jones, K.C., Johnston, A.E. & Rappe, C. (1991) Increases in the polychlorinated dibenzo-*p*-dioxin and -furan content of soils and vegetation since the 1840s. *Environ. Sci. Technol.* **25**, 1619–1627.

Kjeller, L-O., Kulp, S-E., Bergek, S., Boström, M., Bergqvist, P-A., Rappe, C., Jonsson, B., de Wit, D., Jansson, B. & Olsson, M. (1990) Levels and possible sources of PCDD/PCDF in sediment and pike from Swedish lakes and rivers. *Chemosphere* **20**, 1489–1496.

Kleeman, J.M., Olson, J.R., Chen, S.M. & Peterson, R.E. (1986) Metabolism and disposition of 2,3,7,8-tetrachlorodibenzo-*p*-dioxin in rainbow trout. *Toxicol. Appl. Pharmacol.* **83**, 391–401.

Kleeman, J.M., Olson, J.R. & Peterson, R.J. (1988) Species differences in 2,3,7,8-tetrachlorodibenzo-*p*-dioxin toxicity and biotransformation in fish. *Fund. Appl. Technol.* **10**, 206–213.

Knickmeyer, R. & Steinhart, H. (1989) Cyclic organochlorines in the whelks *Buccium undatum* and *Neptunea antiqua* from the North Sea and the Irish Sea. *Mar. Pollut. Bull.* **20**, 433–437.

Kociba, R.J. & Cabey, O. (1985) Comparative toxicity and biologic activity of chlorinated dibenzo-*p*-dioxins and furans relative to 2,3,7,8-tetrachlorodibenzo-*p*-dioxin (TCDD). *Chemosphere* **14**, 649–660.

Kociba, R.J., Keyes, D.G., Beyer, J.E., Carreon, R.M., Wade, C.E., Dittenber, D.A., Kalnins, R.P., Franson, L.E., Parks, C.N., Bernard, S.D., Hummel, R.A. & Humiston, C.G. (1978) Results of a two-year chronic toxicity and oncogenicity study on 2,3,7,8-tetrachlorodibenzo-*p*-dioxin in rats. *Toxicol. Appl. Pharmacol.* **46**, 279–303.

Koester, C.J. & Hites, R.A. (1992) Photodegradation of polychlorinated dioxins and dibenzofurans adsorbed to fly ash. *Environ. Sci. Technol.* **26**, 502–507.

Kubiak, T.J., Harris, H.J., Smith, L.M., Schartz, T.S., Stalling, D.L., Trick, J.A., Sileo, L., Docherty, D.E. & Erdman, T.C. (1989) Microcontaminants and reproductive impairment of the Forester's tern on Green Bay Lake Michigan—1983. *Arch. Environ. Contam. Toxicol.* **18**, 706–727.

Kuehl, D.W., Butterworth, B.C., Libal, J. & Marquis, P. (1991) An isotope dilution high resolution gas chromatographic-high resolution mass spectrometric method for the determination of coplanar polychlorinated biphenyls: application to fish and marine mammals. *Chemosphere* **22**, 849–858.

Kuratsune, M. (1989) Yusho, with reference to Yu-Cheng. In: *Halogenated Biphenyls, Terphenyls, Naphthalenes, Dibenzofurans and Related Products.*

(Eds R.D. Kimbrough & A.A. Jensen), pp. 381–400. Elsevier, Amsterdam.

Kutz, F.W., Barnes, D.G., Bottimore, D.P., Greim, H. & Bretthauer, E.W. (1990) The International Toxicity Equivalency Factor (I-TEF) method of risk assessment for complex mixtures and related compounds. *Chemosphere* **20**, 751–757.

Langer, H.G., Bradey, T.P. & Briggs, P.R. (1973) Formation of dibenzodioxins and other condensation products from chlorinated phenols and derivatives. *Environ. Health Technol.* **5**, 3–7.

Larsen B. & Riego, J. (1990) Interference from 2,3,5,6-3',4'-hexachlorobiphenyl (CB 163) in the determination of 2,3,4-2',4',5'-hexachlorobiphenyl (CB 138) in environmental and technical samples. *Int. J. Environ. Anal. Chem.* **40**, 59–68.

Larsson, P., Okla, L., Ryding, S.-O. & Westöö, B. (1990) Contaminated sediment as a source of PCBs in a river system. *Can. J. Fish. Aquat. Sci.* **47**, 746–754.

Law, R.J., Allchin, C.R. & Harwood, J. (1989) Concentrations of organochlorine compounds in the blubber of seals from eastern and north-eastern England, 1988. *Mar. Pollut. Bull.* **20**, 110–115.

LeBel, G.L., Williams, D.T., Benoit, F.M. & Goddard, M. (1990) Polychlorinated dibenzodioxins and dibenzofurans in human adipose tissue samples from five Ontario municipalities. *Chemosphere* **21**, 1465–1475.

Lemmetyinen, R., Rantamäki, P. & Karlin, A. (1982) Levels of DDT and PCB's in different stages of life cycle of the Arctic tern *Sterna paradisaea* and the herring gull *Larus argentatus*. *Chemosphere* **11**, 1059–1068.

Liem, A.K.D., de Jong, A.P.J.M., Marsman, J.A., den Boer, A.C., Groenemeijer, G.S., den Hartog, R.S., de Korte, G.A.L. & van't Klooster, H.A. (1990) A rapid clean-up procedure for the analysis of polychlorinated dibenzo-*p*-dioxins and dibenzofurans in milk samples. *Chemosphere* **20**, 843–850.

Ligon, Jr, W.V., Dorn, S.B. & May, R.J. (1989) Chlorodibenzofuran and chlorodibenzo-*p*-dioxin levels in Chilean mummies dated to about 2800 years before the present. *Environ. Sci. Technol.* **23**, 1286–1290.

Lindstrom, G. & Rappe, C. (1990) Analytical procedure operating on 0.1–1 ppt level applied in toxicological studies on PCDFs/PCDDs. *Chemosphere* **20**, 851–856.

Lockhart, W.L., Wagemann, R., Tracey, B., Sutherland, D. & Thomas, D.J. (1992) Presence and implications of chemical contaminants in the freshwaters of the Canadian Arctic. *Sci. Total Environ.* **122**, 165–243.

Lower, W.R., Yanders, A.F., Orazio, C.E., Puri, R.K., Hancock, J. & Kapila, S. (1989) A survey of 2,3,7,8-tetrachlorodibenzo-*p*-dioxin residues in selected animal species from Times Beach, Missouri.

Chemosphere **18**, 1079–1088.

Luotamo, M. (1991) Congener specific assessment of human exposure to polychlorinated biphenyls. *Chemosphere* **23**, 1685–1698.

Lutz, R.J. & Dedrick, R.L. (1987) Physiologic pharmacokinetic modeling of polychlorinated biphenyls. In: *Environmental Toxic Series*, Vol. 1 (Eds S. Safe & O. Hutzinger), pp. 111–131. Springer-Verlag, Berlin.

Macdonald, C.R., Norstrom, R.J. & Turle, R. (1992) Application of pattern recognition techniques to assessment of biomagnification and sources of polychlorinated multicomponent pollutants, such as PCBs, PCDDs and PCDFs. *Chemosphere* **25**, 129–134.

Maisel, B.E. & Hunt, G.T. (1990) Background concentrations of PCDDs/PCDFs in ambient air—a comparison of toxic equivalency factor (TEF) models. *Chemosphere* **20**, 771–778.

Manscher, O.H., Heidam, N.Z., Vikelsøe, J., Nielsen, P., Blinksbjerg, P., Madsen, H., Pappesen, L. & Tiernan, T.O. (1990) The Danish incinerator dioxin study. *Chemosphere* **20**, 1779–1784.

Marchand, M., Caprais, J.C. & Pignet, P. (1988) Hydrocarbons and halogenated hydrocarbons in coastal waters of the western Mediterranean (France). *Mar. Environ. Res.* **25**, 131–159.

Marklund, S., Andersson, R., Tysklind, M., Rappe, C., Egebäck, K-E., Björkman, E. & Grigoriadis, V. (1990) Emissions of PCDDs and PCDFs in gasoline and diesel fueled cars. *Chemosphere* **20**, 553–561.

Marple, L., Brunck, R. & Throop, L. (1986) Water solubility of 2,3,7,8-tetrachlorodibenzo-p-dioxin. *Environ. Sci. Technol.* **20**, 180–182.

Mason, C.F. & O'Sullivan, W.M. (1992) Organochlorine pesticide residues and PCBs in otters (*Lutra lutra*) from Ireland. *Bull. Environ. Contam. Toxicol.* **48**, 387–393.

Masuda, Y., Kagawa, R., Kuroki, H., Kuratsune, M., Yoshimura, T., Taki, I., Kusuda, M., Yamashita, F. & Hayashi, H. (1978) Transfer of polychlorinated biphenyls from mothers to foetuses and infants. *Food Cosmet. Toxicol.* **16**, 543–546.

Masuda, Y., Kuroki, H., Haraguchi, K. & Nagayama, J. (1985) PCB and PCDF congeners in the blood and tissues of Yusho and Yu-Cheng patients. *Environ. Health Perspect.* **59**, 53–58.

Masuda, Y., Kuroki, H., Haraguchi, K. & Nagayama, J. (1986) PCDFs and related compounds in humans from Yusho and Yu-Cheng incidents. *Chemosphere* **15**, 1621–1628.

Matsumoto, H., Murakami, Y., Kuwabara, K., Tanaka, R. & Kashimoto, T. (1987) Average daily intake of pesticides and polychlorinated biphenyls in total diets in Osaka, Japan. *Bull. Environ. Contam. Toxicol.* **38**, 954–958.

Matsumura, F. & Benezet, H.J. (1973) Studies on the bioaccumulation and microbial degradation of 2,3,7,8-tetrachlorodibenzo-p-dioxin. *Environ. Health Perspect.* **5**, 253–258.

McConnell, E.E. (1989) Acute and chronic toxicity and carcinogenesis in animals. In: *Halogenated Biphenyls, Terphenyls, Naphthalenes, Dibenzofurans and Related Products* (Eds R.D. Kimbrough & A.A. Jensen), pp. 161–193. Elsevier, Amsterdam.

McCrady, J.K., McFarlane, C. & Gander, L.K. (1990) The transport and fate of 2,3,7,8-TCDD in soybean and corn. *Chemosphere* **21**, 359–376.

McCrea, R.C. & Fischer, J.D. (1986) Heavy metal and organochlorine contaminants in the five major Ontario rivers in the Hudson Bay lowland. *Water Pollut. Res. J. Can.* **21**, 225–234.

McCurvin, D.M.A., Clement, R.E., Taguchi, V.Y., Reiner, E.J., Schellenberg, D.H. & Bobbie, B.A. (1989) A comparison of the capabilities of mass spectral techniques for the detection of chlorinated dibenzo-p-dioxins and dibenzofurans in environmental samples. *Chemosphere* **19**, 205–212.

McKee, P., Burt, A., McCurvin, D., Hollinger, D. Clement, R., Sutherland, D.H. & Neaves, W. (1990) Levels of dioxins, furans and other organic contaminants in harbour sediments near a wood preserving plant using pentachlorophenol and creosote. *Chemosphere* **20**, 1679–1685.

McLeese, D.W., Metcalfe, C.D. & Pezzack, D.S. (1980) Uptake of PCB's from sediment by *Nereis virens* and *Crangon septemspinosa*. *Arch. Environ. Contam. Toxicol.* **9**, 507–518.

Melancon, M.J. & Lech, J.J. (1976) Isolation and identification of a polar metabolite of tetrachlorobiphenyl from bile of rainbow trout exposed to ^{14}C-tetrachlorobiphenyl. *Bull. Environ. Contam. Toxicol.* **15**, 181–187.

Metcalfe, C.D., Balch, G.C., Cairns, V.W., Fitzsimons, J.D. & Dunn, B.P. (1990) Carcinogenic and genotoxic activity of extracts from contaminated sediments in western Lake Ontario. *Sci. Total Environ.* **94**, 125–141.

Metcalfe, D.E., Zukovs, G., Mackay, D. & Paterson, S. (1986) Polychlorinated biphenyls (PCBs) physical and chemical property data. In: *Hazards, Decontamination, and Replacement of PCB* (Ed. J-P. Crine), pp. 3–33. Plenum Press, New York.

Meyer, C., O'Keefe, P., Hilker, D., Rafferty, L., Wilson, L., Conner, S., Aldous, K., Markussen, K. & Slade, K. (1989) A survey of twenty community watersystems in New York state for PCDDs and PCDFs. *Chemosphere* **19**, 21–26.

Miles, W.F., Singh, J., Gurprasad, M.P. & Malis, G.P. (1985) Isomer specific determination of hexachlorodioxins in technical pentachlorophenol (PCP) and its sodium salt. *Chemosphere* **14**, 807–810.

Morrissey, R.E. & Schwetz, B.A. (1989) Reproductive and developmental toxicity in animals. In: *Halogenated Biphenyls, Terphenyls, Naphthalenes, Dibenzofurans and Related Products*. (Eds R.D. Kimbrough & A.A. Jensen), pp. 195–225. Elsevier, Amsterdam.

Morse, R.A., Culliney, T.W., Gutenmann, W.H., Littman, C.B. & Lisk, D.J. (1987) Polychlorinated biphenyls in honey bees. *Bull. Environ. Contam. Toxicol.* **38**, 271–276.

Mouvet, C., Galoux, M. & Bernes, A. (1985) Monitoring of polychlorinated biphenyls (PCBs) and hexachloro-cyclohexanes (HCH) in freshwater using the aquatic moss *Cinclidotus danubicus*. *Sci. Total Environ.* **44**, 253–267.

Mowrer, J., Åswald, K., Burgermeister, G., Machado, L. & Tarradellas, J. (1982) PCB in a Lake Geneva eco-system. *Ambio* **11**, 355–358.

Muir, D.C.G. & Yarechewski, A.L. (1988) Dietary accumulation of four chlorinated dioxin congeners by rainbow trout and fathead minnows. *Environ. Toxicol. Chem.* **7**, 227–236.

Muir, D.C.G., Fairchild, W.L., Yarechewski, A.L. & Whittle, D.M. (1992a) Derivation of bioaccumulation parameters and application of food chain models for chlorinated dioxins and furans. In: *Chemical Dynamics in Freshwater Ecosystems* (Eds F.A.P.C. Gobas & F. MacCorquadale), pp. 185–208. Lewis, Ann Arbor, MI.

Muir, D.C.G., Ford, C.A., Grift, N.P., Stewart, R.E.A. & Bidleman, T.F. (1992b) Organochlorine contaminants in narwhal (*Monodon monoceros*) from the Canadian Arctic. *Environ. Pollut.* **75**, 307–316.

Muir, D.C.G., Norstrom, R.J. & Simon, M. (1988) Organochlorine contaminants in Arctic marine food chains: accumulation of specific polychlorinated biphenyls and chlordane-related compounds. *Environ. Sci. Technol.* **22**, 1071–1079.

Mullin, M.D., Pochini, C.M., McCrindle, S., Romkes, M., Safe, S.H. & Safe, L.M. (1984) High-resolution PCB analysis: synthesis and chromatographic properties of all 209 PCB congeners. *Environ. Sci. Technol.* **18**, 468–476.

Murphy, T.J., Mullin, M.D. & Meyer, J.A., (1987) Equilibration of polychlorinated biphenyls and toxaphene with air and water. *Environ. Sci. Technol.* **21**, 155–162.

Näf, C., Broman, D., Ishaq, R. & Zebühr, Y. (1990) PCDDs and PCDFs in water, sludge and air samples from various levels in a waste water treatment plant with respect to composition changes and total flux. *Chemosphere* **20**, 1503–1510.

Nagayama, J., Tokudome, S. & Kuratsune, M. (1980) Transfer of polychlorinated dibenzofurans to the foetuses and offspring of mice. *Food Cosmet. Toxicol.* **18**, 153–157.

National Oceanic and Atmospheric Administration (1987) National status and trends program for marine environmental quality. A summary of selected data on chemical contaminants in tissues collected during 1984, 1985, and 1986. NOAA Tech. Memo. NOS OMA 38. Office Oceanogr. Mar. Assess., Rockville, MD.

National Research Council of Canada (1981) *Polychlorinated Dibenzo*-p-*dioxins: Criteria for their Effects on Man and his Environment*. NRCC, Ottawa, Canada.

Nau, H., Bass, R. & Neubert, T. (1986) Transfer of 2,3,7,8-tetrachlorodibenzo-*p*-dioxin (TCDD) via placenta and milk and postnatal toxicity in the mouse. *Arch. Toxicol.* **59**, 36–40.

Nebeker, A.V. & Puglisi, F.A. (1974) Effect of poly-chlorinated biphenyls (PCBs) on survival and repro-duction of *Daphnia, Gammarus*, and *Tanytarsus*. *Trans. Am. Fish. Soc.* **103**, 562–568.

Nestrick, T.J., Lamparski, L.L. & Stehl, R.H. (1979) Synthesis and identification of the 22 tetrachlorodibenzo-*p*-dioxin isomers by high per-formance liquid chromatography and gas chroma-tography. *Anal. Chem.* **51**, 2273–2281.

Nestrick, T.J., Lamparski, L.L. & Townsend, D.I. (1980) Identification of tetrachlorodibenzo-*p*-dioxin isomers at the 1 ng level by photolytic degradation and pattern recognition techniques. *Anal. Chem.* **52**, 1865–1874.

Neubert, D., Golor, G. & Neubert, R. (1992) TCDD-toxicity equivalences for PCDD/PCDF congeners: prerequisites and limitations. *Chemosphere* **25**, 65–70.

Nies, L. & Vogel, T.M. (1990) Effects of organic sub-strates on dechlorination of Aroclor 1242 in anaerobic sediments. *Appl. Environ. Microbiol.* **56**, 2612–2617.

Niimi, A.J. (1983) Biological and toxicological effects of environmental contaminants in fish and their eggs. *Can. J. Fish. Aquat. Sci.* **40**, 306–312.

Niimi, A.J. (1985) Use of laboratory studies in assessing the behavior of contaminants in fish inhabiting natural ecosystems. *Water Pollut. Res. J. Can.* **20**, 79–88.

Niimi, A.J. (1990) Review of biochemical methods and other indicators to assess fish health in aquatic eco-systems containing toxic chemicals. *J Great Lakes Res.* **105**, 529–541.

Niimi, A.J. & Oliver, B.G. (1983) Biological half-lives of polychlorinated biphenyl (PCB) congeners in whole fish and muscle of rainbow trout (*Salmo gairdneri*). *Can. J. Fish. Aquat. Sci.* **40**, 1388–1394.

Niimi, A.J. & Oliver, B.G. (1986) Biological half-lives of chlorinated dibenzo-*p*-dioxins and dibenzofurans in rainbow trout (*Salmo gairdneri*). *Environ. Toxicol.*

Chem. **5**, 49–53.

Niimi, A.J. & Oliver, B.G. (1988) Influence of molecular weight and molecular volume on dietary absorption efficiency of chemicals by fishes. *Can. J. Fish. Aquat. Sci.* **45**, 222–227.

Niimi, A.J. & Oliver, B.G. (1989a) Distribution of polychlorinated biphenyl congeners and other halocarbons in whole fish and muscle among Lake Ontario salmonids. *Environ. Sci. Technol.* **23**, 83–88.

Niimi, A.J. & Oliver, B.G. (1989b) Assessment of relative toxicity of chlorinated dibenzo-*p*-dioxins, dibenzofurans, and biphenyls in Lake Ontario salmonids to mammalian systems using toxic equivalent factors (TEF). *Chemosphere* **18**, 1413–1423.

Norback, D.H., Engblom, J.F. & Allen, J.R. (1975) Tissue distribution and excretion of octachlorodibenzo-*p*-dioxin in the rat. *Toxicol. Appl. Pharmacol.* **32**, 330–338.

Norén, K., Lundén, A., Sjövall, J. & Bergman, Å. (1990) Coplanar polychlorinated biphenyls in Swedish human milk. *Chemosphere* **20**, 935–941.

Norström, A., Chaudhary, S.K., Albro, P.W. & McKinney, J.D. (1979) Synthesis of chlorinated dibenzofurans and chlorinated aminodibenzofurans from the corresponding diphenyl ethers and nitrodiphenyl ethers. *Chemosphere* **8**, 331–343.

Norstrom, R.J. (1986) Bioaccumulation of polychlorinated biphenyls in Canadian wildlife. In: *Hazards, Decontamination, and Replacement of PCB* (Ed. J.-P. Crine), pp. 85–100. Plenum Press, New York.

Norstrom, R.J., Muir, D.C.G., Ford, C.A., Simon, M., Macdonald, C.R. & Béland, P. (1992) Indications of P450 monooxygenase activities in beluga (*Delphinapterus leucas*) and narwhal (*Monodon monoceros*) from patterns of PCB, PCDD and PCDF accumulation. *Mar. Environ. Res.* **34**, 267–272.

Norstrom, R.J., Simon, M. & Muir, D.C.G. (1990) Polychlorinated dibenzo-*p*-dioxins and dibenzofurans in marine mammals in the Canadian north. *Environ. Pollut.* **66**, 1–19.

North Atlantic Treaty Organization (1988) Pilot study on international information exchange on dioxins and related compounds. International toxicity equivalency factor (I-TEF) method of risk assessment for complex mixtures of dioxins and related compounds. Report No. 176. Plenum Press, NY.

Nottrodt, I.A. & Ballschmiter, K. (1986) Causes for, and reduction strategies against emissions of PCDD/PCDF from waste incineration plants. Interpretations of recent measurements. *Chemosphere* **15**, 1225–1237.

Nottrodt, I.A., Düwel, U. & Ballschmiter, K. (1990) The influence of increased excess air on the formation of PCDD/PCDF in a municipal waste incineration plant. *Chemosphere* **20**, 1847–1854.

Öberg, T. & Allhammar, G. (1989) Chlorinated aromatics from metallurgical industries — process factors influencing production and emissions. *Chemosphere* **19**, 711–716.

O'Conner, T.P. (1991) Concentrations of organic contaminants in mollusks and sediments at NOAA national status and trend sites in the coastal and estuarine United States. *Environ. Health Perspect.* **90**, 69–73.

Oehme, M., Larssen, S. & Brevik, E.M. (1991) Emission factors of PCDD and PCDF for road vehicles obtained by tunnel experiment. *Chemosphere* **23**, 1699–1708.

Ohlendorf, H.M., Marios, K.C., Lowe, R.W., Harvey, T.E. & Kelly, P.R. (1991) Trace elements and organochlorines in surf scoters from San Francisco Bay, 1985. *Environ. Monitor. Assess.* **18**, 105–122.

Olie, K., Vermeulen, P.L. & Hutzinger, O. (1977) Chlorodibenzo-*p*-dioxins and chlorodibenzofurans are trace components of fly ash and flue gas of some municipal incinerators in the Netherlands. *Chemosphere* **6**, 455–459.

Oliver, B.G. (1984) Uptake of chlorinated organics from anthropogenically contaminated sediments by oligochaete worms. *Can. J. Fish. Aquat. Sci.* **41**, 873–878.

Oliver, B.G. & Niimi, A.J. (1988) Trophodynamic analysis of polychlorinated biphenyl congeners and other chlorinated hydrocarbons in the Lake Ontario ecosystem. *Environ. Sci. Technol.* **22**, 388–397.

Olsen, J.R. (1986) Metabolism and disposition of 2,3,7,8-tetrachlorodibenzo-*p*-dioxin in guinea pigs. *Toxicol. Appl. Pharmacol.* **85**, 263–273.

Olsen, J.R., Gasiewicz, T.A. & Neal, R.A. (1980) Tissue distribution, excretion, and metabolism of 2,3,7,8-tetrachlorodibenzo-*p*-dioxin in the golden syrian hamster. *Toxicol. Appl. Pharmacol.* **55**, 67–78.

Ono, M., Kannan, N., Wakimoto, T. & Tatsukawa, R. (1987a) Dibenzofurans a greater global pollutant than dioxins? Evidence from analyses of open ocean killer whale. *Mar. Pollut. Bull.* **18**, 640–643.

Ono, M., Kashima, Y., Wakimoto, T. & Tatsukawa, R. (1987b) Daily intake of PCDDs and PCDFs by Japanese through food. *Chemosphere* **16**, 1823–1828.

Opperhuizen. A. & Sijm, D.T.H.M. (1990) Bioaccumulation and biotransformation of polychlorinated dibenzo-*p*-dioxins and dibenzofurans in fish. *Environ. Toxicol. Chem.* **9**, 175–186.

Osterhaus, A.D.M.E. & Vedder, E.J. (1988) Identification of virus causing recent seal deaths. *Nature* **335**, 20.

Overcash, M.R., McPeters, A.L., Dougherty, E.J. & Carbonell, R.G. (1991) Diffusion of 2,3,7,8-tetrachlorodibenzo-*p*-dioxin in soil containing organic solvents. *Environ. Sci. Toxicol.* **25**, 1479–1485.

Päpke, O., Ball, M., Lis, Z.A. & Scheunert, K. (1989)

PCDD and PCDF in indoor air of kindergartens in northern W. Germany. *Chemosphere* **18**, 617–626.

Pardue, J.H., Delaune, R.D. & Patrick, Jr, W.H. (1988) Effect of sediment pH and oxidation-reduction potential on PCB mineralization. *Water, Air, Soil Pollut.* **37**, 439–447.

Patil, G.S. (1991) Correlation of aqueous solubility and octanol–water partition coefficient based on molecular structure. *Chemosphere* **22**, 723–738.

Patterson, Jr, D.G., Reddy, V.V., Barnhart, E.R., Ashley, D.L., Lapeza, Jr, C.R., Alexander, L.R. & Gelbaum, L.T. (1989) Synthesis and analytical characterization of all tetra to octachlorodibenzo-p-dioxins. *Chemosphere* **19**, 233–240.

Philippi, M., Schmid, J., Wipf, H.K. & Hutter, R. (1982) A microbial metabolite of TCDD. *Experientia* **38**, 659–661.

Piper, W.N., Rose, J.Q. & Gehring, P.J. (1973) Excretion and tissue distribution of 2,3,7,8-tetrachlorodibenzo-p-dioxin in the rat. *Environ. Health Perspect.* **5**, 241–244.

Piskorska-Pliszczynska, J., Keys, B., Safe, S. & Newman, M.S. (1986) The cytosolic receptor binding affinities and AHH induction potencies of 29 polynuclear aromatic hydrocarbons. *Toxicol. Lett.* **34**, 67–74.

Pohland, A.E. & Yang, G.C. (1972) Preparation and characterization of chlorinated dibenzo-p-dioxins. *J. Agric. Food Chem.* **20**, 1093–1099.

Poiger, H. & Schlatter, C. (1986) Pharmacokinetics of 2,3,7,8-TCDD in man. *Chemosphere* **15**, 1489–1494.

Poland, A. & Glover, E. (1977) Chlorinated biphenyl induction of aryl hydrocarbon hydroxylase activity: a study of the structure-activity relationship. *Mol. Pharmacol.* **13**, 924–938.

Poland, A., Glover, E. & Kende, A.S. (1976) Stereospecific high affinity binding of 2,3,7,8-tetrachlorodibenzo-p-dioxin by hepatic cytosol. *J. Biol. Chem.* **251**, 4936–4946.

Poland, A. & Knutson, J.C. (1982) 2,3,7,8-Tetrachlorodibenzo-p-dioxin and related halogenated aromatic hydrocarbons: examination of the mechanisms of toxicity. *Annu. Rev. Pharmacol. Toxicol.* **22**, 517–554.

Proulx, G., Weseloh, D.V.C., Elliott, J.E., Teeple, S., Anghem, P.A.M. & Mineau, P. (1987) Organochlorine and PCB residues in Lake Erie mink populations. *Bull. Environ. Contam. Toxicol.* **39**, 939–944.

Puhvel, S.M., Sakamoto, M., Ertl, D.C. & Reisner, R.M. (1982) Hairless mice as models for chloracne: a study of cutaneous changes induced by topical application of established chloracnegens. *Toxicol. Appl. Pharmacol.* **64**, 492–503.

Quensen, J.F., III, Tiedje, J.M. & Boyd, S.A. (1988) Reductive dechlorination of polychlorinated biphenyls by anaerobic microorganisms from sediments. *Science* **242**, 752–754.

Ramsey, J.C., Hefner, J.G., Karbowski, R.J., Braun, W.H. & Gehring, P.J. (1982) The *in vivo* biotransformation of 2,3,7,8-tetrachlorodibenzo-p-dioxin (TCDD) in the rat. *Toxicol. Appl. Pharmacol.* **65**, 180–184.

Randerath, K., Reddy, V. & Gupta, R.C. (1982) ^{32}P-Postlabeling test for DNA damage. *Proc. Natl. Acad. Sci.* **78**, 6125–6129.

Rao, C.V. & Banerji, A.S. (1988) Polychlorinated biphenyls in human amniotic fluid. *Bull. Environ. Contam. Toxicol.* **41**, 798–801.

Rapaport, R.A. & Eisenreich, S.J. (1984) Chromatographic determination of octanol-water partition coefficients (K_{ow}'s) for 58 polychlorinated biphenyl congeners. *Environ. Sci. Technol.* **18**, 163–170.

Rappe, C. & Kjeller, L.-O (1987) PCDDs and PCDFs in environmental samples air, particulates, sediments and soil. *Chemosphere* **16**, 1775–1780.

Rappe, C., Andersson, R., Bergqvist, P-A., Brohede, C., Hansson, M., Kjeller, L.-O., Lindström, G., Marklund, S., Nygren, M., Swanson, S.E., Tysklind, M. & Wiberg, K. (1987) Overview of environmental fate of chlorinated dioxins and dibenzofurans. Sources, levels and isometric patterns in various matrices. *Chemosphere* **16**, 1603–1618.

Rappe, C., Buser, H.R., Kuroki, H. & Matsuda, Y. (1979) Identification of polychlorinated dibenzofurans (PCDFs) retained in patients with Yusho. *Chemosphere* **4**, 259–266.

Rappe, C., Glas, B., Kjeller, L.-O., Kulp, S.E., de Wit, C. & Melin, A. (1990a) Levels of PCDDs and PCDFs in products and effluent from the Swedish pulp and paper industry and chloralkali process. *Chemosphere* **20**, 1701–1706.

Rappe, C., Kjeller, L.-O. & Andersson, R. (1989a) Analyses of PCDDs and PCDFs in sludge and water samples. *Chemosphere* **19**, 13–20.

Rappe, C., Lindström, G., Glas, B., Lundström, K. & Borgström, S. (1990b) Levels of PCDDs and PCDFs in milk cartons and in commercial milk. *Chemosphere* **20**, 1649–1656.

Rappe, C., Tarowski, S. & Yrjänheikki, E. (1989b) The WHO/EURO quality control study on PCDDs and PCDFs in human milk. *Chemosphere* **18**, 883–889.

Reggiani, G.M. (1989) The Seveso accident: medical survey of a TCDD exposure. In: *Halogenated Biphenyls, Terphenyls, Naphthalenes, Dibenzofurans and Related Products.* (Eds R.D. Kimbrough & A.A. Jensen), pp. 445–470. Elsevier, Amsterdam.

Reijnders, P.J.H. (1986) Reproductive failure in common seals feeding on fish from polluted coastal waters. *Nature* **324**, 456–457.

Reischl, A., Reissinger, M., Thoma, H. & Hutzinger, O. (1989) Accumulation of organic air constituents by plant surfaces. *Chemosphere* **18**, 561–568.

Rhee, G.-Y. (1982) Overview of phytoplankton contaminant problems. *J. Great Lakes Res.* **8**, 326–327.

Rice, C.P. & White, D.S. (1987) PCB availability assessment of river dredging using caged clams and fish. *Environ. Toxicol. Chem.* **6**, 259–274.

Risebrough, R.W., Rieche, P., Herman, S.G., Peakall, D.B. & Kirven, M.N. (1968) Polychlorinated biphenyls in the global ecosystem. *Nature* **220**, 1098.

Rordorf, B.F., Sarna, L.P., Webster, G.R.B., Safe, S.H., Safe, L.M., Lenoir, D., Schwind, K.H. & Hutzinger, O. (1990) Vapor pressure measurements on halogenated dibenzo-*p*-dioxins and dibenzofurans. An extended data set for a correlation method. *Chemosphere* **20**, 1603–1609.

Rose, J.Q., Ramsey, J.C., Wentzler, T.H., Hummel, R.A. & Gehring, P.J. (1976) The fate of 2,3,7,8-tetrachlorodibenzo-*p*-dioxin following a single and repeated oral doses to the rat. *Toxicol. Appl. Pharmacol.* **36**, 209–226.

Ryhage, R. (1974) Use of mass spectrometer as a detector and analyzer for effluents emerging from high temperature gas liquid chromatographic columns. *Anal. Chem.* **36**, 759–764.

Sabourin, T.D., Stickle, W.B., Michot, T.C., Villars, C.E., Garton, D.W. & Mushinsky, H.R. (1984) Organochlorine residue levels in Mississippi River water snakes in southern Louisiana. *Bull. Environ. Contam. Toxicol.* **32**, 460–468.

Safe, S. (1984) Polychlorinated biphenyls (PCBs) and polybrominated biphenyls (PBBs): biochemistry, toxicology and mechanism of action. *Crit. Rev. Toxicol.* **13**, 319–395.

Safe, S.H. (1986) Comparative toxicology and mechanism of action of polychlorinated dibenzo-*p*-dioxins and dibenzofurans. *Annu. Rev. Pharmacol. Toxicol.* **26**, 371–399.

Safe, S. (1987) Determination of 2,3,7,8-TCDD equivalent factors (TEFs): support for the use of the *in vitro* AHH induction assay. *Chemosphere* **16**, 791–802.

Safe, S. (1990) Polychlorinated biphenyls (PCBs), dibenzo-*p*-dioxins (PCDDs), dibenzofurans (PCDFs), and related compounds: environmental and mechanistic considerations which support the development of toxic equivalency factors (TEFs). *Crit. Rev. Toxicol.* **21**, 51–88.

Safe, S. & Hutzinger, O. (1990) PCDDs and PCDFs: sources and environmental impact. In: *Environmental Toxin Series*, Vol. 3 (Eds S. Safe & O. Hutzinger), pp. 1–20. Springer-Verlag, Berlin.

Safe, S.H. & Safe, L.M. (1984) Synthesis and characterization of twenty-two purified polychlorinated dibenzofuran congeners. *J. Agric. Food Chem.* **32**, 68–71.

Safe, S., Bandiera, S., Sawyer, T., Robertson, L., Safe, L., Parkinson, A., Thomas, P.E., Ryan, D.E., Reik, L.M.,

Levin, W., Denomme, M.A. & Fujita, T. (1985) PCBs: structure–function relationships and mechanism of action. *Environ. Health Perspect.* **60**, 47–56.

Safe, S.H., Gasiewicz, T. & Whitlock, Jr, J.P. (1990) UAREP-Report on health aspects of polychlorinated dibenzo-*p*-dioxins (PCDDs) and polychlorinated dibenzofurans (PCDFs). Mechanism of action. In: *Environmental Toxin Series*, Vol. 3 (Eds S. Safe & O. Hutzinger), pp. 61–91. Springer-Verlag, Berlin.

Saschenbrecker, P.W., Funnell, H.S. & Platonow, N.S. (1972) Persistence of PCB's in milk of exposed cows. *Vet. Rec.* **90**, 100–102.

Schecter, A., Dekin, A., Weerasinghe, N.C.A., Arghestani, S. & Gross, M.L. (1988) Sources of dioxins in the environment: A study of PCDDs and PCDFs in ancient, frozen Eskimo tissue. *Chemosphere* **17**, 627–631.

Schecter, A., Fürst, P., Fürst, C., Groebel, W., Constable, J.D., Kolesnikov, S., Beim, A., Boldonov, A., Trubitsun, E., Vlasov, B., Cau, H.D., Dai, L.C. & Quynh, H.T. (1990a) Levels of chlorinated dioxins, dibenzofurans and other chlorinated xenobiotics in food from the Soviet Union and the south of Vietnam. *Chemosphere* **20**, 799–806.

Schecter, A., Päpke, O. & Ball, M. (1990b) Evidence for transplacental transfer of dioxins from mother to fetus: chlorinated dioxin and dibenzofuran levels in the livers of stillborn infants. *Chemosphere* **21**, 1017–1022.

Schecter, A., Ryan, J.J. & Constable, J.D. (1987) Polychlorinated dibenzo-*p*-dioxin and polychlorinated debenzofuran levels in human milk from Vietnam compared with cow's milk and human breast milk from the North American continent. *Chemosphere* **16**, 2002–2016.

Schimmel, S.C., Habsen, D.J. & Forester, J. (1974) Effects of Aroclor®1254 on laboratory-reared embryos and fry of sheepshead minnows (*Cyrinus variegatus*). *Trans. Am. Fish. Soc.* **103**, 722–727.

Schmid, P., Bühler, F. & Schlatter, C. (1992) Dermal absorption of PCB in man. *Chemosphere* **24**, 1283–1292.

Schulz, C.O., Brown, D.R. & Munro, I.C. (1990) UAREP-Report on health aspects of polychlorinated dibenzo*p*-dioxins and polychlorinated dibenzofurans (PCDFs). Characterization of human health risks. In: *Environmental Toxin Series*, Vol. 3 (Eds S. Safe & O. Hutzinger), pp. 93–139. Springer-Verlag, Berlin.

Schulz, D.E., Petrick, G. & Duinker, J.C. (1988) Chlorinated biphenyls in north Atlantic surface and deep water. *Mar. Pollut. Bull.* **19**, 526–531.

Schulz, D.E., Petrick, G. & Duinker, J.C. (1989) Complete characterization of polychlorinated biphenyl congeners in commercial Aroclor and Clophen mixtures by multidimensional gas chromatography–

electron capture detection. *Environ. Sci. Technol.* **23**, 852–859.

Schwartz, T.R., Stalling, D.L. & Rice, C.L. (1987) Are polychlorinated biphenyl residues adequately described by Aroclor mixture equivalents? Isomer-specific principal components analysis of such residues in fish and turtles. *Environ. Sci. Technol.* **21**, 72–76.

Shain, W., Overmann, S.R., Wilson, L.R., Kostas, J. & Bush, B. (1986) A congener analysis of polychlorinated biphenyls accumulating in rat pups after perinatal exposure. *Arch. Environ. Contam. Toxicol.* **15**, 687–707.

Shelton, D.W., Hendricks, J.D., Coulombe, R.A. & Bailey, G.S. (1984) Effect of dose on the inhibition of carcinogensis/mutagenesis by Aroclor 1254 in rainbow trout fed aflatoxin B1. *J. Toxicol. Environ. Health* **13**, 649–657.

Shiu, W.Y. & Mackay, D. (1986) A critical review of aqueous solubilities, vapor pressures, Henry's law constants, and octanol-water partition coefficients of the polychlorinated biphenyls. *J. Phys. Chem. Ref. Data* **15**, 911–929.

Shiu, W.Y., Doucette, W., Gobas, F.A.P.C., Andren, A. & Mackay, D. (1988) Physical–chemical properties of chlorinated dibenzo-*p*-dioxins. *Environ. Sci. Technol.* **22**, 651–658.

Sijm, D.T.H.M., Wever, H., de Vries, P.J. & Opperhuizen, A. (1989) Octan-1-ol/water partition coefficients of polychlorinated dibenzo-*p*-dioxins and dibenzofurans: experimental values determined with a stirring method. *Chemosphere* **19**, 263–266.

Silberhorn, E.M., Glauert, H.P. & Robertson, L.W. (1990) Carcinigenicity of polychlorinated biphenyls: PCBs and PBBs. *Crit. Rev. Toxicol.* **20**, 440–496.

Sipes, I.G. & Schnellmann, R.G. (1987) Biotransformation of PCBs: metabolic pathways and mechanisms. In: *Environmental Toxin Series*, Vol. 1 (Eds S. Safe & O. Hutzinger), pp. 97–110. Springer-Verlag, Berlin.

Smith, A.H., Kurland, L.T. & Shindell, S. (1990) UAREP-Report on health aspects of polychlorinated dibenzo-*p*-dioxins (PCDDs) and polychlorinated dibenzofurans (PCDFs). Epidemiology. In: *Environmental Toxin Series*, Vol. 3 (Eds S. Safe & O. Hutzinger), pp. 27–59. Springer-Verlag, Berlin.

Smith, L.M. (1981) Carbon dispersed in glass fibers as an adsorbent for contaminant enrichment and fractionation. *Anal. Chem.* **53**, 2152–2154.

Smith, R.M., O'Keefe, P.W., Aldous, K., Connor, S., Lavin, P. & Wade, E. (1990) Continuing atmospheric studies of chlorinated dibenzofurans and dioxins in New York state. *Chemosphere* **20**, 1447–1453.

Smith, R.M., O'Keefe, P.W., Aldous, K.M., Hilker, D.R. & O'Brien, J.E. (1983) 2,3,7,8-Tetrachlorodibenzo-*p*-dioxin in sediment samples from Love Canal storm sewers and creeks. *Environ. Sci. Technol.* **17**, 6–10.

Smith, V.E., Spurr, J.M., Filkins, J.C. & Jones, J.J. (1985) Organochlorine contaminants of wintering ducks foraging on Detroit River sediments. *J. Great Lakes Res.* **11**, 231–246.

Somers, J.D., Goski, B.C. & Barrett, M.W. (1987) Organochlorine residues in northeastern Alberta otters. *Bull. Environ. Contam. Toxicol.* **39**, 783–790.

Sonzogni, W., Maack, L., Gibson, T. & Lawrence, J. (1991) Toxic polychlorinated biphenyl congeners in Sheboygan River (USA) sediments. *Bull. Environ. Contam. Toxicol.* **47**, 398–405.

Startin, J.R., Rose, M., Wright, C., Parker, I. & Gilbert, J. (1990) Surveillance of British foods for PCDDs and PCDFs. *Chemosphere* **20**, 793–798.

Steeves, T., Strickland, M., Frank, R., Rasper, J. & Douglas, C.W. (1991) Organochlorine insecticide and polychlorinated biphenyl residues in martens and fishers from the Algonquin region of south-central Ontario. *Bull. Environ. Contam. Toxicol.* **46**, 368–373.

Strachan, W.M.J. (1990) Atmospheric deposition of selected organochlorine compounds in Canada. In: *Long Range Transport of Pesticides* (Ed. D.A. Kurtz), pp. 233–240. Lewis, Chelsea, MI.

Strek, H.J. & Weber, J.B. (1982) Behaviour of polychlorinated biphenyls (PCBs) in soils and plants. *Environ. Pollut.* **28A**, 291–312.

Subramanian, A., Tanabe, S., Tatsukawa, R. & Miyazaki, N. (1987) Reduction in the testosterone levels by PCBs and DDE in Dall's porpoises of northwestern north Pacific. *Mar. Pollut. Bull.* **18**, 643–646.

Suzuki, M., Aizawa, N., Okano, G. & Takahashi, T. (1977) Translocation of polychlorobiphenyls in soil into plants: a study by a method of culture of soybean sprouts. *Arch. Environ. Contam. Toxicol.* **5**, 343–352.

Swackhamer, D.L. & Armstrong, D.E. (1988) Horizontal and vertical distribution of PCBs in southern Lake Michigan sediments and the effect of Waukegan Harbor as a point source. *J. Great Lakes Res.* **14**, 277–290.

Takizawa, Y. & Muto, H. (1987) PCDDs and PCDFs carried to the human body from the diet. *Chemosphere* **16**, 1971–1975.

Tanabe, S. (1988) PCB problems in the future: foresight from current knowledge. *Environ. Pollut.* **50**, 5–28.

Tanabe, S., Hidaka, H. & Tatsukawa, R. (1983) PCBs and chlorinated hydrocarbon pesticides in Antarctic atmosphere and hydrosphere. *Chemosphere* **12**, 277–288.

Tanabe, S., Kannan, N., Subramanian, A., Watanabe, S. & Tatsukawa, R. (1987a) Highly toxic coplanar PCBs: occurrence, source, persistency and toxic implications to wildlife and humans. *Environ. Pollut.* **47**, 147–163.

Tanabe, S., Kannan, N., Wakimoto, T. & Tatsukawa, R. (1987b) Method for the determination of three toxic non-*ortho*chlorine substituted coplanar PCBs in environmental samples at part-per-trillion levels. *Int. J. Environ. Anal. Chem.* **29**, 199–213.

Tanabe, S., Kannan, N., Fukushima, M., Okamoto, T., Wakimoto, T. & Tatsukawa, R. (1989) Persistent organochlorines in Japanese coastal waters: an introspective summary from a Far East developed nation. *Mar. Pollut. Bull.* **20**, 344–352.

Tanabe, S., Maruyama, K. & Tatsukawa, R. (1982a) Absorption efficiency and biological half-life of individual chlorobiphenyls in carp (*Cyprinus carpio*) orally exposed to Kanechlor products. *Agric. Biol. Chem.* **46**, 891–898.

Tanabe, S., Subramanian, A., Hidaka, H. & Tatsukawa R. (1986) Transfer rates and pattern of PCB isomers and congeners and *p,p'*-DDE from mother to egg in Adelie pengiun (*Pygoscelis adeliae*). *Chemosphere* **15**, 343–351.

Tanabe, S., Tanaka, H. & Tatsukawa, R. (1984) Polychlorobiphenyls, ΣDDT, and hexachlorocyclohexane isomers in the western North Pacific ecosystem. *Arch. Environ. Contam. Toxicol.* **13**, 731–738.

Tanabe, S., Tatsukawa, R., Maruyama, K. & Miyazaki, N. (1982b) Transplacental transfer of PCBs and chlorinated hydrocarbon pesticides from the pregnant striped dolphin (*Stenella coeruleoalba*) to her fetus. *Agric. Biol. Chem.* **46**, 1249–1254.

Tatsukawa, R. (1976) Production, use and properties of PCBs. In: *PCB Poisoning and Pollution.* (Ed. K. Higuchi), pp. 148–152. Academic Press, Tokyo.

Thomann, R.V. (1981) Equilibrium model of fate of microcontaminants in diverse aquatic food chains. *Can. J. Fish. Aquat. Sci.* **38**, 280–296.

Thomas, D.J., Tracey, B., Marshall, H. & Norstrom, R.J. (1992) Arctic terrestrial ecosystem contamination. *Sci. Total Environ.* **122**, 135–164.

Thomas, W., Ruhling, A. & Simon, H. (1984) Accumulation of airborne pollutants (PAH, chlorinated hydrocarbons, heavy metals) in various plants species and humus. *Environ. Pollut.* **36A**, 295–310.

Tiernan, T.O., Wagel, D.J., Vanness, G.F., Garrett, J.H., Solch, J.G. & Harden, L.A. (1989) PCDD/PCDF in the ambient air of a metropolitan area in the U.S. *Chemosphere* **19**, 541–546.

Travis, C.C. & Hattemer-Frey, H.A. (1991) Human exposure to dioxin. *Sci. Total Environ.* **104**, 97–127.

Turner, J.C. (1979) Transplacental movement of organochlorine pesticide residues in desert bighorn sheep. *Bull. Environ. Contam. Toxicol.* **21**, 116–124.

Tysklind, M., Söderström, G., Rappe, C., Hägerstedt, L.-E. & Burström, E. (1989) PCDD and PCDF emissions from scrap metal melting processes at a steel mill. *Chemosphere* **19**, 705–710.

Urey, J.C., Kricher, J.C. & Boylan, J.M. (1976) Bioconcentration of four pure PCB isomers by *Chlorella pyrenoidosa*. *Bull. Environ. Contam. Toxicol.* **16**, 81–85.

US EPA (1976) PCBs in the United States: industrial use and environmental distribution. Task I, Final Report. EPA-560/6–76–005. Washington, DC.

US EPA (1985) Health assessment document for polychlorinated dibenzo-*p*-dioxins. Office of Health and Environmental Assessment. EPA/600/8–84/014F. Washington, DC.

Valentine, R.L., Bache, C.A., Gutenmann, W.H. & Lisk, D.J. (1988) Tissue concentrations of heavy metals and polychlorinated biphenyls in raccoons in central New York. *Bull. Environ. Contam. Toxicol.* **40**, 711–716.

Van den Berg, M., Blank, F., Heeremans, C., Wagenaar, H. & Olie, K. (1987) Presence of polychlorinated dibenzo-*p*-dioxins and dibenzofurans in fish-eating birds and fish from The Netherlands. *Arch. Environ. Contam. Toxicol.* **16**, 149–158.

Varanasi, U., Reichert, W.L. & Stein, J.E. (1989) ^{32}P-Postlabeling analysis of DNA adducts in liver of wild English sole (*Parophrys vetulus*) and winter flounder (*Pseudopleuronectes americanus*). *Cancer Res.* **49**, 1171–1177.

Vodicnik, M.J. & Lech, J.J. (1980) The transfer of 2,4,5,2',4',5'-hexachlorobiphenyl to fetuses and nursing offspring. *Toxicol. Appl. Pharmacol.* **54**, 293–300.

Vogg, H., Metzer, M. & Stieglitz, L. (1987) Recent findings on the formation and decomposition of PCDD/PCDF in municipal solid waste incineration. *Waste Manage. Res.* **5**, 285–294.

Vos, J.G. & Luster, M.I. (1989) Immune alterations. In: *Halogenated Biphenyls, Terphenyls, Naphthalenes, Dibenzofurans and Related Products.* (Eds R.D. Kimbrough & A.A. Jensen), pp. 295–322. Elsevier, Amsterdam.

Wade, R.L. (1986) Development of decontamination guidelines for PCB/PCDF and PCDD decontamination in areas of high explosive potential. In: *Hazards, Decontamination, and Replacement of PCB.* (Ed. J.-P. Crine), pp. 101–114. Plenum Press, New York.

Wakimoto, T., Kannan, N., Ono, M., Tatsukawa, R. & Masuda, Y. (1988) Isomer-specific determination of polychlorinated dibenzofurans in Japanese and American biphenyls. *Chemosphere* **17**, 743–750.

Walker, M.K. & Peterson, R.E. (1991) Potencies of polychlorinated dibenzo-*p*-dioxin, dibenzofuran and biphenyl congeners, relative to 2,3,7,8-tetrachlorodibenzo-*p*-dioxin, for producing early life stage mortality in rainbow trout (*Oncorhynchus mykiss*). *Aquat. Toxicol.* **21**, 219–238.

Walker, M.K., Spitsbergen, J.M., Olson, J.R. & Peterson,

R.E. (1991) 2,3,7,8-Tetrachlorodibenzo-*p*-dioxin (TCDD) toxicity during early life stage development of lake trout (*Salvelinus namaycush*). *Can. J. Fish. Aquat. Sci.* **48**, 875–883.

Ward, C.T. & Matsumura, F. (1979) Fate of 2,3,7,8-tetrachlorodibenzo-*p*-dioxin in a model aquatic ecosystem. *Arch. Environ. Contam. Toxicol.* **7**, 349–357.

Watts, R.J., Smith, B.R. & Miller, G.C. (1991) Catalyzed hydrogen peroxide treatment of octachlorodibenzo-*p*-dioxin (OCDD) in surface soils. *Chemosphere* **23**, 949–955.

Webb, R.G. & McCall, A.C. (1973) Quantitative PCB standards for electron capture gas chromatography. *J. Chromatogr. Sci.* **11**, 366–373.

Weber, R.R. & Montone, R.C. (1990) Distribution of organochlorines in the atmosphere of the south Atlantic and Antarctic oceans. In: *Long Range Transport of Pesticides* (Ed. D.A. Kurtz), pp. 185–197. Lewis, Chelsea, MI.

Weber, H. & Birnbaum, L.S. (1985) 2,3,7,8-Tetrachlorodibenzo-*p*-dioxin (TCDD) and 2,3,7,8-tetrachlorodibenzofuran (TCDF) in pregnant C57B1/6N mice: distribution to the embryo and excretion. *Arch. Toxicol.* **57**, 159–162.

Whittemore, R.C., LaFleur, L.E., Gillespie, W.J. & Amendola, G.A. (1990) US EPA/paper industry cooperative dioxin study: the 104 mill study. *Chemosphere* **20**, 1625–1632.

Whittle, D.M., Sergeant, D.B., Huestis, S.Y. & Hyatt, W.H. (1992) Foodchain accumulation of PCDD and PCDF isomers in the Great Lakes aquatic community. *Chemosphere* **25**, 181–184.

van Wijnen, J., van Bavel, B., Lindström, G., Koppe, J.G. & Olie, K. (1990) Placental transport of PCDD's and PCDF's in humans. In: *Dioxin 90–RPRI Seminar*, Vol. 1 (Eds O. Hutzinger & H. Fiedler), pp. 47–50. Ecoinforma Press, Bayreuth.

Williams, D.T. & LeBel, G.L. (1991) Coplanar polychlorinated biphenyl residues in human adipose tissue samples from Ontario municipalities. *Chemosphere* **22**, 1019–1028.

Williams, D.T., Cunningham, H.M. & Blanchfield, B.J. (1972) Distribution and excretion studies of octachlorodibenzo-*p*-dioxin in the rat. *Bull. Environ. Contam. Toxicol.* **7**, 57–62.

Williams, L.L., Giesy, J.P., DeGalan, N., Verbrugge, D.A., Tillitt, D.E., Ankley, G.T. & Welch, R.L. (1992) Prediction of concentrations of 2,3,7,8-tetrachlorodibenzo-*p*-dioxin equivalents from total concentrations of polychlorinated biphenyls in fish fillets. *Environ. Sci. Technol.* **26**, 1151–1159.

Winter, S. & Streit, B. (1992) Organochlorine compounds in a three-step terrestrial food chain. *Chemosphere* **24**, 1765–1774.

Wipf, H.K. & Schmid, J. (1983) Seveso—an environmental assessment. In: *Human and Environmental Risks of Chlorinated Dioxins and Related Compounds*. (Eds R.E. Tucker, A.L. Young & R. Gray), pp. 255–276. Plenum Press, New York.

Wipf, H., Homberger, E., Neuner, N. & Schenker, F. (1978) Field trials on photodegradation of TCDD on vegetation after spraying with vegetable oil. In: *Dioxin: Toxicological and Chemical Aspects* (Eds F. Cattabeni, A. Cavallaro & G. Galli), pp. 201–207. SP Med. Sci. Books, New York.

Wisk, J.D. & Cooper, K.R. (1990) Comparison of the toxicity of several polychlorinated dibenzo-*p*-dioxins and 2,3,7,8-tetrachlorodibenzofuran in embryos of the Japanese medaka (*Oryzias lapides*). *Chemosphere* **20**, 361–377.

de Wit, C., Jansson, B., Bergek, S., Hjelt, M., Rappe, C., Olsson, M. & Andersson, O. (1992) Polychlorinated dibenzo-*p*-dioxin and polychlorinated dibenzofuran levels and patterns in fish and fish-eating wildlife in the Baltic Sea. *Chemosphere* **25**, 185–188.

de Wit, C., Jansson, B., Strandell, M., Jonsson, P., Bergqvist, P.-A., Bergek, S., Kjeller, L.-O., Rappe, C., Olsson, M. & Slorach, S. (1990) Results from the first year of the Swedish dioxin survey. *Chemosphere* **20**, 1473–1480.

Wolff, M.S., Thornton, J., Foschbein, A., Lilis, R., Selikoff, I.J. (1982) Disposition of polychlorinated biphenyls congeners in occupationally exposed workers. *Toxicol. Appl. Pharmacol.* **62**, 294–306.

World Health Organization (1976) *Polychlorinated Biphenyls and Terphenyls*. WHO, Geneva.

World Health Organization (1989) *Environmental Health Criteria 88. Polychlorinated Dibenzo-para-dioxins and Dibenzofurans*. WHO, Geneva.

Yamashita, F. & Hayashi, M. (1985) Fetal PCB syndrome: clinical features, intrauterine growth retardation and possible alteration in calcium metabolism. *Environ. Health Perspect.* **59**, 41–55.

Yap, T.N., Pond, W.G., Wu, J.F., Stoewsand, G.S., Wszolek, P.C. & Lisk, D.J. (1979) Nutritional and toxicological studies with growing pigs fed aquatic plant rations. *Arch. Environ. Contam. Toxicol.* **8**, 613–619.

Young, A.L., Kang, H.K. & Shepard, B.M. (1983) Chlorinated dioxins as herbicide contaminants. *Environ. Sci. Technol.* **17**, 530A–532A.

Zacharewski, T., Safe, L., Safe, S., Chittim, B., De Vault, D., Wiberg, K., Bergqvist, P.-A. & Rappe, C. (1989) Comparative analysis of polychlorinated dibenzo-*p*-dioxin and dibenzofuran congeners in Great Lakes fish extracts by gas chromatography-mass spectrometry and *in vitro* enzyme induction activities. *Environ. Sci. Technol.* **23**, 730–735.

Zebühr, Y., Näf, C., Broman, D., Lexen, K., Colmsjo, A.

& Östman, C. (1989) Sampling techniques and clean up procedures for some complex environmental samples with respect to PCDDs and other organic contaminants. *Chemosphere* **19**, 39–44.

Zitko, V. (1992) Patterns of 2,3,7,8-substituted chlorinated dibenzodioxins and dibenzofurans in aquatic fauna. *Sci. Total Environ.* **111**, 95–108.

10: Oils and Hydrocarbons

C.I. BETTON

10.1 INTRODUCTION

Crude oil is a material of biological origin obtained from deposits in rock where it has been formed over millennia via a combination of pressure and temperature. The nature of the crude depends upon the original biological source material and the geochemical forces to which it has been subjected. Due to this variability the crude is treated — refined — to give materials of a more consistent nature.

The modern refinery is a complex chemical plant that by distillation, application of vacuum, the use of catalysts and solvent treatment, transforms the various crudes into the variety of performance-based petroleum products we all recognize, and which form the basis of the chemical, energy and transport industries of the late 20th century.

Figures produced by the Institute of Petroleum (UK), in 1992, show that the crude oil entering a refinery is converted primarily to fuel (c. 90%). Only 1% of the crude is converted to lubricating oils (Table 10.1), the basis of the 'oil' that goes

Table 10.1 UK production of petroleum products from crude oil 1991. After Institute of Petroleum, 1992

Products	Tonnes	Percentage
Motor spirit	27 792 971	32.5
Kerosene and jet fuel	9 618 320	11.5
Gas oil/diesel	26 057 467	30.5
Fuel oil	13 204 788	15.5
Gases	1 798 212	2.0
Other	6 030 876	7.0
Lubricating oils and greases	973 333	1.0

into lubricating all forms of machinery in use today. Lubricating oils are complex mixtures of chemicals produced to a performance specification. The foundation of any lubricant is the base 'oil' which may comprise 80–98% of the product. This base oil may be a refinery product — a mineral base — or a synthetic chemical such as a polyalphaolefin or synthetic ester. For some applications it may be a natural oil (triglyceride ester) such as rapeseed oil, sunflower oil or palm oil. The base oil is chosen to give the performance characteristics required of the product, bearing in mind cost. As no base oil is a perfect lubricant on its own, other chemicals are added to improve upon, or add desirable characteristics to, the base oil. These may be antiwear additives, anticorrosion additives, viscosity index improvers, pour-point depressants, antioxidants, etc. They are often complex materials based on metals and hydrocarbons (Mortier & Orszulik, 1992).

This chapter draws heavily on data generated and published by the oil companies' organization for the Conservation of Clean Air and Water in Europe (CONCAWE). Whilst ostensibly a European organization, the oil companies themselves are often multinational, and certainly international in outlook. The work of CONCAWE is therefore carried out by experts in their particular field who are often based outside Europe, and in particular in the USA. For ecological aspects of oil products all available expertise, whether from Europe or the USA, has contributed to the published reports. These can therefore be said to be truly international and not just the 'European view'.

For the remainder of this chapter, unless stated otherwise, 'oil' is used as a generic term to include pure hydrocarbons and performance-based oil products. However, there are several points that all oils have in common.

1 An oil is designed or produced to meet a performance specification; chemical composition can vary in two oils of similar performance.

2 The chemical composition is complex. There may be several hundred, sometimes thousands, of different chemical compounds, mostly hydrocarbons, present in an 'oil'.

3 Most oils have very poor water-solubility, making ecotoxicological assessment difficult.

10.2 ANALYTICAL DETERMINATION

In dealing with a poorly soluble mixture of chemicals, the consideration of what analysis should be performed to confirm exposure concentrations, and what methods should be used for preparation of test media is of particular importance. These two issues are strongly linked, since the way that test media are prepared can affect the composition of the aqueous phase (Bennet *et al.*, 1990).

A standard method of measuring oil in water is to analyse the dissolved total organic carbon (TOC) content of the aqueous phase. This involves taking aqueous samples which are then acidified and purged with nitrogen to remove inorganic CO_2. Total organic carbon can then be measured using commercially available analytical equipment (Somerville *et al.*, 1987). Values for TOC measurement are generally expressed as ppm mass of carbon per volume of extract, and give an overall intergrated figure for exposure concentration.

What such measurements cannot do, however, is to identify the nature of the components present in the aqueous phase. This is of particular relevance when dealing with complex mixtures of hydrocarbons and related compounds characteristic of oils. Analysis of such complex mixtures is normally carried out using gas or liquid chromatography in conjunction with mass spectrometry (GC-MS).

Samples for GC-MS are prepared by extraction of the aqueous test medium with dichloromethane or other suitable solvents (Gough & Rowland, 1990). The solvent containing the extractable oil is then dried and concentrated prior to analysis (Bennet *et al.*, 1990).

Typical GC-MS chromatograms for two formulated oil products are shown in Fig. 10.1. Comparison of the traces for the two products examined, and their respective aqueous phases, reveals distinct differences. The oils themselves are characterized by an unresolved hump (unresolved complex mixture or UCM) starting at 20 min retention time (Gough & Rowland, 1990). This is characteristic of the hydrocarbon base oil.

The UCM has been shown to consist in part of simple monoalkyl and T branched alkanes. As there are 536 possible acyclic T branched alkane structures with carbon numbers between C_{20} and C_{30}, the UCM is a pertinent indication of the complexity of oils and the difficulties inherent in analysing samples from ecotoxicological investigations.

The nature of the water-soluble fraction of the oil can be deduced by examination of fragment ions, searches of and comparison with mass spectral libraries and by judgement of experienced operators. Figure 10.2 shows the result of one such detailed analysis.

Assignment of the various peaks is shown in Table 10.2. It is significant that the identity of the majority of the water-soluble components is attributable to the oil additives used to enhance the base oil properties (i.e. amino succinates, sulphur compounds and methacrylates). Base oil hydrocarbon components in the aqueous phase are present only in minor quantities.

GC-MS analysis is the best means currently widely available to characterize samples. The analytical support that would be expected when dealing with single water-soluble chemicals, e.g. pesticides, would include characterization of parent compound, breakdown products/metabolites and the demonstration that test concentrations have been maintained. For oils, comparison of the traces in Fig. 10.1 shows not only the complexity of the situation, but also that the principal components of the aqueous phase are not even noticeable when looking at the traces for the whole product.

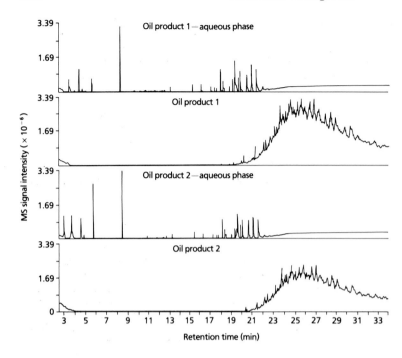

Fig. 10.1 GC/MS chromatograms of two formulated oil products and their aqueous phases at equilibrium. From Bennet *et al.* (1990).

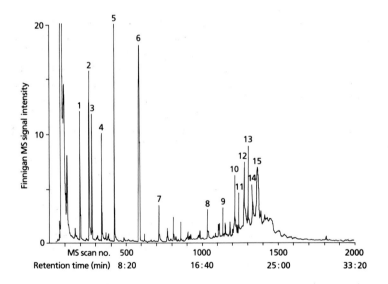

Fig. 10.2 Total ion current chromatogram — aqueous phase for oil product 2. Designated peak numbers refer to Table 10.2. From Bennet *et al.*, 1990.

10.3 ECOTOXICOLOGY

The purpose of carrying out any ecotoxicological investigation is to obtain information upon which to base a prediction of the possible environmental impact of the material under investigation (see Volume 1). A narrower regulatory interpretation of ecotoxicology is to produce data which can be used to rank materials and regulate their use (EEC, 1992). In practice, ecotoxicological tests are carried out to satisfy both scenarios. For the majority of materials, single water-soluble chemi-

Table 10.2 GC/MS characterization of components in the aqueous phase. From Bennet *et al.*, 1990

GC peak no.[a]	Retention time (min)	Major fragment ions (m/z)	Assignment
1	3.22 (not present in product 1)	59, 69, 85, 99, 100	Methyl methacrylate
2	4.24	45, 57, 69, 84, 87, 101	Hexan-2-ol
3	4.38	65, 91, 92	Toluene
4	5.42	43, 61, 69, 84, 104, 118	C_6-Thiol
5	7.04	45, 59, 69, 73, 85, 101, 117, 133	Ethoxyalkane
6	9.53	54, 69, 80, 94, 105, 121, 137	C_5-Cycloalkane
7	12.01	43, 56, 70, 83, 97, 123, 138	C_9-Alcohol or cyclic hydro carbon
8	17.19	57, 89, 145, 201	Amino succinic acid derivative
9	19.01	57, 145, 179, 215, 235	Amino succinic acid derivative
10	20.21	57, 85, 131, 187, 215, 243, 271	Alkyl sulphur compound
11	20.45	57, 131, 187, 201, 257	Alkyl sulphur compound
12	21.24	57, 85, 115, 131, 187, 215, 271, 299	Alkyl sulphur compound
13	21.48	57, 85, 131, 201, 257, 285	Alkyl sulphur compound
14	22.12	57, 71, 131, 201, 271, 327	Alkyl sulphur compound
15	22.48	56, 69, 84, 118	Unsaturated hydrocarbon

[a] GC peak numbers: refer to Fig. 10.2

cals, fulfilling the needs of prediction and regulation, can be accomplished using standard OECD test methods (OECD, 1981) to satisfy both purposes.

With oils, however, which are not only complex mixtures but also have poor water-solubility, very often the way in which the test is performed is determined by the purpose to which the data generated will be put. This is an important consideration, as it is known that the way in which test media are prepared can influence the composition of the aqueous phase (Lockhart *et al.*, 1987; Shiu *et al.*, 1988; Bennet *et al.*, 1990).

10.3.1 Preparation of test media

There are essentially two ways of preparing aqueous test solutions of oils: the use of only those fractions that are soluble in water, the so-called water-accommodated fraction (WAF), or the use of whole product dispersions (CONCAWE, 1992a).

Water-accommodated fractions and water-soluble fractions (WSF) are essentially those fractions of the total product that are present in the aqueous phase following a period of mixing. After mixing, the oil and water are allowed to separate and the oil phase is removed. The term WAF is used in preference to WSF as it is often the case that stable microemulsions or finely dispersed droplets can exist in the aqueous phase that does not represent a true solution (CONCAWE, 1992a).

A WAF may be produced by stirring or shaking a known volume of test material with a known volume of test medium. The amount of test substance per unit volume of test medium is referred to as the loading rate (Girling *et al.*, 1992). Following a preset period of mixing, the mixture is allowed to stand and separate into aqueous and non-aqueous phases. The aqueous phase is then removed and used for toxicity testing (CONCAWE 1992f).

The composition of the aqueous phase thus prepared can be affected by both:

1 Stirring/shaking rate: this should be continuous and rigorous enough to ensure equilibrium conditions are achieved, but not so vigorous as to produce an emulsion. Care should be taken to ensure that, when stirring is employed, a film of oil is not allowed to 'creep' up the side of the

vessel and so avoid contact with the test medium.
2 Mixing time: the longer the mixing time, the
greater the probability of achieving an equilibrium. However, if mixing is prolonged, the
aqueous components can decrease due to volatilization and/or oxidation (Shiu *et al.*, 1988). It is
generally found that a 24-h mixing period is sufficient to produce a stable WAF (Shiu *et al.*, 1988;
Bennet *et al.*, 1990; CONCAWE, 1992f).

In order to produce a range of concentrations
for an assessment of acute toxicity it is standard
practice, when dealing with single water-soluble
chemicals, to prepare a stock solution and then
make serial dilutions of this to give the desired
range of test solutions. However, such a practice
with oils will produce different chemicals in the
aqueous phase, both qualitatively and quantitatively, as compared with separate preparation
of each test dilution by making a series of WAFs
from the original product and testing the resulting
aqueous phase (Shiu *et al.*, 1988). The reason for
this discrepancy is that the least water-soluble
components in the oil may reach their maximum
concentration in the lowest loading rate. This
concentration will not change as the loading rate
increases. Components with increasing water
solubility will reach saturation at increasing
loading rates. For example, consider a poorly
soluble component C, that achieves a maximum
aqueous concentration of 1 ppb at a loading rate
of 1 ppm, and compare this with a more soluble
component D with a maximum achievable concentration of 500 ppm in water. If both C and D
are present in an oil at 10% then the concentration
of components C and D in the aqueous phase of a
series of WAFs would be as shown in Table 10.3.
As can be seen, the composition of the aqueous
phase will change quantitatively and qualitatively

with respect to these two components. These
differences have been demonstrated for a number
of oils (Shiu *et al.*, 1988) and for the hydrocarbons
present in creosote (Benyon & Cowell, 1974).

Other factors, such as some components acting
as co-solvents for others and hence affecting partition coefficients, can also act to modify the
composition of the aqueous phase at each loading
rate. Hence when dealing with complex mixtures
of poorly soluble compounds it is important to
specify the loading rate of test material to water,
to characterize the aqueous phase, and not to rely
on dilutions of a stock mother solution where all
materials will remain in a similar ratio to one
another (CONCAWE, 1992f).

The use of WAFs (i.e. only examining the
aqueous phase) up to a maximum loading rate is
normally preferred when generating data for regulatory purposes. This allows some comparison to
be made with data generated from single water-
soluble chemicals, where it is common practice
to test compounds only as far as the limits of
their water-solubility (OECD, 1981). In order to
examine real-life environmental effects it is more
usual to examine whole-product dispersions
maintained mechanically, as this gives a closer
approximation to the environmental situation
following accidental spillage (Benyon & Cowell,
1974).

A means of maintaining a product dispersed
and distributed in the aqueous phase, but without
the production of an emulsion, places limitations
on the test organisms that can be employed. Small
organisms such as *Daphnia* or unicellular algae
are liable to physical damage from the dispersion
apparatus, and may be of a size where the oil
droplets are liable to cause agglomeration. A
method that has been used successfully for larger

Table 10.3 Aqueous concentrations, achieved at different loading rates, of two components C and D of differing
water solubility, present in an oil

Product load rate (ppm)	Concentration C (ppb)	Concentration D (ppm)
1	1	0.1
10	1	1.0
100	1	10
1000	1	100
10 000	1	500

organisms involves a propeller, housed within a tube, that rotates with sufficient force to maintain good circulation and dispersion as oil and water are passed down the tube and out at the bottom (MAFF, 1984). Factors that can affect the nature of the dispersion are:

1 size and shape of the propeller;

2 rotor speed;

3 nature of any additional chemical dispersant — this is particularly of interest if considering some formulated oil products (G.E. Westlake, personal communication, 1991).

10.3.2 Assessment of methodology

An examination of the recommended OECD guidelines for ecotoxicity, bioaccumulation and biodegradation testing (OECD, 1981), on which regulatory tests are based, reveals that they are not applicable to petroleum products. Experience of tests on oil products has identified some practical difficulties which must be taken into account, both when developing test protocols for a particular product and subsequently interpreting the data. The preparation of test media must produce an aqueous test solution that:

1 adequately represents the nature of a water body in contact with the product under particular conditions; and

2 is definable in terms of reproducible exposure concentrations of the product constituents in the aqueous phase.

These requirements have led to the two different approaches for the preparation of test media, WAFs and dispersions, discussed above. The effects of using these methods on the results generated are discussed below.

Water-accommodated fractions (WAF)

A WAF is a medium containing only that fraction of a product that is retained within the aqueous phase once any mixing energy has been removed. The fraction of the product may be present either in solution or as a stable mixed emulsion.

As previously discussed, the nature of a WAF is determined by the relative volumes of product to aqueous medium and the properties of the product and the mixing environment. If differential par-

titioning effects of product constituents between the product and aqueous phases are ignored, then increasing the product-to-medium ratio will increase the ability for product constituents to enter into solution in the aqueous phase up to the limit of their aqueous solubility. Increasing the energy and time of mixing will increase the potential for products to emulsify. Other factors, such as volatility of the various constituents and water quality (hardness, pH and temperature), may also play a part in determining the properties of a WAF. A diagrammatic representation of a WAF-type test is shown in Fig. 10.3.

For pure substances an 'ideal' WAF is a saturated solution of the substance in water. A single WAF can therefore be produced which, if serially diluted, will be equivalent to a range of concentrations of the substance in water. As discussed previously, for oil products which are mixtures of constituents each with their own water solubility, such a practice is not acceptable, since serial dilutions will not accurately reflect the exposure occurring with different product/water ratios. The alternative in this case therefore is to prepare WAFs at specific product/water ratios.

It is not apparent from much of the literature reporting on ecotoxicity testing, however, whether such effects were taken into consideration. In addition, without analytical data to confirm the stability and durability of test solutions, it is debatable whether a WAF prepared at the start of a 24-h exposure period is the same as the WAF at the end. Quantification of test data is often not provided. Published data on gasolines, where nominal concentrations in open systems are reported to give an EC_{50} of 125–345 mg/l (Westlake, 1991), can be compared to ecotoxicity data generated using measured concentrations in closed systems where equivalent values are orders of magnitude less, e.g. EC_{50}: 2.7–6.25 mg/l (CONCAWE, 1992b). Such differences indicate the importance of analytical support that is often lacking in such tests, and the way in which methodology can affect the results.

Exposure systems based upon WAFs enable the effects of the fractions of a product which are present in a physically stable form in a test medium to be evaluated. The absence of large quantities of undissolved product and the lack of

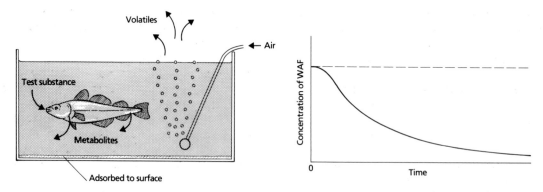

Fig. 10.3 Some of the problems associated with a WAF-type test on oils. Nominal concentration is not the actual aqueous test concentration, due to losses (as shown). Analysis should confirm actual concentration is not less than 80% of the nominal value at the end of the exposure period, for the test to be valid for regulatory purposes.

any requirement for continuous mixing during the exposure period may enable the test protocols used for pure compounds to be employed essentially without modification. However, care must still be taken to avoid problems such as trapping of small test organisms (algal cells or invertebrates) in residual films of oily material on the surface of the test solution, and adhesion of algal cells to undissolved oily material in suspension. Methods for minimizing these effects include additional treatment of the medium (filtering or centrifuging) prior to introducing the test organisms, and discouraging contact with undissolved material. In the latter case it has been found that darkening the top of the test vessel when using *Daphnia*, by covering the neck of a flask with a black cap, discourages the *Daphnia* from coming into contact with any residual surface film.

To achieve consistency in the use of WAFs for determining the effects of a product, it is essential that mixing conditions with respect to method used and duration are well defined. Likewise, the conditions under which separation of the bulk aqueous and product phases is achieved, also require definition.

There are two ways in which the results of tests with WAFs can be expressed:
1 the ratio of product to water used in the preparation of the WAF—the loading rate (Girling *et al.*, 1992);
2 measured concentrations of the product constituents in the aqueous phase.
Loading rates provide a useful basis for expressing

and interpreting the results of tests with products that are poorly soluble. Since the amount of any poorly soluble product spilled into the environment will not be proportional to its water-solubility, expressing the results of tests in terms of loading rate used in the preparation of a WAF provides a common reference point for the interpretation of the results from tests with products of differing solubility. By convention the results of toxicity tests can then be expressed in terms of loading rates:

LL_{50} value, lethal loading value for 50% effect.
EL_{50} value, effective loading value for 50% effect.

Measured concentrations of the product constituents in a WAF may provide a basis for interpreting the results of a test, particularly if known toxicants can be identified. In practice, however, measured concentrations of all of the constituents present are unlikely to be feasible. A degree of judgement is required in identifying those constituents upon which to concentrate analytical effort and resources. A measure such as dissolved organic carbon is of limited value as, whilst it may give an idea of total exposure, it does not identify whether the materials measured are toxicants or metabolites. Some characterization is essential.

Dispersions

A dispersion consists of a quantity of oil distributed uniformly, but not wholly dissolved, in an aqueous phase. To maintain a uniform distribution of the oil throughout an aqueous medium the oil may be dispersed either by continuous mixing or by the use of a chemical dispersant such as a low-toxicity surfactant (G.E. Westlake, personal communication, 1991). While aiding the performance of the test, the use of dispersants cannot easily be compared with other tests where the limit of solubility determines the maximum concentration employed. In practice the limit of 100 mg/l of emulsifier recommended by the OECD (1981) is of little relevance to the testing of oil products (Westlake, 1991). The results obtained for oils with or without dispersant show little variation. A diagrammatic representation of a dispersion type test is shown in Fig. 10.4.

The aqueous phase of a dispersion is enriched by the fraction of the oil which has entered into solution. The relative proportions of each phase will depend upon the solubility of each fraction under the conditions in which the medium is being prepared. For a product which is a mixture, the overall solubility will be a function of the solubility of the constituents.

Unless the material to be tested has a natural tendency to emulsify—as in the case of some cutting oils—exposure systems based upon the dispersion of oils require either the constant input of energy or the use of a chemical dispersing

agent to maintain the dispersed state. These two requirements in turn place limitations on the test methods that can be employed.

The constant input of energy required may preclude the use of small fragile test organisms such as *Daphnia*, and the presence of undissolved material may increase the potential for physical fouling effects. The use of a chemical dispersing agent may not reflect the exposure in the environment, and the presence of the dispersant in the medium may confuse interpretation of the test results.

Despite these shortcomings, dispersion-based tests provide a better assessment of the effects (toxic plus physical) which might result from the introduction of the product into the environment, e.g. via a spill. While the use of dispersants may not be considered valid in relation to other tests carried out only to the limits of solubility, such use could be said to mirror the environmental situation in which, in the case of crude oil spills, dispersants are used, and receiving waters may be subject to sewage input containing domestic-derived surfactants.

Summary of requirements for testing

To avoid confusion when expressing the results of tests with oil products, it is essential that the

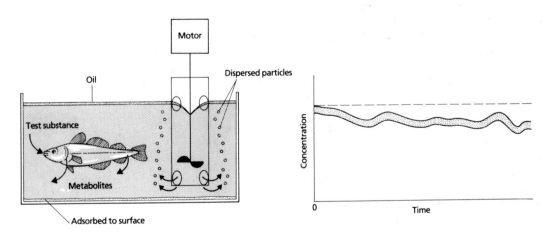

Fig. 10.4 A dispersion-type toxicity test. The concentration will vary in the aqueous phase as different components enter and leave the water. The nominal concentration remains constant, with some loss of volatiles. The concentration and nature of the water-accommodated fraction changes with time due to metabolism and volatilization, enabling different chemicals to move into and out of solution.

conditions under which the test medium is prepared and the exposure is carried out, are accurately described. The following descriptors of the media preparation and test conditions may prove useful:

• type of medium prepared – WAF or dispersion;
• type of mixing system – mass of product and medium;
• geometry of mixing vessel – turbulence and effectiveness of mixing;
• sealed or open mixing vessel – loss of volatiles;
• volume of head space – loss of volatiles and degree of turbulence;
• method of mixing;
• energy input;
• duration of mixing;
• method of phase separation in WAF preparation;
• type of chemical dispersant (if used);
• water quality – pH, hardness, salinity, temperature.

Testing for regulatory purposes may require quantitative chemical analysis to confirm or determine dosing in the test medium, to monitor test medium stability and to characterize the product constituents present.

From the above it is clear that considerable attention must be paid to the preparation, characterization and description of the test media for petroleum products. Failure to do this will not allow an accurate assessment of the toxicity, or meaningful comparisons with other compounds to be made.

10.3.3 Conclusions

A dual approach to the testing of oils and hydrocarbons, utilizing either WAF or whole-product dispersions, is currently in use. In the majority of cases the method employed is chosen depending upon the purpose to which the data are to be put. Data required for regulatory or ranking purposes are generally based on WAF methodologies due to:

1 requirements to test only to the limit of solubility;

2 ability to use standard test organisms such as *Daphnia*;

3 possibility of generating analytical support data.

Whole-oil dispersions have been used predominantly in those cases where data are required to assess potential environmental impact, the effects of physical fouling, losses due to volatilization, and the presence of a large reservoir of undissolved material more closely resembles actual spill situations. A dispersion method, developed by the UK MAFF to assess materials used in the North Sea (MAFF, 1984), has found widespread use in the USA where the interest of regulators is primarily in assessing environmental impact. In Europe, however, where ranking and comparative toxicity and the determination of intrinsic properties are paramount, the WAF methods recommended by the US EPA (1985) are most often used.

The purpose of regulatory toxicology is to enable predictions to be made of potential environmental impacts based on laboratory tests. This being so, is it more realistic to carry out a test on an oil product such as gasoline, when all of the constituents are held in solution by utilizing closed systems (for summary diagram see Fig. 10.5) and obtaining an effect concentration of 3–7 mg/l, or is it more appropriate to use an open vessel and obtain a result of 125–345 mg/l? Moreover, following a gasoline spill in the environment, initially high concentrations of soluble components in water and sediments can return to background values within 48 h (Roubal *et al.*, 1979). As long as regulatory demands are based on the standard methods developed for soluble substances (OECD, 1981) the closed system will be favoured, but this is not necessarily environmentally realistic.

The difference between an acute toxic episode (high concentration which often diminishes rapidly) and a chronic exposure (low concentrations over long periods) is also of importance (see Volume 1). For single chemicals it is possible to relate exposure in laboratory studies to both scenarios, as the materials are quantifiable and generally easily maintained in the aqueous phase and are not volatile. For oils, however, these considerations do not readily apply. They are not present in the aqueous phase at high or constant concentration, they are not easily analysed and the more toxic water-soluble components are often lost rapidly by volatilization (Baker, 1976; Green & Trett, 1989).

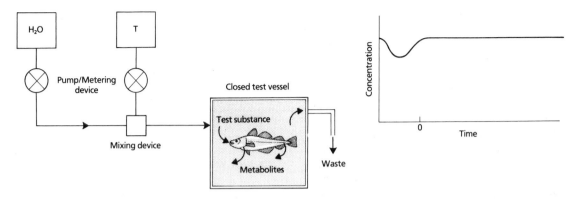

Fig. 10.5 Diagram showing essential elements of a closed flow-through system. Note that, once all surfaces are coated with adsorbed oil, a constant concentration can be maintained. Once set up, it is less labour-intensive and the analytical determination is not time-critical.

10.4 ENVIRONMENTAL CONSEQUENCES

The relatively few incidents, in terms of total tonnages transported, where large volumes of crude oil have entered the environment have provided many opportunities to study the direct environmental impact of crude oil. Accidents such as the Torrey Canyon, Amoco Cadiz, Exxon Valdez and Braer, as well as deliberate pollution such as occurred during the 1991 Gulf War, have led to enormous sums of money being spent on clean-up and scientific investigation, not only at the time of the incident but also for prolonged periods where recovery of ecosystems has been followed. It is not my intention to consider these incidents in detail in this volume; there is an extensive literature on this subject (see Benyon & Cowell, 1974; Baker, 1976; Green & Trett, 1989; Clark, 1982).

In considering the ecotoxicology of oil and hydrocarbons it is important not to forget the purpose of such tests. As mentioned previously, there are tests for regulations which give a ranking of chemicals using comparable methodology, and there are tests to determine potential environmental impacts. These two types of test should be compatible, but unfortunately in the case of oils and hydrocarbons they are frequently not.

Before considering the philosophy of test design we must bear in mind what happens when oil is spilled, as this has implications for the testing strategy finally adopted.

10.4.1 Environmental fate of spilt crude

Crude oil is a variable, complex mixture of chemicals. Once spilt on water, a process known as weathering begins, the progress of which is dependent upon factors such as temperature, wind, degree of sunlight and water turbulence.

The first materials 'lost' from crude oil are, not unexpectedly, the volatile components. It has been demonstrated that under 'normal' conditions C_5 and C_6 compounds are gone in less than 1 h, and components up to C_8 have disappeared after 5 h (Dodd, 1974). The process continues, until by 24 h up to 40% of the original volume may have dissipated. The lower molecular weight hydrocarbons are also generally more water-soluble. The greater the molecular weight, however, the higher the acute toxicity (CONCAWE, 1992b). After several days of weathering, chemical species up to C_{10} are generally absent and even the less volatile fractions are diminished (Dodd, 1974). This process progresses via the production of emulsions commonly known as 'chocolate mousse', until eventually the density of the remnant is such that the material sinks as tar-balls to the ocean floor (AEL, 1967; Ministry of Defence, 1971).

Some oil, however, reaches the shore before it can be effectively weathered, and here it is physical fouling that causes the most visible effects, particularly on seabirds and mammals, but also on intertidal invertebrates. Recovery of such ecosystems can and does occur, however, even in the most inhospitable environment (Engelhardt, 1985; NRC, 1985; Pritchard & Costa, 1991).

Biological activity is primarily responsible for this on both a macro- and micro-scale. Biodegradation by microbes is an essential part of the regenerative process (Cain, 1990; Betton, 1992), and an assessment of biodegradability is central to the ecotoxicological evaluation of all oils and hydrocarbons in terms of both prediction of effects and regulatory assessment (EEC, 1992).

In short, the environmental consequences of crude oils can be catastrophic in the short term; the long-term prognosis for contaminated environments, however, is encouraging. Activities of industry and regulators must be aimed at prevention and, in order to achieve this, sound laboratory data should be available to ensure that appropriate protective measures can be taken to minimize damage. However, even if it were possible to eliminate all oil and hydrocarbon use in the world, the environment would still be subjected to large volumes of such materials from natural seepage. Oil is, after all, a natural product that has leached into the biosphere for many millennia, and it might be anticipated that some species will have evolved adaptations to deal with long-term, low-level exposure to such chemicals.

10.5 LABORATORY DATA ON OILS AND HYDROCARBON ECOTOXICITY

In 1988 CONCAWE published a review of published literature on the ecotoxicology of petroleum products. This review concentrated on refinery products and excluded crude oil. CONCAWE is also submitting all known existing ecotoxicity data, both published and in-house company data from all member companies worldwide, in the form of product dossiers to the European Commission on behalf of the oil industry as a consequence of the EEC Existing Substances Regulations (EEC, 1992). These publications will eventually cover the following groups of petroleum products:
- Liquefied Petroleum Gas (CONCAWE, 1992c)
- Gasolines (CONCAWE, 1992b)
- Kerosenes/Jet Fuels (CONCAWE, in press)
- Gas Oils (diesel fuels/heating oils) (CONCAWE, in press)
- Heavy Fuel Oils (CONCAWE, in press)
- Lubricating Oil base stocks (CONCAWE, in press)
- Aromatic Extracts (CONCAWE, 1992c)
- Waxes and Petrolatum (CONCAWE, in press)
- Bitumens and bitumen derivatives (CONCAWE, 1992d)
- Petroleum Coke (CONCAWE, in press)
- Crude Oil (CONCAWE, in press)

Examination of the data summarized and reported in these dossiers shows that the various oil products can show a range of toxicity in the aquatic environment from highly toxic to non-toxic. The degree of toxicity is not only related to the individual hydrocarbon molecules present, but is also dependent upon whether any performance-enhancing additives are present in the product. Interpretation of these data requires a knowledge of the physicochemical properties of the test material and a knowledge of the test methodology, backed up by chemical analysis of the test media. Very often such data are lacking or incompletely reported. For a summary of data generated using 'standard' test methods see Table 10.4. These data represent the sum of all known literature on this topic.

10.5.1 Review of available information

Following comprehensive literature searches there were no data identified on petroleum products such as liquefied petroleum gas, aromatic extracts, waxes and petrolatum, bitumens and bitumen derivatives and petroleum cokes.

Table 10.4 Summary data on ecotoxicity of petroleum products. From CONCAWE, 1992c,d,e, 1994 in press

Product	$L(E)C_{50}$ Range (mg/l)
Gasoline	10–100
Lubricants/base oils	>1000
Kerosene	1–10
Gas oil/diesel	10–100
Fuel oil	
Nos 1, 2	10–100
Nos 3, 5	100–1000
No. 6	>1000
Crude oil	10–100

Those products for which data have been generated are either those of significant commercial importance or those which possess a significant water-soluble component. The data presented and discussed below for gasolines and crude oil represent the spectrum of reported ecotoxic effects characteristic of oils and hydrocarbons.

Gasoline

Gasoline is a mixture of approximately 280 different hydrocarbons in the range C_4 to C_{12} (Berry & Stein, 1977; API, 1985). Aromatic hydrocarbons account for between 87% and 95% of the WSF derived from gasoline. However, these represent <50% of the volume of the parent gasoline. (Normal composition is approximately 50% aliphatic compounds + 50% aromatic + naphthenic ring compounds).

Assessing the ecotoxicology of gasoline is therefore tantamount to measuring the toxicity of the water-soluble monoaromatic components, particularly benzene, toluene, ethyl benzene and xylenes (BTEX). Less soluble diaromatic components such as naphthalene and methyl-naphthalenes are also of concern.

Toxicity data for the principal components of the WSF are available (Verschueren, 1983) and are summarized in Table 10.5. Examination of these results reveals a wide variation in reported toxicity that probably reflects the methodology employed. It is difficult to comment on this, as these data are often incompletely reported. As is common in ecotoxicological work, nominal concentrations are generally quoted unsupported by analytical information. The figures quoted for the components are of similar orders of magnitude and variability to those reported for gasolines (Table 10.4).

Gasoline and its components (BTEX) clearly exhibit short-term toxic effects to a variety of organisms (Table 10.5), particularly when studied in closed or flow-through systems. The most toxic components are also the most volatile and degradable (CONCAWE, 1992b), and when studied in open systems these components are rapidly lost. This can be said to render the test invalid, but it also reflects the environmental situation where spilt gasoline and its principal water-soluble components are rapidly lost from the aquatic environment, presenting minimal long-term risk and an acute hazard measured in hours not days.

Crude oil

Crude oil is the material most people consider when they think of the environmental consequences of 'oil'. There is a vast literature on environmental contamination by crude. There is also a large literature describing the effects of various crudes and crude fractions on behaviour,

Table 10.5 Summary of toxicity data for principal components of gasoline (CONCAWE, 1992b)

Component	Content in gasoline	Toxicity L(E)C$_{50}$	mg/l
Benzene	2–5%	Fish	5–395
		Crustacea	20–108
		Algae	92–>1000
Toluene	6–7%	Fish	24–1340
		Crustacea	4.3–60
		Algae	105
Ethyl benzene	*c.* 4.5%	Fish	32–97
		Crustacea	—
		Algae	33
Xylenes			
ortho	6.25%	Fish	11–37
meta	12.00%	Crustacea	1.3–1000
para	4.25%	Algae	55

reproduction, physiology, biochemistry and any other measurable parameter to any number of test organisms at various stages of the life cycle. (For a summary see CONCAWE *Product Dossier: Crude Oil* (In press) and Moore & Dwyer, 1974).

Table 10.6 summarizes the acute toxicity data on crude oil and soluble hydrocarbons derived from crude. As stated earlier, the composition of crude oil is both complex and variable. The major constituents are: (i) *n*-alkanes, (ii) isoalkanes, (iii) cycloalkanes (also known as naphthenics); and (iv) aromatics. It is the low-molecular-weight components, and particularly the low-molecular-weight aromatics, that are the most readily water-soluble (Coleman *et al.*, 1984). These are in effect the BTEX components of gasoline (above).

Data reported by CONCAWE (In press) show that crudes from Forties, Kuwait, Prudhoe Bay, South Louisiana and Cook Inlet, as well as a number of unidentified sources, have been studied. Materials studied were generally characterized as 'WSF', although a small number of flow-through tests were performed. Some of the test systems were 'non-aerated', but for the majority no mention of aeration is made; some were 'open' and others 'closed'.

Analytical measurements were made of either total aromatics, aliphatic aromatics, aliphatic hydrocarbons or just 'hydrocarbon' in the 25% of instances where any analytical information is

mentioned. Exposure periods ranged from 24 h to 8 days. It is hardly surprising that results varied from 'no mortality' (e.g. *Menidia berylina*, S. Louisiana Crude, WSF, 48 h) to LC_{50} values of 14 800 mg/l (e.g. *Fundulus similus*, Kuwait crude, shaken mixture, 96-h LC_{50}), and $< 8.4 \mu l/l$ (e.g. *Pemiphales promelas*, 'Mixed blend sweet', flow-through, 48-h LC_{50}).

Approximately 45% of the data in the CONCAWE review indicate a 96-h LC_{50} value between 1 and 10 mg/l. While this could be said to give an idea of some consensus, it may also be argued that over half of the data are outside this range, with the majority of reported being values > 10 mg/l.

10.6 BIOACCUMULATION

Another of the principal considerations in assessing the environmental impact of a chemical is an assessment of its ability to bioaccumulate (Volume 1, Chapter 1). Indications of such potential are given by the partition coefficient between octanol and water (P_{ow} or K_{ow}). The relationship between P_{ow} and actual bioaccumulation potential has been long established (Veith *et al.*, 1980).

The log P_{ow} values can be determined by the classical 'shake flask' method (OECD Method 107) or by the more recently introduced high-

Table 10.6 Summarized acute toxicity data (96 h LC_{50}) for crude oil and soluble hydrocarbons

Species	Test material	Concentration (mg/l)	References
Fish	Fresh crude	88–18 000	
Larvae and eggs	Fresh crude	0.1–100	
Pelagic crustaceans	Fresh crude	100–40 000	Craddock (1977)
Benthic crustaceans	Fresh crude	56	
Bivalves	Fresh crude	1000–100 000	
Aquatic macrophytes	Soluble hydrocarbons	10–100	
Finfish	Soluble hydrocarbons	5–50	
Larvae and eggs (all species)	Soluble hydrocarbons	0.1–1.0	
Pelagic crustaceans	Soluble hydrocarbons	1–10	
Gastropods (snails, etc.)	Soluble hydrocarbons	1–100	Moore & Dwyer (1974)
Bivalves (oysters, clams, etc.)	Soluble hydrocarbons	5–50	
Benthic crustaceans (lobsters, crabs, etc.)	Soluble hydrocarbons	1–10	
Other benthic invertebrates	Soluble hydrocarbons	1–10	

performance liquid chromatography (HPLC) method (OECD, 1981). The latter also allows the possibility of calculating (predicting) the log P_{ow} of a compound from its structure by using the fragment addition method (Hansch & Leo, 1979; see also Chapters 13 and 14).

If the log P_{ow} of a compound exceeds a certain trigger value, normally taken as 3, then consideration should be given to determining the actual bioaccumulation in an aquatic organism, usually a fish (OECD, 1981). However, recent studies of the physiological basis of the bioaccumulation and elimination processes in fish and their relationship to the partitioning process as represented by the log P_{ow} (Hansch & Leo, 1979) have identified three ranges of log P_{ow} associated with different limiting factors in the bioaccumulation process. At low log P_{ow}, bioaccumulation is low and partitioning is not an important factor. Over a middle range of log P_{ow}, bioaccumulation is positively correlated with partitioning unless the compound is rapidly metabolized or steric factors operate. Finally, at high values of log P_{ow}, bioaccumulation is reduced and does not follow partitioning. The limiting factors for this are the small quantity of material available in the water for transfer into organisms, and the comparatively large molecular size of the compound reducing membrane penetration (Barrow, 1990).

As with ecotoxicity tests, the principles and practices of bioaccumulation were developed primarily for chemical products consisting of single well-defined chemicals. Their application to oils presents some technical and interpretive problems.

Determination of the range of log P_{ow} values represented in an oil product cannot be made using the 'shake flask' method becasuse of the analytical complexity of the situation. However, some estimate can usually be made using HPLC, although even this can present problems for the higher log P_{ow} values (see above). However, there is usually sufficient knowledge of the chemical species present in oil products for prediction of log P_{ow} values by the fragment addition method (CONCAWE, 1993).

Study of the actual bioaccumulation of an oil product in fish also presents problems. Again,

each constituent chemical will have its own bioaccumulation characteristic. The analytical problems that this presents are aggravated by the similarity of many of these constituents to compounds naturally present in fish. Many constituents of oil in fact enter the metabolic pathways of the fish, leading to energy production or incorporation into biomass. Such compounds do not bioaccumulate to the degree that might be predicted on the basis of their partitioning characteristics because of this rapid metabolism. However, fish 'tainting' is a recognized problem, and is indicative of some accumulation. Other than effects on taste and smell, the toxicological significance of this is not yet clear.

Currently, there are no known procedures for testing the potential of complex mixtures to bioaccumulate (see Chapters 12 and 13). The HPLC method can be used to separate and quantify components of the mixture with log P_{ow} values below 6. However, GC-MS analyses will be required to identify these compounds. Bioaccumulation studies can then be performed on individual compounds. This approach can be both time-consuming and expensive.

10.7 BIODEGRADATION

The final aspect to be considered when assessing the environmental impact of oils and hydrocarbons is their biodegradability. As mentioned previously, crude oil does eventually degrade in the environment. Oils when tested in standard OECD tests for ready biodegradability (OECD, 1981) do not perform well, however (Cain, 1990; APAVE, 1992).

The standard OECD protocols require either a knowledge of the chemical structure to calculate theoretical values of oxygen uptake, or CO_2 evolution, or a determination of experimental values for these parameters. Information on the purity or the relative proportions of major components of the test material is required to interpret the results obtained. The fundamental characteristics of the ready biodegradation test methods are shown in Table 10.7.

Of the five test methods currently recommended by OECD for assessing ready biodegradation, the Sturm test is the one that has

Table 10.7 Biodegradation test variables (OECD, 1981; CEC, 1982)

Method	Test duration (days)	Measures	Concentration of test substance (mg/l)	Concentration of bacteria (cells/ml)	Pass level (percentage degradation)
1 Closed bottle	28	O_2	2–10	250	60
2 OECD Screening	28	DOC	5–40	50–250	70
3 AFNOR	28	DOC	40	500 000	70
4 Sturm	28	CO_2	10 and 20	10 000–200 000	60
5 MITI	28	O_2	100	200 000–1 000 000	60
6 CEC	21	Hydrocarbon[a]	2500	1 000 000	?(67)

[a] Hydrocarbon is measured using infrared spectroscopy, monitoring IR absorption by the C–H bond. For detail see Section 10.7.6.

gained the most widespread acceptance for examining the biodegradability of oil products. A modified version of the MITI test has also been successfully applied. In addition, the Co-ordinating European Council for the Development of Performance Tests for Lubricants and Engine Fuels (CEC) has published a test method: 'Biodegradability of Two-Stroke Cycle Outboard Engine Oils in Water' (CEC, 1982) which has been used in Europe by both industry and contract test houses for all types of oil products and poorly soluble hydrocarbons (Cain, 1990; Betton, 1992).

A brief outline and assessment of the suitability of these test methods for oils is given below; for a more detailed description of the various methods see Volume 1, Chapter 18.

10.7.1 Closed bottle test

This is the most stringent of the test methods. The theoretical amount of oxygen required to completely oxidize the test material (ThOD) is calculated from its empirical formula. If this is not possible — as with a mixture such as an oil — then a close approximation can be made from elemental analysis. In cases where ThOD cannot be determined, it is recommended that the chemical oxygen demand (COD) is measured by reacting the test material with acid dichromate and carrying out a titration. Many of the hydrocarbons present in oils are resistant to such treatment, however, and reliance on COD can give false results, indicating a higher degree of degradation than has occurred in practice.

10.7.2 Modified OECD test

This method measures dissolved organic carbon as an indication of degradation. The test is applicable only to materials that are soluble in water, have a negligible vapour pressure and do not adsorb to glass surfaces. Oils and oil products do not meet most of these criteria; hence the OECD test is not suitable for use with oils.

10.7.3 AFNOR test

As for the OECD test, this method is not applicable to oils.

10.7.4 Modified Sturm test

This method, developed for use with surfactants, has been used with some oil products. However, it is not suitable for use with volatiles. In addition, presentation of the test material can significantly affect results (CONCAWE, 1992a).

Methods that have been used to present insoluble non-volatile materials in this test include the use of non-degradable emulsifying surfactants, absorbing the material onto glass filter paper, vigorous stirring or shaking. Whilst there has not been any significant difference found between physical and chemical dispersion techniques, the application of the material on a carrier such as glass filter paper (John Boffer, personal communication, 1990) gives significantly lower results than other methods (CONCAWE, 1993).

10.7.5 MITI test

The MITI test requires that an analytical determination of the contents of the test solution is available. As mentioned earlier, this is difficult although not impossible for oils. One of the characteristics of this test is that it uses a mixed inoculum from 10 different sources. This has the advantage of increasing the likelihood of obtaining bacteria that are adapted to the test material (a serious deficiency of the other methods — discussed below). This advantage is subsequently lost, however, by subjecting the mixed inoculum to a prolonged (3-month) acclimatization to synthetic sewage containing glucose as an energy source, thereby diminishing the diversity originally inherent in the mixed inoculum.

10.7.6 CEC test

The CEC have devised a method (L-33-T-82) (CEC, 1982) for testing biodegradable two-stroke outboard engine oils, which has also been used for other oils and oil-based products. The analysis is based on measurement of the disappearance of test material as measured by infrared (IR) analysis of a solvent extract of the test preparation. The CEC test uses IR absorption at $2930\,cm^{-1}$. This wavelength is specific for methylene linkages $(-CH_2-)$ and might need to be modified when dealing with aromatic samples.

The protocol for this method specifies the measurement of biodegradation over a 21-day period, unlike the 28-day period required in the internationally recognized OECD tests. Unlike the other tests described above, the CEC test is a test of 'primary' degradation only in that it measures the disappearance of solvent-extractable hydrocarbons from the aqueous phase. A change in structure of a chemical that changes its water-solubility will be measured as biodegradation, although it may not represent 'ultimate' degradation (mineralization). This can be misleading as, for example, in the case of an ester whose hydrolysis results in two water-soluble fragments. Although these fragments may persist, the test would show loss of solvent-extractable material, which would be interpreted as evidence of complete degradation.

10.7.7 Discussion

A recent review of biodegradation testing carried out by OECD (Painter, 1992) considers in great detail the various test methodologies and how they may be applied, particularly with respect to environmental assessments.

Difficult-to-test materials such as poorly soluble and volatile single chemicals are considered in the review. Appropriate measures can be taken to adapt methodology to deal with such problems. Analytical support, and the use of radio-labelled materials, can be used to monitor breakdown and identify the products of degradation. Not considered, however, are the problems posed by insoluble mixtures of poorly soluble materials such as oils.

The deficiencies of current methods have been recognized by the oil industry, however, and a CONCAWE Task Force is in the process of developing biodegradation test methodology for oil products that more closely reflect the environmental situation (Cain, 1990; Pritchard & Costa, 1991; Verschueren, 1983; CONCAWE, 1993a). Aspects of current methods that are not appropriate for oils and volatile materials are:

Inoculum

The most commonly specified source of bacterial inoculum is sewage effluent from a works treating primarily domestic sewage (OECD, 1981). This requirement is a carry-over from the origins of the tests which were designed initially to assess the degradability of surfactants (Painter, 1992). Surfactants are designed for disposal via treatment works. Such a pattern of environmental exposure is not relevant for oils. Here, exposures are more often directly into the environment via diverse sources and over a prolonged period. In those instances where wastewater from an industrial plant containing hydrocarbons is passed through a treatment works, an adapted bacterial population develops. Use of effluent from such a plant gives significantly 'better' results in a standard biodegradation test when compared with effluent from a works treating domestic sewage only (Dr J. Kung, personal communication).

It has been found, however, that pre-adaptation of any inoculum for a period of 3 weeks prior to starting a test removes not only variability due to source, but also temporal variations in efficiency of the inoculum within the same source (CONCAWE, 1993). The use of a pre-adapted inoculum can be said therefore to:

1 significantly reduce variability; and

2 more accurately reflect the real-life situation for oil products in the same way that current tests reflect the situation for surfactants.

Test concentration

There are two important questions that have to be considered when choosing test concentrations. First, should the concentration be related to the whole product, or the WSF only? Secondly, does the concentration represent the environmental situation and is it able to give a readily measurable result?

Having argued above that the inoculum should be adapted as this more closely represents the real-life situation, the same argument must also call for the testing of whole product, rather than only the water-soluble fraction. It is the whole product that enters the environment and therefore it is this that should be tested. The actual concentration (or loading rate) used should ideally be low to represent environmental conditions and to minimize the problems of toxicity. However, this gives rise to difficulties with regard to measurement, particularly if respirometry is to be used.

Test concentration is therefore a compromise between what is practical and what is desirable. The CONCAWE biodegradation Task Force has suggested that 5–10 ppm is a concentration that could be used successfully in biodegradation tests with oils.

Measured parameters

The parameters commonly used in standard biodegradation tests are dissolved organic carbon (DOC), oxygen uptake (O_2), or carbon dioxide evolution (CO_2).

Removal of DOC, whilst giving an accurate picture of events with a water-soluble compound, is not appropriate for oils and hydrocarbons of poor solubility. One reason for this is that, as materials in the water phase are degraded, other compounds from any undissolved reservoir in the oil phase can diffuse into the water to take their place, thereby masking any degradation that may be occurring. In addition, when dealing with complex mixtures of hydrocarbons, metabolism of one chemical species to another will occur, and it is not possible to determine whether any particular hydrocarbon is original material or an intermediate metabolite.

Ultimate degradation of a hydrocarbon or oil containing N, O, S and P compounds results in the formation of CO_2, water and nitrate, sulphate or phosphate. The amount of oxygen required to carry out this process can be calculated from the elemental analysis of the material, as mentioned previously. Conversely, the amount of CO_2 that should be evolved can be determined from the carbon analysis, and the need to ensure closed systems makes CO_2 evolution the most suitable parameter to measure in assessing the degradation of oils. All CO_2 evolved in a test system must have come from the metabolism of the test compound, and this gives a direct means of assessing degradation.

Biomass generation occurs with many hydrocarbons but is a factor not considered in standard tests. A significant percentage of test material can be converted to biomass instead of being degraded to CO_2. A system where significant bacterial growth occurs could result in apparently low degradation being measured, although a large proportion of material had been removed from the system. This is another reason why the test concentration should be kept as low as possible within the constraints of practicability (CONCAWE, 1993).

10.7.8 Conclusion

The majority of hydrocarbons and oil products are ultimately of biological origin and do not persist in the environment. Current biodegradation methodologies do not necessarily reflect the environmental realities of degradability, due primarily to their original purpose being to predict the effects of single compounds of ready water-solubility. Methods designed to cope with the specific properties of oils and hydrocarbons are needed to enable meaningful predictions of environmental impact to be made. Such methods are currently under development.

10.8 GENERAL CONCLUSIONS

This chapter has lacked detail in the sense that there is little in the way of specific information on the acute or chronic toxicity of a particular hydrocarbon to a particular named species. This is deliberate. Oils and hydrocarbons have poor water-solubility. Water-soluble components tend to be those exhibiting the greatest volatility, while the most toxic of the component parts are increasingly poorly water-soluble. Current test methodologies were not designed for such materials. Their use has been characterized by modifications of various kinds that render comparisons of data generated from one test to another of limited value. It is for this reason that ranges of test data, often varying by orders of magnitude, have been quoted. It is undoubtedly true that different results can be obtained using different test methods. For example, closed systems ensure that the test material stays in solution, while open systems allow the toxic components to escape. Consequently, the test concentrations are not maintained in the latter case and the test is said to be invalid. But is it? It can be argued that the second system mirrors more closely the environmental situation, and as has already been noted, environmental spillages of materials either do not persist or, when they do persist, it is the less bioavailable and hence less toxic, but also ultimately degradable materials that do so.

What we must ultimately consider is why we are doing ecotoxicological studies in the first place if not to predict environmental consequences? This is surely also the point of doing tests for regulatory purposes. Of course, in enforcing regulations there is a need for standardization of tests so that regulators may compare data and establish relative toxicity values with some degree of confidence. It is up to the experienced ecotoxicologist to assist in this, by providing as clear and concise a picture as possible without seeking to mislead. However, it is often too easy to lose sight of the purpose of test data. Determination of 'intrinsic toxicity' can become an end in itself, forming a part of a data package, including information on biodegradability, volatility, bioaccumulation and other 'base set' information. These data can be fed into assessment systems to predict environmental impact, completely losing sight of the fact that it is possible to carry out a similar study that mirrors actual environmental conditions and provides data that are more closely related to real life, requiring less interpretation. This seems particularly true for oils and hydrocarbons.

REFERENCES

AOL (1967) The weathering of *Torrey Canyon* crude oil. AOL Technical Note No. 32.

APAVE (1992) NF-Environment. Huiles et Lubrificants Moteurs. Projet de Reglement Technique. Rapporteur Michel Genesco. May 1992.

API (1985) Literature survey: hydrocarbon solubilities and attenuation mechanisms. API Publication No. 4414. API, Washington, DC.

Baker J.M. (Ed.) (1976) *Marine Ecology and Oil Pollution*. Applied Science Publishers, London.

Barrow M.B. (1990) Bioconcentration: will water-borne organic chemicals accumulate in aquatic animals? *Environ. Sci. Technol.* **24**, 1612–1618.

Bennet, D., Girling, A.E. & Bounds, A. (1990) Ecotoxicology of oil products: preparation and characterisation of aqueous test media. *Chemosphere* **21**, 659–669.

Benyon, L.R. & Cowell, E.B. (Eds) (1974) *Ecological aspects of Toxicity Testing of Oils and Dispersants*. Applied Science Publishers, Essex.

Berry, W.O. & Stein, P.J. (1977) Quantification of water-soluble gasoline fractions using a radio-isotopic standard with direct aqueous injection gas chromatography. *Bull. Environ. Contam. Toxicol.* **18**, 308–316.

Betton, C.I. (1992) Environmental effects of lubricants. In: *Lubricant Technology and Chemistry* (Eds R.M. Mortier & S.T. Orszulik), pp. 282–298. Blackwell Scientific Publications, Oxford.

Cain R.B. (1990) Biodegradation of lubricants. In: *Proceedings of 8th International Biodegradation and Biodegradation Symposium*, Windsor, Ontario, Canada, 25–31 August. (Ed. H.W. Rossmoore), pp. 249–275.

CEC (1982) Tentative test method: biodegradability of 2-stroke cycle outboard engine oils in water. Report CEC L-33-T-82. Coordinating European Council for the Development of Performance Tests for Lubricants and Engine Fuels, London.

Clarke, R.B. (Ed.) (1982) *The Long Term Effects of Oil Pollution on Marine Populations, Communities and*

Ecosystems. Royal Society, London.

Coleman, W.E., Munch, J.W., Streicher, R.P., Ringhand, H.P. & Koffer, F.C. (1984) The identification and measurement of components in gasoline, kerosene and no. 2 fuel oil that partition into the aqueous phase after mixing. *Arch. Environ. Contam. Toxicol.* **13**, 171–178.

CONCAWE (1988) Ecotoxicology of Petroleum Products. A Review of Published Literature. Report No. 88/60. CONCAWE, Brussels.

CONCAWE (1992a) Ecotoxicology Testing of Petroleum Products. A Tier Testing Approach. Report No. 91/56. CONCAWE, Brussels.

CONCAWE (1992b) Product Dossier No. 92/103. CONCAWE, Brussels.

CONCAWE (1992c) Product Dossier No. 92/102. CONCAWE, Brussels.

CONCAWE (1992d) Product Dossier No. 92/104. CONCAWE, Brussels.

CONCAWE (1992e) Product Dossier No. 92/101. CONCAWE, Brussels.

CONCAWE (1992f) Ecotoxicology Testing of Petroleum Products. Test Methodology. Report No. 92/56. CONCAWE, Brussels.

CONCAWE (1993) Biodegradation Task Force on Method Development. Committee Research Programme. CONCAWE, Brussels.

CONCAWE (In press) Product Dossier: Kerosenes/Jet fuels. CONCAWE, Brussels.

CONCAWE (In press) Product Dossier: Gas oils (diesel fuels/heating oils). CONCAWE, Brussels.

CONCAWE (In press) Product Dossier: Heavy fuel oils. CONCAWE, Brussels.

CONCAWE (In press) Product Dossier: Lubricating oil basestocks. CONCAWE, Brussels.

CONCAWE (In press) Product Dossier: Waxes and related products. CONCAWE, Brussels.

CONCAWE (In press) Product Dossier: Petroleum coke. CONCAWE, Brussels.

CONCAWE (In press) Product Dossier: Crude oil. CONCAWE, Brussels.

Craddock, D.R. (1977) Acute toxic effects of petroleum on Arctic and sub-Arctic marine organisms. In: *Effects of Petroleum on Arctic and Sub-arctic Marine Environments and Organisms.* (Ed. D.C. Malins), pp. 1–93. Academic Press. New York.

Dodd, E.N. (1974) Oils and Dispersants: chemical considerations. In: *Ecological Aspects of Toxicity Testing of Oils and Dispersants* (Eds L.R. Benyon & E.B. Cowell), pp. 3–9. Applied Science Publishers, Essex.

EEC (1984) Council Directive 79/831/EEC Annex I, Part C: Methods for the determination of ecotoxicity. Official Journal of the European Communities No. L251/3.

EEC (1992). Council Directive 92/32/EEC amending for the seventh time. Directive 67/548/EEC on the approximation of the laws, regulations and administrative provisions relating to the classification, packaging and labelling of dangerous substances, 7th Amendment. Official Journal of the European Communities No. L154/35.

Engelhardt, F.R. (Ed.) (1985) *Petroleum Effects in the Arctic Environment.* Elsevier Applied Science, London.

Girling, A.E., Markarian, R.K. & Bennett, D. (1992) Aquatic toxicity testing of oil products—some recommendations. *Chemosphere* **24**, 1469–1472.

Gough, M.A. & Rowland, S.J. (1990) Characterisation of unresolved complex mixtures of hydrocarbons in petroleum. *Nature* **344**, 648–650.

Green, J. & Trett, M.W. (Eds) (1989) *The Fate and Effects of Oil in Fresh Water.* Elsevier Applied Science, London.

Hansch, C. & Leo, A.J. (1979). *Substitute Constants for Correlation Analysis in Chemistry and Biology.* Wiley, New York.

Institute of Petroleum UK (1992) *Petroleum Industry Statistics: Consumption and Refinery Production 1990 and 1991.*

Lockhart, W.L. *et al.* (1987) Acute toxicity bioassay with petroleum products: influence of exposure conditions. In: *Oil in Fresh Water: Chemistry, Biology, Counter Measure Technology*, pp. 335–344.

Ministry of Agriculture, Fisheries and Food (1984) *Guidelines for the Toxicity Testing of Substances to be Submitted to the Department of Energy's Notification Scheme for the Selection of Chemicals for Use Offshore.* AEP2 MAFF, Burnham-on-Crouch, Essex.

Ministry of Defence (Navy) (1971) *Effects of Natural Factors on the Fate of Oil at Sea.* Section VII.

Moore, S.F. & Dwyer, D. (1974) Effects of oil on marine organisms: a critical assessment of published data. *Water Res.* **8**, 819–827.

Mortier, R.M. & Orszulik, S.T. (Eds) (1992) *Chemistry and Technology of Lubricants.* Blackie, Edinburgh.

NRC (1985) *Oil in the Sea: Inputs, Fate and Effects.* Natural Academy Press, Washington, DC.

OECD (1981) *Guidelines for Testing of Chemicals. Section 2: Effects on Biotic Systems.* OECD, Paris.

Painter, H. (1992) *Detailed Review of Literature on Biodegradability Testing.* OECD, Paris.

Pritchard, P.H. & Costa, C.F. (1991) EPAs, Alaska Oil Spill Biovenediation Project. *ES and I* **25**, 372–379.

Roubal, G.E., Horowitz, A. & Atlas, R.M. (1979). Disappearance of hydrocarbons following a major gasoline spill in the Ohio River. *Dev. Ind. Microbiol.* **20**, 503–507.

Shiu, W.Y., Maijanen, A., Ng, A.L.Y. & Mackay, D. (1988) Preparation of aqueous solutions of sparingly soluble organic substances. II: Multicomponent systems—hydrocarbon mixtures and petroleum products. *Environ. Toxicol. Chem.* **7**, 125–137.

Somerville, H.J., Bennett, D., Davenport, J.N., Holt, M.S., Lynes, A., Mahieu, A., McCourt, B., Parker, J.G., Stephenson, R.R., Watkinson, R.J. & Wilkinson, T.G. (1987) Environmental effects of water produced from North Sea Oil operations. *Mar. Pollut. Bull.* **18**, 549–558.

US EPA (1985) Toxic Substances Control Act test guidelines: Final results. *Fed. Reg.* **50**, 39333–39342.

Veith, D.T., Macek, K.J., Tetrocelli, S.R. & Carroll, J. (1980) An evaluation of using partition co-efficients and water solubility to estimate bioconcentration factors for organic chemicals in fish. In: *Aquatic Toxicology* (Ed. J.G. Eaton) ASTM STP 707. ASTM, Philadelphia, PA.

Verschueren, K. (Ed.) (1983) *Handbook of Environmental Data on Organic Chemicals*. Van Nostrand Reinhold, New York.

11: Organic Pollution: Biochemical Oxygen Demand and Ammonia

N. ADAMS AND D. BEALING

11.1 INTRODUCTION

This chapter reviews the main effects of general organic pollution of freshwater organisms. It concentrates on the effects of reduced dissolved oxygen and elevated ammonia concentrations; other components of organic pollution are covered elsewhere in the handbook (Chapters 5–10).

Micro-organisms decompose organic compounds when they enter freshwater, and it is this decomposition that is responsible for the decrease in oxygen content. The amount of oxygen that micro-organisms require to degrade compounds 'ultimately' to carbon dioxide and water is assessed by determining the biochemical oxygen demand using methods similar to those for measuring biodegradibility (see Volume 1, Chapter 18). Nitrifying bacteria also consume oxygen when they oxidize the ammonia to nitrite and nitrate.

Sewage treatment works' effluents, storm sewage discharges and agriculture, mainly from discharges of slurries and silage liquor, are the major sources of general organic and ammonia pollution in rivers. Although most work on the impact of these has examined the damage caused by continuous exposure, many discharges are intermittent. Therefore there is a need to establish a relationship between the length and intensity of pollution episodes and their environmental impact. However, these are difficult to establish because many variables are involved and some, such as ammonia content, can have both a direct toxic effect and an indirect action by lowering the dissolved oxygen level.

11.2 AMMONIA

11.2.1 Introduction

There have been many reviews of the toxicity of ammonia to aquatic life, mainly fish, reflecting its importance as a pollutant and the difficulty in setting unambiguous water quality standards for a substance where the toxicity is greatly affected by prevailing conditions. Ammonia can exist in water in two forms: as ionized ammonium ion (NH_4^+) or as un-ionized ammonia (NH_3). The relative concentrations present depend on pH and temperature. For example, raising the pH from 7.0 to 7.3, or the temperature by 10°C, doubles the concentrations of un-ionized ammonia. Ammonia is generally more toxic to fish than to other aquatic organisms. A number of studies, such as those of Wuhrmann & Woker (1948), have demonstrated that the un-ionized form has the direct toxic effect.

Section 11.2.2 reviews recent toxicity data, concentrating mainly on information that has become available since 1988. Information published before 1988 has been reviewed by Alabaster & Lloyd (1982), US EPA (1985) and Seager *et al.* (1988). Section 11.2.3 considers the mode of toxic action of ammonia and Section 11.2.4 examines models that assist in standard setting. Section 11.2.5 reviews the standards that pollution control authorities around the world have set and Section 11.2.6 outlines the methods available for analysing ammonia in water.

11.2.2 The direct toxicity of ammonia

Seager *et al.* (1988) reported LC$_{50}$s for salmonids to be within the range 0.068–0.91 mg/l NH$_3$-N, and since that review there have been few new reports. This section considers those that are available, and earlier key papers relevant to the setting of quality standards. Alabaster & Lloyd (1982) suggested that the lowest lethal concentration of un-ionized ammonia found for salmonid fish was 0.2 mg/l as NH$_3$. Some workers, such as Thurston *et al.* (1984), have supported this, although others have reported lethal effects at < 0.2 mg/l NH$_3$. Calamari *et al.* (1981) established an LC$_{50}$ of 0.056 mg/l NH$_3$-N, for trout early life stages (1-day-old eggs to 72-day-old alevins); newly hatched larvae were most sensitive. This led them to propose a tentative water quality standard of 0.02 mg/l NH$_3$. More recently, Solbé & Shurben (1989) showed significantly higher mortality of unhatched eggs exposed from 24 h postfertilization to 73 days at 0.022 mg/l NH$_3$-N. When exposed from the eyed egg stage, however, mortality was considerably less, and occurred mainly among the early fry with 40% mortality at 0.23 mg/l NH$_3$-N. The study of Solbé & Shurben (1989) showed that severe effects on rainbow trout eggs can be demonstrated in waters containing concentrations of un-ionized ammonia at least as low as 0.027 mg/l NH$_3$ (0.022 mg/l NH$_3$-N). This result caused concern, since it came closer to the EIFAC standard for European fish of 0.025 mg/l NH$_3$-N. This being a 95-percentile, however, some margin of safety still existed. However, to relate laboratory toxicity data directly to effects of ammonia on fisheries in the natural environment requires more information on the toxicity of ammonia to early life stages of fish at low dissolved oxygen, low temperature (especially at spawning time) and high free CO$_2$. It is possible that the increased stress that these conditions could cause may increase ammonia toxicity. Unfortunately, these data are very limited at present.

There is some discussion as to whether cyprinids are more resistant than salmonids to un-ionized ammonia. EIFAC (Alabaster & Lloyd, 1982) concluded that prolonged exposure to ammonia is as toxic to certain cyprinids as to salmonids, though Seager *et al.* (1988) felt that there is evidence of cyprinids being more resistant than salmonids to ammonia over short periods. There is a need for further work on chronic exposure to early life stages of cyprinids to demonstrate the similarity in chronic sensitivity.

Since Seager *et al.* (1988) there has been very little published work on the toxicity of un-ionized ammonia to cyprinids. Hermenutz *et al.* (1987) reviewed and cited a number of studies, reporting sublethal effects on fathead minnows (*Pimephales promelas*) (range 0.091–0.297 mg/l NH$_3$), channel catfish (*Ictalurus punctatus*) (range 0.011–0.153 mg/l NH$_3$), walleye (*Stizostedion vitreum*) (range 0.099–0.268 mg/l NH$_3$) and white sucker (*Catostomus commersoni*) (30-day study showing swim-up delay and length reduction, 0.058–0.068 mg/l NH$_3$). A 30-day early life stage test on channel catfish indicated significant reduction in weight at 0.20 mg/l NH$_3$ and 96-h LC$_{50}$s for white sucker and walleye were in the range 0.36–2.22 mg/l NH$_3$. These data are similar to the range of LC$_{50}$s for salmonids reported by Seager *et al.* (1988) of 0.068–0.91 mg/l NH$_3$-N.

Various studies have shown that ammonia has a range of biochemical and histopathological effects at concentrations lower than those affecting early life stage survival and growth (Herbert & Shurben, 1964; Flis, 1968; Lloyd & Orr, 1969; Alabaster *et al.*, 1979). For example, Thurston *et al.* (1984) exposed rainbow trout (*Oncorhynchus mykiss*) for 5 years, and found that at 0.07 mg/l NH$_3$-N the only observed effect was the common appearance of histopathological lesions in parental and F$_1$ fish. Concentrations up to and including 0.07 mg/l NH$_3$-N did not affect mortality or growth of the F$_1$ generation from egg to 10 months old. These measures of stress have not been used in the development of standards for ammonia; they may not take into account the ability of the stressed organisms to survive in the wild. However, the stress could make individuals easier prey, or predispose them to disease. Smart (1976) found a high incidence of disease among rainbow trout exposed over an extended period to ammonia, and suggested it was related to stress caused by the ammonia exposure. The possibility of increased susceptibility to disease is a complex area that merits further study.

11.2.3 The toxic mode of action of ammonia

There are a number of theories about the mode of toxic action of ammonia to fish. Tomasso *et al.* (1980) reviewed effects reported in the literature, and suggested that the cause of death could be due to one or more of the following:

1 gill damage eventually causing asphyxiation;
2 biochemical mechanisms involving stimulation of glycolysis and suppression of Krebs cycle causing acidosis and eventual asphyxiation;
3 as (2) but causing death due to depletion of ATP in the basilar region of the brain (this has been observed in mammals with high internal ammonia concentrations brought about by hepatic encephalopathy);
4 osmoregulatory disturbance upsetting the kidneys;
5 severe electrolyte imbalance (NH_4^+ reduces internal Na^+ to possibly fatally low levels);
6 inhibition of ATP production by uncoupling oxidative phosphorylation as a direct effect.

Several of these mechanisms could act on a fish simultaneously, and the relative contributions of each to the total stress could depend on the concentration and form (ionized or un-ionized) of the external ammonia. Mechanisms of toxicity at low concentrations over extended periods are extremely important when trying to set water quality standards to maintain the biological quality of a water body in the long term. At the same time it must be remembered that effects on critical stages of the life cycle by transient peaks in pollutant concentration could cause long-term damage to populations.

11.2.4 Modelling the toxic action of ammonia to set quality standards

The toxicity of un-ionized ammonia decreases as the pH of the water increases up to about pH 7.7 (Thurston *et al.*, 1981; McCormick *et al.*, 1984; Mallett, 1990). Data available up to 1980, including the findings of Lloyd & Herbert (1960), led Alabaster & Lloyd (1982) to suggest that this

could be explained by the release of carbon dioxide by respiratory processes at the gill surface causing a localized reduction in pH. Moreover, they suggested that, as the pH of water at the gill surface is unlikely to exceed 8.0, then this level could be used in calculating the amount of un-ionized ammonia present in waters where pH > 8.0. This localized reduction in pH would affect the concentration of un-ionized ammonia and reduce its dependence on ambient water pH. Erickson (1985) reviewed more recent data and used them to derive models to describe the relationship between pH and temperature on ammonia toxicity. He found that the gill pH model did not describe the relationship very well, and that a joint toxicity model provided the best fit. However, this model assumes that un-ionized ammonia and the ammonium ion are jointly toxic, but there is not much evidence either for or against this. Clearly, more experimental work is required.

Lewis (1988) has built on Erickson's work to develop a toxicity-curve modelling approach to set ammonia standards for the US EPA, based on a comprehensive review of the toxicity of ammonia to aquatic life. He argues that this approach is more realistic and better justified than simpler criteria if both environmental protection and cost-effectiveness of wastewater treatment are to be maximized. A model was developed that could predict the LC_{50} to aquatic organisms, given the pH and temperature of the water. The shape of the toxicity curve has four regions: one of near linearity at pH 5–7; one of pronounced curvature at pH 7–8; a plateau over which the LC_{50} remains virtually constant at pH 8–8.5; and a downward inflection suggesting increased toxicity with a further rise in pH at pH 8.5–9.0. A negative relationship between toxicity and temperature was also incorporated.

For a water quality standard to be effective, it is important that all stages of the life cycle of the most sensitive species on which population and community health depend are protected (Chapter 16). Uncertainties in the approach used by the EPA to set ammonia standards in waters were recognized by both Erickson (1985) and Lewis (1988). Their main concern related to the uncertainties with the data on which the models were based, the variations between species and the lack of information on the combined temperature and pH effects. The final curve-fitting, from which the pH model was developed, was based on data from just four species: rainbow trout, fathead minnow, coho salmon (*Oncorhynchus kisutch*) and *Daphnia*.

Fish seem to be the most sensitive group. Where data on sensitive stages in the life cycle are lacking, regulators normally apply 'safety factors' (Chapter 16). Where data on early life stages (ELS), or other long-term effects exist, the safety factors required can be more precisely defined. The US EPA has placed emphasis on defining the LC_{50} under differing conditions, and relies on safety margins to protect sensitive stages. A good deal of uncertainty still exists regarding the latter. McKim (1985) demonstrated the potential of ELS tests to indicate safety margins, or maximum acceptable toxicant concentrations (MATCs). He compared full life-cycle studies with short-term ELS tests and concluded that the MATC could often be quite accurately predicted by exposing the early developmental stages only.

Clearly, a modelling approach to setting environmental quality standards for ammonia must take into account the sensitivity of different stages in the life history of the affected organisms, and not just rely on arbitrary safety factors applied to measures of acute lethality. Ideally, a model should also consider the relationships between the physicochemical properties of the water and ammonia toxicity that have already been discussed. The model should also be able to take into account fluctuating ammonia concentrations in rivers, resulting from episodic discharges from storm sewage overflows, farmyards and the like. Lumbers & Wishart (1989) integrated the effects of temporal variation in ammonia concentration with a compartmental modelling approach. This simulates the uptake and distribution of ammonia in fish. Models of this sort will enable regulators to set more accurate quality standards that relate directly to the conditions prevailing in a water course throughout the year, but they require further experimental validation to make sure that they are calibrated properly.

11.2.5 Existing and proposed quality standards for ammonia

Alabaster & Lloyd (1982) reported a European Inland Fisheries Advisory Commission (EIFAC) working group recommended standard of 0.021 mg/l NH_3-N as a 95-percentile value for protecting salmonid fisheries (not applying to temperatures below 5°C or pH greater than 8). This is also the mandatory limit specified by the EC Directive on the quality of freshwaters needing protection or improvement in order to support fish life (78/659/EEC); a guide value of 0.004 mg/l NH_3-N is given. The Directive also quotes mandatory and guide values as total ammonia of 0.78 and 0.031 mg N/l. Seager *et al.* (1988) proposed an annual average standard of 0.015 mg/l NH_3-N for waters in the EC not designated by this Directive.

Calamari *et al.* (1981) suggested a standard for rainbow trout of 0.061 mg/l NH_3-N, and an early EPA criterion for the protection of salmonids was 0.02 mg/l NH_3-N. Szumski *et al.* (1982) assessed this criterion and proposed another for protecting against effects to warmwater species, which considered pH, temperature and alkalinity. The criterion was defined in terms of total ammonia concentrations to maintain < 0.08 mg/l NH_3-N at the gill surface.

More recent EPA criteria (US EPA, 1985) incorporate pH and temperature, and have been based on a 'final acute value' (the fifth percentile of the distribution of 'genus mean acute values'). They are expressed as maximum permissible 1-h and 4-day average concentrations, with separate criteria for waters with and without salmonids and other sensitive cold-water species present. The EPA set national standards and describe a procedure for calculating site-specific modifications where appropriate. The 1-h criterion for salmonids at 20°C and pH 7.00 is 0.076 mg/l NH_3-N; the 4-day equivalent is 0.048 mg/l NH_3-N. These are comparable with the standards already quoted.

Arthur (1986) reports on a research programme to test the application of the EPA water quality criteria for ammonia and other substances in artificial stream ecosystems. In nearly all cases the site-specific criterion afforded adequate protection for macroinvertebrates (Zischke & Arthur,

1987), microinvertebrates and fish (Hermenutz *et al.*, 1987). There were effects (reduced growth) conclusively observed at un-ionized ammonia concentrations below criteria concentrations in only two cases, for channel catfish and white sucker.

Solbé and co-workers developed water quality standards for short-term exposure to ammonia (Bowden & Solbé, 1987; Whitelaw & Solbé, 1989). Their approach was based on median lethal threshold concentration data and took the form of a time-related LC_{50} standard, for exposure periods up to 1000 min (16.7 h). Separate standards were proposed for salmonids, cyprinids and pike/perch.

Milne & Seager (1991) have extended this work to propose water quality standards for exposure events of 1, 6 and 24 h duration. They also considered the time elapsing before the event returned: 1-month, 3-month and 1-year returns. The standards are plotted as a series of curves (see Fig. 11.1). They derive from data on fish, with the assumption that this also affords adequate protection for other organisms. Experimental evidence indicates that, for ammonia, event concentration, duration and frequency are all important factors in determining toxic effect. Frequency appears to become a dominant factor when it is very high, such as greater than once a week. This suggests initially steep curves at return periods of a few weeks, flattening as they approach the

upper limit to avoid mortality. The authors allow for the influence of pH and temperature on un-ionized ammonia toxicity by including bandings. Standards aimed at conditions of around pH 7 and > 10°C have 2-fold reduction factors, to account for the approximately 2-fold increase in ammonia toxicity caused by a unit drop in pH or a 10°C drop in temperature. The standard is equal to the EIFAC/EC 95 percentile criterion. Table 11.1 describes the proposed reduction factors.

With a maximum frequency of once a year the 1-h standard is likely to cause stress to fish, but will avoid mortality and is thought unlikely to cause permanent damage. It is less than nearly all > 24-h LC_{50} values and well below threshold concentrations for mortality at either 1- or 6-h exposure. The 24-h standard should allow 100% survival of rainbow and brown trout. The 3-monthly-return frequency 1-h standard is likely

Table 11.1 pH and temperature correction factors for ammonia standards for conditions at time of exposure (see also Fig. 11.1)

	Temperature	
pH	< 5°C	> 5°C
> 7	1/2	1
6–7	1/2	1/2
< 6	1/4	1/4

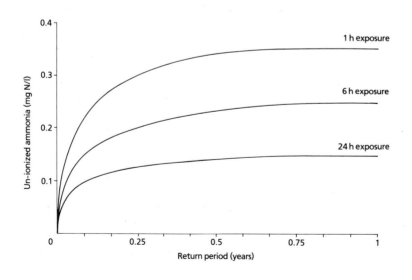

Fig. 11.1 Water quality standards to protect aquatic life from intermittent exposure to ammonia. For each exposure period (1, 6 or 24 h) the concentration of ammonia tolerated increases as the time between exposures increases. The continuous exposure concentration is the lowest, at 0.021 mg/l NH_3-N. There also needs to be correction for prevailing pH and temperature (see Table 11.1).

to cause stress, but long-term, harmful effects are unlikely to occur. The 24-h standard is below concentrations affecting growth in the laboratory and is unlikely to cause any severe stress. The standard for a monthly 1-hour event allows a margin to avoid harmful effects, but it does cause damage to gill tissues at very frequent exposure of longer duration. It is likely to cause stress, but not long-term effects. The 24-h standard at this frequency is less than half the concentration that allows 100% survival of rainbow trout in the laboratory, and should provide a good safety margin against long-term effects.

11.2.6 Ammonia analysis

There are several methods for the analysis of total ammonia, including potentiometric and colorimetric procedures using various chemical techniques and instruments (DoE/NWC, 1980). Flow injection analysis is the usual method: a sample is injected into a flowing stream of alkaline buffer, which converts all of the ammonia to free ammonia, which then diffuses across a gas-permeable polytetrafluoroethylene (PTFE) membrane into an indicator reagent stream to produce a change of absorption which is measured on a spectrophotometer.

The indophenol colorimetric method is a common alternative, especially in conjunction with continuous-flow automatic analysers. Ammonia reacts with sodium hypochlorite and alkaline phenate in the presence of sodium nitroprusside to form a blue complex thought to be related to indophenol blue, and this is measured spectrophotometrically.

Both of these methods can be used for a wide range of sample types including potable and marine waters.

11.2.7 Conclusions

Ammonia toxicity is complex, closely related to pH and temperature levels in water. Regulators must ensure that the models they develop are based on good experimentally derived data, and are tested in each appropriate situation.

There are two areas where data are particularly lacking. Firstly, there is little information on the toxicity of ammonia to invertebrates; most effort has concentrated on fish. Secondly, there is the lack of knowledge about long-term sublethal effects on sensitive stages in the life cycle of cyprinid fish. A further cause for concern is that most studies have considered the effects of continuous exposure to ammonia, whereas the greatest need is for standards that protect aquatic systems from intermittent exposure. This could result from intermittent discharges or significant changes in ammonia concentration that result from changes in the properties of the receiving waters.

11.3 DISSOLVED OXYGEN AND BIOCHEMICAL OXYGEN DEMAND (BOD)

11.3.1 Introduction

The oxygen balance of a water body is very important in maintaining the health and richness of populations and communities of organisms that live in it. This section considers the impact that different dissolved oxygen (DO) levels have on water bodies, and the basis on which various authorities have set quality standards (or criteria) for DO. We shall also consider methods of assessing or predicting the impact of pollutants on DO in receiving waters, with a discussion of the biochemical oxygen demand (BOD) test and possible alternatives. Section 11.3.2 considers the processes that remove oxygen from water bodies and how they relate to their biological quality. Section 11.3.3 reviews the current literature on the effects of DO on aquatic organisms. Section 11.3.4 discusses quality standards or criteria that have been derived from the data discussed in Section 11.3.3, and considers the different approaches that authorities have used to propose them. Finally, the BOD test and its alternatives are considered.

Oxygen is present in the aquatic environment as dissolved, molecular oxygen (DO). It is also part of the water molecule and present in dissolved anions such as carbonate, nitrate, sulphate and phosphate. Some micro-organisms, such as anaerobic bacteria, use these anions as their source of oxygen and some, the obligate anaerobes, may actually be poisoned by DO. However, most aquatic organisms are entirely dependent on molecular oxygen either dissolved in the water or obtained directly from the atmosphere.

The atmosphere is the main reservoir of oxygen in the environment. The principal subsidiary source is photosynthesis. Diffusion of oxygen across the air–water interface is the main input of oxygen to a river or lake. In an estuary the DO concentration in water varies with the state of the tide. Water from the open sea can be expected to be well oxygenated, but freshwater river inputs could be depleted by biological activity. Also, BOD will have a greater effect at low tide because of the smaller volume of water. The National Rivers Authority in the UK has given an example of the range of DO in a tidal stretch of the Yorkshire River Ouse: it ranged from 5–10% of air-saturation at low tide, to 65–75% at high tide (Stiff *et al.*, 1990).

The concentration of oxygen in water is usually described as mass per unit volume (e.g. mg/l) or percentage air saturation value (% ASV). Mass per unit volume is better because it is a measure of the actual amount of oxygen available to biota. The saturation concentration varies with temperature, salinity and atmospheric pressure. When calibrating instruments it is important to make sure that calibrant solutions are at the same temperature and salinity as the medium under test. Table 11.2 gives a selection of DO saturation concentrations at different temperatures and salinities (DoE, 1988). Henry's Law describes the relationship between dissolved oxygen in water and gas in the atmosphere (Perry & Green, 1984; Sukatsch & Dziengel, 1984). The value of Henry's

constant increases with temperature, and so oxygen solubility decreases. DO also decreases as the concentration of dissolved salts increases. Sodium ions in particular are solvated in solution; each ion being surrounded by a shell of water molecules. This is the equivalent of reducing the effective amount of water present, and accounts for the reduced solubility. In hard freshwaters calcium is the usual dominant cation, which is solvated to a lesser degree than sodium. Therefore the solubility/salt content relationship is different from that in seawater; so data in DO/salinity tables are not strictly applicable. In any case most freshwaters have a mineral content of $<0.5\,g/l$, so the difference in the solubility of oxygen between these waters and pure water is negligible.

11.3.2 The removal of DO from water

Oxidation reactions that remove oxygen from water are common, but they usually take place only slowly at ambient temperature. The main mode of removal of oxygen from natural waters is biological oxidation or biodegradation. Microorganisms play an important part in the cycling of organic matter in aquatic ecosystems. Under appropriate conditions they are able to break down almost all naturally occurring organic substances and many industrial chemicals, and convert them into new biomass, carbon dioxide, water and various inorganic substances. Since biodegradation processes are oxidative, complete biodegradation generally occurs only in the presence of oxygen. Under anaerobic conditions, although bound oxygen in the form of anions such as sulphate

Table 11.2 Effect of temperature and salinity on saturated dissolved oxygen concentration (mg/l)

Temperature (°C)	Salinity (grams NaCl per kilogram water)							
	0	5	10	15	20	25	30	35
0	14.6	14.2	13.7	13.3	12.9	12.4	12.0	11.5
5	12.8	12.4	12.0	11.7	11.3	10.9	10.5	10.2
10	11.3	11.0	10.7	10.3	10.0	9.7	9.4	9.0
15	10.1	9.8	9.5	9.3	9.0	8.7	8.4	8.2
20	9.1	8.9	8.6	8.4	8.1	7.9	7.6	7.4
25	8.3	8.0	7.8	7.6	7.4	7.2	7.0	6.8
30	7.6	7.4	7.2	7.0	6.8	6.6	6.4	6.2

may still be available, conversions are slower and often incomplete. Accumulation of the products of incomplete degradation, which could include non-degradable materials, may lead to a reduction in the aesthetic quality of a water or the presence of materials toxic to aquatic organisms. On the other hand, rapid biodegradation uses up dissolved oxygen, possibly at a rate faster than the natural re-aeration rate. This may be a problem especially during the summer months when micro-organisms are more active.

Oxygen uptake by micro-organisms varies with the substrate being utilized, and with the fraction of the substrate that is converted to biomass, to mineral products (largely carbon dioxide and water) or to other organic metabolic by-products. For example, the conversion of glucose ($C_6H_{12}O_6$, $C:O = 1:1$) to biomass with a carbon to oxygen ratio $1:2.4$ will require excess oxygen of about 1.9 g oxygen per gram of carbon. The conversion of glucose completely to carbon dioxide and water ($6CO_2 + 6H_2O$, $C:O = 1:3$) requires 2.7 g oxygen per gram of carbon.

Overall, aerobic biodegradation obeys first-order kinetics (APHA-AWWA-WPCF, 1985), and the rate of re-aeration by diffusion is directly proportional to the oxygen deficit (which is the difference between the saturation concentration and the instantaneous DO concentration). The biodegradation rate constant varies with the nature of the organic matter present, with values in the range 0.05–0.30 per day (DoE, 1988). The re-aeration rate constant is unique to any particular body of water, being a function of many factors including depth, the state of the surface (degree of ripple), velocity and turbulence. Many empirical formulae have been developed, such as the Whipple equation, which give a simple approximation of re-aeration in streams, based on stream velocity and stream depth (Pfafflin & Ziegler, 1976).

Excess amounts of nutrient such as phosphate and nitrate in a body of water stimulate the growth of plants at a rate which cannot be sustained. Overproduction of algae and plants can result in depletion of the DO and the obscuring of the surface of the water, thus limiting re-aeration and photosynthetic activity below the surface. Ultimately, when plant growth is no longer sus-

tainable, bacterial growth, feeding on the dying plants, becomes the dominant process, resulting in depletion of DO. These processes can be seasonal developments in bodies of water that are naturally eutrophic. There is always diurnal fluctuation in DO levels, that results from oxygen production by plants during daylight and uptake at night.

Pollution controllers need to assess or predict the amount of oxygen that a potential pollutant uses up in this way. The traditional method of measurement is the BOD. However, this has various drawbacks (Section 11.3.6).

11.3.3 Dissolved oxygen requirements of aquatic organisms

There have been several comprehensive reviews of the DO requirements of aquatic organisms, especially freshwater fish (Doudoroff & Shumway, 1970; Davis, 1975; Alabaster & Lloyd, 1982; Chapman, 1986; US EPA, 1986; Whitelaw & Solbé, 1989). Stiff *et al.* (1990) and Milne & Seager (1991) bring these reviews up to date and consider a wider range of organisms and more recent data.

Studies have focused on both the effects on aquatic life of continuous exposure to low DO, and the effects of intermittent episodes of low DO. We shall consider continuous exposure first, review the data on intermittent exposure and then, in the next section (11.3.4), consider the standards that have been proposed using this information.

Doudoroff & Shumway (1970) found no convincing evidence that adults of any fish species are killed by DO concentrations above 3 mg/l under otherwise normal conditions. However, it is clear that early life stages of many species are more sensitive than adults, and are killed by DO concentrations > 3 mg/l. To hatch Chum salmon (*Oncorhynchus keta*) eggs required 1 mg/l DO for early stages, and 7 mg/l prior to hatching (Alderdice *et al.*, 1958). Sowden & Power (1985) showed that rainbow trout eggs in redds where the mean intergravel DO concentration was less than 4.3 mg/l did not survive, and there was negligible survival below 5.2 mg/l. This led them to conclude that the lethal threshold for pre-

Table 11.3 Range of oxygen concentrations separating complete survival from complete mortality for three species of fish at three temperatures. After Downing & Merkens, 1957

Fish	Exposure time	Temperature (°C)					
		10		16		20	
Rainbow trout	3.5 h	1.7	(1.2)	1.9	(1.5)	2.1	(1.6)
	3.5 days	1.9	(1.3)	3.0	(2.4)	2.6	(2.3)
Perch	3.5 h	0.7	(0.4)	1.1	(0.6)	1.2	(0.9)
	3.5 days	1.0	(0.4)	1.3	(0.9)	1.2	(1.0)
Roach	3.5 h	0.4	(0.2)	0.6	(0.3)	1.1	(0.5)
	3.5 days	0.7	(0.2)	0.7	(0.7)	1.4	(1.0)

All concentrations are milligrams per litre. First concentration is minimum allowing 100% survival, that in parentheses is maximum causing 100% mortality.

emergent survival was about 4–5 mg/l. There are numerous reports of similar results (Coble, 1961; Turnpenny & Williams, 1980).

Much work has been done on the relationship between duration of exposure and survival. One study showed that pike (*Esox lucius*) larvae required a minimum of 2 mg/l during a 2-h exposure and 4 mg/l for an 8-h exposure (Peterka & Kent, 1976, quoted in Alabaster & Lloyd, 1982). Downing & Merkens (1957) performed a number of studies on the influence of temperature on the survival of adults of several species of fish during short- to medium-term exposure to low DO. Table 11.3 summarizes some of their results, and shows that rainbow trout were more sensitive than the non-salmonids. There was a remarkably small difference between oxygen concentrations that allow 100% survival and those that result in 100% mortality.

There are few data on the lowest DO concentration that fish can survive, for periods of less than about 10 min. Those that are available suggest that non-salmonids are more resistant than salmonids. Work in the 1950s (Shepard, 1955; Alabaster *et al.*, 1957; Downing & Merkens, 1957) suggested that 50–100% of rainbow trout and other salmonids die in less than 1 h at around 1.1 mg/l at 10–16.5°C.

As well as studying the lethal effects of low DO, a number of workers have investigated effects on sublethal measures of stress, such as growth or

reproduction rate. Doudoroff & Shumway (1970) concluded that low DO had little or no effect on salmonid growth down to 5–6 mg/l, and only moderately decreased growth at 3 mg/l. Alabaster & Lloyd (1982) stated that neither reduced DO to 5 mg/l nor wide diel fluctuations about this concentration had much effect on growth of salmonid alevins, and there may be only slight growth reduction at 3 mg/l. According to Itazawa (1971) the minimum DO concentration for maintaining maximum feeding, growth and food conversion efficiency is 4–4.5 mg/l for rainbow trout at 10.5°C and 3 mg/l for carp at 21.5°C. Large diel fluctuations of DO down to minima of 2–3 mg/l have been found to inhibit growth and appetite of brook trout (*Salvelinus fontinalis*) almost as much as continuous low DO conditions. The US EPA reviewed levels of production impairment quoted in the literature (US EPA, 1986; Table 11.7).

For salmonids, the 'no production impairment' concentration was 11 mg/l for embryos and larvae, and 8 mg/l for other life stages. However, for non-salmonids the concentration was 6.5 mg/l for early life stages and 6 mg/l for others. Salmonid eggs are buried in gravel, and the figure of 11 mg/l for salmon embryos assumes an intergravel DO concentration of 8 mg/l.

Whitworth (1968) found that diel DO fluctuations significantly reduced growth of brook trout (*Salvelinus fontinalis*). The experimental regime was 14 h at low DO, reduced from 11 mg/l to 5.3, 3.6, 3.5 or 2.0 mg/l. At 2.0 mg/l few survived, but at the other concentrations growth was severely reduced with no difference between the groups.

Davis (1975) catalogued a wide range of effects induced by low DO, and calculated a mean incipient threshold of 6 mg/l for salmonids and 4 mg/l for non-salmonids. The point below which the fish cannot maintain 100% oxygen saturation of the blood could be an important threshold for physiological stress. An estimate of the DO concentration required to keep 100% blood saturation in rainbow trout at 15°C is 6.5 mg/l. Alabaster & Lloyd (1982), however, did not consider that respiratory and cardiovascular responses to DO changes necessarily showed that ecologically important functions were impaired. They did not favour their use as a basis for judging DO requirements. However, the responses must involve increased energy expenditure, and would be at the expense of some other function, such as growth or reproduction. Under field conditions rainbow trout have increased ventilation when DO has fallen to *c.* 4.5−5 mg/l (Seager & Abrahams, 1988; Milne *et al.*, 1989) due to storm events. However, other factors in the storm event, such as changes in turbidity and temperature, also need to be considered. In the laboratory, with other conditions constant, brown trout showed an increase in ventilation rate of over 20% when DO was reduced to 4 mg/l (Milne & Seager 1990). Fish may continue to swim at near-lethal, low DO concentrations, but maximum sustainable speed and time for which the speed is sustained both decrease (Alabaster & Lloyd, 1982). This may have important implications for migrating salmonids.

Many fish species live at DO concentrations only slightly above lethal levels without showing strong avoidance (Doudoroff & Shumway, 1970), but they do avoid low DO if they can. Chinook (*Oncorhyncus tshawytscha*) and coho salmon strongly prefer water with 9 mg/l DO against 1.5 mg/l, moderately prefer it to 3 mg/l, commonly prefer it to 4.5 mg/l and sometimes prefer it to 6 mg/l (Whitmore *et al.*, 1960). Non-salmonids were less sensitive; large-mouth bass (*Micorpterus salmoides*) showed a slight tendency to avoid 3 and 4.6 mg/l and definitely avoided 1.5 mg/l,

whereas bluegills (*Lepomis macrochirus*) avoided 1.5 mg/l only. Alabaster & Robertson (1961) found that roach showed a burst of activity when DO concentration fell from 10 to 8 mg/l. The concentration at which fish moved away was very variable, but was as high as 7.6 mg/l (roach), 6.7 mg/l (perch) and 1.5 mg/l (bream). Bishai (1962) found that salmonid fry up to 26 weeks old avoided low DO up to 4.6 mg/l at 13.5°C; older stages were less sensitive, though they did avoid less than 3 mg/l.

Avoidance of water of low DO may be beneficial, in that it could prevent possible physiological stress; it can also be detrimental. A plug of low DO in a river may disrupt the up-river migration of salmonids to spawning grounds. Alabaster (1988) found that DO below 5.3 mg/l (19−24°C) stopped most chinook salmon running. Alabaster & Gough (1986) found similar results for salmon in the River Thames estuary in the UK, and estimated that successful passage occurred over short distances (1 km) where the median DO concentration was 3.5 mg/l. As the distance of passage increased so did the DO concentration required, to 4.3 mg/l for 30 km. Priede *et al.* (1988) reported inhibition of running in the River Ribble (Lancashire, UK) below 40% air saturation value (4 mg/l at 15°C) and avoidance of DO below 55% saturation (5.5 mg/l).

Acclimatization to low DO (Doudoroff & Shumway, 1970; Davis, 1975; Alabaster & Lloyd, 1982) may take up to 10 days and result in tolerance thresholds half the levels seen without acclimation. However, acclimation by gradual exposure is probably irrelevant in the case of many rivers affected by storm sewage discharges, where DO is likely to be reduced relatively rapidly. Frequent exposure to low DO episodes also helps fish to acclimatize. Milne & Seager (1991) report work indicating that physiological changes can occur with frequent exposure, but there was no evidence of acclimation in the form of resistance to frequent events. Brown trout exposed to 23-h low DO pulses of 4 or 5.5 mg/l once or twice a week showed changes in blood haemoglobin concentration and the weights of spleen, kidney and liver. DO concentration, rather than frequency, was found to be the important factor. At the end of the experiment, however, the experimental fish showed no increased resistance to acute exposure; in fact there was a trend for them to be more sensitive than controls. Milne _et al._ (1989) note that recovery is very rapid on return to normoxia, and post-event survival is very high. For DO, therefore, post-exposure mortality may not be important in the development of water quality standards.

It is important also to consider invertebrate communities. Maltby & Naylor (1989) showed that _Gammarus pulex_ survives pulses of 24 h of 2 mg/l DO, and the 1 mg/l LC_{50} was > 9.5 h. As with fish, low DO pulses can have sublethal effects, but there is rapid post-exposure recovery. Twenty-four-hour pulses of 3.75 and 2.5 mg/l significantly reduced energy consumption on the day of exposure, but when considered over a 6-day period the effect was no longer apparent (Maltby & Naylor, 1989). Many of the insect species common to salmonid habitats must be more sensitive to low DO than the fish themselves (US EPA, 1986). However, the studies reported must be treated with some caution because the exposure conditions were not ideal.

The National Rivers Authority and River Purification Boards in the UK have recorded salmonids at 95-percentile concentrations of $7-8$ mg/l and coarse fish at 5.3 mg/l (Stiff _et al._, 1990). The biological quality was excellent at 95-percentile concentrations of 9 mg/l and very good at as low as 3.7 mg/l. Various national and international bodies have assessed the available data and proposed quality standards or criteria for DO. The next section reviews these.

11.3.4 Environmental quality standards for dissolved oxygen

Davis (1975) studied documented sublethal effects and proposed the standards shown in Table 11.4. The European Inland Fisheries Advisory Commission (EIFAC) has proposed standards for DO in freshwaters for the protection of fish, as has the US EPA.

EIFAC (Alabaster & Lloyd, 1982) has derived tentative minimum criteria for dissolved oxygen for the protection of freshwater fish (Table 11.5). They proposed that for resident populations of moderately tolerant freshwater species, such as

Table 11.4 Environmental quality standards for DO concentrations (mg/l) proposed by Davis (1975)

Degree of protection	Salmonids		Non-salmonids
	Adults	Eggs/larvae	
A	7.75	9.75	5.50
B	6.00	8.00	4.00
C	4.25	6.50	2.50

roach, the annual 50- and 95-percentile DO values should be > 5 mg/l and 2 mg/l, respectively, and for salmonids the corresponding values should be 9 mg/l and 5 mg/l, respectively. They pointed out that the values were for general guidance; there are circumstances where the seasonal and geographical variation of DO must be considered. For example, for adult migrant salmonids the 50- and 95-percentile values for periods of low water during the summer months in the region of an estuary where the DO is lowest should be 5 mg/l and 2 mg/l, respectively but higher values are recommended if the estuary has an extensive deoxy-

Table 11.5 Tentative minimum sustained DO for maintaining the normal attributes of the life cycle of fish under otherwise favourable conditions. After Alabaster & Lloyd, 1982

Attribute	DO (mg/l)
Survival of juveniles and adults for 1 day or longer	3
Fecundity, hatch of eggs, larval survival	5
Ten per cent reduction in hatched larval weight	7
Larval growth	5
Juvenile growth (could be reduced 20%)	4
Growth of juvenile carp (_Cyprinus carpio_)	3
Cruising swimming speed (maximum sustainable speed could be reduced 10%)	5
Upstream migration of Pacific salmon (_Oncorhynchus_ spp.) and Atlantic salmon (_Salmo salar_)	5
Upstream migration of American shad (_Alosa sapidissima_)	2
Schooling behaviour of American shad	5
Sheltering behaviour of walleye (_Stizostedion vitreum_)	6

genated zone (more than a few kilometres). Further, because the early life stages of fish are especially sensitive, the lower levels of DO should not occur when these are present.

Table 11.6 lists the US national criteria for ambient DO concentrations for the protection of freshwater aquatic life (US EPA, 1986). The EPA criteria for coldwater fish apply to waters where one or more salmonid species are present; they should represent conditions where there is a risk of only a slight impairment of fish productivity. The 1-day minimum values should be achieved at all times. The coldwater minimum is 4 mg/l because the EPA considers many of the insect

species common to salmonid habitats more sensitive than the salmonids, which have acutely lethal levels at or below 3 mg/l. The EPA states that the criteria do not represent assured no-effect levels. If a slight risk is not acceptable, then the assessor should use the values given in Table 11.7 to establish the oxygen required for the protection of the relevant life stages and species.

Bowden & Solbé (1987) and Whitelaw & Solbé (1989) proposed standards based on LC_{50} data for intermittent episodes of low DO. They plotted exposure response curves and derived time-varying standards for periods up to 1000 min

Table 11.6 US national water quality criteria for minimum ambient dissolved oxygen concentration (mg/l)

	Coldwater criteria		Warmwater criteria	
	Early life stages[a]	Other life stages	Early life stages[a]	Other life stages
30-day mean	n.a.[b]	6.5	n.a.	5.5
7-day mean	9.5[c] (6.5)	n.a.	6.0	n.a.
7-day minimum	n.a.	5.0	n.a.	4.0
1-day minimum[d,e]	8.03 (5.0)	4.0	5.0	3.0

[a] Includes all embryonic and larval stages and all juvenile forms to 30 days following hatching.
[b] n.a., Not applicable.
[c] Concentrations of DO in the water column recommended to achieve the required DO concentrations in the intergravel water shown in parentheses. The 3 mg/l differential is discussed in the criteria document. For species that have early life stages exposed directly to the water column, the figures in parentheses apply.
[d] For easily controlled discharges, further restrictions apply.
[e] All minima should be considered as instantaneous concentrations to be achieved at all times.

Table 11.7 Minimum DO concentrations (mg/l) versus quantitative level of effect (US EPA, 1986)

Production impairment	Early life stages		Other life stages		
	Salmonid waters	Non-salmonid waters	Salmonid waters	Non-salmonid waters	Invertebrates
None	11[a] (8)	6.5	8	6	8
Slight	9[a] (6)	5.5	6	5	
Moderate	8[a] (5)	5	5	4	5
Severe	7[a] (4)	4.5	4	3.5	
To avoid acute mortality	6[a] (3)	4	4	3	4

[a] These are concentrations of DO in the water column recommended to achieve the required DO concentrations in the intragravel water shown in parentheses.

(16.7 h). These tended to an asymptote as exposure duration increased. The DO standards (corrected to 17.5°C) were for salmonids: minimum 1.9 mg/l, 1000 min 2.87 mg/l, and for cyprinids minimum 0.96 mg/l, 1000 min 2.39 mg/l.

Danish standards (Danish Engineering Union, 1985) considered event duration and also return period (frequency of episode). The standards (Table 11.8) consist of linear interpolations between a point at the return period where 50% mortality is deemed acceptable and the continuous exposure criterion for very short return (0.1 year). Although a novel and useful approach was used in developing these criteria, there were shortcomings. There were no objectives set for intermediate points (return period between 0.1 and 8–16 years) and no sound biological basis for joining the points by a straight line.

Milne & Seager (1990) reviewed available data and proposed standards based on a no-effect for continuous exposure conditions and no major mortality for 1-year return events. They also considered the protection of invertebrate communities, and 1- and 3-month return events to define the shape of the criteria. Their experimental evidence indicates that fish recover very rapidly following exposure to sublethal low DO concentrations (Milne *et al.*, 1989) and that DO concentration rather than event frequency is the more important factor. Together, these observations suggest an initially steep curve at low return period (high frequency), becoming shallower as return period increases. For 1-month return the standards are around, or just below, the threshold for physiological stress and behavioural responses. For 3-month return the standards allow moderate

physiological stress and behavioural responses including avoidance. The criteria are based on temperatures around 15°C. For very high temperatures ($> 20°C$) they may afford insufficient protection. They also assume otherwise normal conditions for other factors. Fig. 11.2 displays the proposed criteria for events of 1 and 24 h duration.

The continuous exposure level should result in no effect and the 1-year return levels should avoid mortality. The authors suggested that invertebrate drift would be minor for a 24-h event, major for a 1-h event. The 1-month frequency standards might cause some degree of physiological stress (increased ventilation rate, etc.), particularly for salmonids, but would probably not induce strong avoidance reactions in either fish or invertebrates. They believed that the 3-month frequency DO levels would result in greater physiological stress, causing avoidance in fish and moderate drift. The difference between no lethal effect and major fish kill may be a relatively small drop in DO (0.5 mg/l or less), so the standard should permit little or no room for doubt. The Danish approach that determines the frequency at which 50% mortality is acceptable is a sensible one in this regard (Danish Engineering Union, 1985). However, this requires data to determine the maximum frequency to protect long-term population viability.

It is clear that most effort has focused on the effects of DO on fish, and these data have been used to derive standards. The general approach to setting short- and long-term standards is sound, but there is a need to generate good data on DO effects on aquatic invertebrates.

Table 11.8 Danish criteria for DO concentrations (mg/l) for different river types. After Danish Engineering Union, 1985

River type	Exposure (h)	Continuous	50% mortality
Salmon spawning	1	8	1.5
	12	9	2
Salmon fishery	1	6	1.5
	12	7	2
Carp fishery	1	4	1
	12	5	1.5

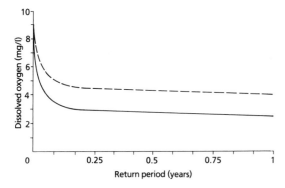

Fig. 11.2 Water quality standards to protect aquatic life during intermittent low DO episodes. For each exposure period (1, 6 or 24 h) the concentration of dissolved oxygen tolerated decreases as the time between exposures increases. Event duration: ●, 24 h; +, 1 h.

11.3.5 The BOD test and its alternatives

Early investigations in the UK into oxygen absorption by sealed water samples date back to the 1870s. The Royal Commission on Sewage Disposal, appointed in 1898, examined methods of assessing oxygen absorption by river water. The test to measure the BOD of a sample over 5 days, the BOD_5 test, was first prescribed in the Royal Commission's 8th Report (1912, 1913) concerning 'standards to be applied to sewage and effluents discharging to rivers and streams and the tests which should be used to determine the standards'. The Commission recognized a relationship between river quality and DO concentration, and hence the requirement to maintain the latter at a level sufficient to prevent development of nuisance. The method was adopted as a means of predicting the DO depletion a river would suffer as a consequence of effluent or sewage discharge, and taken together with other factors such as flow, dilution and reaeration, to determine whether such discharges were permissible in view of the predicted consequences. The BOD_5 method was considered the most trustworthy test, but the Report conceded that it was difficult to perform, and that it gave inconsistent results. In addition, the Royal Commission established a system of river classification, defining

river quality in terms of BOD and a number of other chemical parameters. A qualitative classification scheme was also introduced that correlated visible characteristics of rivers during the summer with their chemical properties. Additionally, the BOD method has become established for assessing design requirements for wastewater treatment processes and for evaluating treatment efficiency. Although the test is universally recognized as being unsatisfactory, it has withstood attempts to replace it. It yields a unique measure of oxygen uptake by microbial metabolism, and is applied to a wider range of situations than any single alternative method examined to date.

The BOD_5 test is performed by preparing a suitable dilution of the sample with a mineral salts medium previously aerated to achieve oxygen saturation at 20°C (Standing Committee of Analysts (SCA), 1983a). A suitable inoculum is added, if appropriate, and the diluted sample divided to fill two narrow-necked, 250-ml glass bottles. Test bottles are fitted with tapered stoppers to force out air bubbles which might otherwise become trapped and provide an additional source of oxygen during incubation. The first bottle is used to obtain an immediate DO measurement; the second is incubated at 20°C and provides a final measurement after 5 days. Incubation takes place in darkness to prevent oxygen production by algal photosynthesis which would otherwise counteract uptake by microbial metabolism. The extent of dilution depends upon the anticipated BOD of the sample; its purpose is to limit oxygen depletion during incubation to between 30% and 70% of the starting oxygen concentration (9.2 mg/l at saturation at 20°C). Ranges of dilutions are prepared according to type for samples whose strengths are unknown but whose BOD is expected to lie within certain limits. Typical dilutions are indicated in Table 11.9.

The method recommended in the 8th report of the Royal Commission on Sewage Disposal (1913) has undergone a succession of modifications in the interest of precision and elimination of interferences. Thus the original incubation temperature of 65°F (18.3°C) was raised to 20°C, a standard mineral salts medium has replaced the tapwater hitherto used as dilution water, and addition of

Table 11.9 Typical dilutions for BOD samples. After Standing Committee of Analysts, 1983a

Sample type	BOD range	Sample vol.	Diluent vol.
Clean river water	<6	1	0
River water	6–20	1	2
Sewage effluent	10–30	1	4
Poor effluent	20–60	1	9
Bad effluent	40–120	1	19
Raw sewage or industrial effluent	100–300	1	49

allylthiourea has become a standard practice in the UK (and elsewhere in Europe) whenever it is considered necessary to eliminate the effects of nitrification. The principle of the test has remained unchanged since 1912.

The SCA method 'Biochemical Oxygen Demand 1981' (SCA, 1983a) discusses graphically the significance of BOD_5 results (see Fig. 11.3), and describes various idealized patterns of oxygen uptake that may be obtained in continuously recording respirometric experiments.

Curve 1 is a common result, typical of domestic wastewaters and uncomplicated by nitrification. It indicates a sample containing readily degradable material and an adequate population of competent bacteria. (Competence is defined as possession of the enzyme(s) required to break down and utilize a particular substrate.) A sample containing a less readily degradable substance, but inoculated with an acclimatized flora, might also produce such a curve.

Curve 1a is identical to Curve 1 up to Day 5 when nitrification causes an additional uptake of oxygen. Nitrogenous oxygen uptake (NOD) has been represented as a discrete step of little significance until after the fifth day of incubation. This is because nitrifiers multiply slowly, and there is a lag before there are sufficient to cause a measurable effect. While this may often be the case, significant nitrification begins earlier in samples inoculated with a nitrified effluent already containing adequate numbers of nitrifiers. In such cases, NOD can make a considerable contribution to the BOD_5 indistinguishable from carbonaceous oxygen uptake.

Curve 2 indicates insufficient competent bacteria were initially present in the sample and there was a lag phase during which cell multiplication occurred before significant oxidation began. Complete oxidation was not achieved until after 5 days.

Curve 2a points to the presence of material which was not readily degraded and required a period of acclimatization before breakdown began. Acclimatized inocula may be used for such samples, but this raises one question, whether results obtained by such means have any significance or validity *in situ*.

Curve 3 may be produced if the sample contains inhibitors or toxicants that stop microbial growth or biochemical oxidation. Absence of oxygen uptake may also be indicative of a sample that is poorly degradable or non-degradable. Presence of toxicants is often exposed by an effect known as 'sliding BOD' where the measured BOD increases with higher dilutions and the corresponding decline in toxicant concentration.

Whereas a significance can be ascribed to the course of oxygen uptake, none of this information can be extrapolated from BOD_5 results. The latter is no more than a number representing a point on a curve, and has no significance if taken out of context.

BOD_5 determinations are used to monitor the quality of effluents and check that they comply with the appropriate emission consent conditions. Whereas the BOD_5 of an incoming sewage effluent is exerted by oxidation of its biodegradable constituents, sewage residues comprising refractory material that resisted oxidation in the treatment process (Heddle, 1984) generally make little contribution to effluent BOD. The BOD of sewage effluents is principally due to endogenous respiration of suspended bacterial flocs washed from the treatment process (Montgomery & Gardiner, 1971; Downing 1983); these could be material stripped from the biofilm of a percolating filter, or activated sludge which remains suspended after final settlement. For this reason, effluent quality standards were originally defined simply in terms of BOD_5 and suspended solids (SS). Thus the BOD test serves only a retrospective function. Measurements of SS are commonly used to obtain a more immediate indication of effluent quality, and may also point to interference with treatment. For example, certain toxicants cause disintegration of the bacterial flocs of activated sludge into extremely fine, non-settleable particles that then produce a turbid solid-laden effluent (Hawkes, 1983).

Suppression of nitrification to obtain a more acceptable assessment of the performance of a treatment works conflicts with the function that the BOD$_5$ test was intended to perform: to give an indication of biochemical oxygen uptake to be expected from an effluent discharge and thus to safeguard against depletion of DO in the receiving water. For a river it is irrelevant whether deoxygenation is caused by carbonaceous oxidation or by nitrification. Clearly, if nitrification occurs under test conditions in a BOD bottle, but not in the water to which an effluent is discharged, the use of inhibitors in the test is legitimate. An estimate of potential NOD can be derived from analysis of ammonia and nitrite; ammonia concentrations are nowadays stated among water quality criteria, and prescribed in emission consent requirements.

11.3.6 Limitations of the BOD test

Dilution

The dilution and the diluent used in the 5-day laboratory test bear no resemblance to real conditions (Montgomery, 1967; Bridié, 1969). In particular, the substrate : micro-organism ratios under test conditions are unlikely to match those of sewage treatment processes (Stover & McCartney, 1984).

Limiting oxygen concentration

Oxidation rates in BOD bottles may be constrained by diminishing DO concentrations. On the other hand, control measures in the aerobic processes at sewage treatment works aim to maintain DO at optimal high concentrations and reaeration occurs naturally or may be encouraged in rivers.

Mismatched microflora

Often the seed flora do not represent the micro-organisms encountered *in situ*. For example,

sewages are typically tested unseeded, as it is assumed they already contain adequate populations of bacteria, though these certainly do not correspond — in terms of number, diversity or range of oxidative competence — to the organisms present in an activated sludge or a percolating filter biofilm.

Irrelevant incubation period

The test duration has neither theoretical nor practical significance; it was adopted because oxygen uptake due to biochemical oxidation of some readily degraded materials was virtually complete by the fifth day. Measured oxygen uptakes were consequently less erratic than those obtained after shorter intervals. The 5-day incubation is also of limited relevance in predicting the effect of effluent oxidation on river DO. Whereas residence periods in many British rivers are less than 5 days, they may be considerably more in others and in oscillating estuarine systems where continuing oxidation of accumulated discharges may cause serious deoxygenation (Downing & Edwards, 1969).

Precision

A cause of dissatisfaction with the BOD$_5$ method has been the poor precision with which results are obtained; variation of around 20% has been reported (Woodward, 1970). Variation of a comparable magnitude was reported in an inter-laboratory exercise which used samples containing a glucose and glutamic acid solution (Ballinger & Lishka, 1962). These constituents are widely used for standardization purposes because of the great ease with which they are biodegraded, and this represents elimination of a major cause of variability. Poor precision can be expected with samples containing substances less amenable to microbial attack. Nitrification is notoriously sporadic (Painter, 1986) often occurring to dissimilar extents in replicate test bottles.

Substrate characteristics and bacterial competence

The extent of biochemical oxidation is determined in part by the nature and properties of organic material present. Is the material readily biodegradable and in a labile form, and does the sample contain toxicants or bacteriostatic inhibitors? The other determining factor is the nature of the microflora that make up the test seed. Do the organisms have the degradative ability to break down sample components, or have they undergone prior acclimatization? Where only very low numbers of competent bacteria are present, an uneven distribution may occur between replicate bottles which is later reflected in dissimilar oxygen uptake.

Dilution effects

The low solubility of oxygen in water requires that samples be diluted to limit oxygen uptake in the test bottles. Educated guesswork is used to decide a range of dilution factors appropriate for a particular sample type, and to obtain valid results at more than one dilution. If guesses are wide of the mark, results must be quoted as greater than the expected maximum or lower than the expected minimum. An effect known as 'sliding BOD' may occur where BOD rises with increasing dilution (Klein, 1959; Standing Committee of Analysts, 1983a) usually caused by reduced inhibition as a consequence of dilution of toxic materials. However, one reported instance of sliding BOD was attributed to contaminated dilution water (Bryan, 1966).

Algal interference

Many effluents contain diatoms and other unicellular algae which, unable to photosynthesize under the test conditions, contribute to oxygen uptake by dark respiration, then die and ultimately disintegrate to release substrates for further microbial oxidation (Klein, 1959; Price & Pearson, 1979).

11.3.7 Improvements and alternatives to the BOD test

Attempts to modify or replace the test have concentrated upon the function of the BOD_5 test as a tool for process control in sewage treatment. Little attention appears to have been paid to the role of the BOD test in defining effluent discharges. Most approaches have sought to reduce the duration of the test so that results can be obtained in a timescale more compatible with sewage treatment processes.

Higher test temperatures have been used to shorten the incubation necessary to obtain results comparable to BOD_5 values. Incubation at 27°C and 35°C has been used to obtain results in 3 and 2½ days, respectively; generally, 3-day BOD values were within 5% of the corresponding 5-day figures (Standing Committee of Analysts, 1983a).

Respirometric methods are considered to have a number of advantages over the standard BOD test. Firstly, it is possible to recharge the test vessel with oxygen. This ensures that test oxygen levels do not fall low enough to limit the rate of oxidation (as they may in the BOD_5 method). It also enables the test conditions to correspond more closely to reality; samples require no dilution, and more realistic inoculum concentrations are used to represent conditions in a treatment plant. More importantly, however, the larger inoculum causes an accelerated oxygen uptake, and indications of oxygen demand are obtained more quickly than with the BOD_5 method. Secondly, the continuous oxygen uptake recordings yield more useful information than BOD_5 values (see Fig. 11.3). Incubation extended to an uptake plateau gives a measure of ultimate oxygen demand (UOD), and enables an estimate to be made of UOD reduction during the retention period of the aerobic processes of sewage treatment. Other types of respirometric equipment which determine the change in oxygen uptake rate of an endogenously respiring inoculum upon addition of a wastewater sample allow treatability assessments to be made, or may be used to give indications of toxicity (Lamb *et al.*, 1964; Standing Committee of Analysts, 1983b; Pagga, 1985).

Respirometry techniques have been reviewed by Montgomery (1967), and subsequent investigations are mentioned in a later paper (Montgomery & Gardiner, 1971). It was concluded that none of the methods offered a single, rapid means of replacing the BOD_5 test in all its applications.

Other methods have employed biochemical reduction of inorganic oxides to circumvent the

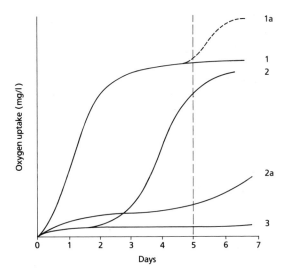

Fig. 11.3 Effects of various influences on the course of oxygen uptake at constant temperature and carbonaceous substrate concentration. The vertical, broken line is the normal endpoint of the BOD_5 test. See text for further explanation.

limitation imposed on oxygen availability by its poor solubility. They depend upon the presence of micro-organisms capable of utilizing the additional oxygen source. A procedure using nitrate was included for a time among the US standard methods. It was thought to have been unsatisfactory because of the poor precision with which nitrite and nitrate could be determined before and after incubation, but other factors were probably also involved. In any event, the method is invalid as a model of oxygen uptake in aerobic systems, as nitrate reduction occurs only when low DO or possibly even anaerobic conditions would develop. Bryan (1966) reported modification of a similar method in which the concentration of chloride formed by reduction of sodium chlorate was measured. Chlorate concentrations equivalent to 1000 mg/l oxygen were used without causing inhibition, but early investigations showed that oxidation comparable with that attained by the standard BOD method was dependent upon sample type, and for strong samples could be achieved only after 10 days incubation at 20°C. Incubation was reduced to 5 days by raising the temperature to 30°C. There was no advantage over the BOD_5 test.

An alternative approach has been to develop a mild chemical oxidation method to match the oxidative power of the biochemical processes occurring in the BOD_5 test. Shriver & Young (1972) described application of an 'oxygen demand index' to estimate biochemical oxygen demand. The method was essentially similar to that employed in chemical oxygen demand (COD) determination, and used acidified dichromate, though at a lower temperature and for only 20 min. Samples are oxidized in a standard, arbitrary manner by refluxing with sulphuric acid and potassium dichromate with a silver salt to catalyse the oxidation of alcohols and low-molecular-weight acids. Chromium III is added as potassium chromium sulphate dodecahydrate which, along with excess silver salt, suppresses chloride interference and with it the effect due to ammonia. The mixture is refluxed for 2 h and the residual dichromate is determined by titration with standardized ferrous ammonium sulphate solution. The amount of dichromate reduced is expressed in the form of milligrams of oxygen consumed per litre of sample; this is the COD (DoE, 1987). Correlation of the results from the Shriver & Young method with those from BOD_5 varied according to sample type, was poorest in the case of treatment works effluents, and was generally not as good as the correlation obtained between BOD_5 and COD for identical samples subjected to all three methods. It is unlikely that any purely chemical oxidation process will yield a satisfactory estimate of biochemical oxidation controlled by a combination of substrate properties and biological factors.

Bourdon *et al.* (1986) have described a method by which BOD_5 is estimated from measurements of ultraviolet light absorption between 250 and 280 nm. This range coincides with an absorption maximum for unsaturated organic compounds, and the method presumes that such compounds form the predominant components of wastewaters. Factors were derived to enable correlation with BOD_5, but the method gives no indication of changes in wastewater composition or the presence of inhibitors.

Another recent development has been the construction of biosensors to obtain rapid estimates of BOD_5 (Strand & Carlson, 1984; Karube

& Tamiya, 1987). These were built by securing a paste of living cells to the exterior face of the gas-permeable membrane of a DO probe. At equilibrium the electrode output indicated a balance between microbial respiration and the rate of oxygen diffusion to the probe membrane. The length of time required to reach equilibrium varied according to sample composition, but occurred within 20 min in all cases examined. In a range corresponding to BOD_5 of 2–22 mg/l, biosensor measurements gave good correlation with BOD values for samples comprising standard glucose/glutamic acid solutions, primary and secondary effluents and certain wastewaters containing no significant oxidizable solids (Strand & Carlson, 1984). Wastewaters needed to be diluted to fall inside the measurement range, and it is likely that the method is therefore prone to the problem of 'sliding BOD' where toxic wastewaters are concerned. The heterogeneous culture described by Strand & Carlson (1984), and derived from activated sludge, can be expected to cover a comparatively wide range of bacterial activity and would therefore appear preferable to the single-species cultures of bacteria or yeast cells reported by Karube & Tamiya (1987).

Riegler (1984), Köhne (1985) and Köhne *et al.* (1986) describe application of an on-line instrument constructed by Siepmann and Teutscher GmbH which determines BOD continually and with a response time of only 3 min. It was possible to follow changes in loading at sewage treatment works to enable effective response by prompt adjustment of process controls. The measured parameter, 'BOD-M3', corresponds fairly closely to BOD_5 determinations carried out on spot samples. However, no result spans an interval of more than 24 h; neither do any two figures cover consecutive days. Consequently it is impossible to tell whether such correspondence is typical or consistent through irregular adverse events such as storm flows. The expectation is that, under such conditions, the acclimatization possible during the 5-day incubation of the standard BOD test might cause BOD_5 values to exceed those derived from BOD-M3 quite significantly. Nevertheless, oxidation events *in situ* are probably better reflected by BOD-M3 than by BOD_5.

Despite general dissatisfaction, the BOD_5 test has been retained for the functions originally introduced in 1912. Its use has also been extended beyond the purpose originally envisaged to application in design and control of sewage treatment processes — the consequence of a desire to match the means of biological treatment to the ends of standards and consent conditions prescribed in terms of BOD_5 (Heddle, 1984). Montgomery & Gardiner (1971) believed it unlikely that a single test could be devised to apply both to untreated wastewaters and to biologically treated effluents.

Methods proposed for replacing the BOD test are unlikely to gain acceptance until the present status of BOD_5 in deciding river quality and discharge consents is either lowered or removed altogether. As explained earlier, BOD_5 is likely to be of little relevance to river quality where substantial and prolonged deoxygenation is not a regular occurrence. However, BOD_5 is the only parameter determined currently and included routinely among discharge consent conditions to involve an element of biological activity. Simply withdrawing BOD_5 as a means of determining water quality and consent requirements, without an alternative, cannot be considered satisfactory.

In the present context a replacement(s) for the BOD_5 test must: (i) give a measure of biological activity; (ii) predict whether a discharge is likely to have a deleterious effect on the quality of the receiving stream as indicated by its biological condition; and (iii) be based on an easily determined parameter. It should also ideally respond rapidly, continuously, and with minimal operator attendance.

REFERENCES

Alabaster, J.S. (1988) The dissolved oxygen requirements of upstream migrant chinook salmon, *Oncorhynchus tshawytscha*, in the lower Whillamette River, Oregon. *J. Fish Biol.* **32**, 635–636.

Alabaster, J.S. & Gough, P.J. (1986) The dissolved oxygen and temperature requirements of Atlantic salmon, *Salmo salar* L, in the Thames estuary. *J. Fish. Biol.* **29**, 613–621.

Alabaster, J.S. & Lloyd, R. (1982) *Water Quality Criteria for Freshwater Fish*, 2nd edn. Butterworths, London.

Alabaster, J.S. & Robertson, K.G. (1961) The effect of changes in temperature, dissolved oxygen and illumination on the behaviour of roach (*Rutilus rutilus* (L)), bream (*Abramis brama* (L)) and perch (*Perca fluvia-tilis* (L)). *Anim. Behav.* **9**, 187–192.

Alabaster, J.S., Hergert, D.W.M. & Hemens, J. (1957) The survival of rainbow trout (*Salmo gairdneri* Richardson) and perch (*Perca fluviatilis* L) at various concentrations of dissolved oxygen and carbon dioxide. *Ann. Appl. Biol.* **45**, 177–188.

Alabaster, J.S., Shurben, D.G. & Knowles, G. (1979) The effect of dissolved oxygen and salinity on the toxicity of ammonia to smolts of salmon, *Salmo salar* L. *J. Fish. Biol.* **15**, 705–712.

Alderdice, D.F., Wickett, W.P. & Brett, J.R. (1958) Some effects of temporary exposure to low dissolved oxygen levels on Pacific salmon eggs. *J. Fish. Res. Bd. Canada* **15**, 229–250.

APHA-AWWA-WPCF (1985) *Standard Methods for the Examination of Water and Wastewater*, 16th edn. American Public Health Association, Washington, DC.

Arthur, R.A.J. (1986) The importance of being absent. *Water Waste Treat.* **29**(11), 50–54.

Ballinger, D.G. & Lishka, R.J. (1962) Reliability and precision of BOD and COD determinations. *J. Water Pollut. Contr. Fed.* **34**, 470–474.

Bishai, H.M. (1962) Reactions of larval and young salmonids to water of low oxygen concentration. *J. Cons. Perm. Int. Explor. Mer.* **27**, 167–180.

Bourdon, F., Jestin, J.-M. & Ravarini, P. (1986) Util-isation de l'absorption ultraviolet pour l'estimation de la DCO et DBO₅ des eaux. *Tech. Sci. Méthodes* **81**, 187–191.

Bowden, A.V. & Solbé, J.F. de L.G. (1987) Effects of intermittent pollution on rivers — derivation of water quality standards. Water Research Centre ER 1227-M/2.

Bridié, A.L.A.M. (1969) Determination of biochemical oxygen demand with continuous recording of oxygen uptake. *Water Res.* **3**, 157–165.

Bryan, E.H. (1966) Application of the chlorate BOD procedure to routine measurement of waste-water strength. *J. Water Pollut. Contr. Fed.* **38**, 1350–1362.

Calamari, D., Merchetti, R. & Vailati, G. (1981) Effects of long-term exposure to ammonia on the develop-mental stages of rainbow trout (*Salmo gairdneri* Richardson). *Rapp P-v Reun. Cons. Int. Explor., Mer.* **178**, 81–86.

Chapman, G. (1986) Ambient water quality criteria for dissolved oxygen. US Environmental Protection Agency Report No. EPA 440/5–86–003, Washington, DC.

Coble, D.W. (1961) Influence of water exchange and dissolved oxygen in the redds on survival of steelhead trout embryos. *Trans. Am. Fish. Soc.* **90**, 469–474.

Danish Engineering Union (1985) Pollution of water-courses from overflow structures. Danish Engineering Union Wastewater Committee Publication No. 22.

Davis, J.C. (1975) Minimal dissolved oxygen require-ments of aquatic life with emphasis on Canadian species: a review. *J. Fish. Res. Bd. Canada* **32**, 2295–2332.

DoE (1987) *Chemical Oxygen Demand (Dichromate Value) of Polluted and Waste Waters 1986*, 2nd edn. HMSO, London.

DoE (1988) *Methods for the Examination of Waters and Associated Materials, 5 Day Biochemical Oxygen Demand (BOD₅)*, 2nd edn, with Dissolved Oxygen in Waters, Amendments 1988. HMSO, London.

DoE/NWC (1980) *Methods for the Examination of Waters and Associated Materials, Dissolved Oxygen in Natural and Waste Waters 1979*. HMSO, London.

Doudoroff, P. & Shumway, D. (1970) Dissolved oxygen requirements of freshwater fishes. Food and Agri-cultural Organization of the United Nations. FAO Technical Paper No. 86. Rome, Italy.

Downing, A.L. (1983) Used-water treatment today and tomorrow. In: *Ecological Aspects of Used-Water Treatment*, Vol. 2: *Biological Activities and Treat-ment Processes* (Eds C.R. Curds & H.A. Hawkes), pp. 1–10. Academic Press, London.

Downing, A.L. & Edwards, R.W. (1969) Effluent stan-dards and the assessment of the effects of pollution on rivers. *Water Pollut. Contr.* **68**, 283–299.

Downing, K.M. & Merkens, J.C. (1957) The influence of temperature on the survival of several species of fish in low tensions of dissolved oxygen. *Ann. Appl. Biol.* **45**, 261–267.

Erickson, R.J. (1985) An evaluation of mathematical models for the effect of pH and temperature on ammonia toxicity to aquatic organisms. *Water Res.* **19**, 1047–1058.

Flis, J. (1968) Anatomicohistopathological changes induced in carp (*Cyprinus carpio* L.) by ammonia water. Part II. Effects of subtoxic concentrations. *Acta Hydrobiol.* **10**, 225–238.

Hawkes, H.A. (1983) Activated sludge. In: *Ecological Aspects of Used-water Treatment*, Vol. 2: *Biological Activities and Treatment Processes* (Eds C.R. Curds & H.A. Hawkes), pp. 77–162. Academic Press, London.

Heddle, J.F. (1984) Communication: the need to reassess use of the BOD₅ test. *J. Water Pollut. Contr. Fed.* **56**, 292–293.

Herbert, D.W.M. & Shurben, D.S. (1964) The suscep-tibility of salmonid fish to poisons under estaurine conditions. 2: Ammonium chloride. *Int. J. Air Water Pollut.* **9**, 89–91.

Hermenutz, R.O., Hedtke, S.F., Arthur, J.W., Andrew, R.W. & Allen, K.N. (1987) Ammonia effects on micro-invertebrates and fish in outdoor experimental streams. *Environ. Pollut. A* **47**, 249–283.

Itazawa, Y. (1971) An estimation of the minimum level of dissolved oxygen in water required for the normal life of fish. *Bull. Jpn. Soc. Sci. Fish.* **37**, 273–276.

Karube, I. & Tamiya, E. (1987) Biosensors for environmental control. *Pure Appl. Chem.* **59**, 545–554.

Klein, L. (1959) *River Pollution*, Vol. 1: *Chemical Analysis.* Butterworths, London.

Köhne, M. (1985) Practical experiences with a new on-line BOD measuring device. *Environ. Technol. Lett.* **6**, 546–555.

Köhne, M., Siepmann, F.W. & Te Heesen, D. (1986) Der BSB5 und der kontinuierliche kurzzeit-BSB (BSB-M3) in Vergleich. *Korres. Abwass.* **33**, 787–793.

Lamb, J.C., Westgarth, W.C., Rogers, J.L. & Vernimmen, A.P. (1964) A technique for evaluating the biological treatability of industrial wastes. *J. Water Pollut. Contr. Fed.* **36**, 1263–1284.

Lewis, Jr, W.J. (1988) Uncertainty in pH and temperature corrections for ammonia toxicity. *J. Water Pollut. Contr. Fed.* **60**, 1922–1929.

Lloyd, R. & Herbert, D.W.M. (1960) The influence of carbon dioxide on the toxicity of un-ionised ammonia to rainbow trout (*Salmo gairdneri* Richardson). *Ann. Appl. Biol.* **48**, 399–404.

Lloyd, R. & Orr, L.D. (1969) The diuretic response of rainbow trout to sub-lethal concentrations of ammonia. *Water Res.* **3**, 335–344.

Lumbers, J.P. & Wishart, S.J. (1989) Ammonia toxicity simulation using a compartmental model. Water Research Centre Draft Report. WRc, Medmenham.

Mallett, M.J. (1990) Review of the toxicity of common pollutants to indigenous species of freshwater fish: ammonia, arsenic and cadmium. WRc Report NR-2541, for the National Rivers Authority. WRc, Medmenham.

Maltby, L. & Naylor, C. (1989) Investigations of the lethal and sub-lethal effects of episodes of low dissolved oxygen on *Gammarus pulex*. WRc Report PRS 2257-M. WRc, Medmenham.

McCormick, J.H., Broderius, S.J. & Fiandt, J.T. (1984) Toxicity of ammonia to early life stages of the green sunfish *Lepomis cyanellus*. *Environ. Pollut. A* **36**, 147–163.

McKim, J.M. (1985) Early life stage toxicity tests. In: *Fundamentals of Aquatic Toxicology* (Eds Rand & Petrocelli), pp. 58–95. Hemisphere, New York.

Milne, I. & Seager, J. (1990) Proposed water quality critera for the protection of aquatic life from intermittent pollution: Dissolved Oxygen. WRc Report PRS 2498-M.

Milne, I. & Seager, J. (1991) Proposed water quality criteria for the protection of aquatic life from intermittent pollution: Ammonia. WRc Report NR-2682/1, for the National Rivers Authority. WRc Medmenham.

Milne, I., Seager, J., Mallett, M. & Holmes, D. (1990) The impact of combined sewer overflow discharges on river quality and the development of water quality criteria for intermittent pollution—interim report.

Water Research Centre Report No. 2572. WRc, Medmenham.

Milne, I., Seager, J., Mallett, M. & Sims, I. (1989) The impact of combined sewer overflow discharges on river quality—progress report on the Pendle Water field study, 1988–1989. Water Research Centre Report PRS 2251-M. WRc, Medmenham.

Montgomery, H.A.C. (1967) The determination of Biochemical Oxygen Demand by respirometric methods. *Water Res.* **1**, 631–662.

Montgomery, H.A.C. & Gardiner, D.K. (1971) Experience with a bacterial inoculum for use in respirometric tests for oxygen demand. *Water Res.* **5**, 147–163.

Pagga, U. (1985) Stoffprüfungen in einem Kläranlagenmodell- Abbaubarkeits- und Toxizitätstests im BASF-Toximeter. *Z Wasser-Abwasser-Forsch.* **18**, 222–232.

Painter, H.A. (1986) Nitrification in the Treatment of Sewage and Waste-waters. In: *Nitrification Special Publications of the Society for General Microbiology*, Vol. 20, (Ed. J.I. Prosser), pp. 185–211. IRL Press, Oxford.

Perry, R.H. & Green, D. (1984) *Perry's Chemical Engineers Handbook*, 6th edn. McGraw Hill, New York.

Peterka, J.J. & Kent, J.S. (1976) Final Report. Dissolved oxygen, temperature, survival of young at fish spawning sites. (EPA/600/3–76–113). Environmental Protection Agency, Washington, DC.

Pfafflin, J.R. & Ziegler, E.N. (Eds) (1976) *Encyclopaedia of Environment Science and Engineering.* Gordon & Breach, New York.

Priede, I.G., Solbé, J.F. de L.G., Nott, J.E., O'Grady, K.T. & Cragg-Hine, D. (1988) Behaviour of Atlantic salmon, *Salmo salar* L, in the estuary of the River Ribble in relation to variations in dissolved oxygen and tidal flow. *J. Fish. Biol.* **33**(A), 133–139.

Price, D.R.H. & Pearson, M.J. (1979) The derivation of quality conditions for effluents discharged to freshwaters. *Water Pollut. Contr.* **78**, 118–131.

Riegler, G. (1984) Kontinuierliche Kurzzeit-BSB-Messung. *Korresp. Abwasser* **31**(5), 369–370, 372, 375–377.

Royal Commission on Sewage Disposal (1912) 8th Report, Vol. 1: Report Cmd 6464. HMSO, London.

Royal Commission on Sewage Disposal (1913) 8th Report, Vol. 2: Appendix. Part II Cmd 6943. HMSO, London.

Seager, J. & Abrahams, R.A. (1988) The impact of combined sewer overflow discharges on river quality—progress report on the Pendle Water field study. Water Research Centre Report PRS 1951-M. WRc, Medmenham.

Seager, J., Wolff, E.W. & Cooper, V.A. (1988) Proposed environmental quality standards for List II substances in water—ammonia. Water Research Centre Report

TR 260. WRc, Medmenham.

Shepard, M.P. (1955) Resistance and tolerance of young speckled trout (*Salvelinus fontinalis*) to oxygen lack, with special reference to low oxygen acclimation. *J. Fish. Res. Bd. Canada* **12**, 387–446.

Shriver, L.E. & Young, J.C. (1972) Oxygen Demand Index as a Rapid Estimate of Biochemical Oxygen Demand. *J. Water Pollut. Contr. Fed.* **44**, 2140–2147.

Smart, G. (1976) The effect of ammonia exposure on gill structure of the rainbow trout (*Salmo gairdneri*). *J. Fish. Biol.* **8**, 471–475.

Solbé, J.F. de L.G. & Shurben, D.G. (1989) Toxicity of ammonia to early life stages of rainbow trout (*Salmo gairdneri*). *Water Res.* **23**, 127–129.

Sowden, T.K. & Power, G. (1985) Prediction of rainbow trout embryo survival in relation to groundwater seepage and particle size of spawning sustrates. *Trans. Am. Fish. Soc.* **114**, 804–812.

Standing Committee of Analysts (1983a) *Biochemical Oxygen Demand 1981. Methods for the Examination of Waters and Associated Materials.* HMSO, London.

Standing Committee of Analysts (1983b) *Methods for Assessing the Treatability of Chemicals and Industrial Waste Waters and their Toxicity to Sewage Treatment Processes 1982. Methods for Examination of Waters and Associated Materials.* HMSO, London.

Stiff, M.J., Cartwright, N.G. & Crane, R.I. (1990) Environmental Quality Standards for dissolved oxygen. WRc Report PRS 2415-M, for the National River Authority.

Stover, E.L. & McCartney, D.E. (1984) BOD results that are believable. *Water/Eng. Manage.* **131**, 37–40, 62, 66.

Strand, S.E. & Carlson, D.A. (1984) Rapid BOD measurement for municipal wastewater samples using a biofilm electrode. *J. Water Pollut. Contr. Fed.* **56**, 464–467.

Sukatsch, D.A. & Dziengel, A. (1984) *Biotechnology: A Handbook of Practical Formulae.* Longman, London.

Szumski, D.S., Barton, D.A., Putnam, H.D. & Polta, R.C. (1982) Evaluation of EPA un-ionized ammonia toxicity criteria. *J. Water Pollut. Contr. Fed.* **54**, 281–291.

Thurston, R.V., Russo, R.C., Luedtke, R.J., Smith, C.E., Meyn, E.L., Chakoumakos, C., Wang, K.C. & Brown, C.J.D. (1984) Chronic toxicity of ammonia to rainbow trout. *Trans. Am. Fish. Soc.* **113**, 56–73.

Thurston, R.V., Russo, R.C. & Vinogradov, G.A. (1981) Ammonia toxicity to fishes. Effect of pH on the toxicity of the un-ionized ammonia species. *Environ. Sci. Technol.* **15**, 837–840.

Tomasso, J.R., Goudie, C.A., Simco, B.A. & Davis, K.B. (1980) Effects of environmental pH and calcium on ammonia toxicity in channel catfish. *Trans. Am. Fish. Soc.* **109**, 229–234.

Turnpenny, A.W.H. & Williams, R. (1980) Effects of sedimentation on the gravels of an industrial river system. *J. Fish. Biol.* **17**, 681–693.

US EPA (1985) Ambient water quality criteria for ammonia—1984. United States Environmental Protection Agency. EPA 440/5–85–001.

US EPA (1986) Ambient water quality criteria for dissolved oxygen. United States Environmental Protection Agency. EPA 440/5–86–003.

Whitelaw, K. & Solbé, J.F. de L.G. (1989) River catchment management: an approach to the derivation of quality standards for farm pollution and storm sewage discharges. *Water Sci. Technol.* **21**, 1065–1076.

Whitmore, C.M., Warren, C.E. & Doudoroff, P. (1960) Avoidance reactions of salmonid and centrarchid fishes to low oxygen concentrations. *Trans. Am. Fish. Soc.* **89**, 17–26.

Whitworth, W.R. (1968) Effects of diurnal fluctuations of dissolved oxygen on the growth of brook trout. *J. Fish. Res. Bd. Canada* **25**, 579–584.

Woodward, G.M. (1970) Determination of Biochemical Oxygen Demand, Chemical Oxygen Demand and Total Organic Carbon. *Proc. Soc. Analyt. Chem.* **7**, 94–95.

Wuhrmann, K. & Woker, H. (1948) Experimentelle Untersuchungen uber die Ammoniak-und Blausaurevergiftung. *Schweiz. Z. Hydrol.* **11**, 210–244.

Zischke, J.A. & Arthur, J.W. (1987) Effects of ammonia levels on the fingernail clam, *Musculium transversum*, in outdoor experimental streams. *Arch. Environ. Contam. Toxicol.* **16**, 225–231.

PART 2
GENERAL

Chapters are collected together in this part that make various contributions to the risk assessment and management exercises described in Chapter 1. The all-important fate models are critically reviewed in Chapter 15; an element of this—estimating the way chemicals are likely to partition between organic and aquatic phases—is described in Chapter 13. Methods of predicting various ecotoxicologically important properties of chemicals from their structures and general properties are addressed in Chapter 14. As well as presenting the general principles involved in understanding the ecotoxicology of mixtures of chemicals (see Introduction to part I), Chapter 12 provides a systematic way of analysing and managing the complex mixtures issuing from specific effluents. On the other hand, Chapter 16 describes how ecotoxicology is used in the development of emission controls in general, and also reviews methods of prioritizing chemicals for regulatory attention.

12: Complex Mixtures

J. DOI

12.1 INTRODUCTION

This chapter deals with how the toxic effects of complex chemical mixtures in aqueous environments are addressed. A definition of a 'complex mixture' is any aquatic matrix that contains a myriad of chemicals existing together, be it produced naturally or by humans. This chapter will focus on complex chemical mixtures produced by humans. Complex mixtures include wastewaters, industrial plant effluents, publicly owned treatment works (POTW) or municipal plant effluents, groundwater, stormwater, ambient waters, sediment porewaters, sediment elutriates, hazardous waste leachates or any other aquatic matrix that contains chemicals. The following questions will be addressed: What techniques and guidelines are available for predicting and assessing the toxicity of chemical mixtures in aqueous environments? What analytical methodologies are available for the separation and quantification of the toxic components in a complex mixture of chemicals? Do compounds behave similarly in mixtures as in isolation? Are there general rules/principles that can be applied to the type of toxicological responses expected when different classes of compounds are together as mixtures in aqueous environments? What is the current state-of-the-art in toxicity reduction evaluations (TREs) and toxicity identification evaluations (TIEs)?

The majority of examples and methodologies come from the USA. There are several reasons for this. In the field of assessing complex mixtures from an ecotoxicological point of view, the USA has taken the technical and regulatory lead. Methodologies and regulations involving TREs and TIEs have been worked on by the US Environmental Protection Agency (US EPA) for over 7 years. In addition, the majority of the references that deal with complex mixtures in the context discussed in this chapter come from work carried out in the USA. Finally, many of the methods in this field used by countries around the world are either US methods or based on them.

Historically, wastewater effluents from industrial or municipal wastewater treatment plants in the USA had to pass general water quality conditions such as TSS (total suspended solids), pH or BOD (biochemical oxygen demand, see Chapter 11) tests. Later, chemical-specific requirements were added to their discharge permits. Guidelines such as 'Priority Pollutants' and 'Criteria Documents' determined the chemical-specific nature of wastewater discharge permits. More recently, emphasis has shifted to also controlling surface-water toxics. Biomonitoring has become an additional requirement for wastewater discharges as regulatory agencies moved to an integrated 'water quality-based' approach consisting of chemical and biological methods for controlling discharges of toxic chemicals. Finally, regulatory agencies have required dischargers to rid their wastewater effluents of unacceptable toxicity through the use of TREs or other similar requirements. It is in the methodologies developed for TREs that the issue of how to deal with the toxicological implications of a complex chemical mixture is addressed.

The following example shows one way in which regulatory agencies formulate policy with regulations and guidance documents. The regulatory driving force behind the focus on a water quality-based approach to aquatic toxicity in the USA is the US EPA. It has issued a variety of regulations and guidelines that have formed the current regulatory framework of controlling aquatic toxicity in wastewater discharges. A description of some of the important regulations and guidance documents is shown in Table 12.1.

12.2 THEORY

This section introduces some terms and assumptions that are made when evaluating complex

Table 12.1 US EPA regulations and guidance documents for addressing aquatic toxicity

1977	*Amendment to the Federal Water Pollution Control Act (Clean Water Act)*
1984	*Development of Water Quality-based Permit Limitations for Toxic Pollutants*
1985	*Technical Support Document for Water Quality-based Toxics Control*
1985	*Methods for Measuring the Acute Toxicity of Effluents to Freshwater and Marine Organisms*
1987	*Permit Writer's Guide to Water Quality-based Permitting for Toxic Pollutants*
1987	*Reauthorization of the Clean Water Act*
1988	*Methods for Aquatic Toxicity Identification Evaluations: Phase I Toxicity Characterization Procedures*
1988	*Short-term Methods for Estimating the Chronic Toxicity of Effluents and Receiving Waters to Marine and Estuarine Organisms*
1989	*Toxicity Reduction Evaluation Protocol for Municipal Wastewater Treatment Plants*
1989	*Generalized Methodology for Conducting Industrial Toxicity Reduction Evaluations (TREs)*
1989	*Methods for Aquatic Toxicity Identification Evaluations: Phase II Toxicity Identification Procedures*
1989	*Methods for Aquatic Toxicity Identification Evaluations: Phase III Toxicity Confirmation Procedures*
1989	*Short-term Methods for Estimating the Chronic Toxicity of Effluents and Receiving Waters to Freshwater Organisms*
1992	*Reissuance of the Technical Support Document for Water Quality-based Toxics Control*

mixtures. It is important to define terminology because different people have individual concepts of what is meant by specific terms, and can use them in dissimilar ways.

When evaluating the ecotoxicology of complex mixtures, the terms additivity, synergism, antagonism and independent action are discussed quite frequently. These define the relationship between the environmentally available concentration of the suspected toxicant(s) and the whole effluent toxicity. When the toxicity of the effluent is equal to the sum of the toxicities of the individual constituents that comprise the effluent toxicity, *additivity* is signified. When the effluent toxicity is greater than the sum of the toxicities of the individual constituents, *synergism* is indicated. When the toxicity of the effluent is less than the sum of the toxicities of the individual constituents that comprise the effluent toxicity, *antagonism* is implied. Finally, when the effluent toxicity correlates with the toxicity of an individual toxicant regardless of the other chemicals in the complex mixture, *independent action* is suggested.

Another term that is used frequently is 'toxic unit' or TU. In many cases it is easier to describe toxicity in terms of toxic units rather than LC_{50}. The TU of an effluent is defined as 100 divided by the effluent LC_{50} (as a percentage). The toxicant concentration is converted to a TU by dividing the toxicant concentration by the LC_{50} of the toxicant. If more than one toxicant is present, the concentration of each is divided by its respective LC_{50}, and the TUs can then be summed.

If additivity were always the mechanism of toxicity for effluents, the TIE process would be significantly easier. The individual TUs of each toxicant could be determined and summed together to match the TU of the whole effluent. If synergistic toxicity occurred in every case, i.e. the TU of the whole effluent was more than the sum of the individual toxicants' TUs, TIE procedures would rarely work. If antagonistic or independent action mechanisms of toxicity were applicable in every case, the TIE process would be much more difficult, but possible.

The factors causing complex mixture toxicity are uncertain. However, it is important to use whatever information is available about a particular wastewater as a starting point when trying to determine the source of its aquatic toxicity. In this light it is easier in most cases to gather information about industrial wastewaters than about municipal effluents. Some of the questions that should be addressed before embarking on a TRE are: how well is the waste treatment plant run? How variable are the inputs to the plant? How consistent or seasonal is the flow to the plant? What manufacturing sites feed into the plant? What type of pretreatment is occurring before the water reaches the plant? In some cases, knowing specific information about the possible nature of the toxicity from the gathering of the above information can point to certain types of toxicant groups and can significantly shorten and lower the cost of the TRE. In other cases, having this background information does no good at all in determining the cause of the wastewater toxicity. In a few cases, knowing the background about a particular effluent can actually increase the time and cost of successfully performing a TRE, because the gathered information could point to possible sources of toxicity and the true sources could be completely different.

It may be the experience of the investigators that is the deciding factor in how successfully the toxicity of a complex mixture can be determined. The methods for assessing and quantifying effluent toxicity, described below, are well laid out. However, accurately determining the factors responsible for wastewater toxicity can be a very complex problem, and many decision points are involved when performing a TRE or TIE. An experienced team of scientists, including toxicologists, chemists and engineers, are in many cases necessary to successfully perform TREs on complex mixtures of chemicals in environmental matrices.

12.3 BACKGROUND

There are many different ways to assess the impact of toxicity on an aqueous environment (Volume 1). The most inexpensive and widely used methods are the short-term bioassay toxicity tests. These include tests such as invertebrate bioassays (e.g. *Ceriodaphnia dubia, Daphnia magna*), fish bioassays (e.g. fathead minnow (*Pimephales promelas*), rainbow trout (*Salmo gairdneri*)) and plant bioassays (e.g. algae

(*Selenastrum capricorutum*), duckweed (*Lemna minor*)). They include both acute and chronic testing as well as static versus flow-through methods. Moving further up in complexity and cost are microcosm tests, mesocosm tests and in-stream testing. Only the short-term bioassay toxicity tests will be discussed here. When biomonitoring requirements began to be put into discharge permits (in the mid to late 1980s), there were controversies surrounding the short-term bioassay toxicity tests. Questions such as the following were posed: are these types of tests reproducible and do these simple tests show a correlation with in-stream community effects? Grothe & Kimerle (1985) and Rue *et al.* (1988) examined the reproducibility of aquatic toxicity tests and concluded that the precision was comparable with commonly accepted analytical test methodologies. Schimel (1981), Broderius (1983) and DeGraeve *et al.* (1991) independently showed that short-term bioassays could be successfully performed by reputable laboratories. Dickson *et al.* (1992) showed that short-term bioassay toxicity tests were a reasonable surrogate of in-stream community effects. These results were important for the acceptance of bio-monitoring and short-term bioassay toxicity tests by the regulated community. Other less expensive and even quicker toxicity tests such as the Microtox and *Daphnia magna* IQ Toxicity Test will be discussed as alternatives to traditional short-term bioassay toxicity tests.

The question beyond biomonitoring is what should be done with the biomonitoring results? The answer is very different depending on the regulatory body making the determination. Some agencies require only single-species toxicity tests quarterly or annually; some require additional and more frequent toxicity testing; some require that additional species be tested; still others require that a TRE be performed on the aquatic discharge in question. It is this last regulatory action that will be the central focus of this chapter, because only TREs address the causative agents of unacceptable effluent toxicity in the complex mixture of chemicals that may be found in wastewater effluents and other aquatic matrices.

There are many definitions of a TRE. One is that a TRE is an investigation conducted within a plant or municipal system to: (i) isolate the sources of pollutant(s) causing effluent toxicity, and (ii) determine the effectiveness of pollution control options in reducing the effluent's toxicity (US EPA, 1984). Another is that a TRE is a stepwise process that combines toxicity testing and analysis of the physical and chemical characteristics of causative agents, to zero in the toxicants causing effluent toxicity, and/or on treatment methods that will reduce the effluent toxicity (US EPA, 1987). The purpose of a TRE is to determine how the aquatic toxicity of an effluent can be reduced to an acceptable level. There are two different approaches for performing a TRE and achieving these aims. The first is to use engineering treatment technologies to reduce the wastewater effluent to an acceptable level of toxicity (i.e. treatability approach). The second is to use a combination of biological and chemical tools to identify the factor(s) responsible for wastewater toxicity (called a TIE) and then design a treatment technology to reduce the toxicity of the causative agent(s) to an acceptable level (i.e. identification approach). There are variants of the above approaches which incorporate parts of both treatability and identification options in a single TRE.

12.4 TECHNIQUES FOR ASSESSING AQUATIC TOXICITY OF COMPLEX MIXTURES

As mentioned in Section 12.3, emphasis will be placed on the short-term aquatic toxicity tests as indicators of wastewater toxicity. There are many different methodologies for short-term toxicity tests. The most common currently used ones are the acute and chronic toxicity tests. This chapter will focus on the acute toxicity test because TIE methods have been worked out utilizing this test. Although chronic toxicity tests are used as regulatory tools, straightforward and proven methods are not as available as with acute toxicity tests for TIEs. The US EPA has however published a guidance document on chronic TIEs (US EPA, 1991a).

The basis for the short-term bioassay toxicity tests is that the critical toxicity limit for protecting the aquatic environment is the assurance that the instream waste concentration (IWC) is equal to or less than the no-observed-effect

concentration or NOEC, determined by chronic toxicity testing or by extrapolation from acute tests. The NOEC is 'the highest measured continuous concentration of an effluent or toxicant that causes no observed effect on a test organism' (US EPA, 1985a). The IWC and NOEC are concepts that allow laboratory toxicity tests to be acceptable regulatory surrogates for instream community effects.

Many organisms in different life stages are acceptable to the different regulatory agencies around the world, based on whether the wastewater effluent and receiving stream are freshwater, estuarine or saltwater environments, and the relative sensitivity of different species to the effluent. However, the discussions to follow will focus on only one organism, *Ceriodaphnia dubia*. This has become a widely accepted test organism in the USA for many biomonitoring purposes, and the methodologies used to assess the impact of the aquatic toxicity of a wastewater or other complex mixture have been worked out in detail (Volume 1, Chapter 4). Reasons for concentrating on this organism are several-fold. First, *Ceriodaphnia dubia* is generally sensitive to many toxicants. Second, it is more cost-effective than many other organisms for toxicity testing. This is because the life period is relatively short and therefore both acute and chronic toxicity tests are very short term relative to most other test species; an acute test lasts 48 h and a chronic test is 7 days. Third, the care and maintenance required for the culturing of *Ceriodaphnia dubia* is more streamlined, and much less space is required than for many larger test species. However, proper culturing still requires dedication, much observation and strict feeding, lighting and temperature controls. Finally, the data base of toxicity tests using *Ceriodaphnia dubia* is rapidly increasing in size. If evidence exists for which toxicant is the causative agent in a complex mixture, information on the response of *Ceriodaphnia dubia* to it is likely to be available for reference. It is important to recognize that this organism should not be used to the exclusion of all others; in fact, most permitting agencies require more than one organism for biomonitoring (i.e. the inclusion of an invertebrate, fish and plant species is common). Nevertheless, the choice of *Ceriodaphnia dubia* as the first organism to test is probably a proper and cost-effective one.

12.5 METHODS FOR IDENTIFYING AND QUANTIFYING TOXICITY OF COMPLEX MIXTURES

As mentioned above, it was the advent of TREs that encouraged development of methods to identify and quantify the causative agents of toxicity in complex mixtures such as wastewaters, groundwater, stormwater, etc. In this section several dissimilar methodologies will be described to address the problem of how to assess the toxicity of wastewaters. No details will be given on what criteria are used to determine an unacceptable level of toxicity in a wastewater. This is a regulatory issue and beyond the scope of this chapter. The methodologies are examples of the various types of procedures that are available for performing TREs. These were chosen because they lead the field in TIEs and TREs, show innovative approaches and form the foundation of many other approaches that have been developed. The Phase I, II and III procedures used by US EPA (see below) have been widely applied all over the world to identify the factors responsible for effluent toxicity. Hence they will be discussed in more detail than other approaches.

The first methods that will be discussed involve the identification of the toxic causative agents, i.e. TIEs. In these methods, both aquatic toxicology and chemistry methodologies are used together to identify the toxicants responsible for wastewater toxicity.

12.5.1 Walsh & Garnas — an early example of chemical fractionation

In one of the earliest publications on this topic, Walsh & Garnas (1983) demonstrated that fractionation procedures could be used to isolate and identify the causative agents responsible for the toxicity of freshwater and saltwater wastewaters. These authors were responsible for pioneering fractionation techniques that are used in virtually all TIE methodologies. In most cases with complex mixtures it is analytically difficult, if not impossible, to isolate, identify and confirm the

toxicant(s) responsible for wastewater toxicity without some way to simplify or separate the complex mixture. Walsh & Garnas used column chromatographic techniques with various exchange resins to fractionate a complex mixture into simpler fractions, in this case organic and inorganic fractions. These more simplified fractions could then be evaluated for acute toxicity. If either fraction exhibited toxicity, further fractionation procedures were implemented using ion-exchange resins for inorganic compounds and extraction techniques for organic compounds. Then a combination of chemical analyses and toxicity tests would be performed on each fraction or subfraction exhibiting toxicity, to determine if the toxic component(s) could be identified.

12.5.2 US EPA methods for aquatic toxicity identification evaluations: Phases I, II and III

In the US EPA guidance documents the TIE approach is divided into three phases. Phase I contains methods to characterize the physical and chemical nature of the factors responsible for environmental water toxicity (US EPA, 1988). A revised version was issued in 1991 (US EPA, 1991b). Such characteristics as solubility, volatility, filterability, chelation and adsorption capacity are determined without specifically identifying the toxicants. Only data regarding the general characterization of the environmental water are generated. However, the information produced may also be used to develop treatment methods to remove the acute toxicity without specific identification of the causative agents. Phase II describes methods to specifically identify the toxic compounds (US EPA, 1989a). However, there are limitations to the techniques proposed. Chemical groups such as non-polar organics, ammonia or metals are amenable to these methods, but other chemical classes such as polar organics, polymers and certain ionic species cannot be identified. There are other methodologies that can deal with these chemical groups and will be described later. Phase III describes methods to confirm the suspected toxicants (US EPA, 1989b). These techniques can be used whether the identification of the toxic agents occurred using Phase I or II methods or any other methodologies. Although complete Phase III confirmations are relatively limited, it is important to note that disregarding these confirmation procedures could be dangerous, because the suspected toxicant(s) may not be the actual toxicant(s).

In Phase I, II and III procedures, as well as most other TIE methodologies, environmental water toxicity is tracked using aquatic organisms and standard toxicity tests. The response of the organisms is the basis of TIE methods and therefore the toxicity tests' results must be dependable and meaningful. System and method blanks must be run extensively throughout these TIE methods in order to detect and eliminate artifactual toxicity during characterization manipulations. The latter is a real concern in TIE procedures and must be thought about and dealt with. In Phase I, sources of toxicity artifacts include excessive ionic strength from additions of acid and/or base during manipulations, formation of toxic products by acids and bases, contaminated air or nitrogen sources, inadequate mixing of test solutions, contaminants leached from filters, pH probes, solid phase extraction (SPE) columns and reagents added. On occasion, toxic artifacts are unknowingly introduced into the test system. For example, pH meters with membrane electrodes can act as a source of silver which can reach toxic levels in solutions where pH measurements are being taken. Randomization techniques, careful observation of organism health and exposure times, and the use of organisms of approximately the same age ensure quality toxicity data. Standard reference toxicant tests should be performed with the aquatic test species on a regular basis, and control charts should be developed (US EPA, 1985b).

In Phase II a more detailed quality control programme is required. Interferences in toxicant analysis are initially unknown but as toxicant(s) identifications are made, interferences can be identified and dealt with. Likewise, analytical separation and analysis procedure particulars can be determined while proceeding through the identification methods in Phase II.

In Phase III of a TIE, detail paid to quality control and verification is at its maximum. Confidence intervals for toxicity tests and chemical measurements must be calculated because only then can correlation between concentration of toxicant(s) and environmental water toxicity be checked for significance based on test and measurement variability. Manipulations of the water sample prior to chemical analysis and toxicity testing are minimized to decrease the chance for appearance of artifacts. Strict attention to detail will determine whether these confirmation procedures produce the correct causative chemicals responsible for toxicity of the environmental water tested or an incorrect conclusion is made.

12.5.3 Phase I procedures

An overview of Phase I effluent characterization tests is shown in Fig. 12.1. Relatively simple wet chemistry manipulations for toxicity removal or reduction are performed on the whole effluent. Acute toxicity tests utilizing aquatic organisms are used to determine whether the toxic chemicals have certain physical or chemical characteristics. Two objectives are accomplished during Phase I: (i) the general characteristics of the toxicant(s) are broadly defined; and (ii) information is gathered to indicate whether the toxicant(s) are similar in different effluent samples taken over time. Several patterns of Phase I results are indicative of certain toxicants but, in general, only evidence of characterization of chemical groups that may be the toxicants is provided. For some environmental waters the Phase I tests will provide few or no clues as to the characteristics of the toxicant(s). For such waters other approaches must be tried. Also, originality and innovation may be used to develop other approaches.

12.5.4 Phase II procedures

In Phase II the major objective is to identify the suspected toxicant(s). Since most wastewater effluents are composed of many constituents in a complex chemical mixture, initial efforts are most productively directed towards separating the toxic constituents from the non-toxic ones. The desire to identify the toxicant(s) quickly without separation techniques is tempting, but in most cases should be avoided. A common though very powerful identifying analytical technique is gas

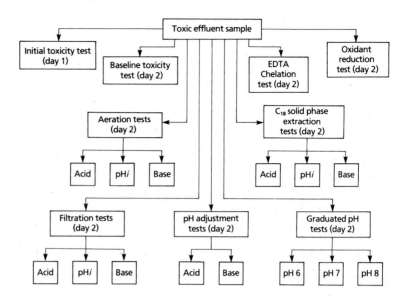

Fig. 12.1 Overview of Phase I effluent characterization tests: pH*i*, initial pH.

chromatography/mass spectrometry (GC-MS). Identification of compounds by GC-MS and subsequent correlation with the associated toxicity is extremely difficult to do on complex mixtures for several important reasons: (i) absolute identification of compounds is rare even with this best of identifying techniques for non-polar organics; (ii) toxicity data for many organic compounds are not usually available; (iii) pure standards of suspected toxicants are often not commercially available in order to measure their toxicity; (iv) the type of interaction (e.g. additivity, synergism, antagonism, independent action; see Section 12.2 p. 290) must be known before correlation of chemical concentrations with toxicological effect can be determined, and this is rarely known quickly for a complex chemical mixture. Because of these factors the separation of toxic constituents from non-toxic ones is the basis for Phase II. A diagram of the general procedures used in Phase II to identify non-polar organic toxicants is shown in Fig. 12.2. The majority of the Phase II protocol is dedicated to the identification of compounds in this class of toxicants. In this procedure, C_{18} SPE columns are used to extract non-polar organic compounds from an effluent sample. If toxicity tests show that the toxicity lies on the SPE column, selective elution of compounds is accomplished by sequentially eluting the column

with solvent mixtures of decreasing polarity. Each fraction is tested for toxicity. The fractions which exhibit toxicity are concentrated on another C_{18} SPE column. The total quantity of organic compounds is extracted off the column, and run through an HPLC fractionation step. The HPLC fractions are collected and tested for toxicity. The

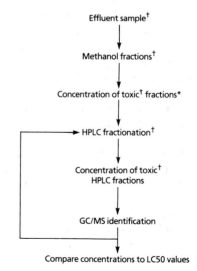

Fig. 12.2 Schematic representation for Phase II identification of C_{18}-removed toxicants. [+], Toxicity tests; [*] in rare cases GC-MS can be useful here.

toxic fractions are concentrated once again and tested for toxicity before analysing by GC-MS. Rough correlation of suspected non-polar organic toxicant(s) concentrations with LC_{50} values for individual compounds are made. If correlations are observed, Phase III procedures may then be used.

Phase II also covers ammonia toxicity identification techniques. Ammonia toxicity can be separated from toxicity of other neutral compounds by the fact that ammonia toxicity increases as the pH increases (Chapter 11). However, some basic compounds also exhibit this trait. Two additional methods can be used to substantiate ammonia toxicity: the use of equitoxic solutions of effluent and use of a zeolite resin to remove ammonia from solution. Ammonia is unique in that the concentration of un-ionized ammonia increases as the pH is raised, and therefore greater toxicity is seen, but the toxicity of un-ionized ammonia increases as the pH is lowered. The equitoxic approach combines these conflicting toxicological effects of pH on ammonia to prepare widely different concentrations of effluent that can be made equally toxic by pH adjustment and dilution (US EPA, 1985c). The zeolite test uses the fact that these naturally occurring or synthetically created crystalline hydrated alkali–aluminium silicates can be employed as ion exchange resins for removal of the ammonium ion from aqueous solutions. Because zeolites are not specific for ammonium ion, but also exchange with other cations such as heavy metals, and can be used as molecular sieves, filter adsorbents and catalysts, they are only effective in Phase II if ammonia is implicated as the causative toxic agent and other groups of toxicants such as organics and metals play no role.

Cationic metals are the third group of toxicants that are specifically covered in Phase II. The reduction of effluent toxicity through the addition of ethylenediaminetetraacetic acid (EDTA) indicates that cationic metals may be present in the effluent at lethal concentrations. The suspect causative agent(s) is(are) chosen based on correlation of effluent toxicity and metals, concentrations, and changes in toxicity observed during manipulation of water quality characteristics. There are many factors that affect the binding strength of cationic metals with EDTA, such as metal electrophilicity, metal–ligand stability constant, amount of other ligands in the complex mixture that compete for the cations, the metal oxidation states, solution pH, temperature and ionic strength, etc. There are several instrumental techniques for analysing for cationic compounds. Specific identification techniques include atomic absorption (AA), inductively coupled plasma–atomic emission spectroscopy (ICP-AES) and inductively coupled plasma/mass spectrometry (ICP-MS). Chromatographic techniques such as ion chromatography (IC) and ion exchange high-performance liquid chromatography can also be used to separate and quantify cations (and anions). Because the concentration of some cationic metals to elicit a measurable effect on an aquatic test organism is so low, i.e. low part-per-billion range is not uncommon, only the AA and ICP/MS instrumental techniques can analyse down to these levels without some sort of concentration step. Once the concentrations of the cationic metals are known, preliminary information about the toxicity of the identified metals can be obtained from data bases such as AQUIRE (US EPA, 1989c) and QSAR (Chapter 14). With literature information, specific metal standards can be made up at effluent concentrations to determine if a correlation between the toxicity of a sample and its concentrations of cationic metals exists.

The groups of possible toxicants that are not covered in Phases I and II include: polar organic compounds, anionic compounds (both organic and inorganic) and high molecular weight compounds such as polymers. The reason for this is that analysis of these groups of chemicals is difficult and none is carried out routinely. Instrumental techniques that show promise are liquid chromatography/mass spectrometry (LC-MS), electrospray MS (ES-MS) and capillary zone electrophoresis/mass spectrometry (CZE-MS). Once routine analytical techniques are available, they will probably be included in future TIE publications.

12.5.5 Phase III procedures

Phase III involves procedures for confirming that suspected toxicants are the true cause of toxicity.

These are applicable and necessary whether Phase I and II or other methodologies have been used. There are two major reasons to require confirmation procedures. First, and as already noted, effluent manipulations can create artifacts that may lead to erroneous conclusions about the toxicants. In Phase III, effluent manipulations are minimal and therefore artifacts are much less likely to occur. Second, there is a definite probability that the factors responsible for toxicity can change from sample to sample, season to season, high flow versus low flow, etc. Since toxicity is a generic measurement, just measuring toxicity will not reveal such variability. Phase III procedures reveal the presence of variable causative agents. This information is essential so that the proper remedial action may be implemented to remove the effluent toxicity. Phase III procedures are applicable to chronic as well as acute toxicity since additives and manipulations are minimal. A section on confirmation is also included for guidance when the treatability approach (to be discussed later), rather than the toxicity identification approach, is taken. The treatability approach requires confirmation at least as much or more so than the toxicity identification approach.

Because, in Phase III, definitive data that constitute the basis for decisions are generated, absolutely clear test methodologies and analytical measurements must be determined. Paying careful attention to test conditions, replicates, quality of test animals, representativeness of the effluent samples tested, and strict Quality Assurance/Quality Control (QA/QC) of analytical procedures, including blanks and recoveries, are mandatory. Because Phase III requires the synthesis of the toxicological and chemical data, keen observation, intuitive insight and extensive knowledge of toxicology and chemistry are keys to the success of confirming the toxicity of an effluent or other complex mixture. The interaction of a well-trained staff of chemists and toxicologists on a routine basis is essential to the dissemination and discussion of diverse information.

Phase III can involve several approaches for confirming the source of toxicity in an effluent. This list is not all-inclusive and will be added to

as experience is gained. A brief summary of the major approaches is given below.

Correlation approach

This is probably the most important of all approaches described in Phase III. The purpose is to show whether there is a consistent relationship between the concentration of the suspected toxicant(s) and effluent toxicity. For the correlation approach to be useful the toxicity of the effluent must be sufficiently variable to provide an adequate range of LC_{50}s over which to do regression analysis. If more than one toxicant is suspected, and each has a different toxicity, the concentration of each must be adjusted for the different toxicity before they can be summed. To prepare the correlation plots, all toxicity data should be converted to toxic units (TUs). If more than one toxicant is present, the concentration of each is divided by its respective LC_{50} and the TUs can then be summed. In Fig. 12.3 an example of the regression from an effluent from a POTW in which the suspected toxicant was the compound, diazinon, is shown. If a perfect correlation between toxicant concentration and effluent toxicity exists, then the expected line should have a slope of 1.00 and a y-intercept of 0.00; i.e. 1 TU of toxicant should show 1 TU for the effluent and

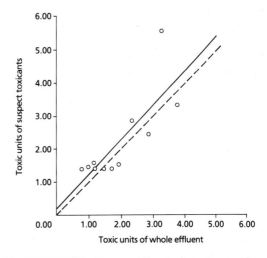

Fig. 12.3 POTW effluent with a single toxicant. After US EPA, 1989b. R2 = 0.63; slope = 1.05; Y-intercept = 0.9. – – –, theoretical; —— observed.

2 TUs of toxicant should show 2 TUs for the effluent, etc. In this example the slope (1.05) and *y*-intercept (0.19) are not much different from theoretical expectations, and indicate that diazinon is probably the major causative toxicant in this effluent.

Experience has shown that there is a strong tendency to subconsciously assume that toxicity is always caused by the same constituents. This assumption can cause possibly very erroneous conclusions.

There are many hindrances that prevent clear-cut correlations between toxicant(s) concentration and effluent toxicity. They include pH, hardness, suspended solids and total organic carbon (TOC) characteristics of the effluent; physical and chemical differences between the effluent and the dilution water when performing toxicity tests; and multiple toxicants that have a minor but measurable effect on the effluent toxicity. Therefore, the correlation approach should be used in conjunction with other Phase III approaches.

Species sensitivity approach

The different sensitivity that various species exhibit with particular toxicants can be used to implicate a causative toxicant. If the suspected toxicants are the true ones, the LC_{50} values of effluent samples with different toxicity to one species will have the same ratio as for a second species of different sensitivity. Further, the ratio for each species should be the same as for known concentrations of the pure toxicant(s). This approach is important because some species may exhibit toxicity from an effluent constituent that the TIE test organism did not. It may then be necessary to revert back to Phase II or even Phase I to characterize the additional toxicant and identify it with the new species.

Spiking approach

In this, suspected toxicants are increased in concentration in the effluent sample, and toxicity is measured to see whether it is increased in proportion to the increase in toxicant concentration. While not conclusive, a proportional increase in effluent toxicity from an increase in toxicant concentration by spiking can give considerable confidence about the identity of the toxicant.

Mass balance approach

This is applicable only to those situations in which the toxicant(s) can be recovered from the effluent. The objective is to account for all toxicity, to ensure that small amounts of toxicity are not being lost in the manipulations.

Symptom approach

Different chemicals may produce very different or similar symptoms in a test organism. It is probable that no symptom is unique to only one chemical. Therefore, while similar symptoms observed between two samples indicate that the toxicants *might* be the same, different symptoms indicate that the toxicants are probably different. By observing symptoms in the effluent, and comparing them to symptoms in organisms exposed to the suspected toxicants, a better understanding of potential causative agents may be obtained.

Deletion approach

In some cases, especially for industrial discharges, suspected toxicants can be removed from a waste stream for short periods by shutting down a process. When this can be done, it offers the most convincing evidence that the suspected toxicants are indeed the true ones. However, the results of such an experiment must be viewed critically, because other compounds could also have been removed, or some effluent characteristic such as pH or ionic strength may also have been changed. These other changes could have caused the removal of effluent toxicity as opposed to the removal of the suspected toxicants.

Often the most laborious and difficult part of the TIE is developing sufficient data to ascertain the cause of effluent toxicity. Frequently, the suspected causative toxicant(s) is(are) found relatively easily. The difficult part is to technically prove that the suspected causative toxic agent(s) is(are) the true one(s).

12.5.6 Doi & Grothe — instrumental approach to performing toxicity identification evaluations

One alternative approach for performing TIEs is the methodology developed by Doi & Grothe (1989). This differs from the US EPA methodology in that much more instrumental analysis is initially performed, and most classes of toxic chemicals may be identified by these procedures. This methodology is close to the conceptual ideas of Walsh & Garnas (1983; p. 293). Column chromatographic techniques such as ion exchange resins, activated carbon and sorbents such as silica gel are used to separate a potentially very complex chemical mixture into organic and inorganic fractions. Each fraction is then evaluated for its acute toxicity to an aquatic organism, and analysed by an appropriate analytical technique. Only the fraction(s) showing toxicity is(are) further evaluated. If the inorganic fraction shows toxicity, then currently available analytical instrumentation can be used to directly identify the inorganic toxicant(s). If the organic fraction contains the acute toxicity of the effluent, another organic separation scheme is used to further fractionate and isolate the toxic components. It is the toxicity of the fractions produced in the frac-

tionation steps that is tracked, identified and eventually correlated with the toxicity of the original sample.

The chemical fractionation scheme used to isolate fractions containing compounds responsible for the toxicity of environmental water samples is shown in Fig. 12.4. Putting the original environmental water through the regime of resins shown in this figure, and submitting each to toxicity and analytical measurements allows a quick determination of which classes of compounds are involved in the toxicity of the original sample. An initial judgement on whether the toxicant(s) is(are) a cation(s), anion(s), metal(s), polar organic(s) or non-polar organic(s) can be made after this simple separation scheme is performed. If the toxicity is associated with the inorganic part of the separation scheme there is usually enough toxicological and analytical information already generated to determine the identity of the inorganic toxic compound(s). If results indicate that the toxicity is associated with the organic part of the separation scheme, another organic fractionation scheme shown in Fig. 12.5 is performed. Performing the organic fractionation scheme provides information on whether the organic toxicant(s) is(are) likely to be

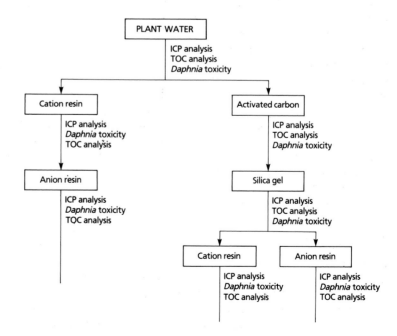

Fig. 12.4 Testing strategy for the evaluation of toxicity in wastewater samples. After Doi & Grothe, 1989.

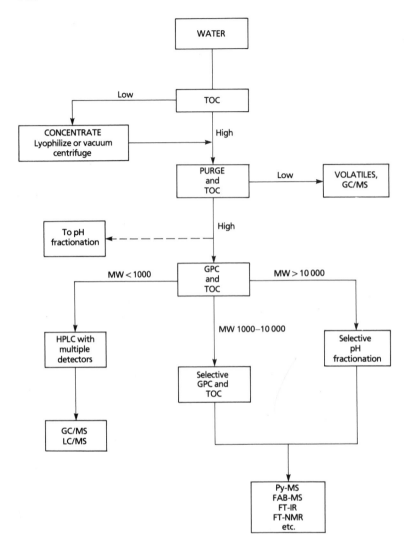

Fig. 12.5 General scheme for the fractionation and analysis of organic compounds in environmental water samples. After Doi & Grothe, 1989.

volatile (purgeable), semi-volatile (extractable), non-volatile (non-purgeable and non-extractable), low or high molecular weight and polar or non-polar. Depending on which compartment(s) the toxicity resides in determines the ease of identifying the compounds causing the environmental water's toxicity. For example, if the toxicity lies in the volatile or semi-volatile compartment, the use of the GC-MS analytical technique allows more routine identification of the toxicant(s) in many cases. If the toxicity resides in the non-volatile, polar or high molecular weight organic fraction, identification of the toxicant(s) will be considerably more difficult, because the analytical methods to routinely identify these classes of organic compounds are not currently available. The type of instrumentation that is available to the analyst, and the amount of expertise in non-routine analytical techniques, will determine whether the toxicant(s) can be identified.

The heavily instrumental analysis methodology discussed above will not be suitable for many permittees. It is intended to be useful for companies and agencies that have the instrumental capabilities in-house and available for use. This scheme offers a quicker and possibly less expensive methodology than other alternative TIE procedures depending on the toxicant(s) found. With proper analytical instrumentation and experienced analysts and toxicologists, most toxicants can be identified and quantified.

In the following two methodologies, the primary emphasis will be on using treatability versus toxicity identification techniques for the methods described above.

12.5.7 Botts — Toxicity Reduction Evaluation protocol for municipal wastewater treatment plants

Botts *et al.* (1989) presented a generalized protocol for conducting a TRE at a municipal wastewater treatment plant (WWTP) or POTW. The overall objectives to be achieved in a municipal TRE are to: (i) evaluate the operation and performance of the POTW in order to identify and correct treatment deficiencies causing effluent toxicity; (ii) identify the toxic compounds causing effluent toxicity; (iii) trace the effluent toxicant(s) and/or toxicity to their sources; and (iv) evaluate, select and implement toxicity reduction methods and technologies to control effluent toxicity.

The overall flowchart for a TRE programme is illustrated in Fig. 12.6. A brief description of major TRE components is presented as follows.

Information and data acquisition

The first step in a TRE is the collection of all information and data pertaining to effluent toxicity. This includes POTW performance data and, if applicable, its pretreatment programme data.

POTW performance evaluation

POTW operating and performance data can be evaluated to indicate possible in-plant sources of toxicity or operational deficiencies that may be allowing toxicity pass-through. In parallel with this, an optional toxicity characterization test (US EPA Phase I) can be performed to indicate the presence of in-plant toxicants caused by incomplete treatment (e.g. ammonia) or routine operating practices (e.g. chlorine). If a treatment deficiency or operating practice is causing effluent toxicity, treatability studies should be conducted to evaluate treatment modifications. If plant performance is not a principal cause of toxicity, or treatment options do not reduce the toxicity, the TRE proceeds to TIE testing.

Toxicity identification evaluations

US EPA Phases I, II and III are performed on the effluent. If specific effluent toxicants are identified, a control method such as local pretreatment limits may be implemented. If additional data are required to determine the nature and sources of the toxicants, a toxicity source evaluation is conducted.

Toxicity source evaluation (Tier I)

The initial stage of a toxicity source evaluation involves sampling the effluent of sewer dischargers or sewer lines and analysing the wastewaters for toxics and/or toxicity. Since influent wastewater is rarely the same as POTW effluent in terms of numbers of chemicals and amounts or toxicity, sewer samples are treated in a simulated biological treatment process prior to toxicity analysis. The choice of chemical-specific analyses versus toxicity tests for source tracking is dependent on the quality of the TIE results. If specific toxicants have been identified with certainty, and can be traced to the responsible sewer dischargers, then chemical-specific analyses are preferred. Toxicity tracking is required when TIE data on specific toxicants are not definitive. If Tier I testing is successful in locating the sources that are contributing to POTW effluent toxicity, a toxicity control method such as local limits can be developed and implemented. If additional information on toxic indirect discharges is needed, further toxicity source testing is conducted.

Toxicity source evaluation (Tier II)

A Tier II evaluation is performed to confirm the suspected sources of toxicity identified in Tier I. Tier II testing involves testing the toxicity of selected sewer dischargers following simulated POTW treatment as in Tier I. Additional characterization steps are used in Tier II to determine

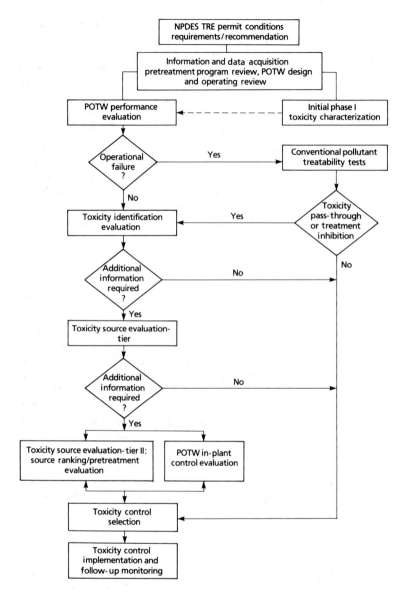

Fig. 12.6 TRE flow diagram for municipal wastewater treatment plant. After Botts *et al.*, 1989.

the relative amount and types of toxicity contributed by each discharger. Tier II information is used to rank the indirect dischargers with respect to their toxicity/toxics loading and to evaluate local limits as a toxicity control option.

POTW in-plant control evaluation

A POTW control evaluation is conducted in parallel with the Tier II assessment to evaluate in-plant options for reducing effluent toxicity. If in-plant control appears to be a feasible approach, treatability testing is used to evaluate methods for optimizing existing treatment processes and to assess options for additional treatment.

Toxicity control selection

Using all the results generated above, alternatives for effluent toxicity reduction are evaluated and the most feasible option(s) is(are) selected for implementation. The choice of a control option(s) is(are) based on several technical and cost considerations.

Toxicity control implementation

The toxicity control method or technology is implemented and follow-up monitoring is conducted to ensure that the control method achieves the TRE objectives and meets permit limits.

The municipal TRE protocol was developed based on results and findings of several TRE and TIE studies. Some of the procedures used in these studies, especially tools for toxicity source evaluations, have not been widely used, and therefore require refinement as experience is gained. Additionally, the feasibility and effectiveness of in-plant and pretreatment toxicity control options have not been well documented and more experience is needed in this area.

12.5.8 Fava — generalized methodology for conducting industrial toxicity reduction evaluations

Fava *et al.* (1989) presented guidance for the performance of TREs at industrial facilities. This generalized methodology is supported by case studies which illustrate various approaches that have been used in the performance of TREs. Because of the numerous differences in operations and complexity of industrial facilities in the characteristics and variability of their effluents (both chemical and toxicological) and in existing wastewater treatment systems, flexibility in the design and performance of a TRE is essential. The approaches must be facility-specific. A generalized flowchart for performing a TRE at an industrial facility is presented in Fig. 12.7. This presents a conceptual overview of the TRE process, illustrating how the steps are linked and indicating when decision points are reached. The framework of the TRE process involves six tiers of action. They are implemented sequentially.

Tier I

The first tier of a TRE process is the acquisition of available data and facility-specific information. The available information can generally be divided into three categories: (i) regulatory information which specifies events leading up to a TRE, defines the regulatory objectives of the study and clearly identifies the target for successful completion; (ii) effluent monitoring data, chemical and biological, which may provide information on the toxicity of the effluent; and (iii) facility and process information which describes the configuration and operation of the facility. These pieces of information are used to define study objectives, identify what is already known and possibly provide clues as to the causes and sources of toxicity.

Tier II

In this tier, an evaluation of remedial actions to optimize the operation of the facility so as to reduce final effluent toxicity is prepared. Three general areas of facility operation are considered: general housekeeping; treatment plant operation; and the selection and use of process and treatment chemicals. This evaluation should be made to identify obvious problem areas, plan and perform remedial actions and determine if these actions reduce the final effluent toxicity to an acceptable level. If the problem appears solved, a monitoring programme is initiated to confirm the solution and to ensure the problem does not recur. If these remedial actions fail to solve the toxicity problem, the study will proceed to the next tier.

Tier III

In this tier, a TIE based on US EPA Phase I, II and III procedures is performed. Because multiple samples are required to perform this tier, a major objective of the TIE is to determine if and how the factors responsible for final effluent toxicity vary over time. Once the TIE has been completed the TRE process can go in two directions. One approach is to evaluate options for treating the final effluent. The other is to identify the source(s) of final effluent toxicity and then evaluate upstream (within-plant) treatment options or process modifications. These two alternatives can be approached separately or a decision can be made

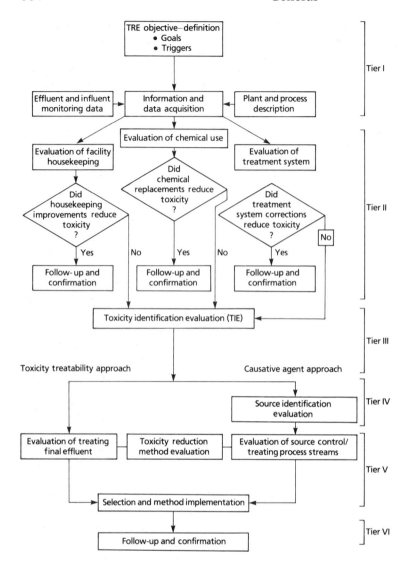

Fig. 12.7 Toxicity reduction evaluation (TRE) flowchart. After Fava *et al.*, 1989.

to pursue both approaches simultaneously and then a selection of the most technically and economically attractive option can be made.

Tier IV

The source identification evaluation (SIE) is performed in this tier. The objective of this evaluation is to identify those process streams that are significant sources of the final effluent toxicity. If a specific toxic chemical had been identified as the causative agent, the SIE would be straight-

forward and have a high probability of success. If, however, no specific causative agent had been found and the cause of toxicity remained in a class of compounds, the SIE would be more complicated. It would include the determination of the characteristics of the toxicity in the process streams feeding into the wastewater treatment. Then a comparison would be made between the characteristics of the process streams and the final effluent, and those process streams which are prime suspects would be identified. In either case the treatability or application of other control methods to these process streams would then be evaluated and the effectiveness confirmed as discussed in Tier V.

Tier V

The evaluation of toxicity reduction methods is performed in this tier. The objective is to identify methods for reducing toxicity in the final effluent and/or source streams. The most technically effective and cost-effective method would be selected and implemented.

Tier VI

Follow-up and confirmation is the sixth and last tier in the TRE process. This becomes operative after the selected method for toxicity reduction has been implemented. Once implemented, continued effluent toxicity testing over time is important to confirm that the toxicity target has been achieved and maintained.

12.6 GENERAL CONCLUSIONS BASED ON TIE AND TRE STUDIES

There are some general conclusions that can be made from the TIE and TRE studies carried out to date. Although these conclusions are based on many TIE and TRE studies, any new sample may show completely different results. The following are therefore only offered as a guide when evaluating new complex mixtures.

1 It is rare for many toxicants to contribute equally to the toxicity of an effluent. It is much more likely for an effluent to have one, two or possibly three causative agents which predominantly produce the toxicity.

2 Relatively few toxicants have a major contribution to the whole effluent toxicity for all effluents which have had the toxicant(s) identified. Certain metals, non-polar organics and ammonia produce the acute toxicity for most effluents that have gone through the TIE process.

3 In many TRE plans, a phase involving a TIE strategy is incorporated. The municipal and industrial TRE plans discussed above are examples of where a TIE is structured into the TRE. Frequently, US EPA's Phase I Guidance Document (Section 12.5.3 p. 294) is used to characterize the toxicity of the effluent. Once the acute toxicity is compartmentalized (i.e. cationic metals, non-polar organics, ammonia, etc.), treatability approaches are used to remove the effluent toxicity. This approach to a TRE is frequently used because of the cost of performing US EPA's Phase II and III procedures. The Phase I procedures are relatively inexpensive; besides being expensive, Phase II and III procedures can be intimidating for many users, and many people who oversee TREs are more comfortable with engineering approaches that are typically used in treatability solutions. Although there is a comfort factor in using this approach to a TRE, less options are available for treatment than when using the full complement of TIE procedures in order to identify the toxicant(s) before treatability options are chosen. A more detailed explanation will be discussed later in this chapter.

4 TIEs and TREs of effluents tend to be site-specific. Although general observations are seen when comparing many sites, it is important to understand that each effluent is unique, and perhaps each different sample of the same effluent over time may also be unique. The data generated from any TIE or TRE must be examined closely in order to make the correct site-specific conclusion.

5 Although many different methodological approaches exist for TIEs and TREs, it is primarily the experience of the person overseeing the TIE or TRE process that will ultimately determine the success of these procedures. The goal of reducing toxicity of a complex chemical mixture cannot be routinely solved by a 'cookbook' approach. Although the methods used in performing TIEs and TREs give exacting procedures and fine detail in many cases, the data generated from these procedures must be reviewed with an experienced eye before a correct interpretation can be made. TIEs and TREs are site-specific, and each mixture may have hidden problems that only an experienced person in the field may be able to find, solve, and then go on to achieve the ultimate goal of reducing aquatic toxicity to an acceptable level.

Below are specific observations based on the TIE and TRE methodological approaches described above.

12.6.1 US EPA Phase I, II and III methods for aquatic TIEs

Mount *et al.* (1992) have made the following observations based on over 100 TIEs carried out in their laboratory.

1 Ammonia toxicity is prevalent in many effluent samples.

2 Organophosphate pesticides play a major role in the acute toxicity of effluents at many sites in the southeastern United States.

3 Metals such as Cu, Cr, Ni, Zn and Cd are among the primary toxicants when metal toxicity is the issue.

4 Where effluent sites are composed of both municipal and industrial inputs, it is more commonly found that the primary sources of the aquatic toxicity are from the municipal, not the industrial, sector.

12.6.2 Doi & Grothe — instrumental approach to performing TIEs (Section 12.5.6, p. 299)

1 In the majority of cases, simple toxicants such as ammonia, metals such as chromium and high levels of salts such as calcium and chloride caused the acute toxicity of effluent samples at various industrial plant sites.

2 Only in a few cases was the toxicity caused by organic chemicals, and the TIE procedures were costly and results were complicated. In these cases, sophisticated instrumentation such as LC-MS, electrospray-MS and capillary zone electrophoresis were required to identify the organic toxicants. It is unlikely that most laboratories would have been able to identify these toxicants without using these sophisticated instrumental analytical techniques.

3 In only one case was the toxicity apparently caused by a combination of organic and inorganic chemicals that separately showed no acute toxicity. It was only in the combination of three compounds that toxicity appeared and correlated with the toxicity of the whole effluent.

12.6.3 Botts — TRE protocol for municipal wastewater treatment plants (Section 12.5.7 p. 301)

1 At one large POTW site, an evaluation of WWTP operations indicated that waste treatment per-

formance was not the cause of effluent toxicity. The TIE portion of the TRE identified non-polar organic compounds as the major constituents causing the effluent toxicity. However, specific analysis did not lead to definitive identification of the specific toxicants. The TIE results showed that the toxic non-polar organic compounds tended to sorb to solids due to their physical and chemical properties. It was the solids that were the major toxic fraction. An evaluation of toxic industrial wastewater samples from selected industries was performed to determine the major contributors of refractory toxicity to the WWTP.

2 At another POTW site the TIE portion of the TRE showed that the non-polar class of compounds was causing the effluent toxicity. Continued toxicity monitoring indicated an abrupt cessation of toxicity before identification of the toxicants or treatment options could be determined.

3 A third POTW site used primarily acute toxicity testing at different inputs to the municipal wastewater treatment plant to identify the particular site of the major contributor of the WWTP toxicity. This screening technique is useful for determining WWTP influent toxicity. However, this approach alone cannot determine the source of the WWTP effluent toxicity. Further tests are required for this determination. This is an inherent shortcoming of this approach. There may or may not be a correlation between influent toxicity and refractory effluent toxicity.

12.6.4 Fava — generalized methodology for conducting industrial toxicity reduction evaluations (Section 12.5.8 p. 303)

1 Of the 10 case studies referenced in this report, three showed non-polar organics to be the toxicants, two showed non-ionic surfactants, one showed a metal and residual chlorine, one indicated only metals, one showed ammonia, one showed various inorganic compounds, and one did not identify any class of toxic compounds. These case studies show the variety of toxicants that may be present in any effluent sample, but commonly, the major causative agents are one or two compounds only.

2 In many cases the actual toxic compounds were not definitively identified, especially when the toxicants were organic. However, once the compound class that caused the effluent toxicity was determined, treatability plans were put into effect and generally solved the toxicity problems.

12.7 CONCERNS AND LIMITATIONS OF TIE AND TRE METHODOLOGIES

The science of TIEs and TREs has been tremendously improved over the past 5 years. However, there are still some serious concerns and limitations when dealing with toxicity issues and complex chemical mixtures. Major ones will be discussed below.

12.7.1 Methodologies for performing TIEs and TREs are not scientifically complete

Even though there is a considerable amount of technical knowledge and experience behind TIE and TRE methodologies, these cannot be treated as scientifically complete. If every procedure in these methodologies were performed exactly as written, there would be cases where the toxicity problem could not be solved. Because of the site-specific nature of effluents, and their variability over time, the identification and/or reduction of the causative toxic agents is difficult to pinpoint with complete certainty. There will be times when the constituents causing effluent toxicity cannot be identified even with the most sophisticated instrumentation and most experienced toxicologists and chemists.

There will be other cases when the solution to the problem is found, but is so costly that it is not economically feasible to implement the proposed solution. Alternatives that are cost-effective which may somewhat compromise the ultimate goal of no toxicity at toxic amounts may have to be developed and agreed by the regulatory agency.

12.7.2 How do the mechanisms of toxicity, i.e. additivity, antagonism, synergism and independent action, complicate TIE procedures and conclusions?

If the toxicity of an effluent were always additive in nature (Section 12.2 p. 290), the TIE process would be significantly easier. If synergistic toxicity happened in every case, TIE procedures

would rarely work, because no correlations could be determined. If antagonistic or independent action toxicity occurred in all cases, the TIE process would be difficult and complex, but solutions would be possible. Fortunately, in many cases, the additivity of toxicity whereby the toxicity of compounds separately is the same as when put together in a mixture, occurs in many instances. In most of the remaining cases, antagonistic and independent action toxicity is usually observed. Rarely has synergistic toxicity in effluents been seen. If antagonistic toxicity is the predominant mechanism of toxicity, only the toxicity confirmation procedures would show this to be the case. Tests must be carried out with extreme care to be able to differentiate between antagonism and artifactual toxicity.

It is important to note that the type of water that the effluent contains may determine the amount of toxicity that a chemical has in that effluent. For example, it is well documented that certain cationic metals such as Cu, Cr, Cd, Zn, Ni, etc., are less toxic in hard than in soft water. The salts mitigate the toxicity of the cationic metals. It is also known that the toxicity of some organic compounds is affected by certain water quality parameters such as pH or ionic strength. This makes it very important to carefully choose the dilution water to be as similar as possible to the effluent as measured by water quality parameters when performing the toxicity tests, so that these types of results are not artificially produced. In general the following statements regarding compound classes can be made:

1 With regard to metals, whether they are cationic or anionic (e.g. chromate), water quality parameters such as hardness and pH, as well as certain organic classes of compounds such as chelators (e.g. EDTA, nitrilotriacetic acid (NTA)) will affect toxicity.

2 With respect to ammonia, pH will dramatically affect its toxicity. In addition, certain metals may bind the ammonia and make it biologically unavailable, and therefore reduce its toxicity.

3 For non-polar organic compounds, few other classes of compounds or water quality parameters affect their toxicity. However, physical binding to sediment, sludge or soil could make these compounds biologically unavailable and therefore attenuate their toxicity.

4 In the case of polar organic compounds, pH and ionic strength can affect toxicity. Because the toxicity of organic ions is generally much lower than their neutral forms, pH, which controls whether the organic acid or base is in its neutral or ionic form, has a major effect on toxicity of these compounds.

12.7.3 What compound classes tend not to be amenable to TIEs?

In general, inorganic toxicants are simpler to identify than organic ones. Part of the reason is that there are many fewer inorganic than organic toxicants. Second, the analytical techniques for the identification and quantification of inorganic compounds are much more developed than those for organic compounds. With ICP-MS, AA or IC instrumentation, most inorganic compounds can be identified and quantified. There is no corresponding analytical technique that can uniquely identify the majority of organic compounds. Although GC-MS is used successfully for some TIEs, only 10–20% of all organic compounds are analysable by this technique. LC-MS has the potential to analyse for the majority of organic compounds, but the current state-of-the-art is not close to the point where many organic compounds can be routinely identified and quantified by this instrumental analytical technique. The types of organic compounds that are more likely to be identified by current analytical techniques are volatile and semi-volatile non-polar organic compounds. Non-volatile and polar organic compounds are difficult to analyse by current techniques. In the future, LC-MS, ES-MS and CZE-MS may provide the solution for analysing these classes of organic compounds.

12.7.4 What are the costs associated with TIEs and TREs?

This is a difficult question to answer because of the variability of the problem. Performing TIEs can be very simple to extremely complex. The cost of TIEs obviously will reflect the amount of effort to solve the problem. In US dollars (1993 prices), the cost of a TIE programme could be as little as several thousand dollars to hundreds of

thousands of dollars for a particularly difficult and variable complex mixture. TREs are considerably more expensive than TIEs.

12.8 ALTERNATIVE RELATIONSHIPS BETWEEN COMPLEX MIXTURES AND TOXICOLOGICAL EFFECT

For most of this chapter the discussion has centred on the relationship between complex chemical mixtures and ecotoxicological effects using simple acute aquatic toxicity testing methods. There are alternative short-term toxicity tests relative to invertebrates, algae and fish that are both less expensive and much shorter-term (Volume 1). These include Microtox and the *Daphnia magna* IQ Toxicity Test. They are measured in minutes, rather than days for traditional short-term acute toxicity tests.

12.8.1 Microtox

This has been available for over 10 years (see also Volume 1, Chapter 2). It measures the amount of light generated by living bioluminescent marine bacteria, *Photobacterium phosphoreum*. The Microtox instrument measures the light output of the bacteria before and after they are exposed to a test chemical. The degree of light loss is indicative of metabolic inhibition of the test organism. This inhibition is related to the toxicity of the test chemical. Microlitre volumes only of test solution are needed. Multiple tests can be carried out in its 30-sample wells, and the test only takes a few minutes. However, Microtox has gained only limited acceptance because no regulatory agency has agreed to its use as a routine acute toxicity testing species. Microtox has been used effectively as a screening tool for effluents of known consistency, and when a correlation between Microtox results and an accepted test species has been previously determined. In this case Microtox can be a quick, inexpensive screening tool to monitor effluent upsets or problems.

12.8.2 *Daphnia magna* IQ Toxicity Test or IQ TEST

This test has been available in kit form for the last three years. The IQ TEST exposes *Daphnia magna* to a test chemical for a 1-h period. Then, the IQ additive is introduced to the test system and allowed to incubate for 15 min. After this time, the test system is illuminated with ultraviolet light and scored visually. An organism in the test concentration series that does not fluoresce as brightly as the control organisms is adversely affected and indicates toxicity at that concentration. Based on some comparison tests with agricultural pesticides carried out by the manufacturer, the IQ TEST appears to be more sensitive and accurate at predicting 48-h *Daphnia magna* values than Microtox. There is no guarantee that the sensitivity and accuracy observed for agricultural pesticides would hold for other chemical classes of compounds.

Alternative toxicity parameters can be followed other than acute toxicity measurements. The use of chronic toxicity has been discussed earlier. Sediment toxicity is another element that can be followed for complex mixtures (US EPA, 1991c). Another issue that is of concern is the assessment and control of bioconcentratable materials in complex mixtures (US EPA, 1991d).

12.9 CONCLUSIONS

This chapter has reviewed the current thinking behind the assessment and control of toxicity in environmental complex chemical mixtures. It has focused on the use of standard acute toxicity tests combined with physical and chemical separation techniques and analytical identification approaches to deal with identifying sources of toxicity in complex mixtures as well as toxicity treatment techniques. Several different approaches were evaluated in order to show the broad diversity of carrying out toxicity identification and toxicity reduction evaluations. These approaches are not the only methodologies that are available but are useful as examples of how TIEs and TREs can be performed. The US EPA Phase I, II and III TIE Guidance Documents (Section 12.5.2, p. 293) are especially noteworthy in that they are widely used, have a strong technical backing, and are accepted as standard methods by many regulatory agencies. Results, observations and conclusions based on TIE and TRE schemes have been discussed. Finally, other

options for dealing with complex chemical mixtures were briefly considered.

The manner in which complex mixtures are dealt with in the future will change as we discover more about chemicals and their interactions, bioavailability of chemicals in environmental matrices, the mechanisms through which organisms are physiologically affected by chemicals, and better analytical techniques become available to allow more definitive identification and quantification of chemicals.

REFERENCES

Botts, J.A., Braswell, J.W., Zyman, J., Goodfellow, W.L. & Moore, S.B. (1989) *Toxicity Reduction Evaluation Protocol for Municipal Wastewater Treatment Plants.* US EPA. Project Contract Number 68–03–3431. EPA/600/2–88/062. Cincinnati, OH.

Broderius, S.J. (1983) *Analysis of an Interlaboratory Comparative Study of Acute Toxicity Tests with Freshwater Organisms.* US EPA. Duluth, MN.

DeGraeve, G.M., Cooney, J.D., McIntyre, D.C., Pollock, T.L., Reichenbach, N.G., Dean, J.H. & Marcus, M.D. (1991) Variability in the performance of the seven-day fathead minnow (*Pimephales promelas*) larval survival and growth test: an intra- and interlaboratory study. *Environ. Toxicol. Chem.* **10**, 1189–1203.

Dickson, K.L., Waller, W.T., Kennedy, J.H. & Ammann, L.P. (1992) Assessing the relationship between ambient toxicity and instream biological response. *Environ. Toxicol. Chem.* **11**, 1307–1322.

Doi, J. & Grothe, D.R. (1989) Use of fractionation and chemical analysis schemes for plant effluent toxicity evaluations. In: *Aquatic Toxicology and Environmental Fate,* Vol. 11 (Eds G.W. Suter II and M.A. Lewis) pp. 123–138. ASTM STP 1007. Philadelphia.

Fava, J.A., Lindsay, D., Clement, W.H., Clark, R., DeGraeve, G.M., Cooney, J.D., Hansen, S.R., Rue, W.J., Moore, S.B. & Lankford, P.W. (1989) *Generalized Methodology for Conducting Industrial Toxicity Reduction Evaluations (TREs).* US EPA. Project Contract Number 68–03–3248. EPA/600/2–88/070. Cincinnati, OH.

Grothe, D.R. & Kimerle, R.A. (1985) Inter- and intralaboratory variability in *Daphnia magna* effluent toxicity test results. *Environ. Toxicol. Chem.* **4**, 189–192.

Mount, D.I., Amato, J.R., Durhan, E.J., Lukasewyez, M.T., Ankley, G.T. & Robert, E.D. (1992). An example of the identification of diazinon as a primary toxicant in an effluent. *Environ. Toxicol. Chem.* **11**, 209–216.

Rue, W.J., Fava, J.A. & Grothe, D.R. (1988) A review of inter- and intralaboratory toxicity test method variability. In: *Aquatic Toxicology and Hazard Assessment*, Vol. 10. (Eds W.J. Adams, G.A. Chapman and W.G. Landis) pp. 190–203. ASTM STP 971. Philadelphia.

Schimel, S.C. (1981) *Results Interlaboratory Comparison—Acute Toxicity Tests Using Estuarine Animals*. US EPA, Gulf Breeze, FL.

US EPA (1984) Policy for the development of water quality-based permit limitations for toxic pollutants. *Fed. Reg.* **49**, FR9016. US EPA Washington.

US EPA (1985a) *Technical Support Document for Water Quality-based Toxics Control*. EPA 440/4–85–032.

US EPA (1985b) *Methods for Measuring the Acute Toxicity of Effluents to Freshwater and Marine Organisms*, 3rd edn. EPA-600/4–85/013. Cincinnati, OH.

US EPA (1985c) *Ambient Water Quality Criteria for Ammonia*. EPA-440/5–85–001. Duluth, MN.

US EPA (1987) *Permit Writer's Guide to Water Quality-based Permitting for Toxic Pollutants*. EPA-440/4–87–005. US EPA Washington.

US EPA (1988) *Methods for Aquatic Toxicity Identification Evaluations: Phase I Characterization Procedures*. EPA-600/3–88/034. Duluth, MN.

US EPA (1989a) *Methods for Aquatic Toxicity Identification Evaluations: Phase II Identification Procedures*. EPA-600/3–88/035. Duluth, MN.

US EPA (1989b) *Methods for Aquatic Toxicity Identification Evaluations: Phase III Confirmation Procedures*. EPA-600/3–88/036. Duluth, MN.

US EPA (1989c) *Aquatic Toxicity Information Retrieval (AQUIRE) Database*. Environmental Research Laboratory—Duluth, Scientific Outreach Program. Duluth, MN.

US EPA (1991a) *Toxicity Identification Evaluation: Characterization of Chronically Toxic Effluents, Phase I*. EPA-600/6–91/005. Duluth, MN.

US EPA (1991b) *Methods for Aquatic Toxicity Identification Evaluations: Phase I Characterization Procedures*, 2nd edn. EPA-600/6–91/003. Duluth, MN.

US EPA (1991c) *Toxicity Identification Evaluations for Sediment: Phase I (Characterization), Phase II (Identification), Phase III (Confirmation) Modifications of Effluent Procedures*. EPA-600/6–91/007. Duluth, MN.

US EPA (1991d) *Assessment and Control of Bioconcentratable Contaminants in Surface Waters*. Draft. Duluth, MN.

Walsh, G.E. & Garnas, R.L. (1983) Determination of bioactivity of chemical fractions of liquid wastes using freshwater and saltwater algae and crustaceans. *Env. Sci. Technol.* **17**, 180–182.

13: The Octanol–Water Partition Coefficient

D.W. CONNELL

13.1 INTRODUCTION

In the late 1800s various researchers discovered that a pure compound partitioned between two phases in a constant ratio essentially independent of the concentration of the compound. This constant ratio was considered to be a physicochemical property of the compound relative to the properties of the two phases involved. It was described as the 'distribution ratio', later to be called the 'partition coefficient'.

Charles Overton and Hans Meyer put this characteristic to good use at about the turn of the century. They found that when the two phases were lipid and water then the partition coefficients were related to the narcotic effect of the chemicals. This relationship was found to be consistent with a range of biota including tadpoles, fish, crustaceans and daphnids. The work culminated in Overton's book *Studien Uber die Narkose* (Studies of Narcosis), published in 1901.

Later Hansch *et al.* (1962) suggested that *n*-octanol would provide a more satisfactory lipoid phase than the oily fats previously used by Overton and Meyer and other researchers. Octanol resembled a fat in having a polar group together with a long alkyl chain. It also had the practical advantages that it was readily available and could be prepared in a highly pure form. Since then a considerable volume of data has been accumulated on the octanol–water partition coefficient for a wide range of compounds. This information has been principally used in the development of relationships to evaluate pharmacological properties. As a general rule the octanol–water partition coefficient has proved to be a valuable predictor of the biological properties

of chemicals. It is now the most widely available and used physicochemical property for this purpose.

In recent years the environmental properties of chemicals have been under intensive investigation. For example, Hamelink *et al.* (1971) proposed that the bioconcentration of chemicals in water by fish was a partition process. As a result of this a range of relationships has been established between the octanol–water partition coefficient and bioconcentration (Mackay, 1982; Connell, 1988a). Könemann (1981) also established a relationship between toxicity to fish and the octanol–water partition coefficient for a range of environmental pollutants. Many important environmental processes have been shown to be related to the octanol–water partition coefficient, e.g. the sediment–water partition coefficient and water-solubility. This value is now the key parameter in evaluating the environmental properties of organic compounds. The objectives of this chapter are to review the nature and use of this important characteristic.

13.2 NATURE OF THE OCTANOL–WATER PARTITION COEFFICIENT

The octanol–water partition coefficient is principally used for hydrophobic organic compounds and is considered to be a measure of their hydrophobicity. These substances, by definition, are soluble in non-polar solvents, such as hexane and octanol, and sparingly soluble in water. The 'hydrophobic effect' causes the partitioning between water and octanol, and originates from the polarity of the water molecule and the non-polarity of octanol and the hydrophobic com-

311

pound itself. Water is a polar molecule with a moderately high dielectric constant and dipole moment (see Table 13.1). Polar solutes dissolve in water, and the strong attractive forces between the polar solvent and the polar water molecules stabilize the matrix which forms on solution. An example of this is the ready solution of ethanol in water (see Table 13.1). On the other hand, compounds such as alkanes, aromatic hydrocarbons and chlorinated hydrocarbons are non-polar, as illustrated by the data in Table 13.1 where benzene, carbon tetrachloride and cyclohexane have low dielectric constants, and dipole moments of zero. The solution of these molecules into water does not result in bonding between solute molecules and the water solvent; in fact the non-polar molecules distort and displace the overall matrix of water molecules which are held together by strong intermolecular forces. A surface is formed by water around the non-polar molecule. This results in the water molecules orientating to form hydrogen bonds between adjacent molecules and a consequent intermolecular attractive force is generated. The forces generated by the insertion of a non-polar molecule into the water solvent tend to reduce the water-solubility of non-polar substances. Therefore it would be expected that non-polar substances would be sparingly soluble in water and soluble in non-polar octanol. The 1-octanol–water partition coefficient (K_{ow}) is a defining characteristic of the hydrophobic class of organic compounds and is defined as

$$K_{ow} = C_o/C_w$$

where C_o and C_w are the concentrations of the solute in 1-octanol and water, respectively, when the octanol–water system is at equilibrium.

The octanol–water partition coefficient is often represented by other symbols, particularly P and P_{ow}. The values of C_o and C_w are expressed in the same units so that K_{ow} is unitless. Compounds readily soluble in fat are usually described as lipophilic compounds and have values of K_{ow} ranging from about 100 to about 1 000 000. This range of data is difficult to present effectively on a linear graph, and so the logarithms of the K_{ow} values are commonly used; thus log K_{ow} values usually range from about 2 to about 6 for lipophilic compounds.

It is important to note that octanol and water are mutually soluble. At equilibrium, octanol contains 2.3 mol/l of water and water contains 4.5×10^{-3} mol/l of octanol. An estimation of the K_{ow} value can be obtained as the ratio of the solubility of the test compound in octanol and in water. However, the value obtained takes no account of the alteration in the characteristics of both the octanol and the water, due to the mutual solubility of these two solvents. Thus the properties of pure octanol and pure water cannot be accurately applied to the octanol–water system (Lyman et al., 1990). Niimi (1991) has evaluated the relationship between K_{ow} and the solubility ratio of compounds in octanol and water ($K_{oct/wat}$) and found the following regression equation

$$K_{ow} = 0.761 + 0.851 \ (\pm 0.122) \ K_{oct/wat} \quad (n = 45; \ r = 0.91)$$

In the measurement of the octanol–water partition coefficient the test compound should be in the same chemical form in both octanol and

Table 13.1 Dipole moments and dielectric constants of some representative organic compounds

Compound	Dielectric constant (ε)	Dipole moment (Debyes)	Aqueous solubility (mg/l)
Water	80.4	1.85	∞
Ethanol	25.7	1.69	∞
Benzene	2.3	0	1780
Carbon tetrachloride	2.2	0	1160
Cyclohexane	2.0	0	56

water. Most lipophilic compounds are non-polar neutral non-electrolytes such as hydrocarbons and chlorohydrocarbons. However, some compounds of interest are weak electrolytes. This means that the proportion of compound in the ionic form could be different in water and in octanol in the same system at equilibrium. As a result, the K_{ow} values would not be representative of either the un-ionized or ionized form. In addition the apparent K_{ow} values could vary depending on the pH of the water used in the experiment. The OECD *Guidelines for Testing of Chemicals* (OECD, 1981), which describe the measurement of the octanol—water partition coefficient, recommend that the measurement should be made on the non-ionized form by adjusting the pH above or below the pK_a of the compound. In addition the presence of other electrolytes in the water should be controlled, since this may have an influence on the ionization of the solute.

The octanol—water partition coefficient is constant under defined conditions, particularly temperature. K_{ow} values are usually measured at 20°C or 25°C, and there is a dependence on temperature. Lyman (1990) has reported this to be 0.001 to 0.01 log K_{ow} units per degree and may be either positive or negative depending on the solute. Lyman (1990) has also reported that there may be a dependence on the concentration of the solute, particularly at concentrations greater than 0.01 mol/l, but for practical purposes in predicting environmental properties the concentration dependence is trivial.

13.3 MEASUREMENT OF K_{ow} VALUES

Partition coefficients have been measured experimentally since the last century. The traditional methods involved placing the two immiscible solvents together in a vessel, adding a small concentration of the solute (below the maximum solubility) and shaking the vessel for a period of time followed by analysis of the two phases to yield the concentrations contained and subsequently the partition coefficient. The OECD *Guidelines for Testing of Chemicals* (OECD, 1981) have detailed experimental procedures for measurement of the K_{ow} values by this procedure.

A number of more sophisticated techniques have been developed based on this simple 'shake flask' method, which have been described by James (1986a). There are disadvantages with the 'shake flask' method and related techniques. The lipophilic compounds dissolve in very low concentrations in the water phase at equilibrium. The true concentration can be extremely difficult to measure accurately, due to the presence of very small aggregates or micelles of the solute (Tanford, 1973). In order to overcome these problems chromatographic methods and generator columns are often used.

Generator columns are columns containing Chromosorb, or similar material, packed into a column, and onto which an appropriate hydrophobic solvent is coated. This system presents a large interface surface area between the lipophilic and water phases, and so allows equilibrium to be established relatively rapidly. Water is passed slowly through the column, and thus is equivalent to the aqueous phase in octanol—water partitioning. Wasik *et al.* (1983) have found that there is good agreement between K_{ow} values determined using this experimental technique and those determined by the traditional 'shake flask' method.

James (1986a) has described several chromatographic techniques that have been used to estimate K_{ow} values. Paper, thin-layer and gas—liquid chromatography have been commonly used for the estimation of K_{ow}. Probably the most successful technique is high-pressure or high-performance liquid chromatography (HPLC), since this system consists of stationary and mobile phases that are liquids, and the nature of the phases can be most closely arranged to resemble the octanol—water system. As a general rule, with these techniques the chromatographic characteristic is measured for several reference compounds and a relationship established between the chromatographic characteristic and the known K_{ow} values of the reference compounds. The K_{ow} value of an unknown compound is then estimated by the measurement of its appropriate chromatographic characteristic.

Chessels *et al.* (1991) have evaluated the accuracy of experimental measurements of log K_{ow}. These authors plotted the mean experimental

$\log K_{ow}$ values based on five experimental methods of determination versus the standard error for a set of 40 chlorohydrocarbons, hydrocarbons and related compounds. This plot is shown in Fig. 13.1 and shows that compounds having $\log K_{ow}$ values less than about 6 usually give a reasonably high level of accuracy with the various experimental methods available. But the standard error increases substantially with compounds having values greater than about 6, and have errors of almost one $\log K_{ow}$ unit with compounds having $\log K_{ow}$ values in the range from 10 to 12. The various experimental techniques for determining K_{ow} values, referred to above, were found to give similar results for compounds in the range of $\log K_{ow}$ values from 2 to approximately 6. However, above this range deviations between the results obtained with the various techniques were apparent.

13.4 COMPOUND TYPES

There are many organic compounds which are environmental contaminants. Within this wide range of compounds there are some that are particularly appropriate for evaluation of their environmental characteristics using the K_{ow} value. Connell (1990) has summarized the characteristics of organic compounds which result in bioaccumulation, as shown in Table 13.2. The chemical structure characteristics shown in Table 13.2 are most applicable to persistent lipophilic

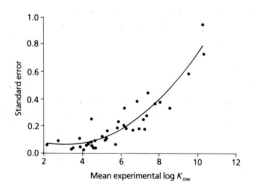

Fig. 13.1 Plot of the mean experimental $\log K_{ow}$ value against the standard error for 40 chlorohydrocarbons, hydrocarbons and related compounds with the regression line. After Chessels *et al.* (1991).

Table 13.2 Some general characteristics of organic chemicals which exhibit bioaccumulation. Adapted from Connell (1990), p. 56

Characteristic	Features giving bioaccumulation
Chemical structure	*High capacity*: high proportion of C−C (aliphatic), C−C (aromatic), C−H and C−halogen bonds
Molecular weight	> 100 giving a maximum capacity at about 350, then declining to very low capacity about 600
Stability	Resistant to degradation reflected in soil persistence in the order of years
Log K_{ow}	>2 giving a maximum capacity of about 6 and a decline to very low capacity at about 10−12
Water solubility (mol/m³)	< 18 giving a maximum at about 0.002 with declining capacity at lower values
Degree of ionization	Very low

compounds such as the chlorohydrocarbons, and to a lesser extent to the polyaromatic hydrocarbons. Stability in the environment must be relatively high for bioaccumulation to occur, and this is a reflection of the stability represented by the chemical structure characteristics mentioned above. However, other groups may give consistent patterns of behaviour in relation to K_{ow}, but relationships may exhibit specific characteristics which reflect the chemical nature and properties of the compounds involved. Anliker *et al.* (1988) have outlined problems in evaluating the K_{ow} value with disperse dyes and its application to bioaccumulation with this group.

13.5 CALCULATION OF K_{ow} VALUES

Hansch & Leo (1979), Lyman (1990) and James (1986a) have reviewed the various methods that are available to calculate K_{ow} values. An overview of these is provided in Table 13.3. They have differing reliability depending on the particular

Table 13.3 Overview of methods for calculation of K_{ow} values

Method	Information required
Substituent constants	Parent compound and additive values for common functional groups
Fragmental constants	Values for structural fragments within a molecule
Parachor	Values for atoms and structural features of a molecule
Molecular connectivity	Regression equation relating molecular connectivity to K_{ow} and molecular connectivity of the test compound
Other partition coefficients	Regression equation relating other partition coefficients to K_{ow} and corresponding partition coefficient of the test compound
Water-solubility	Regression equation relating water-solubility to K_{ow} and water-solubility of test compound

compound involved and the reliability of the relationships used in that particular situation.

One of the earliest methods to estimate K_{ow} was that based on 'substituent constants'. Leo et al. (1971) defined the substituent constant π_x as the $\log K_{ow}$ value due to the addition of a functional group x added to a parent molecule. Thus

$$\pi_x = \log P_x - \log P_H$$

where P_H is the K_{ow} value of the parent molecule and P_x the K_{ow} value of the derivative of P_H with a substituent x.

Thus the $\log K_{ow}$ value of the parent compound is required, as well as the substituent constant. A parent compound that is often used with this approach is benzene, which has a $\log K_{ow}$ value of 3.46. Substituent constants are then added to this value to obtain the $\log K_{ow}$ of other aromatic compounds. The method has found wide application, and with many compounds yields accurate results. However, there are a number of limitations and objections to this approach (Rekker, 1977).

Rekker (1977) and Hansch & Leo (1979) have devised an approach that addresses these difficulties and limitations. The method requires a set of fragmental constants (f) that represent structural fragments within molecules. An evaluation of the $\log K_{ow}$ is obtained as the sum of the structural fragments for the whole molecule. Some fragmental constants are shown in Table 13.4. Using these the $\log K_{ow}$ of a compound can be calculated from the following equation

$$\log K_{ow} = \Sigma \, a_n f_n$$

where a is the numerical factor indicating the incidence of the fragment within the molecule, and f the fragmental constant value. Additionally, corrections are necessary in some situations to allow for the proximity of hydrophilic groups (Table 13.4). Lyman (1990) has assessed this method, and with a variety of different compounds found that the error involved was ± 0.12 $\log K_{ow}$ unit.

As outlined before, the solubility in water, and consequently the K_{ow} value, are related to the size of the cavity formed by the molecule in the water mass. Thus it would be expected that there is a relationship between molar volume and the partition coefficient. The 'parachor' was developed as a measure of molar volume, and has been related to K_{ow} by the following equation (McGowan, 1952)

$$\log K_{ow} = 0.012 \, [P] + E_a$$

where $[P]$ is the parachor and E_a is a correction to account for the interaction of solute and solvent molecules. Briggs (1981) found the following relationship between parachor and $\log K_{ow}$

$$\log K_{ow} = 0.011 \, [P] - 1.2n - 0.18 \quad (n = 26; \, r = 0.95)$$

where n is taken as one for each oxygen atom not bonded or conjugated to an aromatic ring, one for each singly bonded nitrogen atom, one for each heterocyclic aromatic ring and 0.025 for each

Table 13.4 Some fragmental constants for calculation of K_{ow} values. Adapted from James (1986a), pp. 317 and 319

Fragment	Constant	Fragment	Constant
Aliphatic compounds			
H	0.21	OH	−1.440
CH	0.236	Br	0.24
Cl	0.06	$CH = CH_2$	0.93
CH_3	0.702	CH_2	0.527
NH_2	−1.380	O	−1.536
C_6H_5	1.896	C (quat)	0.14
Aromatic compounds			
H	0.21	C_6C_4	1.719
Cl	0.943	COO	−0.43
OH	−0.359	NH	−0.93
C_6H_5	1.90	N	−1.06
COOH	0.00	C_6H_3	1.440

Proximity effects:
(pe1) Two hydrophilic groups separated by one CH_2; +0.80.
(pe2) Two hydrophilic groups separated by two CH_2; +0.46.
Primary and secondary constants:
Those expressed to three decimal places are primary constants and those expressed to two decimal places are secondary constants.

halogen attached to a saturated carbon atom. Some [P] values are shown in Table 13.5.

Molecular descriptors have considerable potential as predictors of K_{ow} values. These include a range of topological indices as well as molecular characteristics such as the molecular surface area and molecular volume. Murray *et al.* (1975) have demonstrated that the first-order connectivity index is a good predictor of the log K_{ow} of sets of related compounds such as alcohols, esters, ketones and so on. Warne *et al.* (1990) have evaluated the utility of a set of 39 molecular descriptors and physicochemical properties to model the octanol−water partition coefficient of lipophilic compounds. They found that satisfactory relationships were expressed by multiparametric linear regression equations utilizing molecular surface area (SA) and sum of absolute values of all sigma charges (SAVC) as follows

$$\log K_{ow} = 0.030\ \text{SA} + 3.473\ \text{SAVC} - 1.491 \quad (n = 25;\ r^2 = 0.944)$$

Table 13.5 Some simplified parachor values (see Table 13.3)

C-9	S-49
O-20	Cl-55
H-15.5	Br-68
N-17.5	F-26

Double bond, 18; triple bond, 40.

Warne *et al.* (1990) also found the following simple linear regression equation

$$\log K_{ow} = 0.035\ \text{SA} - 2.486 \quad (n = 25;\ r^2 = 0.876)$$

where SA is the molecular surface area.

In many cases the most readily available physicochemical characteristic of a compound may be its solubility in water. The use of this property to calculate the K_{ow} value thus considerably extends the range of K_{ow} values available. Hansch *et al.* (1968) investigated the relationship between water solubility and K_{ow} using the following general equation

$$\log (1/S) = a \log K_{ow} + b$$

where a and b are empirical constants for different groups of chemicals and S is the aqueous solubility. Miller *et al.* (1985) have reviewed the relationship between aqueous solubility and the K_{ow} value. A summary of the investigations of this relationship carried out by Chiou *et al.* (1982) is shown in Table 13.6.

To assess calculated log K_{ow} values Chessels

Table 13.6 Correlations of octanol−water partition coefficients log K_{ow} with aqueous solubility log S. Data from Chiou *et al.* (1982)

Type of compound	x	y	Number	r
Liquid organics				
Alkylhalides	−0.819	0.681	20	0.928
Alkenes	−0.773	0.192	12	0.985
Aromatics	−1.004	0.340	16	0.975
Alkanes	−0.808	−0.201	16	0.953
All compounds	−0.747	0.730	156	0.935
Solid organics				
Organophosphates	−0.747	0.472	10	0.969
PCBs and DDT	−0.518	2.222	7	0.997
Ideal line	−1.000	0.92	−	−

Note: $\log K_{ow} = x \log S + y$; where S is in moles/litre.

et al. (1991) have plotted the mean experimental value for a number of $\log K_{ow}$ values measured separately on 40 chlorohydrocarbons, hydrocarbons and related compounds against the calculated $\log K_{ow}$ value. This was obtained using the method described by Hansch & Leo (1979) and referred as $C \log P$. The plot is shown in Fig. 13.2, where the individual data points are indicated, as well as the line with a slope of unity, which would indicate agreement between the calculated and experimental values. Good agreement is observed with compounds having $\log K_{ow}$ values between 2 and 5.5, but calculated values for some compounds having $\log K_{ow}$ values greater than 5.5 show an increasing deviation from the experimental values, while with others there is excellent agreement. Thus the calculation method does not give reliable results with compounds having $\log K_{ow}$ values greater than 5.5. Chessels *et al.* (1991) have suggested that the overestimation by this method results from the use of fragmental constants generally derived from relatively smaller molecules. This suggests that a degree of caution should be attached to all methods of calculation for superhydrophobic compounds which have $\log K_{ow}$ values greater than about 6.

13.6 APPLICATIONS IN ECOTOXICOLOGY

Many quantitative structure–activity relationships (QSARs) used in ecotoxicology are based on

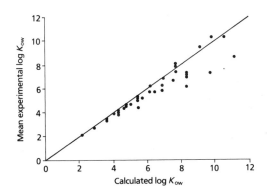

Fig. 13.2 Plot of the calculated $\log K_{ow}$ values ($C \log P$ calculated by the Hansch & Leo method) against the mean experimental value for 40 chlorohydrocarbons, hydrocarbons and related compounds with the line having a slope of unity. After Chessels *et al.* (1991).

the K_{ow} value (see Chapter 14). Some of the most commonly used relationships are employed in the prediction of bioconcentration, non-specific toxicity and the sediment–water partition coefficient. These QSARs relate the logarithm of the environmental factor to the $\log K_{ow}$ value in a direct linear relationship (Table 13.7). The relationships apply for persistent lipophilic compounds, i.e. those having $\log K_{ow}$ values between about 2 and about 6.

Compounds having $\log K_{ow}$ values greater than 6 are usually referred as superhydrophobic compounds. These exhibit deviant relationships from the linear forms shown in Table 13.7, and this has been particularly demonstrated with the bioconcentration relationship. With bioconcentration of superhydrophobic compounds the bioconcentration factor, K_B value declines while the K_{ow} value continues to increase, leading to a relationship having a general parabolic shape (Connell & Hawker, 1988). Thus the following equation has been found for the relationship between $\log K_B$ and $\log K_{ow}$ for compounds over the $\log K_{ow}$ range from 2 to 10:

$$\log K_B = 6.9 \times 10^{-3} (\log K_{ow})^4 - 1.85 \times 10^{-1} (\log K_{ow})^3 + 1.55 (\log K_{ow})^2 - 4.18 \log K_{ow} + 4.79$$

This is a somewhat similar general form to the equation suggested by Hansch (1969) to describe the general relationship between biological response and the K_{ow} value.

The observed decline in activity while K_{ow} continues to increase has been attributed to a variety of factors. Some of these include the declining solubility of the substances in water, inaccurate calculation of K_{ow} values, steric chemical structure, non-attainment of equilibrium, and resistance to transfer of large molecules across membranes. Recently, Chessels *et al.* (1993) have provided evidence to suggest that the decline in activity is due to declining solubility of the superhydrophobic compounds in biota lipid. This factor suggests that the use of K_{ow} is most appropriate when octanol mimics the lipid phase in the environmental system most closely.

13.7 SOURCES OF K_{ow} VALUES

It is important to obtain reliable K_{ow} values for use in ecotoxicology, otherwise the environmen-

Table 13.7 Some QSARs for persistent lipophilic compounds used in ecotoxicology. Data from Connell (1988b), p. 221

Ecotoxicology factor	Relationship to K_{ow}
Bioconcentration factor, K_B, with fish	$\log K_B = 0.94 \log K_{ow} - 1.00$
Bioconcentration Factor, K_B, with molluscs	$\log K_B = 0.84 \log K_{ow} - 1.23$
Non-specific toxicity to fish LC_{50}	$\log(1/LC_{50}) = 0.907 \log K_{ow} - 4.94$
Sediment to water partition coefficient, K_{oc}	$\log K_{oc} = 1.029 \log K_{ow} - 0.18$

tal characteristics derived will reflect inaccuracies in the original K_{ow} values. The reliability of K_{ow} values is difficult to determine, and relatively few investigations have been undertaken into this, e.g. Chessels *et al.* (1991). The literature contains a variety of K_{ow} values (Table 13.8). As a general rule, techniques for the measurement of K_{ow} have improved, and the reliability of the values obtained has increased, over recent years. Many techniques have a consistent error, (e.g. Chessels *et al.*, 1991) as a result of the basic mechanism involved, or possibly the equipment used. This

Table 13.8 Some sources of log K_{ow} values

Reference	Compound types
Leo *et al.* (1971)	Wide range
Hansch & Leo (1979)	Wide range
Banerjee *et al.* (1980)	Mainly hydrocarbons, chlorohydrocarbons and substituted phenols
Mackay *et al.* (1980)	Mainly aromatic hydrocarbons
Nahum & Horvath (1980)	Wide range
Ellegehausen *et al.* (1981)	Pesticides
Eadsforth (1983)	Wide range
Wasik *et al.* (1983)	Mainly aromatic hydrocarbons
Woodburn *et al.* (1984)	PCBs
Miller *et al.* (1984)	Chlorinated benzenes and biphenyls
Rapaport & Eisenreich (1984)	PCBs
Coates *et al.* (1985)	Alkanes
Miller & Wasik (1985)	Alkanes, alkenes, halogenated hydrocarbons, aromatic and polyaromatic hydrocarbons, PCBs
Chiou (1985)	Mainly hydrocarbons, chlorohydrocarbons and PCBs
Shiu *et al.* (1988)	PCDDs
Connell & Hawker (1988)	PCBs
Doucette & Andren (1988)	Mainly halogenated aromatic compounds
Klein *et al.* (1988)	Wide range
Hawker (1989)	Chlorohydrocarbons and aromatic hydrocarbons
Warne *et al.* (1990)	Various compounds
Gusten *et al.* (1991)	Polycyclic aromatic hydrocarbons and their alkyl derivatives
Chessels *et al.* (1991)	Chloroaromatic hydrocarbons, PCBs and PCDDs

suggests that values derived by aggregating sets of results may be more accurate than those derived from a single set of experiments. In many cases K_{ow} values have been calculated using one of the methods previously described. A large evaluated set of K_{ow} values is available from the 'MedChem' Software Package developed at Pomona College, California (MedChem, 1989). This is useful for obtaining evaluated experimental as well as calculated values.

13.8 CONCLUSIONS

The octanol–water partition coefficient has been demonstrated to be a very useful physicochemical characteristic of a chemical for the prediction of its environmental properties. Applications in environmental chemistry and ecotoxicology are extensive, and range from predicting toxicity to predicting the sediment–water partition coefficient. The data base of K_{ow} values is large and growing, but the reliability of values for some compounds can be difficult to evaluate. The techniques for experimental determination have improved in recent times, and can be successfully applied to superhydrophobic compounds. The calculation of K_{ow} values has been successful with lipophilic compounds, but needs further development with other compounds, particularly the superhydrophobic compounds.

REFERENCES

Anliker, R., Moser, P. & Poppinger, D. (1988) Bioaccumulation of dyestuffs and organic pigments in fish. Relationships to hydrophobicity and steric factors. *Chemosphere* **17**, 1631–1636.

Banerjee, S., Yalkowski, S.H. & Valvani, S.C. (1980) Water solubility and octanol/water partition coefficients of organics. Limitations of the solubility–partition coefficient correlation. *Environ. Sci. Technol.* **14**, 1227–1235.

Briggs, G.G. (1981) Theoretical and experimental relationships between soil adsorption, octanol–water partition coefficients, water solubility, bioconcentration factors and the parachor. *J. Agric. Food Chem.* **29**, 1050–1053.

Chessels, M., Hawker, D.W. & Connell, D.W. (1991) Critical evaluation of the measurement of the 1-octanol/water partition coefficient of hydrophobic compounds. *Chemosphere* **22**, 1175–1190.

Chessels, M., Hawker, D.W. & Connell, D.W. (1993) Influence of solubility in lipid on bioconcentration of hydrophobic compounds. *Ecotoxicol. Environ. Safety* **23**, 260–273.

Chiou, C.T. (1985) Partition coefficients of organic compounds in lipid–water systems and correlations with fish bioconcentration factors. *Environ. Sci. Technol.* **19**, 57–63.

Chiou, C.T., Schmedding, D.W. & Mains, M. (1982) Partitioning of organic compounds in octanol–water systems. *Environ. Sci. Technol.* **16**, 4–10.

Coates, M., Connell, D.W. & Barron, D.M. (1985) Aqueous solubility and octan-1-ol to water partition coefficients of aliphatic hydrocarbons. *Environ. Sci. Technol.* **19**, 628–632.

Connell, D.W. (1988a) Bioaccumulation behaviour of persistent organic chemicals with aquatic organisms. *Rev. Environ. Contam. Toxicol.* **101**, 117–154.

Connell, D.W. (1988b) Quantitative structure–activity relationships and the ecotoxicology of chemicals in aquatic systems. *ISI Atlas of Science: Plants and Animals* **1**, 221–225.

Connell, D.W. (1990) *Bioaccumulation of Xenobiotic Compounds*, CRC Press, Boca Raton, FL.

Connell, D.W. & Hawker, D.W. (1988) Use of polynomial expressions to describe the bioconcentration of hydrophobic chemicals by fish. *Ecotoxicol. Environ. Safety* **16**, 242–257.

Doucette, W.J. & Andren, A.W. (1988). Estimation of octanol/water partition coefficients: evaluation of six methods for highly hydrophobic aromatic hydrocarbons. *Chemosphere* **17**, 345–352.

Eadsforth, C.B. (1983) Assessment of reverse phase chromatographic methods for determining partition coefficients. *Chemosphere* **12**, 1459–1466.

Ellegehausen, H., D'Hont, C. & Fuerer, R. (1981) Reversed phase chromatography as a general method for determining octan-1-ol/water partition coefficients. *Pestic. Sci.* **12**, 219–225.

Gusten, H., Horvatic, D. & Sabljic, A. (1991) Modelling n-octanol/water partition coefficients by molecular topology: polycyclic aromatic hydrocarbons and their alkyl derivatives. *Chemosphere* **23**, 199–204.

Hamelink, J.L., Waybrant, R.C. & Ball, R.C. (1971) A proposal: exchange equilibria control the degree chlorinated hydrocarbons biologically magnify in lentic environments. *Trans. Am. Fish. Soc.* **100**, 207.

Hansch, C. (1969) A quantitative approach to biochemical structure–activity relationships. *Accounts Chem. Res.* **2**, 232–238.

Hansch, C. & Leo, A. (1979) *Substituent Constants for Correlation Analysis in Chemistry and Biology*, John Wiley & Sons, New York.

Hansch, G., Maloney, P., Fujita, T. & Muir, R.M. (1962) Correlation of biological activity of phenoxyacctic acids with Hammette substituent constants and partition coefficients. *Nature* **192**, 178–179.

Hansch, C., Quinlan, J.E. & Lawrence, G.L. (1968) The linear free energy relationship between partition coefficients and aqueous solubility of organic liquids. *J. Org. Chem.* **33**, 347–350.

Hawker, D.W. (1989) The relationship between octan-1-ol/water partition coefficient and aqueous solubility in terms of solvatochromic parameters. *Chemosphere* **19**, 1585–1591.

James, K.C. (1986a) *Solubility and Related Properties* (Ed. K.C. Jones). Marcel Dekker, New York, p. 341.

James, K.C. (1986b) The distribution law. In: *Solubility and Related Properties* (Ed. K.C. Jones), p. 279. Marcel Dekker, New York.

Klein, W., Kordel, W., Weiss, M. & Poremski, H.J. (1988) Updating of the OECD test guideline 107 'partition coefficient n-octanol/water': OECD laboratory intercomparison test on the HPLC method. *Chemosphere* **17**, 361–369.

Könemann, H. (1981) Quantitative structure–activity relationships in fish toxicity studies. Part I. A relationship for 50 industrial pollutants. *Toxicology* **19**, 209.

Leo, A., Hansch, C. & Elkins, D. (1971) Partition coefficients and their uses. *Chem. Rev.* **71**, 525–553.

Lyman, W.J. (1990) Octanol/water partition coefficient. In: *Handbook of Chemical Property Estimation Methods, Environmental Behaviour of Organic Compounds*, (Eds W.J. Lyman, W.P. Reehl & D.H. Rosenblatt), p. 1.1. American Chemical Society, Washington, DC.

Lyman, W.J., Reehl, W.P. & Rosenblatt, D.H. (1990) *Handbook of Chemical Property Estimation Methods, Environmental Behaviour of Organic Compounds*. American Chemical Society, Washington,

Mackay, D. (1982) Correlation of bioconcentration factors. *Environ. Sci. Technol.* **16**, 274–282.

Mackay, D., Bobra, A. & Shiu, W.Y. (1980) Relationships between aqueous solubility and octanol–water partition coefficients. *Chemosphere* **9**, 701–708.

McGowan, J.C. (1952) The physical toxicity of chemicals. II. Factors affecting physical toxicity in aqueous solutions. *J. Appl. Chem.* **2**, 323–329.

MedChem, (1989) *MedChem Software Manual Release 3.54*. Daylight Chemical Information Systems, Inc., Pomona CA.

Miller, M.M., Ghodbane, S., Wasik, S.P., Tewari, Y.B. & Martire, B.E. (1984) Aqueous solubilities, octanol/water partition coefficients and entropies of melting of chlorinated benzenes and biphenyls. *Chem. Eng. Data* **29**, 184–187.

Miller, M.M., Wasik, S.P., Huang, G.L., Shiu, W.Y. & McKay, D. (1985) Relationship between octanol–water partition coefficients and aqueous solubility. *Environ. Sci. Technol.* **19**, 522–528.

Murray, W.J., Hall, L.H. & Kier, L.B. (1975) Molecular connectivity. III. Relationship to partition coefficients. *J. Pharm. Sci.* **64**, 1978–1981.

Nahum, A. & Horvath, S. (1980) Evaluation of octanol/water partition coefficients by using high performance liquid chromatography. *Chromatography* **192**, 315–321.

Niimi, A.J. (1991) Solubility of organic chemicals in octanol, triolein and cod liver oil and relationships between solubility and partition coefficients. *Water Res.* **25**, 1515–1522.

OECD (1981) *Guidelines for Testing of Chemicals, Subject No. 107, Partition Coefficient (n-octanol/water)*. Organization for Economic Cooperation and Development, Paris.

Rapaport, R.A. & Eisenreich, S.J. (1984) Chromatographic determination of octanol–water partition coefficients (K_{ow}'s) for 57 polychlorinated biphenyl congeners. *Environ. Sci. Technol.* **18**, 163–170.

Rekker, R.F. (1977) *The Hydrophobic Fragmental Constant*. Elsevier, Amsterdam.

Shiu, W.Y., Dasette, W., Gobas, F.A.P.C., Andren, A. & MacKay, B. (1988) Physical–chemical properties of chlorinated dibenzo-p-dioxins. *Environ. Sci. Technol.* **22**, 651–659.

Tanford, C. (1973) *The Hydrophobic Effect: Formation of Micelles and Biological Membranes*. John Wiley & Sons, New York, p. 2.

Warne, M.StJ., Connell, D.W., Hawker, D.W. & Schüürmann, G. (1990) Prediction of aqueous solubility and the octanol–water partition coefficient for lipophilic organic compounds using molecular descriptors and physicochemical properties. *Chemosphere* **21**, 877–888.

Wasik, S.P., Miller, M.M., Tewari, Y.B., May, W.E., Sonnefeld, W.J., DeVoe, H. & Zoller, W.H. (1983) Determination of the vapour pressure, aqueous solubility and octanol/water partition coefficient of hydrophobic substances by coupled generator column/liquid chromatographic methods. *Residue Rev.* **85**, 29–35.

Woodburn, K.B., Dasette, W.J. & Andren, A.W. (1984) Generator column determination of octanol/water partition coefficients for selected polychlorinated biphenyl congeners. *Environ. Sci. Technol.* **18**, 457–562.

14: Quantitative Structure–Activity Relationships

P. DONKIN

14.1 INTRODUCTION

14.1.1 Predictive ecotoxicology: a complex problem

The main aims of ecotoxicology are to predict and to monitor the impact of potentially harmful chemicals on natural ecosystems (Volume 1). The levels of toxic chemicals in the environment have increased as a consequence of human activity, most obviously in the case of synthetic chemicals such as pesticides and plasticizers, but in addition there has been increased contamination from combustion products, crude oils and algal toxins. It has been estimated that over 100 000 chemical substances are in commercial use, and that they are being added to at the rate of between 500 and 5000 per year (Hansch *et al.*, 1989; Blum & Speece, 1990). Most of these substances are compounds of carbon, or complex mixtures of such compounds. The potential scale of contamination by organic compounds has been illustrated by Nirmalakhandan & Speece (1988a,b), who calculated that if the industrial chemical benzene is substituted on three of its six carbon atoms with a choice of 90 common constituents, 729 000 derivatives are possible.

In order to predict and to monitor the impact of these organic chemicals on ecosystems, it is necessary to know how they are dispersed from their sources, how they partition into various environmental compartments, how they are degraded or modified by physical and biological processes, and what is their mode of toxic action and that of their degradation products once they enter the myriad of organisms that comprise an ecosystem (Jaffé, 1991). Experimentally it would be impossible to handle the problem on a compound-by-compound, species-by-species, habitat-by-habitat basis.

However, although by definition all organic compounds are unique, many share common structural features and/or physicochemical properties that allow them to be grouped. The environmental and toxicological behaviour of the different compounds in the group can then be shown to change in a systematic and predictable way in relation to these features or properties, often to the extent of fitting precise mathematical relationships. The utility of such structure–activity relationships (SAR) and property–activity relationships (PAR) in enhancing our understanding of the toxicological behaviour of organic compounds has been appreciated for over 100 years, but their application has now been extended to all aspects of the environmental behaviour of organic chemicals. In much of the literature the term SAR is used to describe SARs and PARs. Except where there is a need to be more specific, the same convention is used throughout this review.

14.1.2 The history of quantitative SARs (QSARs)

We have long been aware that different chemicals produce different toxicological responses (Nirmalakhandan & Speece, 1988a,b), but the first reports of systematic studies on organic compounds date from the middle of the nineteenth century. These pioneering investigations have been comprehensively reviewed by Lipnick (1989, 1990). The stimulus for much of this research was an interest in the mode of action of anaesthetics/narcotics, though other compounds

321

of pharmacological interest such as antipyretic and antiseptic drugs were also studied. The nineteenth-century scientists drew attention to the important influence that the water-solubility and molecular weight of organic chemicals have on the biological response produced. This culminated in the proposal, made independently by Meyer and by Overton in 1899 (see Lipnick, 1989), that narcosis was the result of partition into lipid phases of the cell; a process that could be accurately modelled by the partition ratio between olive oil and water (see Chapter 13).

The remarkable contribution of Overton and his contemporaries to our understanding of the pharmacology and toxicology of organic compounds can be seen from the frequency with which concepts discussed in writings published in 1901 (Overton, 1901), appear in the modern literature. Overton considered that narcosis may be related to molecular volume; noted that excess toxicity of esters could be correlated with their rates of hydrolysis; and reported a narcotic 'cut-off' which he thought was due to a combination of limited aqueous and lipid phase solubility. He observed that the effects induced by different narcotic molecules could be concentration additive; recognized the importance of both lecithin and cholesterol in controlling toxicant— lipid phase interactions; discussed pigment loss from cells as a means of detecting narcotic effects; and recognized that many species differences in response to narcotics were reflections of different sensitivity, not selectivity, and that high sensitivity could probably be associated with the presence of an advanced nervous system. Overton thought that the excess toxicity of compounds such as aldehydes could be due to reactivity; discussed his results in terms of ratios to saturated solution concentration; noted that branching of chains reduces narcotic potency; and considered interaction with proteins to be an important mechanism of alkaloid action. His overriding theme, of narcosis related to partitioning into lipids, was discussed in terms of lipid boundaries to the cell, at a time when the existence of these was disputed, and many years before they could be visualized. However, despite the generality of his theory, Overton recognized that there were subtle differences in the responses of organisms

to different groups of narcotics, a subject of active research today.

Overton's primary interest was in pharmacology, but he considered that the situation of aquatic organisms bathed in toxicant solutions was in some respects analogous to that of human patients inhaling anaesthetic vapours. Consequently, aquatic organisms, particularly tadpoles, were favoured experimental animals, so these nineteenth-century 'pharmacological' studies are of direct relevance to modern aquatic ecotoxicology.

For many years following this period of intense activity, progress was slow, though important contributions were made by Ferguson (1939), who proposed that narcosis occurred at a constant thermodynamic activity of the pharmacological agent (i.e. constant ratio to aqueous solubility), and Mullins (1954), who developed the idea that the fraction of the volume of the membrane occupied by the narcotic was the key correlate of efficacy. Practical use was made of QSARs in risk assessment as early as the 1930s by the Russian scientist Lazarev (see Lipnick & Filov, 1992). Lazarev was concerned with the health of industrial workers exposed to organic chemicals. Studies of ecotoxicological relevance are more recent. For example, Corner & Sparrow (1957) showed that the relative toxicity of a series of alkyl mercuric chlorides to marine crustaceans could be related to lipophilic behaviour, and 10 years later Crisp *et al.* (1967) applied Ferguson's thermodynamic approach to explain the narcotic action of a diverse group of organic compounds on barnacle larvae. However, probably the greatest stimulus to modern QSAR development was the work of Hansch and his colleagues (e.g. Hansch & Fujita, 1964), who combined into a single linear free energy equation, electronic (σ) and steric (E_s) substituent constants developed from studies of chemical reactivity (Hammett, 1937; Taft, 1953), and a hydrophobicity substituent constant (π). The equation is generally presented in the form:

$$\log 1/C = -K_1\pi^2 + K_2\pi + K_3\sigma + K_4E_s + K_5$$

where C is the concentration of chemical required to produce a particular biological response.

Today, such a multiparameter regression approach is probably the most widely used QSAR technique in environmental toxicology, and examples of it appear throughout this chapter. Nevertheless, the increased availability of powerful computers has stimulated the application of alternative approaches using multivariate projection methods as a means of establishing relationships within complex data sets.

As a result of these developments, publications containing biological QSARs are appearing at the rate of 400–500/year (Hansch *et al.*, 1989); the proceedings of a recent conference on 'QSAR in Environmental Toxicology' (Hermens & Opperhuizen, 1991) included 59 papers. A comprehensive account of the field cannot be provided in a single short chapter, but the subject can be conveniently divided into distinct specializations, several of which have been extensively reviewed in recent years. Frequent reference to these reviews will be made in this chapter. A general overview of the use of QSARs in ecotoxicology has been provided by Calamari & Vighi (1990).

14.2 MOLECULAR DESCRIPTORS AND PROPERTY–PROPERTY/ STRUCTURE RELATIONSHIPS

14.2.1 Characteristics of molecular descriptors

Before QSARs can be established, a means must be found to define the properties of the molecules of interest in a quantifiable way. The behaviour of molecules can be described in terms of bulk properties such as aqueous solubility or olive-oil–water partitioning favoured by early workers in the field, or deduced from an understanding of the influence of the constituent parts of the molecule on the properties of the whole.

For biological effects' QSARs, the most important molecular properties are those that relate to solubility in aqueous and lipid components of the organism, since these determine transport to and interaction with active sites (Chapter 13). However, if the mode of action is specific, particular steric and electronic factors can also be important (Dearden, 1990). The distribution of contaminants in the environment is also strongly influenced by hydrophilic/hydrophobic properties, but there is an additional need to consider

factors such as evaporation rates and photolysis (Govers, 1990).

Van de Waterbeemd & Testa (1987) list more than 200 physicochemical and structural descriptors that have been used in QSAR studies, though a relatively small number appear frequently in the ecotoxicology literature. Descriptors may be divided into two broad categories: those that can be calculated from a knowledge of molecular structure, and those that are based on experimental observations.

Calculated descriptors

These have the advantage that they should be free from experimental error. Calculated bulk properties such as molecular volume and surface area have been used successfully in bioaccumulation QSARs (Connell & Schüürmann, 1988), but the mathematically derived parameters which have generated most interest in the environmental sciences to date are topological, especially connectivity indices (Basak, 1990; Sabljic, 1990). These express the shape and size of the molecule in numerical form and so correlate well with characteristics such as molecular surface area and molecular volume. However, since valence electrons are included in the calculations, electronic properties are also described.

A common criticism of connectivity indices is that their precise physical and chemical meaning is not clear. A less empirical way of calculating parameters for QSARs is to start from a fundamental understanding of the quantum chemistry of the molecule. Many successful QSARs have been generated by this means, though mostly in the field of pharmacology (Dearden, 1990; Lewis, 1990).

Experimentally determined descriptors

The majority of QSARs developed in environmental science use descriptors of this type. Their advantage is that the influence they have on the environmental behaviour of the chemical is often easy to understand; e.g. aqueous solubility relates to dispersion and transport processes and hydrophobicity relates to partition into organisms and onto particles.

Hydrophobicity is undoubtedly the most important molecular characteristic determining the behaviour of organic chemicals in aquatic systems, and it also influences the partitioning of chemicals in soil. The pioneering pharmacologists tried olive oil and organic solvents (Overton, 1901; Meyer, 1937) to model lipid–water partitioning, but the *n*-octanol/water system has now become an almost universally accepted 'standard' (Dearden, 1985; this volume, Chapter 13). However, it is not a perfect model for biological membranes, the presumed site of action for many toxicants, since the partition ratio is generally lower into membranes (Seeman, 1972), an effect which is particularly pronounced for larger, more hydrophobic molecules (Abernethy *et al.*, 1988; Gobas *et al.*, 1988). Nevertheless, since the partitioning properties of membranes vary with composition (Miller, 1985), a search for a universal model would be futile.

Despite these reservations, the *n*-octanol/water partition system has proved to be of outstanding practical utility in QSAR research (Chapter 13). Its main disadvantage is that experimental determination can be subject to very large errors (Garten & Trabalka, 1983; Sabljic, 1987; Dearden, 1990). This problem is partly overcome by the fragment method of estimation, where results are derived from experiments with many compounds (Hansch & Leo, 1979). Similarly, the key parameter, water-solubility, is subject to large experimental errors, particularly for the very hydrophobic molecules of greatest environmental interest. Furthermore, the experimental determination of all the physicochemical parameters of interest in environmental modelling for a high proportion of the organic contaminants in or likely to be released into the environment is impractical. For these reasons, considerable effort has been directed towards the establishment of property–property and property–structure relationships, as a potential way of estimating one parameter from another.

14.2.2 Property–property/structure relationships

Property–property (PP) and property–structure (PS) relationships are particularly useful if a par-

ameter which is difficult to determine experimentally can be estimated either from one which is easy to determine, or from one that can be calculated. There is also the possibility of cross-checking values derived through different methods of estimation, and concentrating effort on data points shown to be anomalous. Similarly, outliers from the individual QSARs can point to a need for further investigation.

PP and PS relationships have been discussed in detail by Nirmalakhandan & Speece (1988a). Most interest has centred on measures of hydrophobicity (usually the *n*-octanol/water partition coefficient (K_{ow})) and aqueous solubility. K_{ow} has been correlated with numerous variables (Dearden, 1985) including aqueous solubility (Miller *et al.*, 1985), total molecular surface area (Doucette & Andren, 1987; see Table 14.1) and capacity factors on high-performance liquid chromatography (HPLC; Dearden, 1990; Coates, Connell & Barron, 1985). Mailhot & Peters (1988) developed relationships between the K_{ow} of over 250 compounds and nine physicochemical parameters. Aqueous solubility was the best single correlate, but addition of a second descriptor, either melting point or the first-order connectivity index, gave equations with improved statistics. Aqueous solubility itself has been modelled with high precision using either descriptors which were all calculated (connectivity indices and a polarizability term; Nirmalakhandan & Speece, 1989), or 'solvatochromic' parameters (Kamlet *et al.*, 1986a).

Parameters of particular importance in environmental fate models can be estimated using a similar approach. Henry's Law constants, a measure of partitioning between air and water, can be predicted with quantitative structure–property relationship (QSPR), equations including connectivity indices (Nirmalakhandan & Speece, 1988b; Brunner *et al.*, 1990) or specifically for polychlorinated biphenyls (PCBs), including number and position of chlorine atoms or total planar surface area (Brunner *et al.*, 1990). Total planar surface area has also been used to predict the vapour pressure of PCBs (Hawker, 1989).

Table 14.1 Property–property/structure relationships

Compounds and equation	n	r^2	References
Aromatic			
$\log K_{ow} = 0.0238 \, (TSA) - 0.142$	32	0.92	Doucette & Andren (1987)
TSA, total surface area			
Polar and non-polar aliphatic			
$\log S = 0.54 - 3.32 V/100 + 0.46\pi^a + 5.17\beta$	105	0.99	Kamlet *et al.* (1986a)
S, aqueous solubility; V, solute molar volume; π^a and β, solvatochromic parameters			
Hydrocarbons, esters, alcohols			
$\mathrm{Log}\, H = -1.29 + 1.005\Phi - 0.468\,^1\chi^v - 1.258I$	180	0.98	Nirmalakhandan & Speece (1988a,b)
H, Henry's constant; Φ, polarizability parameter; χ, connectivity index; I, hydrogen bonding indicator variable			

14.3 QSARs AND PREDICTION OF THE PHYSICAL FATE OF CHEMICALS IN THE ENVIRONMENT

Which organisms in an ecosystem are exposed to which chemicals and at what concentrations is determined to a large extent by the way various physical processes distribute the contaminants (Chapter 15). Processes of major importance are volatilization, solubilization, interaction with dissolved organic matter, partitioning into soil or sediments and, depending on the chemical, transformations such as photo-oxidation and hydrolysis (Govers, 1990; Nendza *et al.*, 1990; Jaffé, 1991).

Very volatile chemicals have the potential for rapid, long-distance dispersion in the atmosphere, and can be taken up by terrestrial animals by skin absorption or inhalation, and by terrestrial plants through the cuticle or stomatal pores. Volatility is modelled using vapour pressure or Henry's Law constants, both of which can be estimated using QSAR methods (see Section 14.2.2).

If a chemical is significantly soluble in water it can be transported directly to aquatic organisms which can accumulate it across their body surfaces. This can be a major route of uptake, particularly for gill-breathing animals. Solubility can be estimated using QSARs (Section 14.2.2).

Interaction of organic contaminants with dissolved or colloidal macromolecules often reduces their bioavailability. Partitioning into soil or sediments (suspended or settled) potentially puts at risk organisms that live in, or feed on, these matrices. These processes have been modelled using linear free energy equations with K_{ow}, connectivity indices and other established QSAR parameters (Dobbs *et al.*, 1989; Govers, 1990; Nendza *et al.*, 1990; Sabljic, 1990; Jaffé, 1991). Usually, account must be taken of the organic matter content of the particulate matrices.

Chemical fate models based on the fugacity concept depend upon accurate modelling of partitioning processes with K_{ow} (Mackay & Paterson, 1990; Chapter 15). The capability to estimate K_{ow} using PP and PS relationships (Section 14.2.2) is therefore of considerable significance.

Chemicals which are transformed by photochemical or hydrolysis reactions can be converted into products that are more toxic or less toxic than the parent molecule. Furthermore, their physicochemical properties can be changed such that their partition between the various environmental compartments differs from that of the original molecule, changing the organisms which are exposed. Some of the earliest studies from which electronic (Hammett, 1937) and steric (Taft, 1953) descriptors were developed, were directed towards understanding the factors which influence the reaction rates of organic chemicals (Nirmalakhandan & Speece, 1988a). QSARs modelling, for example, the environmental hydrolysis of esters and carbamates (Drossman et al., 1988), have been established using these parameters. Realistic QSARs for photodegradation are less common, though attempts have been made (Nendza et al., 1990).

14.4 BIOACCUMULATION QSARs

Before a chemical can induce a response in an organism, it must at least be adsorbed on to its surface, but more commonly it must be accumulated into its tissues. An understanding of the process of bioaccumulation is therefore of fundamental importance to the goal of predictive ecotoxicology (Volume 1, Chapter 19). The application of the QSAR approach to this problem has been most extensive in aquatic toxicology, particularly in the study of animals.

14.4.1 Bioaccumulation by aquatic animals

The factors influencing bioaccumulation have been discussed in recent reviews by Esser (1986) and by Barron (1990), with an emphasis on fish, by Donkin & Widdows (1990), who considered invertebrates, and by Connell (1988), Farrington (1989) and Nagel & Loskill (1991), who describe results obtained with a variety of organisms. Aquatic organisms can accumulate contaminants from their food or by direct absorption from water across body surfaces, particularly gills. QSARs have mostly been applied to rationalize laboratory data on uptake from water, generally termed bioconcentration (Ernst, 1985).

The simplest interpretation of bioconcentration is that the *dissolved* contaminants partition between water and hydrophobic components of the organism (largely lipids) until the rate of uptake into, and the rate of loss from the body are in equilibrium, resulting in a steady-state body burden. By using a parameter that accurately mimics the partitioning of the organic compounds into lipid, a linear relationship with bioconcentration factor (BCF; concentration of compound in organism/concentration in water) can be

anticipated, with the log/log plot having a slope of one (Mackay, 1982). Many such relationships have been reported for fish and invertebrates, with K_{ow} as the hydrophobicity/lipophilicity descriptor (Table 14.2). Slopes are, however, often less than one, a result which is in part due to the application of regression analysis without accounting for errors in the K_{ow} term (McCarty, 1986).

Such linear relationships have been shown to apply to compounds with $\log K_{ow}$ values in the approximate range of 2–6, though even within this range, scatter of data points around a regression line can amount to a 10–100-fold variation in estimates of BCF (Barron, 1990). Larger deviations may occur for compounds of $\log K_{ow}$ < 2 because their behaviour is no longer dominated by partition into lipid phases, and small molecules may be able to enter cells by diffusion through membrane pores (Esser, 1986).

The most frequently observed systematic deviation from linear bioconcentration QSARs is the lower-than-anticipated accumulation of compounds with $\log K_{ow}$ > 6. Combining this behaviour with that of the compounds with $\log K_{ow}$ < 2 can produce a sigmoid curve for the $\log$BCF/ $\log K_{ow}$ relationship. However, as compounds with $\log K_{ow}$ values considerably in excess of 6 are included, the upper portion of the curve becomes more parabolic (Esser, 1986; Connell & Hawker, 1988). Several explanations for the curvature of the relationship at high $\log K_{ow}$ have been advanced:

1 Experimental BCF data were not derived from steady-state measurements. Hawker & Connell (1986) and Connell & Hawker (1988) have shown that for certain compound (very lipophilic)– organism combinations, a steady-state body burden can take months to achieve, or may never be achieved within the lifetime of an organism. Furthermore, experimentation with chemicals of very low water-solubility is technically difficult, so the quality of BCF data may be inadequate for QSAR studies.

2 n-Octanol/water partitioning is a poor model for lipid/water partitioning. Support for this view based on thermodynamic considerations was presented by Opperhuizen et al. (1988), although these authors did indicate that octanol–water

Table 14.2 Bioaccumulation QSARs for animals

Animal, compounds and equation	n	r^2	References
Bioconcentration QSARs			
Daphnids; aromatic, heterocyclic and alicyclic			
$\log \text{BCF} = 0.898 \log K_{ow} - 1.315$	22	0.93	Hawker & Connell (1986)
BCF, bioconcentration factor			
Fish; aromatic, aliphatic and alicyclic			
$\log \text{BCF} = -1.13 + 1.02 \log K_{ow} + 0.84 \log S_{octanol} + 0.0004 \, (\text{mp-}25)$	36	0.90	Banerjee & Baughman (1991)
S, solubility in *n*-octanol; mp = melting point			
Fish; aromatic, aliphatic, alicyclic			
$\log \text{BCF} = -2.13 + 2.12 = {}^2\chi^v - 0.16 = ({}^2\chi^v)^2$	84	0.93	Sabljic (1990)
${}^2\chi^v$, second-order valence connectivity index			
Fish; chlorinated hydrocarbons			
$\log \text{BCF} = 0.0069[\log K_{ow}]^4 - 0.185[\log K_{ow}]^3 + 1.55[\log K_{ow}]^2 - 4.18 \log K_{ow} + 4.79$	45	—	Connell & Hawker (1988)
Bioaccumulation QSAR			
Birds (into fat); pesticides, chlorinated hydrocarbons			
$\log \text{BAF} = 0.542 \log K_{ow} - 2.743$	47	0.54	Garten & Trabalka (1983)
BAF, bioaccumulation factor			

partitioning may be a better model for partitioning into storage lipids than into the membranes through which lipophilic contaminants must pass to reach storage sites. Studies of phospholipid vesicles (Gobas et al., 1988) have clearly demonstrated that for chemicals with $\log K_{ow}$ values between 1 and 5.5 membrane/water and n-octanol/water partition coefficients are approximately equal and linearly related, but compounds with a $\log K_{ow} > 5.5$ (corresponding to a molar volume $> 230 \, cm^3/mol$ for the chemicals in this study) partition less into the structured environment of a membrane than into n-octanol. However, even when unstructured lipid (triolein) is used, similar disparities between lipid–water and n-octanol–water partitioning are observed (Chiou, 1985).

Banerjee & Baughman (1991) have recently suggested adding a term to bioconcentration QSARs which corrects for the inadequacies of n-octanol as a lipid model. This equation (Table 14.2) predicted with satisfactory accuracy the low bioaccumulation of certain hydrophobic dyes by fish (Anliker & Moser, 1987), though it failed to predict the zero bioaccumulation of octachloronaphthalene reported by Opperhuizen et al. (1985). It was suggested that hydrophobic molecules such as octachloronaphthalene with widths in excess of 0.95 nm cannot permeate fish gills.

An alternative approach to non-linear bioconcentration modelling is to use calculated parameters such as connectivity indices (Sabljic, 1990; Table 14.2). However, the predictability of this equation has not been tested with such a structurally diverse group of chemicals as that described by Banerjee & Baughman (1991).

3 Metabolism. QSARs based on partitioning can provide an accurate model of bioconcentration only if the chemicals of interest are not metabolically transformed by the organism to a significant extent. For many aquatic invertebrates the rate of metabolism of a wide range of organic contaminants is small in relation to rates of diffusional uptake and depuration (Livingstone, 1991; Donkin & Widdows, 1990), so hydrophobicity-based QSARs accurately reflect experimental observations. However, fish can effectively metabolize a wide range of organic

compounds (Barron, 1990), so successful fish QSARs are usually developed either with compounds that are resistant to metabolism (e.g. halogenated hydrocarbons), or with molecules that are rapidly bioaccumulated to high levels, overwhelming the metabolic capabilities of the animal. Furthermore, the time required to induce maximum metabolic activity may be shorter than the duration of some bioconcentration experiments. Attempts to modify ecotoxicology QSARs to account for metabolism are few, though electronic terms have been proposed for this purpose (Tulp & Hutzinger, 1978). Walker (1985) has suggested using in-vitro metabolism data to predict bioaccumulation factors in birds. As more in-vitro data become available for a wider range of organisms, this may become a practical way of increasing the precision of bioconcentration QSARs.

4 Kinetics. The BCF reflects a balance between the rate of uptake of chemical into the organism and the rate of loss from it. By understanding the factors that influence the first-order rate constants for these kinetic processes (K_1 and K_2 respectively), a better understanding of the mechanism of bioconcentration can be achieved. Using this approach, Gobas et al. (1986) developed a model from which they deduced that for moderately hydrophobic chemicals ($\log K_{ow} < 3-4$), $\log K_1$ should be linearly related to $\log K_{ow}$ but that for more hydrophobic chemicals, $\log K_2$ should be inversely related to $\log K_{ow}$. Similar linear relationships over restricted $\log K_{ow}$ ranges have been demonstrated experimentally. The transitions from non-linear to linear relationships with $\log K_{ow}$ were considered to reflect transitions from membrane-permeation controlled kinetics for the less hydrophobic compounds, to aqueous diffusion-layer controlled kinetics for the more hydrophobic ones. The combination of the linear sectors of the two separate rate–constant relationships with $\log K_{ow}$, results in a linear relationship between $\log K_{ow}$ and $\log BCF$ over a wide range of hydrophobicities. More recently Connell & Hawker (1988) have developed a kinetic approach to explaining bioconcentration in fish, which takes into account the limited solubility of very large hydrophobic molecules in lipid membranes.

Table 14.3 Bioaccumulation QSARs for plants and micro-organisms

Organism, compounds and equation	n	r^2	References
Aquatic macrophyte; chlorinated hydrocarbons			
$\log K_{pw} = 0.98(\pm 0.15) \log K_{ow} - 2.24(\pm 0.61)$	9	0.97	Gobas *et al.* (1991)
$\log K_{pw}$, plant/water bioconcentration factor			
Terrestrial plant cuticles; structurally diverse group			
$\log K_{cw} = 0.37(\pm 0.16) + 1.31(\pm 0.05)^3\chi^v - 1.49(\pm 19)$ (no. OH$_{aliph}$)	47	0.98	Sabljic *et al.* (1990)
K_{cw}, cuticle/water partition coefficient; $^3\chi^v$, valence third-order connectivity index; no. OH$_{aliph}$, number of aliphatic hydroxy groups			
Microorganisms; aromatic, heterocyclic, aliphatic			
$\log BCF = 0.907 \log K_{ow} - 0.361$	14	0.954	Baughman & Paris (1981)
BCF, bioconcentration factor			

The uptake and elimination rate constants were related to $\log K_{ow}$ by polynomial expressions, as was $\log BCF$ (Table 14.2).

In the discussion above, the main reasons for non-linearity of hydrophobicity-related BCF QSARs derived from laboratory experimentation have been considered. However, other factors contribute to the variance throughout the range of such QSARs. For example, PCBs often bioconcentrate in a way that cannot be fully explained in terms of hydrophobicity and molecular size. Shaw & Connell (1984) suggested that adsorption and desorption processes might be involved in bioconcentration, and that these processes are influenced by the planarity of the molecule. Ionization of molecules such as dyes (Esser, 1986) and phenols (Saarikoski & Viluksela, 1982; Pärt, 1990) also influences bioconcentration.

Although there are many reasons for deviations from the simple linear partitioning theory for the uptake of hydrophobic organic molecules, significant progress has been made in recent years in understanding and modelling these anomalies. However, bioconcentration experiments from water under constant laboratory conditions do not realistically reflect the environment that many aquatic animals inhabit.

Natural marine and freshwater habitats contain dissolved and colloidal humic materials which can bind hydrophobic organic contaminants and reduce their bioavailability (Barron, 1990; Donkin & Widdows, 1990). The quantitative significance of this effect depends largely upon the type and concentration of the humic material in the water column and the hydrophobicity of the contaminant. McCarthy *et al.* (1985) developed a QSAR for the effect of dissolved humic acid on the bioaccumulation of polycyclic aromatic hydrocarbons by the freshwater cladoceran *Daphnia magna* (Fig. 14.1). Humic acid concentrations as high as 50 mg/l had little effect on the bioaccumulation of naphthalene ($\log K_{ow}$ 3.3), but benzo(a)pyrene, the most hydrophobic compound tested ($\log K_{ow}$ 6.1), bound so strongly to the humic acid that concentrations less than 1 mg/l were sufficient to significantly reduce its bioaccumulation. Several studies have, however, shown that the commercially available humic acid (Aldrich) used in this work binds hydrophobic organics more strongly than the humic materials which occur naturally in the aquatic environment (Chiou *et al.*, 1987; Whitehouse, 1985; Servos *et al.*, 1989). Consequently, the impact of dissolved humic materials on bioaccumulation in the aquatic environment, though undoubtedly significant in some situations (Larsson *et al.*,

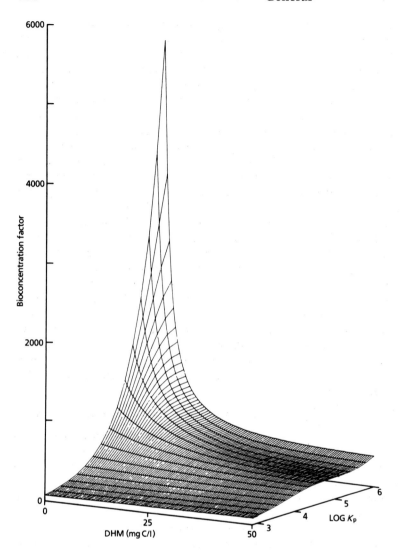

Fig. 14.1 Structure–activity relationship between K_p, concentration of dissolved humic material (DHM) and predicted bioconcentration factor (BCF) for the accumulation of polycyclic aromatic hydrocarbons (PAH) by the freshwater cladoceran *Daphnia magna*. K_p is an association coefficient expressing the affinity of binding of PAH to DHM. BCF was calculated from uptake and elimination rate constants for the compounds benzo(*a*)pyrene, benzanthracene, anthracene and naphthalene. From McCarthy *et al.*, 1985).

1992), may not be as quantitatively important as McCarthy's results suggest.

Aquatic animals are also exposed to contaminated suspended particulate materials of heterogeneous composition. These particulates have little influence on the bioaccumulation of compounds of low-to-moderate hydrophobicity, but can act as a source of very hydrophobic chemicals, which desorb into aqueous solution, then partition into the animal (Pruell *et al.*, 1986; Opperhuizen & Stokkel, 1988). Consequently, a BCF QSAR with log K_{ow} can still be linear in the presence of particulates if the BCF is calculated from aqueous, not total, contaminant concentrations (Pruell *et al.*, 1986).

Where particles accumulate to form sedimentary habitats, their role as a source of contaminants attains greater significance. Again, desorption into interstitial water followed by diffusive partitioning into organisms is a mechanistic explanation which fits much experimental data (Oliver, 1987; Shaw & Connell, 1987). Based on a comprehensive series of investigations of the bioaccumulation kinetics of a freshwater benthic amphipod, it has been concluded that uptake from interstitial water was of great importance, but that the concentration of contaminants therein, and hence the importance of this route of uptake, could be controlled by the rate of desorption from the particles (Landrum, 1988; Landrum *et al.*, 1989; Landrum *et al.*, 1991). However, assimilation of ingested particulate material was also a major route of uptake for very hydrophobic materials such as benzo(*a*)pyrene (log K_{ow} 6.3) and hexachlorobiphenyl (log K_{ow} 6.8), an observation also supported by recent studies on a sediment-dwelling mollusc (Boese *et al.*, 1990). A further complicating factor in fitting partition bioconcentration QSARs to observations with invertebrates is that contaminant depuration can sometimes be much longer than predicted, suggesting that the animal behaves as a multi-compartment system (Farrington, 1989; Livingstone, 1991; Tanacredi & Cardenas, 1991).

The recognition that bioaccumulation can be a complex process has led to the recent development of several models of the process in fish. Although based fundamentally on partitioning, these models are more sophisticated than the simple QSAR correlation approach, since they describe the properties of the major routes of uptake via gills and the digestive system in detail, and take account of the physiological characteristics of the fish (Barron, 1990; Clark *et al.*, 1990; Erickson & McKim, 1990; Barber *et al.*, 1991; Clark & Mackay, 1991). Their broad conclusions are consistent with invertebrate studies described above, indicating that for compounds of low to intermediate hydrophobicity (< log K_{ow} *c.* 5) uptake through gills will be an important route, but as the hydrophobicity of the contaminants increases, the dietary route increases in significance. Nevertheless, even where dietary uptake is the major route, partitioning across gills is of considerable importance, as a route for excretion (Clark *et al.*, 1990; Barber *et al.*, 1991).

The objective of bioaccumulation modelling is to enable predictions to be made of the biotic concentrations of contaminants in the environment. Several studies with small invertebrates have shown that partitioning of hydrophobic chemicals between aqueous *solution* and organism provides a satisfactory model for environmental observations (Donkin & Widdows, 1990). This is particularly the case for filter-feeding organisms such as mussels that pump large volumes of water, have a high exposed body surface to volume ratio, and a very limited capacity to degrade organic xenobiotics. Even for fish, environmental data can indicate an important role for water–organism partitioning when the data set is dominated by compounds of low to moderate hydrophobicity (log K_{ow} *c.* 2–5.5; Gossett *et al.*, 1983; Connolly & Peterson, 1988; Herbert & Haffner, 1991). However, the bioaccumulation of more hydrophobic compounds into natural fish populations cannot be explained by such a simple model, particularly when sediment interactions are likely (Connolly & Peterson, 1988; Bierman, 1990; Herbert & Haffner, 1991). These observations are consistent with experimental and modelling conclusions.

The difficulty of predicting bioaccumulation in organisms of such commercial importance as fish is unfortunate. However, the majority of animal biomass in most aquatic ecosystems is made up of small, relatively simple invertebrates (Warwick & Joint, 1987), for which steady-state partitioning is likely to be a more satisfactory model.

14.4.2 Bioaccumulation by terrestrial animals

Terrestrial animals can accumulate contaminants by inhalation, by skin absorption and through the diet. All of these processes are influenced by the physicochemical properties of the contaminant, but few QSARs have been developed with ecological objectives in mind.

However, because of the interest of the pioneering QSAR researchers in inhalation anaesthesia, it has long been realized that uptake and loss of these chemicals from lungs can be mimicked by gas/oil (lipid) partitioning (Lipnick, 1989). The same should apply to inhaled hydrophobic environmental pollutants. Recent studies on laboratory rats (Bond *et al.*, 1985), have shown that the long-term half-life of polycyclic aromatic compounds in lungs increases linearly with log K_{ow}. Garten & Trabalka (1983) used log K_{ow} to model the dietary accumulation of a structurally diverse group of chemicals into the fat of birds and mammals. Positive linear correlations were obtained for ruminants, non-ruminants and birds, though the predictive accuracy of the equations was poor ($r^2 < 0.55$). Nevertheless, there were some highly significant correlations between the log-transformed bioaccumulation factors for sheep, poultry, rodents, dogs, cattle and pigs, perhaps indicating that particular compounds gave rise to most of the errors in the QSAR equations. These QSARs took no account of metabolism, though mammals are known to readily transform some xenobiotics. There is clearly a need for further investigation of the potential usefulness of QSARs in terrestrial animal bioaccumulation.

At the cellular level, however, an innovative SAR approach has been applied to the rat. Cultured rat cells were used to study the physico-chemical factors controlling the bioaccumulation of organic chemicals into organelles such as mitochondria (Rashid & Horobin, 1990) and lysosomes (Rashid *et al.*, 1991). Although the compounds investigated were fluorescent dyes, this work has implications for the study of the site of action of toxic compounds.

14.4.3 Bioaccumulation by plants and microorganisms

Comprehensive investigations of the permeability of algal cells to non-electrolytes were reported by Collander in 1954. He demonstrated positive relationships with diethyl ether–water and olive oil–water partition coefficients, though the relationships were complicated by a tendency for compounds of higher molecular weight to have lower than predicted permeation power; complex branching of molecules further reduced permeation rate.

More recently, the BCFs of hydrophobic organic molecules into unicellular green algae have been positively correlated with $\log K_{ow}$. The slopes of the log/log relationships were less than generally observed with aquatic animals, suggesting that octanol–water partitioning could be a less satisfactory model for algal bioaccumulation (Casserly *et al.*, 1983; Mailhot, 1987). However, Gobas *et al.*, (1991), using BCF values determined from uptake and elimination rate constants into the freshwater macrophyte *Myriophyllum spicatum*, obtained a linear relationship to $\log K_{ow}$ with a slope of 0.98 (Table 14.3). In contrast to some fish QSARs, this relationship was linear up to $\log K_{ow}$

values of 8.3, indicating that lipid–water partitioning is an adequate representation of bioaccumulation in such plants. Systematic changes in the uptake and elimination rate constants were interpreted in terms of transfer of chemical through lipid and aqueous barriers in the organism.

Contaminant uptake into terrestrial plants is more complex, since several routes are possible. Topp *et al.* (1986) showed that, for barley seedlings, root concentration factors (based on contaminant concentration in soil) were negatively correlated with carbon-normalized soil sorption coefficients, but when based on concentration in soil-water, the root concentration factor was positively correlated with $\log K_{ow}$. In this experimental study, uptake from the air into leaves was strongly influenced by the rate of volatilization from the soil.

In the environment the foliar route has been recognized to be of global significance in the cycling of atmospheric pollutants (Calamari *et al.*, 1991). Cuticle–water partitioning has been accurately modelled using the valence third-order connectivity index combined with a variable defining the number of aliphatic hydroxy groups in the molecule (Table 14.3; Sabljic *et al.*, 1990). This study clearly indicates that care is needed with the application of these indices to QSARs, since the valence *first*-order connectivity index ($^1\chi^v$) gave a poor correlation, apparently because it did not adequately model the behaviour of polar groups.

Recently, more complex models of foliar bioaccumulation have been published (Reiderer, 1990; Paterson *et al.*, 1991). Both are based on the fugacity concept and use a variety of predictable properties including *n*-octanol–water, cuticle–water, air–water and *n*-octanol–air partition coefficients.

There have been few reports of BCF QSARs for microorganisms, but Baugham & Paris (1981) have derived a good-quality relationship between microbial sorption and the hydrophobicity of a structurally diverse group of organic compounds with $\log K_{ow}$ values in the range 3–7 (Table 14.3). This is indicative of uptake by partitioning.

14.5 BIODEGRADATION QSARs

Chemical contaminants which have been accumulated into organisms can be metabolically transformed into structures with very different physicochemical properties and environmental behaviour from the parent molecule. Virtually all organisms can effect some form of metabolic modification, but quantitatively the most important are micro-organisms, particularly bacteria, that can degrade organic compounds to carbon dioxide. Since microbial metabolism controls the environmental fate of many chemicals, considerable research has been directed towards its prediction. This has been comprehensively reviewed (Kuenemann *et al.*, 1990; Parsons & Govers, 1990; Mani *et al.*, 1991).

Biodegradation of organic chemicals to carbon dioxide is a complex multi-step process involving uptake, intracellular transport, and enzymic reactions. The type of structure–activity relationship obtained depends upon which step is rate-limiting. Furthermore, in the environment, and under the conditions of the commonly used biochemical oxygen demand (BOD) test for biodegradation (Chapter 11), chemicals are degraded by populations of microorganisms with diverse substrate specificities, and their metabolic capabilities can increase with time as the population adapts to a new chemical. Despite these complications, numerous structure–biodegradability relationships (SBRs) have been reported.

The simplest form of biodegradation data are those that classify chemicals as biodegradable or non-biodegradable, or provide a rank order of biodegradability (Kuenemann *et al.*, 1990). Such data have been related to physicochemical properties using multivariate statistical procedures, and a predictive procedure developed that is capable of classifying compounds as biodegradable or non-biodegradable with over 90% accuracy (Niemi *et al.*, 1987). Investigations of this type can lead to an understanding of structural characteristics which enhance or reduce the potential of a molecule to be biodegraded. For example, aerobic biodegradation is generally reduced by branching, halogenation, by heterocycles and large molecular size, but facilitated by the presence of hydrolysable groups (esters, amides). Hydrolysis can

also occur rapidly under anaerobic conditions (Boethling *et al.*, 1989).

While such qualitative studies are of great practical value for environmental protection, they provide only limited information about the processes involved in microbial biodegradation. Consequently, many researchers have sought to develop quantitative relationships and a wide range of descriptors have been employed (Kuenemann *et al.*, 1990; Parsons & Govers, 1990). Simple linear relationships between biodegradation rate constants and hydrophobicity parameters have been reported (e.g. for 12 phthalate esters, Boethling, 1986; Table 14.4) but parabolic or bilinear equations more frequently fit the data (e.g. also for phthalates, Urushigawa & Yonezawa, 1979). Results of this type are often considered to indicate that uptake and transport processes are rate-limiting (Parsons & Govers, 1990).

However, a considerable body of evidence suggests that the primary degradation step is often rate-limiting (Parsons & Govers, 1990). For example, Wolfe *et al.* (1980) showed that the second-order biodegradation rate constants of a group of compounds, which included pesticides, could be linearly correlated with their second-order alkaline hydrolysis rate constants. A linear (though quantitatively different) relationship was also obtained with phthalate esters. Probably the most successful attempt to model the biodegradation of a structurally diverse group of chemicals is the work of Dearden & Nicholson (1987a). They correlated 5-day BOD values with the difference in the modulus of atomic charge across a functional group bond $(X-Y)$ for 112 compounds comprising alcohols, phenols, ketones, carboxylic acids, ethers and sulphonates (Table 14.4). An equation of similar quality has since been derived for a 197-compound data set (Dearden & Nicholson, 1987b). Furthermore, good correlations could also be obtained with the biodegradation of alcohols when electrophilic superdelocalizability on the carbon atom to which the hydroxyl group is attached was used as a descriptor (Table 14.4). Superdelocalizability is a measure of the reactivity of an atom. Thus, despite BOD values representing a complex multi-step process, the rate of this process is limited by the reactivity of a particular atomic centre in the compound, which must relate only to the initial degradation step.

Parsons & Govers (1990) concluded their review by writing: 'There is no general relationship between biodegradability and chemical structure.' While this is no doubt true, considerable progress has been made in both the qualitative and quantitative understanding of the factors that control microbial biodegradation. All reviewers seem to agree that further progress depends upon more high-quality experimental data becoming available, and advances in understanding the mechanisms of biodegradation.

Very little progress has been made in developing QSARs for biotransformation in higher organisms, although electronic parameters have been used to linearize BCF QSARs thought to be influenced by metabolic transformation, and the structural requirements for the formation of reactive carcinogenic metabolites in mammals are well understood for polyaromatic compounds (see Section 14.6).

14.6 TOXICITY QSARs

Once chemicals have been bioaccumulated they can interact with the biochemical processes of the cell in a manner that is detrimental to the organism. Furthermore, some organisms can convert relatively innocuous compounds to highly toxic metabolites. Environmental toxicologists must be able to predict these processes.

14.6.1 Chemicals acting by non-specific narcosis

The mode of toxic action most easily predicted using QSARs is non-specific narcosis. General anaesthetics, and many of the other compounds studied by the pioneering workers in the field, are thought to act in this way (Lipnick, 1990). When toxicity is expressed in terms of concentration in an aqueous phase to which the organism is exposed, the characteristic non-specific narcotic QSAR has a single descriptor for hydrophobicity (usually $\log K_{ow}$), is linear over a $\log K_{ow}$ range of approximately 2–5.5, and has a slope equal to or approaching one (Hansch et al., 1989; Lipnick, 1990). Relationships of this type have been reported for a multitude of biological systems including bacteria (Warne et al., 1989; Blum & Speece, 1991), algae (Nendza & Seydel, 1988), invertebrates (Donkin & Widdows, 1990), fish (Hansch et al., 1989; Hermens, 1990a) and isolated cells from fish and mammals (Babich & Borenfreund, 1987); typical equations are given in Table 14.5. The intercept of the QSAR line varies with the sensitivity of the correlated biological

Table 14.4 Biodegradation QSARs

Compounds, equation	n	r^2	References
Phthalate esters			
$RC = -24.308 \log K_{ow} + 394.84$	12	0.87	Boethling (1986)
RC, primary biodegradability			
Alcohols, phenols, ketones, carboxylic acids, ethers, sulphonates			
$BOD = 1105\ \Delta/\delta/_{x-y} + 1.906$	112	0.98	Dearden & Nicholson (1987a)
BOD, biological oxygen demand; $\Delta/\delta/_{x-y}$, difference in the modulus of atomic charge across a functional group bond $x-y$			
Alcohols			
$BOD = 0.093\ S_E - 3.163$	19	0.96	Dearden & Nicholson (1987a)
S_E, electrophilic superdelocalizability			

response, but both sublethal and lethal endpoints give rise to the same general relationship.

The universality of such QSARs indicates that narcosis (or anaesthesia) is a response of fundamental importance in ecotoxicology, particularly since a large proportion of industrial chemicals have the physicochemical properties ($\log K_{ow}$ in the approximate range 2–5.5) required to produce narcosis (Veith *et al.*, 1984/85). The observation that there are no specific molecular features required to produce narcosis (hence the term non-specific) and that the response can be completely reversible (Schultz, 1989), provides a clue to the mechanism of action.

Meyer and Overton (see Meyer, 1937) suggested that narcosis resulted from membrane perturbations produced when the toxicant/narcotic partitioned into this lipid layer, and that the degree of narcosis was a function only of the molar concentration of the compound in the lipid phase. This theory was modified by Mullins (1954), who suggested that membrane disturbance (hence narcosis) should be more closely related to the volume occupied by the narcotic, rather than its concentration. This approach can provide a satisfactory interpretation of environmental toxicology data for a range of organisms from bacteria (Warne *et al.*, 1991) to fish (Abernethy *et al.*, 1988). These latter authors (1988) suggested that acute lethality to aquatic animals results from a 0.6% volume fraction of narcotic in membranes, whereas 0.06% produces chronic effects.

A considerable body of evidence is now available to support the lipid membrane-based theories

Table 14.5 Toxicity QSARs for non-specific narcosis

Organism, compounds and equation	n	r^2	References
Fish (Poecelia); aromatic, aliphatic and alicyclic			
$\log 1/LC_{50} = 0.871 \log K_{ow} - 4.87$	50	0.98	Könemann (1981a)
LC_{50}, concentration lethal to 50% of fish			
Mixed bacterial culture; hydrocarbons, phenols, pyridines			
$\log EC_{50} = 2.256 - 0.799 \log K_{ow}$	—	0.94	Warne *et al.* (1989)
EC_{50}, concentration inhibiting growth by 50%			
Yeast (Saccharomyces) membrane ATPase; phenols, anilines, alcohols			
$\log EC_{50} = -0.78 \log K_{ow} - 1.09$	—	—	Ahlers *et al.* (1991)
EC_{20}, concentration inhibiting H^+-ATPase activity by 20%			
Fish (Pimephales); alcohols, ketones, ethers, alkyl halides, benzenes			
$\log LC_{50} = -0.94 \log K_{ow} + 0.94 \log(0.000068 \, K_{ow} + 1) - 1.25$	—	—	Veith *et al.* (1984–85)
Fish (Pimephales); phenols, anilines			
$\log LC_{50} -0.65(\pm 0.07) \log K_{ow} - 2.29(\pm 0.22)$	39	0.90	Veith & Broderius (1990)
Bacterium (Photobacterium); structurally diverse aliphatics and benzenes			
$\log EC_{50} = 7.61 - 4.11V/100 - 1.54\pi^a + 3.94\beta - 1.51\alpha_m$	38	0.974	Kamlet *et al.* (1986b)
EC_{50}, concentration inhibiting luminescence by 50%; V, solute molar volume; π^a, β, α, solvatochromic parameters for polarizability, H-bond acceptor basicity, and H-bond donor acidity of toxicant			

of narcosis, though the most-favoured current interpretation is that alterations in the lipid matrix adversely influence the functioning of membrane proteins (Miller, 1985, 1986). The fact that there is a correlation between inhibition of the activity of membrane-bound ATPases by organic compounds, and log K_{ow}, lends support to this interpretation (Table 14.5, Ahlers *et al.*, 1991).

The membrane-based theories predict that narcosis should occur at a constant body burden of chemical (molar concentration corrected for volume for Mullins's theory). Data for fish (McCarty, 1986; Van Hoogen & Opperhuizen, 1988) and mussels (Donkin *et al.*, 1989, 1991) support this conclusion. However, the studies on mussels also showed that some chemicals could be accumulated to considerable concentrations without causing narcotic effects on filter feeding by the animal (Fig. 14.2). Response 'cut-offs' are an established characteristic of narcotic toxicity, giving rise to parabolic or bilinear QSARs where the response data are presented in terms of dose or exposure concentration (Veith *et al.*, 1984/85;

Table 14.5). Part of this non-linearity may result from factors influencing bioaccumulation (Section 14.4), but the primary explanation provided by the membrane theory of narcosis is that large hydrophobic molecules are often very insoluble in water, and of limited solubility in lipid membranes, so are unable to achieve a narcotic concentration in the membrane by partitioning (Miller, 1985; Gobas *et al.*, 1988). This effect may be accentuated for some compounds by intracellular sequestration processes (De Bruijn *et al.*, 1991; Donkin *et al.*, 1991).

An alternative theory to explain both the cut-off and narcosis has been advocated by Franks & Lieb (1990). Based on studies of the enzyme luciferase, they suggest that narcosis is the result of direct interaction between proteins and xenobiotic chemical, and that the toxicity cut-off occurs when the size of the chemical exceeds that of a specific binding pocket in the protein. Such a pocket would probably be too small to account for the narcotic toxicity of large surfactant molecules, so multiple sites of interaction need to be

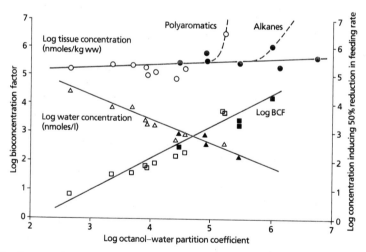

Fig. 14.2 Relationship between log *n*-octanol–water partition coefficient of hydrophobic compounds and the log of their bioconcentration factor into the marine mussel (*Mytilus edulis*), and the log of the concentration of these compounds in water (nmol/l) and mussel tissue (nmol/kg wet weight) which reduce filter feeding by 50%. The symbols on the figure represent the following parameters: water-based EC$_{50}$, ($\triangle$) aromatics, ($\blacktriangle$) alkanes and substituted alkanes; tissue-based EC$_{50}$ ($\circ$) aromatics, ($\bullet$) alkanes and substituted alkanes; BCF ($\square$) aromatics, ($\blacksquare$) alkanes and substituted alkanes. The near vertical lines from the tissue-based EC$_{50}$ plot indicate the approximate position of the toxicity cut-off for polyaromatics and unsubstituted alkanes. From Donkin *et al.*, 1991.

invoked (Schüürmann, 1991). There is increasing evidence that the most sensitive manifestations of narcosis in multicellular animals can be related to specific regions of the nervous system, and perhaps to specific proteins (Evers *et al.*, 1987; Franks & Lieb, 1990; Wafford *et al.*, 1991; McKenzie *et al.*, 1992), but the association between membrane and protein can be so intimate (McCrea *et al.*, 1988), that an unequivocal mechanistic answer may be elusive. However, although cellular narcosis can be observed in all organisms, plants (Overton, 1901; Hutchinson *et al.*, 1979), bacteria (Blum & Speece, 1991), protozoans (Lipnick, 1990) and cultured animal cells (Babich & Borenfreund, 1987) tend to be less sensitive than animals with nervous systems. This may be due to the greater responsiveness of nerve cells to a common mode of toxic action, or perhaps indicates that narcosis is not a single toxicological phenomenon.

Evidence for the latter view comes from studies of the behavioural and physiological responses of fish to acute poisoning by different classes of toxicants. Non-specific narcotic chemicals such as many alcohols, ethers, ketones, hydrocarbons and halogen-substituted hydrocarbons produce a characteristic response in fish (Bradbury *et al.*, 1990). These are the types of compounds that give rise to the typical narcotic QSAR described above, which is thought to represent the minimum toxicity of hydrophobic molecules. The toxicity of chemicals that fit such a QSAR is concentration-additive (Könemann, 1981a,b; Deneer *et al.*, 1988a). However, phenols, anilines, and some other compounds which have polar groups, produce a distinctive narcotic response in fish, are more toxic than predicted by a 'baseline' narcotic QSAR, and their toxicity is not concentration-additive with chemicals that fit such a QSAR (Bradbury *et al.*, 1989; Veith & Broderius, 1990; Table 14.5). These compounds, which generally have a $\log K_{ow} < 2.9$, have been termed polar or type II narcotics. Their important characteristic is the presence of a strong hydrogen-bond donating group, a property also known to increase the response of isolated nerves to halogenated ether anaesthetics (Urban & Haydon, 1987). However, the toxicological distinction between compounds with and without polar hydrogen atoms is not

confined to organisms with nervous systems, since it has clearly been demonstrated in protozoans (Schultz *et al.*, 1987).

Although polar and non-polar narcotics are considered to act by distinctive mechanisms, they can be incorporated successfully into a single QSAR if solvatochromic parameters, which include descriptors for hydrogen bonding and dipolarity, are used. Equations that predict toxicity to bacteria (Kamlet *et al.*, 1986b) and to fish (Kamlet *et al.*, 1987) with high precision, have been developed using this technique (e.g. Table 14.5).

14.6.2 Chemicals acting by specific mechanisms

Despite the effectiveness of solvatochromic parameters at modelling the toxicity of compounds of differing polarity, Kamlet *et al.* (1986a) noted several compounds that were outliers from the relationship. These have also been highlighted by predicting the non-specific narcotic toxicity of a compound from a $\log K_{ow}$-based regression, and calculating a ratio between the predicted and experimentally determined toxicity values (Hermens, 1990a). Such outliers are toxic by means of a specific mode of action in addition to their baseline narcotic activity. Some of the common reasons for specific toxicity and approaches to developing predictive QSARs for them are summarized below.

1 *Neurotoxic compounds.* Many of the pesticides used in agriculture are potent neurotoxins. Although the molecular features which give rise to their toxicological properties are in many cases well established (Coats, 1990; Fukuto, 1990), the diversity of their modes of action precludes the development of multicompound-class QSARs of ecotoxicological value. However, within pesticide classes such as organophosphates, QSARs for toxicity to fish (Schüürmann, 1990), *Daphnia* and honeybee (*Apis mellifera*; Vighi *et al.*, 1991; Table 14.6) have been developed. These include terms to describe the hydrophobicity and reactivity of the molecule.

2 *Esters.* Those with $\log K_{ow}$ values < 4 are more toxic to fish than predicted by a non-specific narcotic QSAR (Veith *et al.*, 1984/85). Although the reason for this additional toxicity is unclear

Table 14.6 Toxicity QSARs for specific mechanisms

Organism, compounds and equation	n	r^2	References
Bees (Apis); organophosphorus pesticides			
$\log 1/LD_{50} = 1.14 \log K_{ow} - 0.28 (\log K_{ow})^2 + 0.28\,^1\chi - 0.76\,^2\chi_{ox}{}^v - 1.09\gamma_3 + 0.096(\gamma_3)^2 + 12.29$	14	0.908	Vighi *et al.* (1991)
LD, lethal dose; K_{ow}, partition coefficient of oxon metabolite; χ, connectivity index; γ_3, electronic parameter			
Protozoan (Tetrahymena); phenols, anilines			
$\log BR = 0.438 \log K_{ow} + 0.157$	27	0.93	Cajina-Quezada & Schultz (1990)
BR, concentration inhibiting population growth by 50%			
Fish (Poecilia); phenols			
$\log 1/LC_{50} = 0.38 \log K_{ow} + 0.16\, pK_a + 0.07$	21	0.81	Lipnick *et al.* (1986)
pK_a, acid dissociation constant			
Fish (Poecilia); epoxy compounds			
$-\log LC_{50} = (0.39 \pm 0.05) \log K_{ow} + (3.0 \pm 0.4) \log K_1 - 2.25$	12	0.89	Deener *et al.* (1988b)
K_1, rate constant for reaction of toxicant with 4-nitrobenzylpyridine			
Alcohol proelectrophiles (general equation)			
$\log 1/LC_{50} = a \log K_{ow} + bE^0 + c \log K + d$	–	–	Veith *et al.* (1989)
E^0, redox potential; K, descriptor for reactivity of metabolite			

(Veith & Broderius, 1990), Kamlet *et al.* (1987) have suggested that an accurate prediction of ester toxicity demands inclusion of a descriptor for hydrolysis rate.

3 *Respiratory uncouplers.* These break the link between electron transport and adenosine triphosphate synthesis in mitochondria. The molecular properties common to most uncouplers are sufficient hydrophobicity to allow penetration of lipoprotein membranes, and a weakly acidic proton (Terada, 1990). Many phenols and anilines have these characteristics. QSAR studies on plants and animals show that hydrophobicity is the most important predictive parameter. It alone can provide high-quality relationships, though their low slopes clearly distinguish them from narcotic QSARs (Cajina-Quezada & Schultz, 1990; Table 14.6; Shannon *et al.*, 1991). When a structurally very diverse set of phenols is used to develop the QSAR, descriptors of the electronic properties of the molecule must be included before an accurate fit to the data is achieved (Lipnick, 1986; Table 14.6; Tissut *et al.*, 1987; Jaworska & Schultz, 1991).

4 *Herbicides.* Herbicides act by a variety of different mechanisms (Duke, 1990). QSARs have been derived for restricted compound classes but these are of limited value in ecotoxicology.

5 *Reactive compounds.* Many industrial organic chemicals are electrophiles, which can react with amino, hydroxy and sulphydryl groups on proteins and nucleic acids, adversely affecting their biological function. A variety of reactions are possible, depending on the nature of the chemical and of the group on the macromolecule which is attacked (Hermens, 1990b). In order to derive QSARs of high predictive capability it is generally necessary to include a descriptor of reactivity. For example, the rate constants for the reaction of alkyl halides (Hermens *et al.*, 1985) and epoxides (Deneer *et al.*, 1988b; Table 14.6) with 4-nitrobenzyl pyridine have been used for this purpose. In contrast, however, Deneer and co-workers (1988c) reported that the toxicity of aldehydes to fish could be modelled as accurately using $\log K_{ow}$ as the only descriptor, as when the rate constant for reaction with cysteine was included in the equation. This was thought to indicate that transport processes were the rate-limiting step in the response of these fish to aldehydes.

The slope of the aldehyde QSAR line was different from that expected for a narcotic mechanism, supporting the idea that reactive toxicity was involved. However, the data showed that as the hydrophobicity of the compounds tested changed from very low ($\log K_{ow}$ 0.53) to high ($\log K_{ow}$ 3.7) values, the toxicity of the compounds approached that anticipated for a narcotic mechanism (Deneer *et al.*, 1988). Similar observations have been reported by Overton (1901) and by many other workers since (Veith *et al.*, 1989; Hermens, 1990b), whenever the compounds tested act by a mechanism more biologically damaging than non-specific 'baseline' narcosis. This includes polar (or Type II) narcotics which are only distinguishable from non-polar narcotics at $\log K_{ow}$ values < approximately 2.7 (Veith & Broderius, 1990).

Some unreactive chemicals can be converted to reactive forms by metabolism within the organism. Perhaps the best known are polyaromatic hydrocarbons which can be converted to carcinogenic metabolites by monooxygenase enzymes; several QSARs have been developed for mammals (Lewis, 1987; Govers, 1990). The toxicity of some nitroaromatics appears to be related to ease of reduction to reactive intermediates; their toxicity to fish was adequately described by a QSAR with $\log K_{ow}$ and electrochemical reduction potential as the only parameters (Deneer *et al.*, 1987). The excess toxicity of some acetylenic alcohols has also been attributed to reduction (by alcohol dehydrogenase) to reactive aldehydes and ketones (Lipnick, 1989; Veith *et al.*, 1989). A suitable QSAR model included redox potential and a descriptor for the reactivity of the reduction product (Table 14.6).

Some metals may be toxic because of their reactivity towards $-SH$ groups in proteins (Hermens, 1990b) in addition to other mechanisms. Successful QSARs have been derived for the toxicity of metals to aquatic animals (Kaiser, 1980), mammals (Turner *et al.*, 1988; Hoeschele *et al.*, 1991), and cultured cells (Babich & Borenfreund, 1987). Generally, equations include more than one descriptor (e.g. for size and electronic properties).

14.7 WHAT USE ARE QSARs?

Hansch *et al.* (1989) have suggested that 'a long-term objective of toxicology must be to devise a computerized data base of numerically defined, statistically valid, structure–activity relationships'. For predictive ecotoxicology this suggestion should be extended to include property–property or property–structure relationships. Why are these worthwhile objectives?

The research reviewed in the preceding sections has shown that QSAR and related approaches can be used to predict the physical and chemical properties of environmental contaminants, their bioaccumulation, biodegradation and toxicity. Most biological QSARs are derived for the response of a single species, so may seem to be of limited value in predicting the behaviour of complex ecosystems.

However, probably the most important finding of QSAR research is that non-specific narcosis is a universal biological response, caused by chemicals with definable physicochemical properties. This response provides a baseline against which the toxicity of other chemicals can be assessed. Outliers from the non-specific narcotic QSAR which exhibit enhanced toxicity, must act by an additional mechanism which needs to be understood (Lipnick, 1991). It could be argued that claims that a biological response reflects a specific mode of action should always be tested against the response to known non-specific narcotics. Some outliers are less toxic than predicted by baseline narcosis, often because their physicochemical properties (hydrophobicity, solubility, molecular size) lie outside the range required to produce this response (Donkin *et al.*, 1991).

QSARs which result from narcosis can be highly correlated between organisms as diverse as bacteria and fish (Moulton & Schultz, 1986; Nendza & Klein, 1990; Schultz *et al.*, 1990) so the toxicity of a compound to a species can often be predicted from data derived using a different species (Zaroogian *et al.*, 1985; Cronin *et al.*, 1991). However, this approach also shows that the toxicity of compounds that act by a specific mechanism can vary dramatically between species (references above and Deneer *et al.*, 1989). This can highlight particularly vulnerable species.

Narcotics research, then subsequent studies of other chemicals, have shown that the toxicity of mixtures of compounds acting by a common mechanism (i.e. they fit a single QSAR) is usually concentration-additive, at least to aquatic animals (Könemann, 1981b; Deneer *et al.*, 1988a,b; Deneer *et al.*, 1988; Widdows & Donkin, 1991; Chapter 12). This observation is of crucial importance in ecotoxicology, since environmental contaminants almost always occur as complete mixtures (e.g. Hardy *et al.*, 1987). Further research is required to establish the generality of additive toxicity, since some exceptions have been observed in studies with multi-species bacterial cultures (Warne *et al.*, 1989).

Practical application of the basic understanding derived from QSAR studies is most advanced in the field of aquatic toxicology. A QSAR approach has been either advocated or used to design toxicity tests, reducing the need for range-finding (Turner *et al.*, 1987); to interpret the toxicological significance of bioaccumulated contaminant residues (Donkin & Widdows, 1986; Widdows & Donkin, 1991); to design organic contaminant monitoring programmes (Donkin *et al.*, 1991), and to carry out environmental risk/hazard assessments (Passino & Smith, 1987; Auer *et al.*, 1990; Nendza *et al.*, 1990; Van Leeuwen *et al.*, 1991). Hickey *et al.* (1990) have developed a QSAR-based expert system using linear solvation energy parameters to predict the toxicity of non-polar narcotics to *Photobacterium*, *Daphnia* and a fish (*Pimephales*), and intend to extend the system to other organisms and modes of action.

Progress in the development and use of QSARs in predictive ecotoxicology has been rapid in recent years, but they do have limitations. Perhaps the greatest of these is recognizing to which of many established QSARs the chemical of interest belongs (Turner *et al.*, 1987). There is a need to improve our understanding of which structural characteristics give rise to particular specific modes of action. QSARs are not a panacea, but they are already an essential tool in predictive ecotoxicology, and their use will certainly increase in the future.

REFERENCES

Abernethy, S.G., Mackay, D. & McCarty, L.S. (1988) 'Volume fraction' correlation for narcosis in aquatic organisms: the key role of partitioning. *Environ. Toxicol. Chem.* **7**, 469–481.

Ahlers, J., Cascorbi, I., Forê, M., Gies, A., Köhler, M., Pauli, W. & Rösick, E. (1991) Interaction with functional membrane proteins—a common mechanism of toxicity for lipophilic environmental chemicals? *Comp. Biochem. Physiol.* **100C**, 111–113.

Anliker, R. & Moser, P. (1987) The limits of bioaccumulation of organic pigments in fish: their relation to the partition coefficient and the solubility in water and octanol. *Ecotoxicol. Environ. Safety* **13**, 43–52.

Auer, C.M., Nabholz, J.V. & Baetcke, K.P. (1990) Mode of action and the assessment of chemical hazards in the presence of limited data: use of structure—activity relationships (SAR) under TSCA, section 5. *Environ. Health Perspect.* **87**, 183–197.

Babich, H. & Borenfreund, E. (1987) Structure—activity relationship (SAR) models established *in vitro* with the neutral red cytotoxicity assay. *Toxicol. in Vitro* **1**, 3–9.

Banerjee, S. & Baughman, G.L. (1991) Bioconcentration factors and lipid solubility. *Environ. Sci. Technol.* **25**, 536–539.

Barber, M.C., Suárez, L.A. & Lassiter, R.R. (1991) Modelling bioaccumulation of organic pollutants in fish with an application to PCBs in Lake Ontario salmonids. *Can. J. Fish. Aquatic Sci.* **48**, 318–337.

Barron, M.G. (1990) Bioconcentration. *Environ. Sci. Technol.* **24**, 1612–1618.

Basak, S.C. (1990) A nonempirical approach to predicting molecular properties using graph-theoretic invariants. In: *Practical Applications of Quantitative Structure—Activity Relationships (QSAR) in Environmental Chemistry and Toxicology* (Eds W. Karcher & J. Devillers), pp. 83–103. Kluwer, Dordrecht.

Baughman, G.L. & Paris, D.F. (1981) Microbial bioconcentration of organic pollutants from aquatic systems—a critical review. *CRC Crit. Rev. Microbiol.* **8**, 205–228.

Bierman, V.J. (1990) Equilibrium partitioning and biomagnification of organic chemicals in benthic animals. *Environ. Sci. Technol.* **24**, 1407–1412.

Blum, D.J.W. & Speece, R.E. (1990) Determining chemical toxicity to aquatic species. *Environ. Sci. Technol.* **24**, 284–293.

Blum, D.J.W. & Speece, R.E. (1991) Quantitative structure-activity relationships for chemical toxicity to environmental bacteria. *Ecotoxicol. Environ. Safety* **22**, 198–224.

Boese, B.L., Lee, H., II, Specht, D.T., Randall, R.C. & Winsor, M.H. (1990) Comparison of aqueous and solid-phase uptake for hexachlorobenzene in the tellinid clam *Macoma nasuta* (Conrad): a mass balance approach. *Environ. Toxicol. Chem.* **9**, 221–231.

Boethling, R.S. (1986) Application of molecular topology to quantitative structure—biodegradability relationships. *Environ. Toxicol. Chem.* **5**, 797–806.

Boethling, R.S., Gregg, B., Frederick, R., Gabel, N.W., Campbell, S.E. & Sabljic, A. (1989) Expert systems survey on biodegradation of xenobiotic chemicals. *Ecotoxicol. Environ. Safety* **18**, 252–267.

Bond, J.A., Baker, S.M. & Bechtold, W.E. (1985) Correlation of the octanol/water partition coefficient with clearance halftimes of intracheally instilled aromatic hydrocarbons in rats. *Toxicology* **36**, 285–295.

Bradbury, S.P., Carlson, R.W. & Henry, T.R. (1989) Polar narcosis in aquatic organisms. In: *Aquatic Toxicology and Hazard Assessment*, Vol. 12 (Eds U.M. Cowgill & L.R. Williams), pp. 59–73. American Society for Testing and Materials, Philadelphia, PA.

Bradbury, S.P., Henry, T.R. & Carlson, R.W. (1990) Fish acute toxicity syndromes in the development of mechanism-specific QSARs. In: *Practical Applications of Quantitative Structure—Activity Relationships (QSAR) in Environmental Chemistry and Toxicology* (Eds W. Karcher & J. Devillers), pp. 295–315. Kluwer, Dordrecht.

Brunner, S., Hornung, E., Santi, H., Wolff, E., Piringer, O.G., Altschuh, J. & Brüggemann, R. (1990) Henry's Law constants for polychlorinated biphenyls: experimental determination and structure—property relationships. *Environ. Sci. Technol.* **24**, 1751–1754.

Cajina-Quezada, M. & Schultz, T.W. (1990) Structure—toxicity relationships for selected weak acid respiratory uncouplers. *Aquat. Toxicol.* **17**, 239–252.

Calamari, D. & Vighi, M. (1990) Quantitative structure—activity relationships in ecotoxicology; value and limitations. *Rev. Environ. Toxicol.* **4**, 1–112.

Calamari, D., Bacci, E., Focardi, S., Gaggi, C., Morosini, M. & Vighi, M. (1991) Role of plant biomass in the global environmental partitioning of chlorinated hydrocarbons. *Environ. Sci. Technol.* **25**, 1489–1495.

Casserly, D.M., Davis, E.M., Downs, T.D. & Guthrie, R.K. (1983) Sorption of organics by *Selenastrum capricornutum*. *Water Res.* **17**, 1591–1594.

Chiou, C.T. (1985) Partition coefficients of organic compounds in lipid-water systems and correlations with fish bioconcentration factors. *Environ. Sci. Technol.* **19**, 57–62.

Chiou, C.T., Kile, D.E., Brinton, T.I., Malcolm, R.L., Leenheer, J.A. & MacCarthy, P. (1987) A comparison of water solubility enhancements of organic solutes by aquatic humic materials and commercial humic acids. *Environ. Sci. Technol.* **21**, 1231–1234.

Clark, K.E. & Mackay, D. (1991) Dietary uptake and biomagnification of four chlorinated hydrocarbons by guppies. *Environ. Toxicol. Chem.* **10**, 1205–1217.

Clark, K.E., Gobas, F.A.P.C. & Mackay, D. (1990) Model

of organic chemical uptake and clearance by fish from food and water. *Environ. Sci. Technol.* **24**, 1203–1213.

Coats, J.R. (1990) Mechanisms of toxic action and structure-activity relationships for organochlorine and synthetic pyrethroid insecticides. *Environ. Health Perspect.* **87**, 255–262.

Coates, M., Connell, D.W. & Barron, D.M. (1985) Aqueous solubility and octanol-1-ol to water partition coefficients of aliphatic hydrocarbons. *Environ. Sci. Technol.* **19**, 628–632.

Collander, R. (1954) The permeability of nitella cells to non-electrolytes. *Physiol. Plant.* **7**, 420–445.

Connell, D.W. (1988) Bioaccumulation behaviour of persistent organic chemicals with aquatic organisms. *Rev. Environ. Contam. Toxicol.* **101**, 117–154.

Connell, D.W. & Hawker, D.W. (1988) Use of polynomial expressions to describe the bioconcentration of hydrophobic chemicals by fish. *Ecotoxicol. Environ. Safety* **16**, 242–257.

Connell, D.W. & Schüürmann, G. (1988) Evaluation of various molecular parameters as predictors of bioconcentration in fish. *Ecotoxicol. Environ. Safety* **15**, 324–355.

Connolly, J.P. & Pedersen, C.D. (1988) A thermodynamic-based evaluation of organic chemical accumulation in aquatic organisms. *Environ. Sci. Technol.* **22**, 99–103.

Corner, E.D.S. & Sparrow, B.W. (1957) The modes of action of toxic agents. II. Factors influencing the toxicities of mercury compounds to certain Crustacea. *J. Mar. Biol. Assoc. UK* **36**, 459–472.

Crisp, D.J., Christie, A.O. & Ghobashy, A.F.A. (1967) Narcotic and toxic action of organic compounds on barnacle larvae. *Comp. Biochem. Physiol.* **22**, 629–649.

Cronin, M.T.D., Dearden, J.C. & Dobbs, A.J. (1991) QSAR studies of comparative toxicity in aquatic organisms. *Sci. Total Environ.* **109/110**, 431–439.

Dearden, J.C. (1985) Partitioning and lipophilicity in quantitative structure–activity relationships. *Environ. Health Perspect.* **61**, 203–228.

Dearden, J.C. (1990) Physico-chemical descriptors. In: *Practical Applications of Quantitative Structure–Activity Relationships (QSAR) in Environmental Chemistry and Toxicology* (Eds W. Karcher & J. Devillers), pp. 25–59. Kluwer, Dordrecht.

Dearden, J.C. & Nicholson, R.M. (1987a) Correlation of biodegradability with atomic charge difference and superdelocalizability. In: *QSAR in Environmental Toxicology–II* (Ed. K.L.E. Kaiser), pp. 83–89. Reidel, Dordrecht.

Dearden, J.C. & Nicholson, R.M. (1987b) QSAR study of the biodegradability of environmental pollutants. In: *QSAR in Drug Design and Toxicology* (Eds D. Hadzi & B. Jerman-Blazic), pp. 307–312. Elsevier, Amsterdam.

De Bruijn, J., Yedema, E., Seinen, W. & Hermens, J. (1991) Lethal body burdens of four organophosphorus pesticides in the guppy (*Poecilia reticulata*). *Aquat. Toxicol.* **20**, 111–122.

Deneer, J.W., Sinnige, T.L., Seinen, W. & Hermens, J.L.M. (1987) Quantitative structure–activity relationships for the toxicity and bioconcentration factor of nitrobenzene derivatives towards the guppy (*Poecilia reticulata*). *Aquat. Toxicol.* **10**, 115–129.

Deneer, J.W., Sinnige, T.L., Seinen, W. & Hermens, J.L.M. (1988a) The joint acute toxicity to *Daphnia magna* of industrial organic chemicals at low concentrations. *Aquat. Toxicol.* **12**, 33–38.

Deneer, J.W., Sinnige, T.L., Seinen, W. & Hermens, J.L.M. (1988b) A quantitative structure–activity relationship for the acute toxicity of some epoxy compounds to the guppy. *Aquat. Toxicol.* **13**, 195–204.

Deneer, J.W., Seinen, W. & Hermens, J.L.M. (1988) The acute toxicity of aldehydes to the guppy. *Aquat. Toxicol.* **12**, 185–192.

Deneer, J.W., van Leeuwen, C.J., Seinen, W., Maas-Diepeveen, J.L. & Hermens, J.L.M. (1989) QSAR study of the toxicity of nitrobenzene derivatives towards *Daphnia magna*, *Chlorella pyrenoidosa* and *Photobacterium phosphoreum*. *Aquat. Toxicol.* **15**, 83–98.

Dobbs, R.A., Wang, L. & Govind, R. (1989) Sorption of toxic organic compounds on wastewater solids: correlation with fundamental properties. *Environ. Sci. Technol.* **23**, 1092–1097.

Donkin, P. & Widdows, J. (1986) Scope for growth as a measurement of environmental pollution and its interpretation using structure–activity relationships. *Chem. Ind.* **21**, 721–752.

Donkin, P. & Widdows, J. (1990) Quantitative structure–activity relationships in aquatic invertebrate toxicology. *Rev. Aquat. Sci.* **2**, 375–398.

Donkin, P., Widdows, J., Evans, S.V., Worrall, C.M. & Carr, M. (1989) Quantitative structure–activity relationships for the effect of hydrophobic organic chemicals on rate of feeding by mussels (*Mytilus edulis*). *Aquat. Toxicol.* **14**, 277–294.

Donkin, P., Widdows, J., Evans, S.V. & Brinsley, M.D. (1991) QSARs for the sublethal responses of marine mussels (*Mytilus edulis*). *Sci. Total Environ.* **109/110**, 461–476.

Doucette, W.J. & Andren, A.W. (1987) Correlation of octanol/water partition coefficients and total molecular surface area for highly hydrophobic aromatic compounds. *Environ. Sci. Technol.* **21**, 821–824.

Drossman, H., Johnson, H. & Mill, T. (1988) Structure–activity relationships in environmental processes. 1. Hydrolysis of esters and carbamates. *Chemosphere* **17**, 1509–1530.

Duke, S.O. (1990) Overview of herbicide mechanisms of action. *Environ. Health Perspect.* **87**, 263–271.

Erickson, R.J. & McKim, J.M. (1990) A model for exchange of organic chemicals at fish gills: flow and diffusion limitations. *Aquat. Toxicol.* **18**, 175–198.

Ernst, W. (1985) Accumulation in aquatic organisms. In: *Appraisal of Tests to Predict the Environmental Behaviour of Chemicals* (Eds P. Sheenan, F. Korte, W. Klein & Ph. Bordeau), pp. 243–255. John Wiley & Sons, Chichester.

Esser, H.O. (1986) A review of the correlation between physicochemical properties and bioaccumulation. *Pestic. Sci.* **17**, 265–276.

Evers, A.S., Berkowitz, B.A. & d'Avignon, D.A. (1987) Correlation between the anaesthetic effect of halothane and saturable binding in brain. *Nature* **328**, 157–160.

Farrington, J.W. (1989) Bioaccumulation of hydrophobic organic pollutant compounds. In: *Ecotoxicology: Problems and Approaches* (Eds S.A. Levin, M.A. Harwell, J.R. Kelly & K.D. Kimball), pp. 279–314. Springer-Verlag, New York.

Ferguson, J. (1939) Use of chemical potentials as indices of toxicity. *Proc. R. Soc. Lond. B* **127**, 387–404.

Franks, N.P. & Lieb, W.R. (1990) Mechanisms of general anaesthesia. *Environ. Health Perspect.* **87**, 199–205.

Fukuto, T.R. (1990) Mechanism of action of organophosphorus and carbamate insecticides. *Environ. Health Perspect.* **87**, 245–254.

Garten, C.T. & Trabalka, J.R. (1983) Evaluation of models for predicting terrestial food chain behaviour of xenobiotics. *Environ. Sci. Technol.* **17**, 590–595.

Gobas, F.A.P.C., Opperhuizen, A. & Hutzinger, O. (1986) Bioconcentration of hydrophobic chemicals in fish: relationship with membrane permeation. *Environ. Toxicol. Chem.* **5**, 637–646.

Gobas, F.A.P.C., Lahittete, J.M., Garofalo, G., Shiu, W.Y. & Mackay, D. (1988) A novel method for measuring membrane–water partition coefficients of hydrophobic organic chemicals: comparison with 1-octanol-water partitioning. *J. Pharm. Sci.* **77**, 265–272.

Gobas, F.A.P.C., McNeil, E.J., Lovett-Doust, L. & Haffner, G.D. (1991) Bioconcentration of chlorinated aromatic hydrocarbons in aquatic macrophytes. *Environ. Sci. Technol.* **25**, 924–929.

Gossett, R.W., Brown, D.A. & Young, D.R. (1983) Predicting the bioaccumulation of organic compounds in marine organisms using octanol/water partition coefficients. *Mar. Pollut. Bull.* **14**, 387–392.

Govers, H.A.J. (1990) Prediction of environmental behaviour and effects of polycyclic aromatic hydrocarbons by PAR and QSAR. In: *Practical Applications of Quantitative Structure-Activity Relationships (QSAR) in Environmental Chemistry and Toxicology* (Eds W. Karcher & J. Devillers), pp. 411–432. Kluwer, Dordrecht.

Hammett, L.P. (1937) The effect of structure upon the reactions of organic compounds. Benzene derivatives. *J. Am. Chem. Soc.* **59**, 96–103.

Hansch, C. & Fujita, T. (1964) ρ-σ-π Analysis. A method for the correlation of biological activity and chemical structure. *J. Am. Chem. Soc.* **86**, 1616–1626.

Hansch, C. & Leo, A. (1979) *Substituent Constants for Correlation Analysis in Chemistry and Biology*. John Wiley, New York.

Hansch, C., Kim, D., Leo, A.J., Norvellino, E., Silipo, C. & Vittoria, A. (1989) Toward a quantitative comparative toxicology of organic compounds. *CRC Crit. Rev. Toxicol.* **19**, 185–226.

Hardy, J.T., Crecelius, E.A., Antrim, L.D., Broadhurst, V.L., Apts, C.W., Gurtisen, J.M. & Fortman, T.J. (1987) The sea-surface microlayer of Puget Sound: Part II. Concentrations of contaminants and relation to toxicity. *Mar. Environ. Res.* **23**, 251–271.

Hawker, D. (1989) Vapour pressures and Henry's Law constants of polychlorinated biphenyls. *Environ. Sci. Technol.* **23**, 1250–1253.

Hawker, D.W. & Connell, D.W. (1986) Bioconcentration of lipophilic compounds by some aquatic organisms. *Ecotoxicol. Environ. Safety* **11**, 184–197.

Herbert, C.E. & Haffner, G.D. (1991) Habitat partitioning and contaminant exposure in cyprinids. *Can. J. Fish. Aquat. Sci.* **48**, 261–266.

Hermens, J.L.M. (1990a) Quantitative structure–activity relationships for predicting fish toxicity. In: *Practical Applications of Quantitative Structure–Activity Relationships (QSAR) in Environmental Chemistry and Toxicology* (Eds W. Karcher & J. Devillers), pp. 263–280. Kluwer, Dordrecht.

Hermens, J.L.M. (1990b) Electrophiles and acute toxicity to fish. *Environ. Health Perspect.* **87**, 219–225.

Hermens, J.L.M. & Opperhuizen, A. (Eds) (1991) QSAR in environmental toxicology. *Sci. Total Environ.* **109/110**, 706.

Hermens, J.L.M., Busser, F., Leeuwangh, P. & Musch, A. (1985) Quantitative correlation studies between the acute lethal toxicity of 15 organic halides to the guppy (*Poecilia reticulata*) and chemical reactivity towards 4-nitrobenzylpyridine. *Toxicol. Environ. Chem.* **9**, 219–236.

Hickey, J.P., Aldridge, A.J., Passino, D.R.M. & Frank, A.M. (1990) An expert system for prediction of aquatic toxicity of contaminants. In: *Expert Systems for Environmental Applications*, ACS Symposium Series 431 (Ed. J.M. Hushon), pp. 90–107. American Chemical Society, Washington, DC.

Hoeschele, J.D., Turner, J.E. & England, M.W. (1991) Inorganic concepts relevant to metal binding, activity, and toxicity in biological systems. *Sci. Total Environ.* **109/110**, 477–492.

Hutchinson, T.C., Hellebust, J.A., Mackay, D., Tam, D. & Kauss, P. (1979) Relationship of hydrocarbon solubility to toxicity in algae and cellular membrane effects. In: *Proceedings, 1979 Oil Spill Conference* (Ed. J.O. Ludwigson), pp. 541–547. American Petroleum Institute, Washington, DC.

Jaffé, R. (1991) Fate of hydrophobic organic pollutants in the aquatic environment: a review. *Environ. Pollut.* **69**, 237–257.

Jaworska, J.S. & Schultz, T.W. (1991) Comparative toxicity and structure-activity in *Chlorella* and *Tetrahymena*: monosubstituted phenols. *Bull. Environ. Contam. Toxicol.* **47**, 57–62.

Kaiser, K.L.E. (1980) Correlation and prediction of metal toxicity to aquatic biota. *Can. J. Fish. Aquat. Sci.* **37**, 211–218.

Kamlet, M.J., Doherty, R.M., Abboud, J.-L. M., Abraham, M.H. & Taft, R.W. (1986a) Linear solvation energy relationships. 36. Molecular properties governing solubilities of organic nonelectrolytes in water. *J. Pharm. Sci.* **75**, 338–349.

Kamlet, M.J., Doherty, R.M., Veith, G.D., Taft, R.W. & Abraham, M.H. (1986b) Solubility properties in polymers and biological media. 7. An analysis of toxicant properties that influence inhibition of bioluminescence in *Photobacterium phosphoreum* (the Microtox test). *Environ. Sci. Technol.* **20**, 690–695.

Kamlet, M.J., Doherty, R.M., Taft, R.W., Abraham, M.H., Veith, G.D. & Abraham, D.J. (1987) Solubility properties in polymers and biological media. 8. An analysis of the factors that influence toxicities of organic nonelectrolytes to the golden orfe fish (*Leuciscus idus melanotus*). *Environ. Sci. Technol.* **21**, 149–155.

Könemann, H. (1981a) Quantitative structure-activity relationships in fish toxicity studies. Part 1: relationship for 50 industrial pollutants. *Toxicology* **19**, 209–221.

Könemann, H. (1981b) Fish toxicity tests with mixtures of more than two chemicals: a proposal for a quantitative approach and experimental results. *Toxicology* **19**, 229–238.

Kuenemann, P., Vasseur, P. & Devillers, J. (1990) Structure–biodegradability relationships. In: *Practical Applications of Quantitative Structure-Activity Relationships (QSAR) in Environmental Chemistry and Toxicology* (Eds W. Karcher & J. Devillers), pp. 343–370. Kluwer, Dordrecht.

Landrum, P.F. (1988) Toxicokinetics of organic xenobiotics in the amphipod, *Pontoporeia hoyi*: role of physiological and environmental variables. *Aquat. Toxicol.* **12**, 245–271.

Landrum, P.F., Faust, W.R. & Eadie, B.J. (1989) Bioavailability and toxicity of a mixture of sediment-associated chlorinated hydrocarbons to the amphipod *Pontoporeia hoyi*. In: *Aquatic Toxicology and Hazard Assessment*, Vol. 12 (Eds. U.M. Cowgill & L.R. Williams), pp. 315–329. American Society for Testing and Materials, Philadelphia, PA.

Landrum, P.F., Eadie, B.J. & Faust, W.R. (1991) Toxicokinetics and toxicity of a mixture of sediment-associated polycyclic aromatic hydrocarbons to the amphipod *Diporeia* sp. *Environ. Toxicol. Chem.* **10**, 35–46.

Larsson, P., Collvin, L., Okla, L. & Meyer, G. (1992) Lake productivity and water chemistry as governors of the uptake of persistent pollutants in fish. *Environ. Sci. Technol.* **26**, 346–352.

Lewis, D.F.V. (1987) Molecular orbital calculations and quantitative structure–activity relationships for some polyaromatic hydrocarbons. *Xenobiotica* **17**, 1351–1361.

Lewis, D.F.V. (1990) MO-QSARs: a review of molecular orbital-generated quantitative structure-activity relationships. In: *Progress in Drug Metabolism*, Vol. 12 (Ed. G.G. Gibson), pp. 205–255. Taylor and Francis, London.

Lipnick, R.L. (1989) Narcosis, electrophile and proelectrophile toxicity mechanisms: application of SAR and QSAR. *Environ. Toxicol. Chem.* **8**, 1–12.

Lipnick, R.L. (1990) Narcosis: fundamental and baseline toxicity mechanism for nonelectrolyte organic chemicals. In: *Practical Applications of Quantitative Structure–Activity Relationships (QSAR) in Environmental Chemistry and Toxicology* (Eds W. Karcher & J. Devillers), pp. 281–293. Kluwer, Dordrecht.

Lipnick, R.L. (1991) Outliers: their origin and use in the classification of molecular mechanisms of toxicity. *Sci. Total Environ.* **109/110**, 131–153.

Lipnick, R.L. & Filov, V.A. (1992) Nikolai Vasilyevich Lazarev, toxicologist and pharmacologist, comes in from the cold. *Trends Pharmacol. Sci.* **13**, 56–60.

Lipnick, R.L., Bickings, C.K., Johnson, D.E. & Eastmond, D.A. (1986) Comparison of QSAR predictions with fish toxicity screening data for 110 phenols. In: *Aquatic Toxicology and Hazard Assessment: Eighth Symposium* (Eds R.C. Bahner & D.J. Hansen), pp. 153–176. American Society for Testing and Materials, Philadelphia, PA.

Livingstone, D.R. (1991) Organic xenobiotic metabolism in marine invertebrates. In: *Advances in Comparative and Environmental Physiology*, Vol. 7 (Ed. R. Gilles), pp. 46–185. Springer Verlag, Berlin.

Mackay, D. (1982) Correlation of bioconcentration factors. *Environ. Sci. Technol.* **16**, 274–278.

Mackay, D. & Paterson, S. (1990) Fugacity models. In: *Practical Applications of Quantitative Structure–Activity Relationships (QSAR) in Environmental Chemistry and Toxicology* (Eds W. Karcher & J. Devillers), pp. 433–460. Kluwer, Dordrecht.

Mailhot, H. (1987) Prediction of algal bioaccumulation and uptake rate of nine organic compounds by ten physicochemical properties. *Environ. Sci. Technol.* **21**, 1009–1013.

Mailhot, H. & Peters, R.H. (1988) Empirical relationships between the 1-octanol/water partition coefficient and nine physicochemical properties. *Environ. Sci. Technol.* **22**, 1479–1488.

Mani, S.V., Connell, D.W. & Braddock, R.D. (1991) Structure–activity relationships for the prediction of biodegradability of environmental pollutants. *Crit. Rev. Environ. Contr.* **21**, 217–236.

McCarthy, J.F., Jimenez, B.D. & Barbee, T. (1985) Effect of dissolved humic material on accumulation of polycyclic aromatic hydrocarbons: structure–activity relationships. *Aquat. Toxicol.* **7**, 15–24.

McCarthy, L.S. (1986) The relationship between aquatic toxicity QSARs and bioconcentration for some organic chemicals. *Environ. Toxicol. Chem.* **5**, 1071–1080.

McCrea, P.D., Engelman, D.M. & Popot, J.-L. (1988) Topography of integral membrane proteins: hydrophobicity analysis vs immunolocalization. *Trends Biol. Sci.* **13**, 289–290.

McKenzie, J.D., Calow, P., Clyde, J., Miles, A., Dickinson, R., Lieb, W.R. & Franks, N.P. (1992) Effects of temperature on the anaesthetic potency of halothane, enflurane and ethanol in *Daphnia magna* (Cladocera: Crustacea) *Comp. Biochem. Physiol.* **101C**, 15–19.

Meyer, K.H. (1937) Contributions to the theory of narcosis. *Trans. Faraday Soc.* **33**, 1062–1064.

Miller, K.W. (1985) The nature of the site of general anesthesia. *Int. Rev. Neurobiol.* **27**, 1–61.

Miller, K.W. (1986) Are lipids or proteins the target of general anaesthetic action? *Trends Neurosci.* **10**, 49–51.

Miller, M.M., Wasik, S.P., Huang, G.-L., Shiu, W.-Y. & Mackay, D. (1985) Relationships between octanol–water partition coefficient and aqueous solubility. *Environ. Sci. Technol.* **19**, 522–529.

Moulton, M.P. & Schultz, T.W. (1986) Comparisons of several structure–toxicity relationships for chlorophenols. *Aquat. Toxicol.* **8**, 121–128.

Mullins, L.J. (1954) Some physical mechanisms in narcosis. *Chem. Rev.* **54**, 289–323.

Nagel, R. & Loskill, R. (Eds) (1991) *Bioaccumulation in Aquatic System*, pp. 239. VCH Publishers, Weinheim.

Nendza, M. & Klein, W. (1990) Comparative QSAR study on freshwater and estuarine toxicity. *Aquat. Toxicol.* **17**, 63–74.

Nendza, M. & Seydel, J.K. (1988) Quantitative structure–toxicity relationships for ecotoxicologically relevant biotestsystems and chemicals. *Chemosphere* **17**, 1585–1602.

Nendza, M., Volmer, J. & Klein, W. (1990) Risk assessment based on QSAR estimates. In: *Practical Applications of Quantitative Structure–Activity Relationships (QSAR) in Environmental Chemistry and Toxicology* (Eds by W. Karcher & J. Devillers), pp. 213–240. Kluwer, Dordrecht.

Niemi, G.J., Veith, G.D., Regal, R.R. & Vaishnav, D.D.

(1987) Structural features associated with degradable and persistent chemicals. *Environ. Toxicol. Chem.* **6**, 515–527.

Nirmalakhandan, N. & Speece, R.E. (1988a) Structure–activity relationships. *Environ. Sci. Technol.* **22**, 606–615.

Nirmalakhandan, N. & Speece, R.E. (1988b) QSAR model for predicting Henry's constant. *Environ. Sci. Technol.* **22**, 1349–1357.

Nirmalakhandan, N. & Speece, R.E. (1989) Prediction of aqueous solubility of organic chemicals based on molecular structure. 2. Application to PNAs, PCBs, PCDDs etc.. *Environ. Sci. Technol.* **23**, 708–713.

Oliver, B.G. (1987) Biouptake of chlorinated hydrocarbons from laboratory-spiked and field sediments by oligochaete worms. *Environ. Sci. Technol.* **21**, 785–790.

Opperhuizen, A. & Stokkel, R.C.A.M. (1988) Influence of contaminated particles on the bioaccumulation of hydrophobic organic micropollutants in fish. *Environ. Pollut.* **51**, 165–177.

Opperhuizen, A., Velde, E.W.v.d., Gobas, F.A.P.C., Liem, D.A.K., Steen, J.M.D.v.d. & Hutzinger, O. (1985) Relationship between bioconcentration in fish and steric factors of hydrophobic chemicals. *Chemosphere* **14**, 1871–1896.

Opperhuizen, A., Serné, P. & Van der Steen, J.M.D. (1988) Thermodynamics of fish/water and octanol-1-ol/water partitioning of some chlorinated benzenes. *Environ. Sci. Technol.* **22**, 286–292.

Overton, E. (1901) *Studien über die Narkose*, Gustav Fischer, Jena. Available as English translation: Overton, E. (1991) *Studies of Narcosis* (Ed. R.L. Lipnick). *Chapman and Hall/Wood Library–Museum of Anesthesiology, London.*

Parsons, J.R. & Govers, H.A.J. (1990) Quantitative structure–activity relationships for biodegradation. *Ecotoxicol. Environ. Safety* **19**, 212–227.

Pärt, P. (1990) The perfused fish gill preparation in studies of the bioavailability of chemicals. *Ecotoxicol. Environ. Safety* **19**, 106–115.

Passino, D.R.M. & Smith, S.B. (1987) Quantitative structure–activity relationships (QSAR) and toxicity data in hazard assessment. In: *QSAR in Environmental Toxicology—II* (Ed. K.L.E. Kaiser), pp. 261–270. Reidel, Dordrecht.

Paterson, S., Mackay, D., Bacci, E. & Calamari, D. (1991) Correlation of the equilibrium and kinetics of leaf-air exchange of hydrophobic organic chemicals. *Environ. Sci. Technol.* **25**, 866–871.

Pruell, R.J., Lake, J.L., Davis, W.R. & Quinn, J.G. (1986) Uptake and depuration of organic contaminants by blue mussels (*Mytilus edulis*) exposed to environmentally contaminated sediment. *Mar. Biol.* **91**, 497–507.

Rashid, F. & Horobin, R.W. (1990) Interaction of molecular probes with living cells and tissues. Part 2.

Histochemistry **94**, 303–308.

Rashid, F., Horobin, R.W. & Williams, M.A. (1991) Predicting the behaviour and selectivity of fluorescent probes for lysosomes and related structures by means of structure-activity models. *Histochem. J.* **23**, 450–459.

Riederer, M. (1990) Estimating partitioning and transport of organic chemicals in the foliage/atmosphere system: discussion of a fugacity-based model. *Environ. Sci. Technol.* **24**, 829–837.

Saarikoski, J. & Viluksela, M. (1982) Relation between physicochemical properties of phenols and their toxicity and accumulation in fish. *Ecotoxicol. Environ. Safety* **6**, 501–512.

Sabljic, A. (1987) On the prediction of soil sorption coefficients of organic pollutants from molecular structure: application of molecular topology model. *Environ. Sci. Technol.* **21**, 358–366.

Sabljic, A. (1990) Topological indices and environmental chemistry. In: *Practical Applications of Quantitative Structure–Activity Relationships (QSAR) in Environmental Chemistry and Toxicology* (Eds W. Karcher & J. Devillers), pp. 61–82. Kluwer, Dordrecht.

Sabljic, A., Güsten, H., Schönherr, J. & Riederer, M. (1990) Modelling plant uptake of airborne organic chemicals. 1. Plant cuticle/water partitioning and molecular connectivity. *Environ. Sci. Technol.* **24**, 1321–1326.

Schultz, T.W. (1989) Nonpolar narcosis: a review of the mechanism of action for baseline aquatic toxicity. In: *Aquatic Toxicology and Hazard Assessment: 12th Volume* (Eds U.M. Cowgill & L.R. Williams), pp. 104–109. American Society for Testing and Materials, Philadelphia, PA.

Schultz, T.W., Appelhans, F.M. & Riggin, G.W. (1987) Structure–activity relationships of selected pyridines. III. Log K_{ow} analysis. *Ecotoxicol. Environ. Safety* **13**, 76–83.

Schultz, T.W., Arnold, L.M., Wilkie, T.S. & Moulton, M.P. (1990) Relationship of quantitative structure–activity for normal aliphatic alcohols. *Ecotoxicol. Environ. Safety* **19**, 243–253.

Schüürmann, G. (1990) QSAR analysis of the acute fish toxicity of organic phosphorothionates using theoretically derived molecular descriptors. *Environ. Toxicol. Chem.* **9**, 417–428.

Schüürmann, G. (1991) Acute aquatic toxicity of alkyl phenol ethoxylates. *Ecotoxicol. Environ. Safety* **21**, 227–233.

Seeman, P. (1972) The membrane actions of anaesthetics and tranquilizers. *Pharmacol. Rev.* **24**, 583–655.

Servos, M.R., Muir, D.C.G. & Webster, G.R.B. (1989) The effect of dissolved organic matter on the bioavailability of polychlorinated dibenzo-p-dioxins. *Aquat. Toxicol.* **14**, 169–184.

Shannon, R.D., Boardman, G.D., Dietrich, A.M. &

Bevan, D.R. (1991) Mitochondrial response to chlorophenols as a short-term toxicity assay. *Environ. Toxicol. Chem.* **10**, 57–66.

Shaw, G.R. & Connell, D.W. (1984) Physicochemical properties controlling polychlorinated biphenyl (PCB) concentrations in aquatic organisms. *Environ. Sci. Technol.* **18**, 18–23.

Shaw, G.R. & Connell, D.W. (1987) Comparative kinetics for bioaccumulation of polychlorinated biphenyls by the polychaete (*Capitella capitata*) and fish (*Mugil cephalus*). *Ecotoxicol. Environ. Safety* **13**, 84–91.

Taft, R.W.J. (1953) The general nature of the proportionality of polar effects of substituent groups in organic chemistry. *J. Am. Chem. Soc.* **75**, 4231–4238.

Tanacredi, J.T. & Cardenas, R.R. (1991) Biodepuration of polynuclear aromatic hydrocarbons from a bivalve mollusc, *Mercenaria mercenaria* L. *Environ. Sci. Technol.* **25**, 1453–1461.

Terada, H. (1990) Uncouplers of oxidative phosphorylation. *Environ. Health Perspect.* **87**, 213–218.

Tissut, M., Taillandier, G., Ravanel, P. & Benoit-Guyod, J.-L. (1987) Effects of chlorophenols on isolated class A chloroplasts and thylakoids: a QSAR study. *Ecotoxicol. Environ. Safety* **13**, 32–42.

Topp, E., Scheunert, I., Attar, A. & Korte, F. (1986) Factors affecting the uptake of ^{14}C-labelled organic chemicals by plants from soil. *Ecotoxicol. Environ. Safety* **11**, 219–228.

Tulp, M.Th.M. & Hutzinger, O. (1978) Some thoughts on aqueous solubilities and partition coefficients of PCB, and the mathematical correlation between bioaccumulation and physicochemical properties. *Chemosphere* **10**, 849–860.

Turner, L., Choplin, F., Dugard, P., Hermens, J., Jaeckh, R., Marsmann, M. & Roberts, D. (1987) Structure–activity relationships in toxicology and ecotoxicology: an assessment. *Toxicol. in Vitro* **1**, 143–171.

Turner, J.E., England, M.W., Hingerty, B.E. & Hayden, T.L. (1988) Correlations between pairs of simple physicochemical parameters of metal ions and acute toxicity in mice. *Sci. Total Environ.* **68**, 275–280.

Urban, B.W. & Haydon, D.A. (1987) The actions of halogenated ethers on the ionic currents of the squid giant axon. *Proc. R. Soc. Lond. B* **231**, 13–26.

Urushigawa, Y. & Yonezawa, Y. (1979) Chemico-biological interactions in biological purification systems. V. Relation between biodegradation rate constants of di-n-alkyl phthalate esters and their retention times in reversed phase partition chromatography. *Chemosphere* **5**, 317–320.

Van de Waterbeemd, H. & Testa, B. (1987) The paramaterization of lipophilicity and other structural properties in drug design. *Adv. Drug Res.* **16**, 85–225.

Van Hoogen, G. & Opperhuizen, A. (1988) Toxicokinetics of chlorobenzenes in fish. *Environ. Toxicol. Chem.* **7**, 213−219.

Van Leeuwen, C.J., Van Der Zandt, P.T.J., Aldenberg, T., Verhaar, H.J.M. & Hermens, J.L.M. (1991) The application of QSARs, extrapolation and equilibrium partitioning in aquatic effects assessment for narcotic pollutants. *Sci. Total Environ.* **109/110**, 681−690.

Veith, G.D. & Broderius, S.J. (1990) Rules for distinguishing toxicants that cause type I and type II narcosis syndromes. *Environ. Health Perspect.* **87**, 207−211.

Veith, G.D., De Foe, D. & Knuth, M. (1984/85) Structure−activity relationships for screening organic chemicals for potential ecotoxicity effects. *Drug Metab. Rev.* **15**, 1295−1303.

Veith, G.D., Lipnick, R.L. & Russom, C.L. (1989) The toxicity of acetylenic alcohols to the fathead minnow, *Pimephales, promelas*: narcosis and proelectrophile activation. *Xenobiotica* **19**, 555−565.

Vighi, M., Garlanda, M.M. & Calamari, D. (1991) QSARs for the toxicity of organophosphorus pesticides to *Daphnia* and honeybees. *Sci. Total Environ.* **109/110**, 605−622.

Wafford, K.A., Burnett, D.M., Leidenheimer, N.J., Burt, D.R., Wang, J.B., Kofuji, P., Dunwiddie, T.V., Harris, R.A. & Sikela, J.M. (1991) Ethanol sensitivity of the GABA$_A$ receptor expressed in *Xenopus* oocytes requires 8 amino acids contained in the γ2L subunit. *Neuron* **7**, 27−33.

Walker, C.H. (1985) Bioaccumulation in marine food chains−a kinetic approach. *Mar. Environ. Res.* **17**, 297−300.

Warne, M.StJ., Connell, D.W., Hawker, D.W. & Schüürmann, G. (1989) Prediction of the toxicity of mixtures of shale oil components. *Ecotoxicol. Environ. Safety* **18**, 121−128.

Warne, M.StJ., Connell, D.W. & Hawker, D.W. (1991) Comparison of the critical concentration and critical volume hypotheses to model non-specific toxicity of individual compounds. *Toxicology* **66**, 187−195.

Warwick, R.M. & Joint, I.R. (1987) The size distribution of organisms in the Celtic Sea. *Oecologia* **73**, 185−191.

Whitehouse, B. (1985) The effects of dissolved organic matter on the aqueous partitioning of polynuclear aromatic hydrocarbons. *Est. Coastal Shelf Sci.* **20**, 393−402.

Widdows, J. & Donkin, P. (1991) Role of physiological energetics in ecotoxicology. *Comp. Biochem. Physiol.* **100C**, 69−75.

Wolfe, N.L., Paris, D.F., Steen, W.C. & Baughman, G.L. (1980) Correlation of microbial degradation rates with chemical structure. *Environ. Sci. Technol.* **14**, 1143−1144.

Zaroogian, G., Heltshe, J.F. & Johnson, M. (1985) Estimation of toxicity to marine species with structure−activity models developed to estimate toxicity to freshwater fish. *Aquat. Toxicol.* **6**, 251−270.

15: Fate Models

D. MACKAY

15.1 INTRODUCTION

15.1.1 The incentive

One of the fascinations and frustrations of ecotoxicology is the wide variety of situations which must be addressed in which chemicals impact organisms and ecosystems. As other chapters in these volumes have discussed, there is a large number of chemicals of ecotoxicological concern that differ greatly in their properties. There is a diversity of environments such as soils, rivers and sediments that must be considered with varying physical, chemical and climatic conditions. Chemicals present in these environments are subject to many processes of transport and transformation. There are numerous organisms ranging from bacteria to whales that are worthy of consideration individually and collectively. There are many manifestations of toxicity ranging from subtle behavioural effects to immediate death.

A comprehensive assessment of the impact of chemicals on an ecosystem thus requires a collaborative effort of many disciplinary skills ranging from analytical chemists to engineers, physicists, mathematicians and animal physiologists. Accordingly, there is a need for a synthesis of the available information on chemical fate and effects into a comprehensive and quantitative statement of chemical impact. A significant component of this task is gathering information on chemical properties, sources, transport and transformation into a quantitative statement of fate. Such statements usually invoke the concept of the mass or material balance, and are thus referred to as mass balance models of chemical fate. The product is usually a set of equations containing numerous parameters that purport to describe the environment, and the chemical as it experiences transport and transformation. The equations are often solved using a computer programme yielding estimates of chemical quantities and concentrations in various locations and at various times. Ideally the results should be validated by comparing the model's assertions with actual observations, usually in the form of concentrations.

This modelling effort can be justified because it formalizes a test that deduced chemical fate is in accord with reality. The model can be used to suggest experiments which can test its validity, and thus test our depth of understanding of chemical fate. There are also more practical reasons for modelling. A reliable model can be used to explore the effects of interventions; e.g. to reduce the rate of discharge of a chemical into an estuary. It can often demonstrate where and when monitoring should take place, avoiding expensive and futile 'non-detects'. It can show the primary processes to which a chemical is subject, and elucidate which parameters, such as a biodegradation rate, are required with greatest accuracy. Models provide the only method of estimating, in advance, the behaviour of new chemicals for which there is, as yet, no environmental experience. They are particularly valuable when the concentrations are dictated by a number of factors or processes which may be understood individually, but combine and interact in ways which are not intuitively obvious to determine the chemical's fate, and hence its effects.

The aim of fate models is thus usually to predict concentrations which may then be compared with concentrations known to cause defined responses (such as death) in organisms. The lethal concentration in water to 50% (LC_{50}) of a population of fish is a common method of expressing chemical toxicity (Volume 1). The concentration may be combined with an intake rate such as a respiration or feeding rate to give a dosage in units such as grams per hour or grams over a defined time interval. The effect can then be related to this dosage. The product of concentration and time can also be regarded as a form of dosage, since it may reflect exposure to a quantity of toxic material.

It is important to emphasize that whereas models can give various estimates of *exposure*, it is a separate and usually more difficult task to estimate *effects* or response. Environmental concentrations often vary with time, and response may not be a linear function of concentration or time or dosage. Thus calculating 'average' or 'total' values can be difficult, since the nature of the exposure integration is not clear. The reader is thus alerted to the existence of this complementary body of scientific knowledge and effort in which the output of fate models must be used with discretion.

15.1.2 Types of models

A wide range of models is available of varied structure applicable to a diversity of situations of ecotoxicological interest. Fig. 15.1 gives a general picture of chemical sources, fate and routes of exposure. Models can be devised to treat the whole system, or some more detailed component process within the larger whole.

Multimedia models attempt to describe the behaviour of chemicals as they migrate throughout an entire ecosystem consisting of soil, air, water, sediments and biota, usually on the scale of a region or country hundreds or thousands of kilometres across. They provide a broad picture of fate (Fig. 15.2), and are invaluable for assessing the extent to which a local discharge of chemical (e.g. SO_2 from the UK) impacts a distant ecosystem (e.g. acidifies a lake in Norway). By focusing on the 'big picture' they necessarily ignore local detail. Models assessing the biogeochemical cycles of natural chemicals such as CO_2 and anthropogenic chemicals such as freons fall into this class, and may even be applied on a global scale.

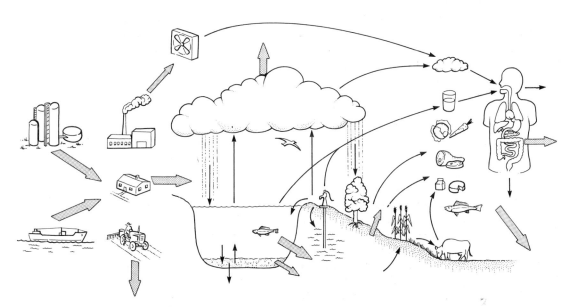

Fig. 15.1 Representation of chemical sources (industrial, import, domestic, and agricultural), environmental fate, accumulation in exposure media of air, water and foodstuffs and transport into, and within humans. Reproduced from Mackay, 1991 with permission.

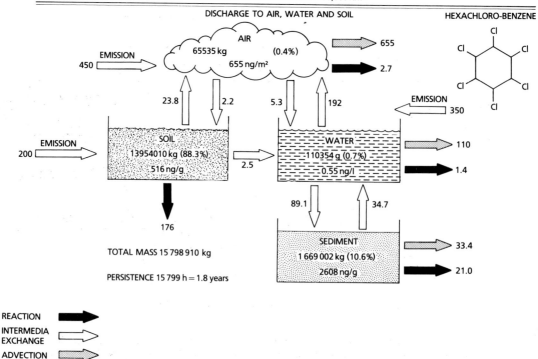

Fig. 15.2 Typical multimedia mass balance diagram of a chemical (hexachlorobenzene) showing sources, reactions, advective loss and transport rates between media. The rates are in units of kilograms/hour and the environment considered is a generic or evaluative area of $100\,000\,km^2$.

Single-media models provide a more detailed assessment of how a chemical is transported and transformed in a phase such as air, water or soil. Examples include models of chemical fate in lakes (Fig. 15.3) or atmospheric dispersion models used to relate stack emission rates to ground-level concentrations in air downwind. River models are used to track chemical fate as the discharged material flows and is subject to sedimentation, evaporation, dilution and degradation. There are numerous models of chemical fate in lakes, rivers and estuaries. Soil models are widely used to assess the behaviour of agrochemicals. Recently groundwater models have assumed greater importance as a result of numerous incidents in which potable well-waters have become contaminated from leaking tanks or leachate from dumps or agricultural practices. The scales range widely from regions of hundreds of kilometres to local areas such as municipal jurisdictions, to single fields and even to small experimental ponds or mesocosms which are used for controlled studies of chemical fate.

Biotic models express mathematically the processes by which an organism interacts with its environment, absorbing a chemical during respiration and feeding, releasing the chemical by respiration or egestion and transforming it metabolically. Fish bioaccumulation models, as illustrated in Fig. 15.4, fall into this class and may be designed to treat a single fish, or an entire food web of plankton, invertebrates, benthos, herbivorous, carnivorous and omnivorous fish, each class consuming other classes according to its food preferences. Such models can be applied to aquatic, marine and terrestrial animals and to plants. An important outcome of these models is the ability to relate chemical concentrations in organisms to the usually much lower concentrations in the environment. This is invaluable

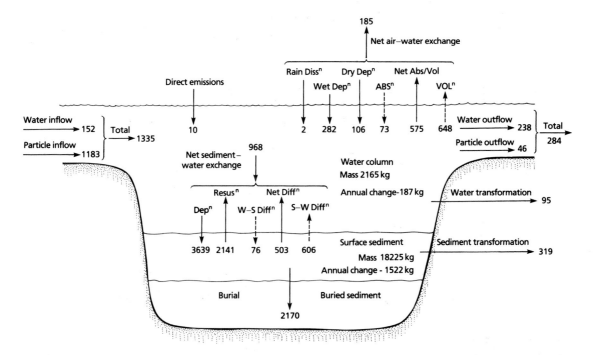

Fig. 15.3 Estimated mass balance diagram of PCBs in the water column and sediments in Lake Ontario about 1985. Rates are in kilograms/year. From *Multimedia Environmental Models: The Fugacity Approach*, by Donald Mackay, Copyright 1991, Lewis Publishers, Chelsea MI. Used with permission.

because it enables biota to be used as biomonitors, exploiting the fact that they may have concentrated and integrated contaminant over a long period of time. For example, analysis of gulls' eggs provides a convenient method of assessing the state of contamination of a lake from which they feed.

Pharmacokinetic models illustrated in Fig. 15.5 continue the process of tracking chemical fate into the organisms, calculating the chemical's disposition between various tissues as it is transported by diffusion, blood flow, or in sap in the case of plants. Such models are most widely used for assessment of the fate of therapeutic drugs in

Fig. 15.4 Structure of a fish bioaccumulation model showing input and output processes for which rate expressions can be written either in rate constant (K) or fugacity (D) value form. Reprinted with permission from *Multimedia Environmental Models: The Fugacity Approach*, by Donald Mackay, Copyright 1991, Lewis Publishers, Chelsea, MI. Used with permission.

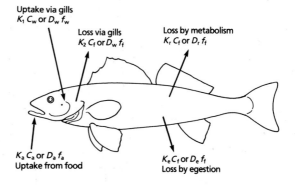

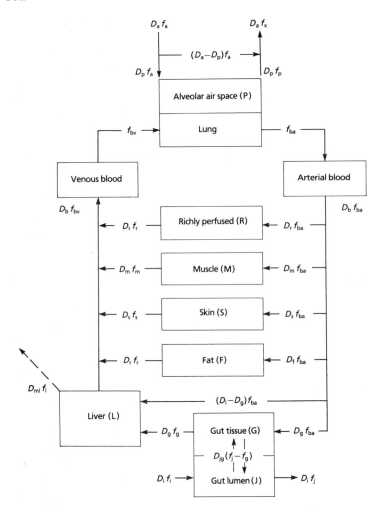

Fig. 15.5 Pharmacokinetic model of chemical distribution in mammalian tissues with chemical introduced in inhaled air and in food. The Ds represent transport and transformation rate coefficients; the fs represent fugacities which are related to concentrations. Reprinted with permission from *Environmental Toxicology and Chemistry*, **6**, Paterson & Mackay. A steady-state fugacity-based pharmacokinetic model with simultaneous multiple exposure routes, Copyright (1987), SETAC.

humans, often using rodents as a smaller model, but they can be applied to contaminants originating in the environment, present in foodstuffs, and encountered in occupational settings and in accidental chemical releases.

Exposure models are usually multimedia in nature, but they are primarily concerned with estimating the exposure (in units such as micrograms per day) experienced by humans (and other organisms) as a result of environmental contamination. Estimates of environmental concentrations are used to deduce concentrations in respired air, ingested water, vegetable, dairy and wild and domestic animal foods, especially fish. Essentially, the aim is to quantify the exposure routes depicted in Fig. 15.1. It often transpires

that the primary route of exposure is not appreciated until the model calculations are complete. For hydrophobic chemicals, consumption of fish and consumption of green vegetation are often the key processes. For volatile organic chemicals such as chloroform present in potable water, consumption by drinking water may be less important than inhalation during showering or washing in hot water. These models make no attempt to predict concentrations in defined locations, they merely estimate generally prevailing levels. They are invaluable for gaining insights into how, and to what extent, humans become exposed to chemicals, and how these exposures can be reduced.

Emergency models

Public concern about the environmental and human health impact of incidents such as large marine oil spills or the gas release at Bhopal has resulted in the development of models that can be applied rapidly to assessing the behaviour of chemical releases under emergency conditions. Although such models are mainly intended for use by emergency personnel under 'real-time' conditions, they are also valuable for use in contingency planning to alert personnel to the likely situations which they will encounter. Models have been developed for spills or releases of oil and other hazardous materials at sea, in harbours and rivers and on land, including urban areas of potential high human exposure. Sophisticated models now exist to describe the behaviour of various gases, especially dense gases such as chlorine, or 'cold' gases such as methane or ammonia, in specified regions as controlled by the current meteorological conditions.

Although most models apply to *real* environments with measurable dimensions and properties, there is a parallel application to *evaluative* environments in which no attempt is made to simulate fate in an actual system (Baughman & Lassiter, 1978). The environment selected is hypothetical, but reasonable, in that it has broad similarities to real environments. It usually comprises homogeneous phases such as soil or water at a constant temperature. By ignoring the multitude of complexities and idiosyncrasies of real environments, the modeller can focus on the fate of the chemical rather than on the state of the environment. Often when assessing the fate of a chemical in, for example, a river, more time, effort and money is devoted to characterizing the river than to understanding the chemical's fate in that river. The use of 'standard' rivers of constant temperature, velocity, depth and width avoids this problem. The disadvantage is, of course, that an assessment of chemical fate in such a setting cannot be validated, and it may thus lack credibility.

The primary focus of this chapter is on *deterministic* or *mechanistic* models in which an attempt is made to write and solve equations that express the 'laws of nature' as we understand them, as they apply to chemical fate in the environment. Often, environmental conditions are so complex and variable in time and space that such simulations become impossible, or at least very inaccurate. The classic case is weather forecasting, in which predictive techniques are limited to a fairly close time horizon. Another is the concentration of suspended solids or oxygen in a river which is subject to fluctuations in flow and temperature. There is also uncertainty about parameter values in models, such as rate constants or flows. The variable or *stochastic* nature of these processes can be treated by allowing the input parameters to vary within prescribed limits, and by running the model repeatedly. This Monte-Carlo approach gives an impression of the variance of the final result, and it can be applied in the form of a sensitivity analysis to determine which parameters introduce the greatest sensitivity in the final result. Often such an analysis reveals that the results are unexpectedly sensitive to one parameter and unexpectedly insensitive to another.

The outcome of a stochastic model is a distribution of results rather than a single value, often plotted as the familiar bell-shaped distribution curve or the S-shaped cumulative distribution function which shows directly the percentage of time that the result is expected to be above or below a certain value. This is particularly important when assessing the response of organisms or an ecosystem to stresses such as high toxicant concentrations or low dissolved oxygen concentrations. Even a very occasional, say once per year, fluctuation to zero dissolved oxygen in a river is sufficient to destroy a fish population.

Monte-Carlo techniques may require excessive computation time. If a model contains five key parameters, each of which is to be varied 10 times, a total of 10^5 runs is needed. A more sophisticated alternative is to transform the governing equations from a purely deterministic form to a stochastic form which includes the variance. This gives an analytical expression for the probability distribution function. Such models are commonly applied to water quality, examples being the studies of Di Toro (1984), Zielinski (1988) and Tumeo & Orlob (1989).

15.1.3 Computers

Models have become more widely used and accepted in ecotoxicology, largely because there has developed in the past two decades a much deeper understanding of chemical fate and effects in the environment. Accordingly there is a social incentive to understand, quantify and control chemical fate. Another important factor is the availability of low-cost computation and a marked increase in computer literacy. Most models can now be transmitted from user to user on diskette, and can be run on personal computers. Many are sufficiently 'user-friendly' that, with moderate training, an ecotoxicologist can run programmes without remembering or even fully understanding the detail contained in the programme. Results can be presented in attractive graphical form as plots, bar charts, pie charts, mass balance diagrams and even as narrative statements. Spreadsheet programmes have become particularly popular because of these features. Models can be run rapidly and repeatedly to determine sensitivities and to explore the effects of changing variables. Monte-Carlo simulations can be run in which parameters are allowed to vary within prescribed limits, thus giving results as a distribution of concentrations instead of a single value. The model thus simulates not only chemical fate in the environment but the variability of that fate.

15.1.4 Concentration and fugacity

Most models are written, as is discussed in Section 15.3, in terms of concentrations of a chemical (e.g. grams per cubic metre) in the various environmental phases. Expressions are written to deduce amounts (e.g. kilograms) and process rates (e.g. evaporation in grams/hour) using concentrations as the basic descriptor of quantity of chemical present. Equilibrium partitioning between phases such as air and water is usually expressed in terms of partition coefficients which are ratios of concentration. Mass balance equations are then written and solved in terms of concentrations, process rate parameters, partition coefficients, phase volumes and flow rates.

An alternative formalism is to use fugacity as the descriptor of chemical quantity. Fugacity (f) is an equilibrium criterion related to chemical potential. It is essentially the chemical's partial pressure and can be viewed as an 'escaping tendency or pressure'. It has units of pressure (Pa) and can usually be linearly related to concentration (C) through a proportionality constant Z, or Z value; i.e. C is Zf. Values of Z depend on the chemical, on the nature of the dissolving or sorbing medium and on temperature, as described by Mackay (1991) or Mackay & Paterson (1982). Equal fugacities prevail when two phases are in equilibrium with respect to chemical transfer. The use of partition coefficients is thus avoided.

Process rates are expressed as Df, where D is a transport or transformation rate parameter deduced from quantities such as rate constants, mass transfer coefficients, diffusivities or flow rates. The advantage of this approach is that D values can be compared and summed when they apply to a common phase. The mass balance equations become much simpler, and are more easily interpreted.

It is, however, important to emphasize that models written in concentration or fugacity format are (or may be) ultimately algebraically identical; thus the benefit of using fugacity is purely one of convenience. Both systems have their place and their proponents. Later in this chapter an example is given in which both approaches are used.

15.1.5 Summary

In summary, there are scientific and social incentives for developing models of chemical fate, as one component of the science of ecotoxicology. It is often an exercise in synthesis, bringing together information from a variety of sources to express the chemical's fate in quantitative terms. There is emerging a variety of models addressing chemical fate on a variety of scales from global biogeochemical assessments to pharmacological descriptions of fate within an organism. The increase in computer literacy and availability has been an added stimulus to this increasingly important aspect of ecotoxicology.

15.2 MODELLING CONCEPTS

15.2.1 The mass balance equations

The fundamental mass balance concept is illustrated in Fig. 15.6, in which a volume of space within the environment is identified as a compartment, and a mass or material balance equation is written around this volume. The volume may be a lake or a section of a river, or a region of the atmosphere. It could be soil to some specified depth, or it could even be an organism such as a fish. Physical boundaries are defined, and conditions should be fairly homogeneous within the phase envelope. The mass balance equation, which is accepted as axiomatic, states that the change in inventory of the chemical in the phase envelope in units such as grams per hour, will equal the sum of the rate of inputs to the phase envelope, again in grams per hour, less the sum of the rate of outputs. Input terms normally include flow in air and water, diffusion from other compartments, direct discharges or even formation from other chemical compounds. The outputs may include flows in air or water, degrading reactions, and diffusion to another compartment. The modeller's task is to develop expressions, equations or quantities for each of the terms in the mass balance equation.

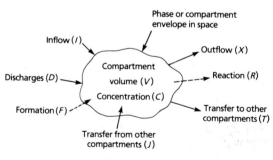

Inventory change = Inputs − Outputs
$V_{dC/dt} = I + D + F + J - X - R - T$ kg/year
At steady state $dC/dt = 0$
$I + D + F + J = X + R + T$ kg/year

Fig. 15.6 Illustration of the mass balancing concept for an environmental compartment showing the basic differential mass balance equation and the steady-state version, when terms in the equation are defined in the figure.

The modeller must make a number of subjective decisions when compiling this mass balance. Herein lies the art of modelling as the intuitive ability to make the 'right' decisions when faced with a number of alternatives.

The first stage is definition of the phase boundary or envelope, usually referred to as the process of segmentation. The volume within the compartment should be well mixed and thus fairly homogeneous in concentration. Examples are river or lake water or surface sediments. If more detail is desired, a water column may be segmented into two layers, a surface and a deep layer. In some cases it is preferable to treat particles in the water column as a separate phase, or they may be lumped with the water column. Obviously, as the number of compartments increases, the fidelity of the model to reality increases, but the mathematics become more complex, more input data are needed, and the results may become more difficult to interpret. Complex models are less likely to be used because there may be a suspicion that there are errors, or processes are misrepresented.

A second stage is to write equations for the various transport and transformation processes such as evaporation, sedimentation, biodegradation and photolysis. This may be difficult because temperature and other conditions such as meteorology, microbial activity, sunlight intensity and pH may vary diurnally and seasonally. Reliable data can be obtained for inclusion in these equations by bench-scale experiments involving exposure of chemical in a beaker to a well-defined transport or reactive environment under controlled conditions. Translating this physical/chemical information to environmental conditions is often difficult. Experiments may also be conducted in larger microcosms or mesocosms to simulate environmental conditions more closely, but there is an inevitable loss of control over these conditions.

A third stage is to define inputs, including discharges to the system, which may be from adjacent compartments, industrial or municipal sources, spills, deliberate application of pesticide chemicals, leaching from groundwater or disposal of consumer products. The total discharge rate, or loading rate, can often only be estimated approximately, at least for large regions in which there are many sources, including deposition from the atmosphere. The magnitude of the discharge rate usually drives the model, and establishes the magnitude of the chemical concentrations, and thus its fate.

When the rates, or the rate expressions, are defined for the various terms in the mass balance equation around the phase envelope, the equations can be solved to obtain estimates of the desired concentrations. Several solution strategies are available.

15.2.2 Solutions to the mass balance equation

A simple strategy is to obtain the solution describing the steady-state condition of the system. The inventory change terms on the left are ignored and the material balance is written as one or more algebraic equations. This solution yields the concentrations which would apply throughout the system, when the input quantities remain constant for a prolonged period of time, such that steady-state or constant conditions apply. The set of algebraic equations is solved either by hand or by matrix techniques. Although this solution may be artificial, it is useful because it can demonstrate which processes and parameters are most important as controlling the chemical's concentrations. Extra effort can be devoted to obtain more accurate values of the more important parameters. Less important processes need only be quantified approximately. For example, an evaporation rate may be so small that it is unimportant compared to a reaction rate by biodegradation.

Another more rigorous, but more demanding strategy is to solve the set of differential equations, usually by numerical integration techniques. The initial or boundary conditions are first set, then the change in inventory is deduced for a specified time period, and the new concentrations are calculated. This process is then repeated. The output from such a calculation is the time-course of concentration changes in the system as a result of inputs which may be constant, or may change with time. Such results are not easily generalizable to other conditions, since the solution is specific to the initial conditions selected and, of course, to the inputs. It may be possible in some simple situations to solve the set of differential equations analytically. This is

usually feasible when discharge rates are constant and there are only two compartments. For time-varying inputs and more compartments, the solutions become very cumbersome and mistakes are likely.

Regrettably, there is a tendency to respond to the complexity of environmental conditions by assembling complicated models of chemical migration in complex multi-media environments with expressions for numerous transport and transformation processes, containing a large number of adjustable parameters. The model then contains so many parameters and assumptions that users are rightly suspicious that there may be some erroneous assumptions, some hidden sensitivity or even a hidden mistake. When a model contains a multitude of adjustable and uncertain parameters, any agreement with reality may be entirely fortuitous. Key processes may be controlled by one, or a few, expressions or parameters that contain considerable error. A frequent issue is that the total rate of loss of chemical from a compartment such as a pond is known fairly accurately from experimental data, and can be modelled accurately, but this rate is the sum of several contributing processes such as degrading reactions, evaporation and sedimentation; thus the magnitudes of the contributing processes to this total are not well known. To an experienced ecotoxicologist the model results should be in intuitive accord with what is expected for that chemical in that environment. Often the primary function of the model is merely to assign the disparate fate processes into an order of importance and identify the key processes. A simple 'back of the envelope' model can then be written, containing only those key processes, and this will give a satisfactory and simple description of the reality of chemical fate.

These concepts are best illustrated by an example, as given in the next section.

15.3 AN ILLUSTRATIVE MASS BALANCE MODEL OF CHEMICAL FATE IN A POND

15.3.1 Environmental conditions

To illustrate a simple one-compartment model we treat the fate of a chemical in a pond (Fig. 15.7)

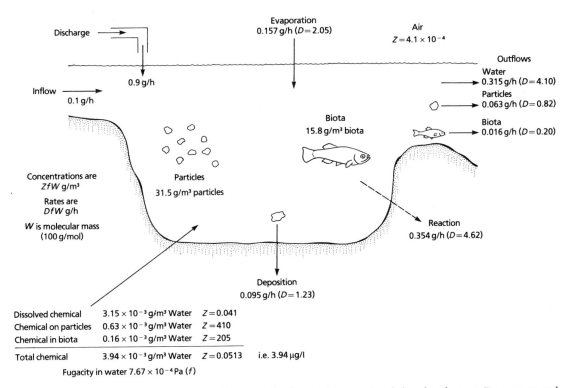

Fig. 15.7 Illustrative mass balance diagram of the fate of a chemical in a pond as deduced in the text. D, transport and transformational coefficients; Z, fugacity capacity.

which is 3 m deep and has an area of $10\,000\,\text{m}^2$, the volume of water being $30\,000\,\text{m}^3$. Water flows into and out of this pond at a rate of $100\,\text{m}^3/\text{h}$ and thus the residence time of the water is 300 h. There is inflow of 5 l/h of suspended sediment, deposition of 3 l/h to the bottom, and the remaining 2 l/h flows out of the system. The bottom (which we ignore here) consists of sediment. A chemical reacts in the water column with a half-life of 231 h, i.e. a rate constant of $0.003\,\text{h}^{-1}$. It evaporates with an overall water-side mass transfer coefficient of $0.005\,\text{m/h}$.

The chemical (molecular mass 100 g/mol) has an air–water partition coefficient of 0.01, a particle–water partition coefficient K_{pw} of 10 000 and a biota–water partition coefficient K_{bw} or bioconcentration factor of 5000. The concentration of particles is 20 ppm and that of biota (including fish) is 10 ppm, both by volume. There is a constant discharge of chemical of 0.9 g/h or 0.009 mol/h, and chemical is present in the inflowing water at a concentration of $0.001\,\text{g/m}^3$ or $10^{-5}\,\text{mol/m}^3$.

The aim of the model is to calculate how the chemical partitions between the water, particles and biota, the steady-state or constant concentration in the system of water, particles and fish, and all the loss rates.

We can undertake this calculation in two ways: first a conventional concentration calculation, then a fugacity calculation.

15.3.2 Concentration calculation

We let the total concentration of chemical in the water (including chemical present in particles and fish) be an unknown $C_w\,\text{g/m}^3$, and the dissolved concentration $C_d\,\text{g/m}^3$. C_d is lower than C_w because some of the chemical is sorbed to particles and biota. The various process rates are expressed in terms of C_d, summed and equated to the total input rate. We then solve for C_d, and finally deduce the process rates. We use units of grams per hour for the mass balance calculation.

Input rate

The discharge is 0.9 g/h. The inflow rate is the product of $100 \, m^3/h$ and $0.001 \, g/m^3$ (i.e. 0.001 mg/l) or 0.1 g/h. This gives a total chemical input of 1.0 g/h.

Partitioning between water, particles and biota

The total amount in the water is $V_w C_w$ g where V_w is the volume of water $(30\,000 \, m^3)$. But this contains 20 ppm of particles; i.e. a volume of $0.6 \, m^3$ and similarly 10 ppm of biota or $0.3 \, m^3$. If the dissolved chemical concentration in water is C_d, then the concentrations in the particles will be $K_{pw} C_d$ and in the biota $K_{bw} C_d$, thus the amounts are respectively:

$30\,000 \; C_d$ in solution in water
$0.6 \; K_{pw} C_d$ in particles and
$0.3 \; K_{bw} C_d$ in biota

Since K_{pw} is 10 000 and K_{bw} is 5000 these add to give

$$C_d(30\,000 + 0.6 \times 10\,000 + 0.3 \times 5000) = 37\,500 C_d$$

But this must equal $30\,000 C_w$, thus C_d is $0.8 C_w$, i.e. 80% of the chemical is dissolved, 16% is sorbed to particles and 4% is bioconcentrated in biota. Note that we use dimensionless partition coefficients K_{pw} and K_{bw}, which are ratios such as (mg/l)/(mg/l). Often K_{pw} is reported as a ratio of (mg/kg)/(mg/l) and thus has dimensions of litres/kilogram.

Outflow

Since the outflow rate is $100 \, m^3/h$, the outflow rate of dissolved chemical must be $100 C_d$ or $80 C_w$ g/h. In addition there is outflow of 2 l/h $(0.002 \, m^3/h)$ of particles containing $10\,000 C_d$ g/m^3, giving $20 C_d$ g/h. The outflow of biota is 0.001 m^3/h (i.e. 10 ppm $\times$ 100 m^3/h) at a concentration of $5000 C_d$ or $5 C_d$ g/h.

Reaction

The reaction rate is expressed as the product of water volume, total concentration and rate constant, i.e. it is $30\,000 \times C_w \times 0.003$ or $90 C_w$ g/h or

$112.5 C_d$ g/h. It is also possible to express the rate constant on the basis of the dissolved concentration if it is known that reaction occurs only in the dissolved phase.

Deposition

The concentration on the particles is $10\,000 C_d$ g/m^3 particle. Since the particle deposition rate is 3 l/h or $0.003 \, m^3/h$, the chemical deposition rate will be $10\,000 \times 0.003 C_d$ or $30 C_d$ g/h.

Evaporation

The evaporation rate is the product of the mass transfer coefficient $(0.005 \, m/h)$, the water area $(10\,000 \, m^2)$ and the concentration dissolved in water; i.e. it is $50 C_d$ g/h. Note that we assume here that the air contains no chemical to create a 'back-pressure' or diffusion from air to water. If this was the case it would be included as another input term.

Combining the process rates we argue that at steady state, when there is no inventory change of chemical in the pond, the output rate will equal the discharge rate of 1.0 g/h, thus

$$1.0 = 100 C_d + 20 C_d + 5 C_d + 112.5 C_d + 30 C_d + 50 C_d = 317.5 C_d$$

so: $C_d = 1/317.5 = 0.00315$ g/m^3 or mg/l or 3.15 µg/l

The dissolved concentration is thus 0.003 15 g/m^3, that on the particles is 10 000 times this or 31.5 g/m^3 particle, and if the particle density is 1.5 g/cm^3, approximately 21 mg/kg. This is also 0.000 63 g/m^3 water. The biotic concentration will be 15.75 g/m^3 fish or mg/kg fish, and is equivalent to 0.0001 575 g/m^3 water. The total water concentration C_w is 0.003 94 g/m^3 or 3.94 µg/l. The various process rates are thus

Outflow in water	0.3150 g/h	31.5%
Outflow in particles	0.0630 g/h	6.3%
Outflow in biota	0.0158 g/h	1.6%
Reaction	0.3544 g/h	35.4%
Deposition	0.0945 g/h	9.5%
Evaporation	0.1575 g/h	15.7%

These total to the input of 1.0 g/h, thus satisfying the mass balance as depicted in Fig. 15.7. The relative importance of the processes becomes immediately obvious, and there now emerges a clear picture of the chemical's fate. The key processes are outflow, reaction and evaporation. The parameters controlling these processes should be established most accurately.

15.3.3 Fugacity calculation

In the fugacity approach, partitioning of a chemical between phases is expressed in terms of the equilibrium criterion of fugacity, as discussed earlier. Fugacity can be used in the calculation as a surrogate for concentration, and is related to concentration through a fugacity capacity or Z value. These Z values are calculated from the chemical's partition coefficients as controlled by its physical properties of molecular weight, vapour pressure, solubility and octanol–water partition coefficient, as well as environmental properties such as density and fraction organic content of the phases present. Definition starts in the air phase in which Z is $1/RT$ where R is the gas constant $(8.314 \, \text{Pa} \, \text{m}^3/\text{mol} \, \text{K})$ and T is absolute temperature. Z values in other phases are calculated from the partition coefficients using the relationship that K_{12} is Z_1/Z_2. A fuller account is given by Mackay (1991).

This calculation can be repeated in fugacity format by first calculating Z values, then D values, then equating input and output rates as before. It is now preferable to use units of moles per hour.

For air, Z is assumed to be $4.1 \times 10^{-4} \, \text{mol/m}^3$ Pa. For water Z is then as Z_a/K_{aw} or 4.1×10^{-2}, for particles Z_p is $K_{pw}Z_w$ or 410, and for biota Z_b is $K_{bw}Z_w$ or 205. The total Z value for water, particles and biota is the sum of these Z values, weighted in proportion of their volume fractions, i.e.

$$Z_{wt} = Z_w + 20 \times 10^{-6} Z_p + 10 \times 10^{-6} Z_b$$
$$= 5.13 \times 10^{-2}$$

Note that the three terms represent 80%, 16% and 4% of the total as before.

The D values (in units of mol/Pa per hour) are calculated from flow rates $(G \, \text{m}^3/\text{h})$, volume $(V \, \text{m}^3)$, area $(A \, \text{m}^2)$, reaction rate constant $(k \, \text{h}^{-1})$ and the mass transfer coefficient $(k_m \, \text{m/h})$ as follows.

Outflow in water	$D_1 = G_w Z_w$	
	$= 100 \times 4.1 \times 10^{-2}$	
	$= 4.10$	
Outflow in particles	$D_2 = G_p Z_p$	
	$= 0.002 \times 410$	
	$= 0.82$	
Outflow in biota	$D_3 = G_b Z_b$	
	$= 0.001 \times 205$	
	$= 0.205$	
Reaction	$D_4 = V Z_{wt} k$	
	$= 30\,000 \times 5.13$	
	$\times 10^{-2} \times 0.003$	
	$= 4.62$	
Deposition	$D_5 = G_d Z_p$	
	$= 0.003 \times 410$	
	$= 1.23$	
Evaporation	$D_6 = k_m A Z_w$	
	$= 0.005 \times 10^4 \times 4.1$	
	$\times 10^{-2}$	
	$= 2.05$	

The overall mass balance is then expressed in terms of the fugacity of the chemical in the water f_w. The input rate of 0.01 mol/h is equated to the sum of the output rates as before.

$$\text{input} = f_w D_1 + f_w D_2 + f_w D_3 + f_w D_4 + f_w D_5 + f_w D_6$$

$$\text{or} \quad 0.01 = f_w \Sigma \, D_i = f_w \, 13.03$$
$$f_w = 7.67 \times 10^{-4}$$
$$C_d = Z_w f_w$$
$$= 3.15 \times 10^{-5} \, \text{mol/m}^3 \text{ or } 0.00315 \, \text{g/m}^3$$
$$C_p = Z_p f_w$$
$$= 0.315 \, \text{mol/m}^3 \text{ or } 31.5 \, \text{g/m}^3$$
$$C_b = Z_b f_w$$
$$= 0.157 \, \text{mol/m}^3 \text{ or } 15.7 \, \text{g/m}^3$$
$$\text{or } 15.7 \, \text{mg/kg}$$
$$C_w = Z_{wt} f_w$$
$$= 3.94 \times 10^{-5} \, \text{mol/m}^3$$
$$\text{or } 0.00394 \, \text{g/m}^3$$

The individual rates are Df, which are identical to those calculated earlier. The D values give a useful direct expression of the relative importance of diverse processes by expressing them in a common system of units. The algebraic simplicity of this approach favours its extension to more phases, processes and compartments with no real increase in complexity.

15.3.4 Discussion

A simple steady-state model such as this yields invaluable information on the status and fate of the chemical in this pond system. Obviously it is desirable to measure concentrations in particles, water and biota to determine if the model assertions are correct. If (as is likely) discrepancies are observed, the model assumptions and parameter values can be examined to test if some process has not been included, or if the assumed parameter values are correct. When reconciliation is successful, the ecotoxicologist can with some justification claim that the system is well understood. It is then possible to explore the effect of changing input rates to the system.

For example, if the discharge of 0.9 g/h was eliminated, the water, particle and biotic concentrations would eventually fall to one-tenth of the values calculated above, because the input would now be 0.1 g/h instead of 1.0 g/h. The biota would now contain 1.57 mg/kg instead of 15.7 mg/kg. To achieve a 'target' of say 1 mg/kg would require reduction of the input rate to about 0.06 g/h. The nature of regulatory measures necessary to achieve this desired environmental quality can thus be determined.

Another issue is how long it will take for such measures to become effective. This requires solution of the differential mass balance equation which in this case is

$$V_w dC_w/dt = \text{Inputs} - \text{Outputs}$$
or $V_w dC_w/dt = 1.0 - 317.5 C_d = 1.0 - 254\, C_w$
or $dC_w/dt = 3.3 \times 10^{-5} - 8.47 \times 10^{-3}\, C_w$
$= A - B C_w$

If the initial water concentration is C_{w0} at time t of zero this equation can be solved by separation of variables to give

$$C_w = C_{wf} + (C_{w0} - C_{wf})\exp(-Bt)$$

where C_{wf} is the final value of C_w at t of infinity, and is A/B. C_w thus changes from C_{w0} to C_{wf} with a rate constant of B or a half-time of $0.693/B$ h. Here B is $8.47 \times 10^{-3}\,h^{-1}$; thus the half-time is 82 h. This is illustrated in Fig. 15.8 using the pond condition in Fig. 15.7 as a starting point, i.e. C_{w0} is 3.94 µg/l.

In reality the biota would respond more slowly

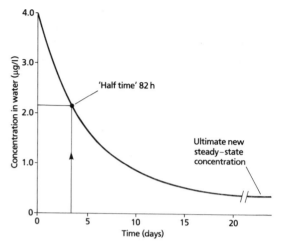

Fig. 15.8 Unsteady-state response of the concentration in the pond to a reduction in total discharge rate from 1.0 to 0.1 g/h. The initial condition is that of Fig. 15.7 of 3.94 µg/l. The concentration falls with a half-time of 82 h to an ultimate 0.394 µg/l.

because of the delay in bio-uptake or release, but the important finding is that within a week the pond system would be well on its way to a new steady-state condition. In fugacity terms the equivalent value of B is $\Sigma D/V_w Z_{wt}$ and the individual processes contributing to this overall rate constant are $D/V_w Z_{wt}$. It is thus apparent that water outflow and reaction have about equivalent influences on the rate or time of response, with evaporation also being important, while deposition and particle and biota outflow are relatively unimportant.

Having set up the model it is possible to explore more complex situations. The effect of varying input rates can be investigated. If the temperature dependence of reaction and other rates are known, the effect of temperature can be tested. The consequences of hydrologic changes can be examined. Experiments can be designed in which the pond is subjected to a pulse of chemical and the time response of concentrations monitored and compared with predictions.

It is possible to include a sediment compartment, or another water compartment to give more detail. If human or wildlife exposure to chemicals from fish consumption is important, a separate bioconcentration or food chain model may be developed. Concentrations can be compared with levels known to cause toxic effects, thus providing an estimate of effects resulting from the chemical discharge. The nature and extent of the modelling activity can be thus tailored to the ecotoxicological needs.

15.4 MODELS

15.4.1 Models and availability

In this section some available environmental fate models are described briefly. No attempt is made to provide a comprehensive list. Reviews of available models are found in the texts by Cohen (1986), OECD (1989), Swann & Eschenroeder (1983) and Dickson *et al.* (1982). Other useful sources are the reports by Thibodeaux (1979), Jorgensen (1984) Hites & Eisenreich (1987). Bonazountas *et al.* (1988) and Jorgensen & Gromiec (1989). Some agencies, notably the US EPA, maintain and update a number of models which are available on request. Table 15.1 lists some of these EPA models. Other models are available commercially. Occasional courses are run to introduce users to these models. Since models are in a continual state of modification and upgrading, the reader is advised to make personal contact with individuals in the appropriate agency to obtain the latest versions.

Models are now sufficiently complex, and are used so widely for decision-making in the regulatory sphere, that it is essential that they be well documented and tested, and fully supported by individuals able and willing to provide advice and assistance. Regrettably, few agencies have the foresight to provide these much-needed resources. Sonzogni *et al.* (1987) discuss these and other issues relating to the use and possible abuse of models.

15.4.2 Multimedia models (primarily for organic chemicals)

Fugacity models

A series of fugacity-based models has been developed by Mackay and co-workers (Mackay, 1991, Mackay & Paterson, 1982, Mackay *et al.*, 1985;

Paterson & Mackay, 1990). They vary in complexity from a simple equilibrium distribution of a conservative chemical, to steady-state and time-varying descriptions of the fate of reactive compounds.

The Level III model (Mackay *et al.*, 1985 and Mackay & Paterson, 1991) is most useful in estimating environmental fate of persistent organic chemicals which are continuously emitted to the environment over a prolonged period. This model treats four bulk compartments: air, water, soil and sediment, each of which is considered to consist of defined fractions of air, water, mineral and organic matter. Chemical equilibrium is assumed to exist within, but not between, compartments. The steady-state distribution and concentration of a chemical, and its persistence in the environment, can be estimated. Media of accumulation and dominant pathways of transfer are determined. Fig. 15.2 is an illustration of the fate of a chemical in such a system.

Geotox

Geotox (McKone & Layton, 1986; McKone & Kastenberg, 1986) is a comprehensive multimedia compartmental fate and exposure model developed under contract from the US Government, which calculates chemical partitioning, degrading reactions and diffusive and non-diffusive interphase transport. The estimated concentrations are subsequently combined with appropriate human inhalation and ingestion rates, and absorption factors to calculate exposure.

Geotox treats an environment representative of a region of the United States, consisting of air, aerosols, biomass, upper and lower soil, groundwater, surface water and sediments. These media are made up of sub-phases of gas, liquid and solid. Chemical partitioning between compartments, interphase transport, reaction and advective loss rates are described by first-order rate constants. The model can be applied to defined regions to constant or time-varying chemical sources. Output is in the form of environmental concentrations, human intake by various exposure pathways, and total intake. Relative health risks for a number of chemicals can be calculated.

Table 15.1 Selected environmental models available from, and supported by, the US Environmental Protection Agency (with EPA report number)

Acronym	Model	EPA Report number
EXAMS	Exposure Analysis Modeling System	EPA/600/3−82/023
FGETS	Food and Gill Exchange of Toxic Substances. A Simulation Model for Predicting Bioaccumulation of Non-polar Organic Pollutants by Fish	EPA/600/3−87/038
GETS	Simulation Model for Dynamic Bioaccumulation of Non-polar Organics by Gill Exchange	EPA/600/3−86/057
HSPF	Hydrologic Simulation Program	EPA/600/3−84/066
MINTEQ	An Equilibrium Metal Speciation Model	EPA/600/3−87/012
PRZM	Pesticide Root Zone Model	EPA/600/3−84/109
QUAL	Enhanced Stream Water Quality Models	EPA/600/3−87/007
RUSTIC	Risk of Unsaturated/ Saturated Transport and Transformation of Chemical Concentration	EPA/600/3−89/048a
SARAH	Surface Water Assessment Model for Back-calculating Reductions in Abiotic Hazardous Wastes	EPA/600/3−86/058
SWMM	Storm Water Management	EPA/600/2−84−109a
WASP	Hydrodynamic and Water Quality Model	EPA/600/3−87/039
WQA	Water Quality Assessment: A Screening Procedure for Toxic and Conventional Pollutants in Surface and Ground Waters	EPA/600/6−85/002a

SMCM (Spatial Multimedia Compartment Model)

This model, developed by the National Center for Intermedia Transport at UCLA by Cohen and co-workers (1990; Cohen & Ryan, 1985), describes the fate of chemicals in a conventional air−water−soil−sediment system under steady-state or unsteady-state conditions. It has the unusual feature that it allows for concentration variation with depth in the soil and sediment, i.e. these compartments are not treated as well-mixed 'boxes'. This hybrid approach results in more complexity, but it yields greater fidelity. The model is 'user-friendly' with help menus and the capability of presenting data in tabular or graphical form. Particular strengths of the model are its treatment of deposition from the atmosphere and volatilization from soil.

Enpart (Environmental Partitioning Model)

Enpart (OECD, 1989) is one of a set of models developed by the US EPA as a first-level screening tool for new and existing organic chemicals. It is a fugacity-based model which estimates the steady-state equilibrium or dynamic partitioning of organic chemicals among environmental compartments. It identifies dominant pathways and data gaps, and estimates the chemical's persistence and bioconcentration potential.

Toxscreen

Toxscreen (Hetrick & McDowell-Boyer, 1983) is a time-dependent multimedia model, developed by the US EPA to assess the potential for environmental transport accumulation of chemicals released to the air, surface water or soil. It is modular in concept and incorporates intermedia transfer processes. It is intended as a screening tool to assess the human exposure potential of organic chemicals. Files of data on climatic and soil conditions for various regions of the USA are included.

Environmental Exposure Potentials (EEP)

EEP (Klein *et al.*, 1988) is a simple fugacity-based equilibrium multicompartment model used by some member states of the EC to determine exposure potential of new organic chemicals. The methodology is applied to chemicals being imported or produced in quantities of over 1 tonne per year. It treats multiple or diffuse sources of continuous emissions and calculates environmental partitioning, quantities in the environment, degradation, and accumulation potential in air, water and soil. It employs weighting factors to produce a 'fingerprint' of environmental fate.

Multiphase Non-Steady State Equilibrium Model (MNSEM)

MNSEM (Yoshida *et al.*, 1987) is a simplified kinetic model designed to predict the fate of organic chemicals under steady-state conditions of continuous loading to the Japanese environment. It calculates the chemical's distribution, persistence and concentrations in an environment representative of Japan. It also estimates recovery time after emissions are terminated.

There are four major compartments: air, water, soil and sediment, with subcompartments defined similarly to the fugacity model. The model has been applied to the fate of chloroform, aniline, chlorobenzene, nitrobenzene and naphthalene in Japan. Estimated concentrations are generally within an order of magnitude of monitored values.

RAPS — Remedial Action Priority System

The RAPS system (Whelan *et al.*, 1987) was developed for use by the US Department of Energy to set priorities for investigation and possible clean-up of chemical and radioactive waste disposal sites. It is intended to be used in a comparative rather than predictive mode. Estimated concentrations in the air, soil, sediments and water media are used to assess exposure to neighbouring populations and determination of a Hazard Potential Index.

15.4.3 Water quality models

EXAMS — Exposure Analysis Modeling System, and other US EPA models

EXAMS (Burns & Cline, 1985) is a very well-used and well-supported interactive mass balance model developed at the US EPA Research Laboratory in Athens, GA, which predicts the fate of organic contaminants in various surface waters as a result of continuous or intermittent releases.

The water body is subdivided into zones or segments, the mass balance of each segment being described by a differential equation. The resulting set of equations, which describes the mass balance of the entire system, incorporates comprehensive transport and transformation processes. EXAMS allows for loadings by point or non-point sources, wet and dry deposition or aerial drift, and ground-water seepage to selected segments.

The user can choose one of three operating modes, depending on the complexity of the problem being studied. The modes range from a steady-state solution for continuous release of a contaminant to the dynamic solution of a time-varying source.

WASP is an EPA-supported hydrodynamic and water quality model, as are QUAL, a stream model and RUSTIC, a groundwater model. Examples of applications of these and related models can be found in the reports by Mackay *et al.* (1983), Thomann & DiToro (1983), Connolly & Winfield (1984), Holysh *et al.* (1986), Schnoor *et al.* (1987), O'Connor (1988a,b) and Mackay (1989). Booty & Lam (1989) have reviewed many of these models. Schnoor (1981) gives an interesting example of the application of a water quality model to ecosystem recovery after a pesticide ban.

EXWAT

EXWAT (OECD, 1989) is a steady-state model, developed in Germany, to describe chemical fate in water bodies. It is a simple approach suitable for continuous single-point sources. It is intended for use as a screening tool to assess comparative hazards of existing chemicals in the Rhine River. It is also a submodel of the multimedia E4CHEM system (Exposure and Ecotoxicity Estimation for Environmental Chemicals), developed in Germany for priority setting within OECD.

EXWAT considers water and sediment, and processes of transport, degradation and advection. Variation of environmental properties along the river and stratification are not taken into account. The output includes a graphic description of sediment clearance as a function of time following cessation of the input.

15.4.4 Inorganic chemical models

Modelling of inorganic compounds proves to be more difficult in the sense that the chemical properties and speciation tend to be element- or compound-specific, thus the generalizations

which apply to organic chemicals do not usually exist. Most models are applied to aquatic or marine situations and explore the fate of metals or inorganic chemicals such as arsenic or phosphorus.

An example is the work of Dolan & Bierman (1982), who developed a dynamic mass balance model of heavy metals in Saginaw Bay, Lake Huron. Extensive environmental data were collected, and attempts were made to relate mass input of trace metals to observed water column concentrations.

Phosphorus models

Many models describing phosphorus loadings in lakes have been developed in an attempt to elucidate and solve eutrophication problems. In Canada and the USA the focus has been on the Great Lakes, whose water quality has deteriorated as the result of rapid development of urban populations.

One of the earliest phosphorus models was developed by Vollenweider (1969a,b). It was an empirical, one-compartment model based on extensive data from lakes in Europe and North America. It defined the relationship between phosphorus loading, mean lake depth and trophic state. It is steady-state and assumes removal of phosphorus is by sedimentation described by a first-order rate constant. This model was later improved (Vollenweider, 1976) to include terms for water residence time and relative residence time of phosphorus in the lake system. Lam *et al.* (1982) have reviewed these models.

MINTEQA1

MINTEQA1 (Brown & Allison, 1987) is an equilibrium metal speciation model applicable to metallic contaminants in surface and ground-waters. It is thus quite different in purpose from the mass-balance models discussed earlier. It calculates the equilibrium aqueous speciation, adsorption, gas-phase partitioning, solid-phase saturation states, and precipitation−dissolution of 11 metals (As, Cd, Cr, Cu, Pb, Hg, Ni, Se, Ag, Tl and Zn). Some expertise regarding kinetic limitations is required for proper application. The output is a description of the major metal species in the system. An extensive thermodynamic data base is included.

15.4.5 Soil models

Several models have been developed to describe chemical fate in soils. Notable are SESOIL (A Seasonal Soil Compartment Model, OECD, 1989); PRZM (Pesticide Root Zone Model) by Carsel *et al.* (1984), PESTAN (Pesticide Analytical Model) by Enfield *et al.* (1982), and the Behavior Assessment Model by Jury *et al.* (1983).

These models generally treat degradation, evaporation, diffusion in water and air, advective flow in water and, in some cases, transport in colloidal organic carbon. They are mainly applied to pesticides in an attempt to describe chemical fate and persistence, and especially transfer to groundwater. Melancorn *et al.* (1986) have evaluated some of these models by comparison of predictions with experimental soil core data.

15.4.6 Fish uptake and food chain models

Because of the importance of the human exposure route via fish consumption, considerable effort has been devoted to estimating chemical concentrations in fish. Recent examples include the work by Thomann (1989), Clark *et al.* (1990) and the text by Connell (1989). The US EPA supports two such models, FGETS (fish and gill exchange of toxic substances) and GETS, which applies only to gill exchange.

15.4.7 Air models

Numerous air dispersion models with general or limited geographic applicability have been developed with the objective of deducing ground-level concentrations, and hence exposures, from stack emissions. Most texts on air pollution contain full descriptions of such models. An example of a comprehensive model of airborne contaminants is ALWAS (Tucker *et al.*, 1982).

A subclass of these models treats dispersion under emergency release conditions resulting from an accident. Particularly challenging is the 'heavy gas' problem, in which the released gas such as chlorine is denser than air and tends to flow as a near-intact plume over the ground surface.

15.5 CONCLUSIONS

In this chapter an attempt has been made to outline the objectives of chemical fate modelling, to discuss some of the benefits gained and to illustrate the underlying principles. The partial list of models will, it is hoped, provide the reader with an entry to this large and growing literature. With the advent of faster, cheaper and more user-friendly computing systems, with attractive outputs in graphical and pictorial form, and growing computer literacy, it is likely that there will continue to be extensive developments in chemical fate modelling. It is important to appreciate that models are only as good as the quality of the input data and the validity of the expressions used for describing the various partitioning, transport and transformation processes. They should be viewed as merely tools, not as an end in themselves. They must therefore be designed to satisfy a stated need. There is often overconfidence that the computed results are accurate, and modellers are frequently guilty of failing to convey a full appreciation of the sensitivity of the results to errors or variation in the numerous parameters used to build the model.

Despite these pitfalls, it is likely that models will play an increasingly important role in ecotoxicology by providing a method of estimating quantities which have not been, or cannot be, measured, and in synthesizing a diversity of information about chemical behaviour into a coherent, quantitative statement of chemical fate in the environment.

REFERENCES

Baughman, G.L. & Lassiter, R.R. (1978) Prediction of environmental pollutant concentration. In: *Estimating the Hazard of Chemical Substances to Aquatic Life* (Eds J. Cairns, Jr, K.L. Dickson & A.W. Maki), pp. 35–54. ASTM STP 657. American Society for Testing and Materials, Philadelphia, PA.

Bonazountas, M., Brecher, A. & Vranka, R.G. (1988) Mathematical environmental fate modeling. In: *Environmental Inorganic Chemistry* (Eds I. Bodek, W.J. Lyman, W.F. Reehl & D.H. Rosenblatt), pp. 5.1–

5.6.10. Pergamon Press, New York.

Booty, W.G. & Lam, D.C.L. (1989) Freshwater Eco-system Water Quality Modelling. Report NWRI 89–63, National Water Research Institute, Burlington, Ontario.

Brown, D.S. & Allison, J.D. (1987) *MINTEQA1 Equilibrium metal speciation model: A user's manual.* US EPA, Athens, GA.

Burns, L.A. & Cline, D.M. (1985) *Exposure Analysis Modeling System: Reference Manual for EXAMS II.* EPA-600/3-85-038, US EPA, Athens, GA.

Carsel, R.F., Smith, C.N., Mulkey, L.A., Dean, J.D. & Jowise, P. (1984) *User's Manual for the Pesticide Root Zone Model (PRZM).* EPA-600/3-84-109. US EPA, Athens, GA.

Clark, K.E., Gobas, F.A.P.C. & Mackay, D. (1990) Model of organic chemical uptake and clearance by fish from food and water. *Environ. Sci. Technol.* **24**, 1203–1213.

Cohen, Y. (1986) *Pollutants in a Multimedia Environment.* Plenum Press, New York.

Cohen, Y., Tsai, W., Chetty, S.L. & Mayer, G.J. (1990) Dynamic partitioning of organic chemicals in regional environments: a multimedia screening-level modeling approach. *Environ. Sci. Technol.* **24**, 1549–1558.

Cohen, Y. & Ryan, P.A. (1985) Multimedia modeling of environmental transport: trichloroethylene test case. *Environ. Sci. Technol.* **19**, 412–417.

Connell, D.W. (1989) *Bioaccumulation of Xenobiotic Compounds.* CRC Press, Boca Raton, FL.

Connolly, J.P. & Winfield, R.P. (1984) *A User's Guide for WASTOX, A Framework or Modeling the Fate of Toxic Chemicals in Aquatic Environments.* Part I: Exposure Concentration, and Part II: Food Chains. Report EPA-600/3-84-077, US EPA, Athens, GA.

Dickson, K.L., Maki, A.W. & Cairns, J. (1982) *Modeling the Fate of Chemicals in the Environment.* Ann Arbor Science Publishers, Ann Arbor, MI.

Di Toro, D.M. (1984) Probability model of stream quality due to runoff. *J. Environ. Eng.* **ASCE 118**(3), 687.

Dolan, D.M. & Bierman, Jr, V.J. (1982) Mass balance modeling of heavy metals in Saginaw Bay, Lake Huron. *J. Great Lakes Res.* **8**(4), 676.

Enfield, G.C., Carsel, R.F., Cohen, S.Z., Phon, T. & Walters, D.M. (1982) Approximating pollutant transport to groundwater. *Ground Water* **20**, 711–727.

Hetrick, D.M. & McDowell-Boyer, L.M. (1983) *User's Manual for TOXSCREEN: A Multimedia Screening-level Program for Assessing the Potential Fate of Chemicals Released to the Environment.* Draft report, Oak Ridge National Laboratory, ORNL/TM8570, Office of Pesticides and Toxic Substances. US EPA, Washington DC, Contract No. W-7405-eng-26.

Hites, R.A. & Eisenreich, S.J. (Eds) (1987) *Sources and Fates of Aquatic Pollutants.* American Chemical Society Symposium Series 216, Washington, DC.

Holysh, M., Paterson, S., Mackay, D. & Bandurraga, M.M. (1986) Assessment of the environmental fate of linear alkylbenzenesulphonates. *Chemosphere* **15**, 3–20.

Jorgensen, S.E. & Gromiec, M.J. (Eds) (1989) *Mathematical Submodels in Water Quality Analysis.* Elsevier, Amsterdam.

Jorgensen, S.E. (Ed.) (1984) *Modelling the Fate and Effect of Toxic Substances in the Environment.* Elsevier, Amsterdam.

Jury, W.A., Spencer, W.F. & Farmer, W.J. (1983) Behavior assessment model for trace organics in soil. *J. Environ. Qual.* **12**, 558–564.

Klein, W., Kordel, W., Klein, A.W., Kuhnen-Clausen, D. & Weiss, M. (1988) Systematic approach for environmental hazard ranking of new chemicals. *Chemosphere* **7**, 1445–1462.

Lam, D.C.L., Schwertzer, W.M. & Fraser, A.S. (1982) Mass balance models of phosphorus in sediments and water. *J. Hydrobiol.* **217**, 91–92.

Mackay, D. (1989) An approach to modelling the long term behaviour of an organic contaminant in a large lake: application to PCBs in Lake Ontario. *J. Great Lakes Res.* **15**, 283–297.

Mackay, D. (1991) *Multimedia Environmental Models: The Fugacity Approach.* Lewis, Chelsea, MI.

Mackay, D. & Paterson, S. (1982) Fugacity revisited. *Environ. Sci. Technol.* **16**, 654A–660A.

Mackay, D. & Paterson, S. (1991) Evaluating the multimedia fate of organic chemicals: a level III fugacity model. *Environ. Sci. Technol.* **25**, 427–436.

Mackay, D., Joy, M. & Paterson, S. (1983) A quantitative water, air, sediment interaction (QWASI) fugacity model for describing the fate of chemicals in lakes. *Chemosphere* **12**, 981–997.

Mackay, D., Paterson, S., Cheung, B. & Neely, W.B. (1985) Evaluating the environmental behavior of chemicals with a level III fugacity model. *Chemosphere* **14**(3/4), 335–374.

McKone, T.E. & Kastenberg, W.E. (1986) Application of multimedia pollutant transport models to risk analysis. In: *Pollutants in a Multimedia Environment* (Ed. Y. Cohen), pp. 167–190. Plenum Press, New York and London.

McKone, T.E. & Layton, D.W. (1986) Screening the potential risks of toxic substances using a multimedia compartment model: estimation of human exposure. *Regul. Toxicol. Pharmacol.* **6**, 359–380.

Melancorn, S.M., Pollard, J.E. & Hern, S.C. (1986) Evaluation of SESOIL, PRZM and PESTAN in a laboratory leaching experiment. *Environ. Toxicol. Chem.* **5**, 865–878.

O'Connor, D.J. (1988a) Models of sorptive toxic substances in freshwater systems. I: Basic equations. *J. Environ. Eng.* **4**(3), 507.

O'Connor, D.J. (1988b) Models of sorptive toxic substances in freshwater systems. II: Lakes and reservoirs. *J. Environ. Eng.* **4**(3), 533.

OECD (1989) Compendium of Environmental Exposure Assessment Methods for Chemicals, Environment Monographs, No. 27. OECD, Paris.

Paterson, S. & Mackay, D. (1987) A steady-state fugacity-based pharmacokinetic model with simultaneous multiple exposure routes. *Environ. Toxicol. Chem.* **6**, 395–408.

Paterson, S. & Mackay, D. (1990) Models of environmental fate and human exposure to toxic chemicals. In: *Reviews in Environmental Toxicology 4* (Ed. E. Hodgson), pp. 241–265. Toxicology Communications, Raleigh, NC.

Schnoor, J.L. (1981) Fate and transport of dieldrin in Coralville Reservoir: residues in fish and water following a pesticide ban. *Science* **211**, 840–842.

Schnoor, J.L., Sato, C., McKechnie, D. & Sahoo, D. (1987) Processes, coefficients, and models for simulating toxic organics and heavy metals in surface waters. Report EPA/600/3-87/015.

Sonzogni, W.C., Canole, R.P., Lam, D.C.L., Lick, W., Mackay, D., Minns, C.K., Richardson, W.L., Scavia, D., Smith, R. & Strahan, W.M.J. (1987) Large lake Models—uses, abuses and future. Report of the Modeling Task Force to the Science Advisory Board of the International Joint Commission, Windsor, 1986. Also published in summary form in *J. Great Lakes Res.* **13**, 387–396.

Swann, R.L. & Eschenroeder, A. (Eds) (1983) *Fate of Chemicals in the Environment*. American Chemical Society Symposium Series 225, Washington, DC.

Thibodeaux, L.J. (1979) *Chemodynamics*. John Wiley and Sons, New York.

Thomann, R.V. & DiToro, D.M. (1983) Physico-chemical model of toxic substances in the Great Lakes. *J. Great Lakes Res.* **9**, 474–496.

Thomann, R.V. (1989) Bioaccumulation model of organic chemical distribution in aquatic food chains. *Environ. Sci. Technol.* **23**, 699–707.

Tucker, W.A., Eschenroeder, A.G. & Magil, G.C. (1982) Air, land, water analysis systems (ALWAS): a multimedia model for assessing the effect of airborne toxic substances on surface quality. First draft report, prepared by Arthur D. Little, Inc. for Environmental Research Laboratory. EPA, Athens, GA.

Tumeo, M.A. & Orlob, G.T. (1989) An analytic technique for stochastic analysis in environmental models. *Water Resour. Res.* **25**(12), 2417.

Vollenweider, R.A. (1969a) Possibilities and limits of elementary models concerning the budget of substances in lakes. *Arch. Hydrobiol.* **66**, 1.

Vollenweider, R.A. (1969b) Moglichkeiten und Grenzen elementares Modelle der Stoffbilanz von Seen. *Arch. Hydrobiol.* **66**, 1–36.

Vollenweider, R.A. (1976) Advances in defining critical loading levels for phosphorus in lake eutrophication. *Mem. 1st. Ital. Idrobiol.* **33**, 53–83.

Whelan, G., Strange, D.L., Droppo, Jr, J.G., Steelman, B.L. & Buck, J.W. (1987) The remedial action priority system (RAPS): Mathematical formulations, prepared for US Department of Energy, Office of Environment, Safety and Health.

Yoshida, K., Shigeoka, T. & Yamauchi, F. (1987) Multiphase non-steady state equilibrium model for evaluation of environmental fate of organic chemicals. *Toxicol. Environ. Chem.* **15**, 159.

Zielinski, P.A. (1988) Stochastic dissolved oxygen model. *J. Environ. Eng.* ASCE **114**(1), 74.

16: Prioritization and Standards for Hazardous Chemicals

S. HEDGECOTT

16.1 INTRODUCTION

16.1.1 Why prioritize?

Estimates suggest that up to 100 000 chemicals are used by humans, increasing by several thousand per year, each with an unknown number of by-products associated with their manufacture, use, degradation or destruction (Schmidt-Bleek & Haberland, 1980; Agg & Zabel, 1990; Toft & Hickman 1990). Each has the potential to cause damage to some part of the environment, whether overtly as in the case of crude oil slicks, or more subtly as in the case of imposex in gastropod molluscs caused by organotins (Smith & McVeagh, 1991; Chapter 6). Obviously it is desirable that the environmental impact of chemicals is kept to a minimum. Thus there is a need to identify those substances that are liable to have the most adverse effects: only through effective control over their manufacture, use and release can we hope to protect our environment. With such a vast list of chemicals there are two issues to be addressed.

1 Some of the 'worst offenders' such as DDT, chlorofluorocarbons and mercury have already been identified, and controls are being implemented. This process needs to be continued in order that the most hazardous chemicals are, one by one, regulated or eliminated.

2 We must ensure that we do not repeat our mistakes. We must fully assess the implications of new chemicals finding their way onto the marketplace and thus into our air, our land, our food and our water.

Most of this chapter considers the first issue, that of existing chemicals, but it should not be forgotten that the principles used apply equally to new chemicals.

In the past, action on existing chemicals has been carried out on a reactive basis, when problems were already manifest. Now more effort is being devoted to identifying harmful substances before their effects become so obvious. However, with thousands of chemicals and only limited resources it is important that we identify priorities for control, and systematically work from the most hazardous to the least. The rationale behind schemes used to create such priority lists, and the results that have been produced, comprise the bulk of this chapter.

The principles applied when setting priorities for existing chemicals can also be used to estimate the potential hazard of those not yet on the market (Toft & Hickman, 1990). The base set of bioaccumulation and toxicity data required for new chemical registration can be used in conjunction with data on production and amounts released to give an indication of hazard ranking (Klein et al., 1984). Therefore we now have the ability to be proactive when considering new chemicals, to ensure that only those which are environmentally acceptable (or, in a few cases, do not have a more viable alternative) can be marketed.

16.1.2 Standards for single substances

Rarely are chemicals routinely used and released in isolation, and even those that are become mixed with other contaminants once in the environment (Chapter 12). These chemicals might form complexes or react to release more hazardous products, or they may have additive or even synergistic toxic effects on biota. Certainly each chemical has the potential to cause stress, and simultaneous stresses that are unimportant in isolation may be very significant in combination. So what is the point in setting priorities and exercising controls on single substances? It is not ideal, but there are three very important reasons why it is the approach used:

1 we know that some chemicals are harmful in isolation, whether or not they may also be harmful in combination with others;
2 the interactions are so diverse and so complex in mixtures that we could never hope to fully understand and predict them, and any mixture of chemicals may be unique to a particular situation;
3 we have limited resources and have to start somewhere!

It is relatively easy and widely beneficial to control releases of a chemical known to be hazardous in isolation, and this should also reduce the number of hazardous combinations in the environment. Thus the role of prioritization schemes for single substances is essentially a regulatory one, but does also have some practical basis.

16.1.3 International agencies

All environmental contaminants show geographical variation, and the problems are not always associated with the point of origin. For example, in Europe 'acid rain' in Scandinavia arises largely from Germany, France and the UK. Thus in many situations the release of harmful chemicals is not stringently controlled because the consequences are seen elsewhere. Similarly, restrictions may be waived if national interests associated with trade and industry take precedence over environmental concerns. For example, although the manufacture and use of DDT is banned in most western countries some still permit the export of existing stocks to developing nations, and thus, although banned in the UK, DDT may still be released during the treatment of imported contaminated fleeces.

The best way to overcome such problems is to have agreement between nations through international agencies and commissions, conferring the following advantages:

1 ensuring that trans-boundary pollution is kept to a minimum;
2 facilitating the control of shared problems by adjacent nations;
3 enabling harmonization of controls between countries, facilitating trade between them;
4 preventing the duplication of effort on priority setting and hazard assessment, thus making the best use of limited resources;
5 raising the profile of such issues in the eyes of both the public and politicians, encouraging the development of good environmental policies.

Thus international agencies play an important role in setting priorities and standards for hazardous chemicals.

16.1.4 Analytical and technical challenges

Prioritization schemes select hazardous substances on the basis of the available data on their inherent properties, and normally some sort of quality standard is derived from these same data. This often incorporates an extrapolation or safety factor, and therefore the resulting allowable concentration may be below the current analytical capability. This begs the question: should we attempt to provide complete protection or should we establish standards that we can actually monitor? It would be defeatist to apply only half-measures to the control of chemicals known to be hazardous, suggesting that we should establish standards independently of analytical capability and use these as the driving force for improving our ability to monitor them.

The control of chemical emissions is in a similar situation, with current technology not always able to meet quality standards. Again we must attempt to give priority to the environment, and the 'polluter pays principle' (OECD, 1975) has been adopted in part to facilitate this, by making it more cost-effective for industry to clean up discharges. Similarly the principle of best available technology (DoE, 1990) has been established to ensure that industry strives to improve treatment technology. If controls are not technically or financially achievable, then an alternative to that chemical should be found.

16.2 SELECTING PRIORITY POLLUTANTS

16.2.1 Elements of prioritization schemes

Prioritization was 'born' out of negotiations on the Oslo Convention during 1971–2, which classified substances onto so-called 'black' and 'grey' lists, according to the extent of control which was required (i.e. prohibition and reduction of release, respectively) (Bjerre & Hayward, 1984). There are now numerous prioritization schemes in operation throughout the world, varying according to the role of the organization using them. They all share a common basis in hazard assessment, which involves classifying a substance as 'of concern' or 'not of concern'. These are the central questions common to such classifications schemes.

1 What volume or amount of this chemical are we concerned with?
2 How persistent is this chemical?
3 What is the likelihood of this chemical causing harm to living organisms?

This final question generally has two components: what level of the chemical would have adverse effects (i.e. what is the environmental concern level), and how close are measured or predicted environmental concentrations to this level? If the concern level is higher than the environmental concentrations the risk (of environmental damage) may be considered to be low, but there are a number of problems inherent in making this comparison.

1 There is uncertainty in defining what margin is acceptable, and this varies from a factor of 1 to 1000 (de Nijs *et al.*, 1988).
2 Reliable, widely applicable environmental concentration data are rare (Branson, 1980) and predictions can only approximate actual concentrations and may therefore underestimate risk. This can be overcome to a certain extent by quantifying the uncertainty in the predictions, and thus having an associated uncertainty in the overall assessment (Slob & de Nijs, 1989).
3 A toxic chemical may not be prioritized if it has low environmental levels, but a single uncontrolled release (such as a spillage) may have a significant impact.
4 If a chemical is not prioritized because of a wide margin between the predicted and concern levels, it may not be controlled and its environmental levels may increase, simply deferring the problem.

For these reasons some prioritization schemes consider the volume of a chemical produced or used and the innate characteristics that give some measure of its 'escapability' to be more appropriate than predicted or a few measured environmental levels.

16.2.2 Required outcome of prioritization

All prioritization schemes aim to characterize chemicals into one of the following groups (Branson, 1980; Klein *et al.*, 1984; Slob & de Nijs, 1989):

1 those with no, or a low, likelihood of environmental hazard;
2 those with a high likelihood of environmental hazard;
3 those with an uncertain likelihood of environmental hazard and which require further study.

Some schemes go a step further and rank individual chemicals according to their relative environmental hazard.

16.2.3 Prioritization methods

Many early selection processes were not rigorous, but did not need to be, since the initial priorities for control were often obvious. Once these initial concerns had been identified there was a need for a more defined approach. Current prioritization schemes may be specific to one particular situation, or may be used to determine priorities for the whole environment. For example, Dieter *et al.* (1990) reported a hazard ranking scheme developed only for a consideration of the human health hazards associated with contaminated groundwater, while Bro-Rasmussen (1985) outlined a screening scheme that takes account of all environmental partitions. In fact, for adequate chemical control both types are needed (Schulze & Mücke, 1985).

Approaches used

Current prioritization schemes may essentially be split into three types.

1 Those in which a score is assigned to each particular property of the chemical, and these individual scores are combined in some way to provide an overall priority score.

2 Those in which a 'pass or fail' or 'decision tree' system is used to assess the importance of each property of the chemical, and the overall priority assessment is determined by the specific combinations of 'passes' and 'fails'.

3 Those in which candidate chemicals are compared with one another and assigned to hierarchical priority position. However, besides being difficult to computerize, such schemes have the major disadvantage of needing complete reruns should the list of candidate chemicals expand, and are therefore not widely used.

The main advantage of scoring systems is that they provide an absolute ranking of priority for any range of chemicals, since each has a discrete score independent of all others. Additionally, scores may be manipulated to adapt the system to particular situations. For example: the range of scores available can be varied for different parameters, and thus greater emphasis can be given to those considered to be more important; the actual numerical value for a property can be translated into a score (e.g. by taking the log), giving a score that is both fully representative and infinitely variable (within the bounds of that parameter); and scores for individual properties may be multiplied (rather than added) to give the total score, thus increasing the range of scores possible.

In 'decision tree' schemes each chemical is assigned to a final priority category (e.g. high, medium or low) but the order of priority within each category is not defined. Thus there is somewhat less sensitivity than with scoring systems. The argument in support of this approach is that the amount and quality of the data currently available, and our present understanding of chemical hazard, are so poor that a 'broad-brush' approach is more appropriate, and any absolute ranking provided by a scoring system will probably be wrong anyway! Scoring systems also have the disadvantage of the scores themselves being misleading, because they may be interpreted as measurements of hazard potential, which they are not. Thus a chemical scoring 50 is not (necessarily) twice as hazardous as one scoring

25. Indeed, if total scores are derived by multiplication (rather than addition) of individual scores then a relatively small difference in hazard may translate into a substantial difference in final scores. Therefore, it may be preferable, if using a scoring system, to remove the scores from any resulting published priority list, in order to prevent confusion and misinterpretation.

Interest in the harmonization of prioritization schemes in Europe has meant that several reviews of existing schemes have recently been carried out, with particularly good ones by Jackson & Peterson (1989) and Crookes & Nielsen (1989). Table 16.1 summarizes the type of scheme (i.e. whether it is based on a 'decision tree' process or a scoring process for each parameter considered), and the data that are used in the selection process, for a variety of priority-setting schemes currently in operation in Europe and elsewhere.

Although the details of various prioritization schemes vary, the principles used are broadly the same in all cases. Therefore, rather than attempt to consider the full range of schemes the rest of Section 16.2 concentrates on one representative scheme. Since most effort has been directed towards prioritization for the aquatic environment a water-based scheme is described. One of the more recent and refined examples of such schemes is that used in the UK to develop a 'Red List' of substances of concern to water, described by Hedgecott & Cooper (1991), and this is considered in the rest of Section 16.2 below. Although this is a 'decision tree' type scheme, it has many elements of a scoring scheme because the decisions allocate particular features into 'high', 'medium' and 'low' categories, which in itself is a kind of scoring system. However, no final numerical score is derived, so prioritization involves allocation into broad priority categories rather than absolute ordering.

'Red List 2'

At the Third International Conference on the Protection of the North Sea (1990) each participating state agreed to develop a national priority list based upon a common reference list of chemicals. The UK Department of the Environment had earlier devised a scheme (DoE, 1986) for identifying priority chemicals (the 'Red List')

Table 16.1 Data requirements of a selection of priority setting schemes

	Can.	CEFIC	Fra.	Ger.	Italy	NL	N. Sea	SRI	UK
Decision process[a]	S	T	T	S	S	T	T	S	T
Mammalian toxicity[b]	CMT	—	C	—	CMT	C	CM	—	CMT
	Acu	—	—	Acu	Acu	—	—	Acu	Acu
	—	—	Chr	—	Chr	—	—	Chr	Chr
Aquatic toxicity[b]	Acu	Acu	Acu	Acu	Acu	Acu	Acu	Acu	Acu
	Chr	—	—	—	—	Chr	—	Chr	Chr
Partitioning models	—	—	—	—	—	—	—	Yes	—
Bioaccumulation or K_{ow}	Yes	Yes	Yes	(Yes)	Yes	—	Yes	Yes	Yes
Persistence	Yes	Yes	Yes	(Yes)	Yes	Yes	Yes	Yes	Yes
Physicochemical	Yes	—	Yes	(Yes)	Yes	—	—	Yes	Yes
Sources (point v. diffuse)	—	—	—	—	—	—	—	Yes	Yes
Environmental concentrations	—	—	—	—	—	Yes	Yes	—	—
Production or usage volume[c]	—	—	—	—	P	P	PU	PU	PU

[a] S, scoring system; T, decision tree system.
[b] C, carcinogenicity data used; M, mutagenicity data used; T, teratogenicity data used; Acu, acute data used; Chr, chronic data used.
[c] P, production data used; U, usage data used.
(Yes), not integral part, but may be used.
References: Can. (Canada), Environment Ontario (1989); CEFIC, CEFIC (1987); Fra. (France), CEESC (1986); Ger. (Germany), BMU (1987); Italy, Sampaolo & Binetti (1989); NL (Netherlands), BCW (1989), N. Sea (North Sea), Second International Conference on the Protection of the North Sea (1987); SRI, SRI (1980); UK (United Kingdom), Hedgecott & Cooper (1991).

(DoE, 1988), and this was refined for use on this new list (DoE, 1989a). This new selection procedure, here termed 'Red List 2' for ease of reference, contains all the elements of a reasoned, practical prioritization scheme.

Red List 2 is a computerized selection procedure consisting of four decision trees which select those substances that pose the greatest hazard to the aquatic environment on the basis of a combination of their toxicity, bioaccumulation, carcinogenicity, persistence and likelihood of entering water. The categorization of values as 'high', 'medium' or 'low' avoids problems experienced with other prioritization schemes based on algorithms where the various parameters are scored

and these are then combined. Such a scheme can easily be biased by extreme values or by the weightings allocated to particular parameters (Agg & Zabel, 1990). The definitions of 'high' and 'low' are given in Table 16.2. These values are somewhat arbitrary, but Red List 2 was designed so that they can be altered for any particular run of the programmes to see what effects this has on the results, and to identify how sensitive the scheme is to particular parameters.

The short-term scenario (Fig. 16.1) and the long-term scenario (Fig. 16.2) consider whether the substance is likely to reach a concentration in water at which acute or chronic toxic effects have been reported, based on the substance's reported

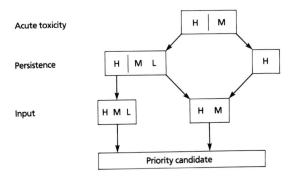

Fig. 16.1 The 'Red List 2' short-term scenario. High (H), medium (M), low (L) are defined in Table 16.2.

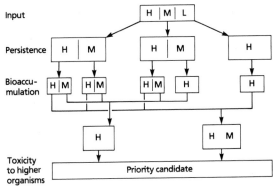

Fig. 16.3 The 'Red List 2' food-chain scenario. High (H), medium (M), low (L) are defined in Table 16.2.

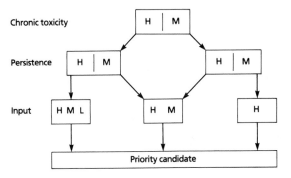

Fig. 16.2 The 'Red List 2' long-term scenario. High (H), medium (M), low (L) are defined in Table 16.2.

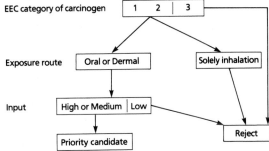

Fig. 16.4 The 'Red List 2' carcinogenicity scenario. High (H), medium (M), low (L) are defined in Table 16.2.

toxicity, persistence in surface waters and estimated level of input. The food chain scenario (Fig. 16.3) considers whether the substance is likely to reach a toxic level as a result of bioconcentration through the food chain, based on the estimated level of input, persistence in surface waters, bioaccumulation in aquatic organisms and toxicity to mammals. The carcinogenicity scenario (Fig. 16.4) considers whether the substance's carcinogenic, mutagenic or teratogenic (CMT) properties may result in an increased risk to human health following exposure in or through water. The decision is based on the substance's European Commission CMT classification, the exposure risk route (those effective by the topical or oral route) and the estimated level of input to water.

16.2.4 Data requirements for prioritization

As mentioned in Section 16.2.1 there are three elements that are essential to all prioritization schemes: quantity, persistence and toxicology. There are a number of ways in which these can be considered, but the required data are similar in each case and are summarized in Table 16.3.

Persistence

Chemicals that are ephemeral in the environment are less likely to have adverse effects, and some value for persistence is included in most prioritization schemes. Red List 2 uses an estimated half-life for the predominant removal processes from surface waters as a measure of persistence.

Table 16.2 Numerical values for high and low thresholds for chemical property or toxic endpoints

Property	Toxic endpoint or parameter	Unit	High	Low	Insignificant
Acute aquatic toxicity	96-h LC_{50} (fish) or 48-h EC_{50}/LC_{50} (invert)	mg/l	< 1.0	> 100	—
Chronic aquatic toxicity	NOEC	mg/l	< 0.01	> 1	—
Persistence in water	Half-life	days	> 100	< 10	< 2
Bioaccumulation potential	Bioconcentration factor	—	> 1000	< 100	—
	$\log K_{ow}$	—	> 3.5	< 2	—
Toxicity to higher organisms	Significant toxic effect or oral LD_{50}	mg/kg	< 50	> 500	—
Production and use		tonnes/year	> 10 000	< 1000	—
Solubility		mg/l	> 1000	< 1	—
Volatility		Pascals	> 0.133	< 0.000133	—

Table 16.3 Types of data required for prioritization schemes

Volume significant
Volume produced and/or used and/or released
Routes of release
Input medium
Environmental fate parameters
Physicochemical characteristics — a minimum of solubility and volatility

Sometimes leading to predicted or measured environmental levels

Persistence
Measured or predicted environmental half-life and/or rate constants for removal processes
Fate during treatment of emission/effluent and during sewage treatment
Environmental mobility

Toxicology
Aquatic and/or terrestrial toxicity — extrapolated to environment
Bioaccumulation/bioconcentration — may be estimated via K_{ow}
Mammalian toxicity — extrapolated to humans
Carcinogenicity/teratogenicity/mutagenicity

Amount significant to water

A basic tenet when determining priorities is that chemicals that are frequently present in the environment, whether spatially or temporally, require particular attention (de Nijs *et al.*, 1988). Red List 2 incorporates data on the total estimated discharge of a chemical significant for water (tonnes/year) (SRI International, 1980). If no such figure is available an estimate of 'high', 'medium' or 'low' input is made, based on solubility, volatility, production volume and expected sources (i.e. point or diffuse).

Aquatic toxicity

There is no such thing as the most sensitive aquatic species, with sensitivity depending largely on the nature of the chemical being considered (Cairns, 1986; Volume 1). Nevertheless, salmonid fish tend to be particularly sensitive to a wide range of contaminants, and various species of *Daphnia* are often considered broadly representative of aquatic invertebrates. Therefore, to provide consistency when comparing different chemicals Red List 2 is biased towards acute and chronic toxicity data for rainbow trout (*Oncorhynchus mykiss*) and *Daphnia magna*, but if there are data for more sensitive species these can be used instead.

The actual toxicity data used are the lowest reported $\leq$ 96-h LC_{50}s (or EC_{50}s for invertebrates) for the acute scenario, and the lowest reported chronic toxicity values (e.g. no observed effect concentration NOEC; maximum acceptable toxicant concentration, MATC) for the chronic scenario. Chronic data require careful consideration because of the variety of different test types (such as lethality, inhibition of reproduction, biochemical changes, and so on), and endpoints (ranging from no effect through a percentage incidence to a total effect). In prioritization it is important that toxicity data are comparable, otherwise different substances will not be considered equitably, and that the toxic effects considered are biologically meaningful. Therefore Red List 2 only considers chronic effects on survival, reproduction and growth, and only no or minimal effect endpoints are considered.

Aquatic bioaccumulation

An increasing body burden of a substance that is readily taken up but not readily depurated may have significant consequences. The highest reported bioaccumulation factors for aquatic fish and invertebrates (based on wet weight) are therefore incorporated into Red List 2 (Volume 1; Chapter 19). In the absence of reliable bioaccumulation factors the octanol–water partition coefficient (K_{ow}; Chapter 13) of a chemical is taken as an indicator of bioaccumulation potential, based on the following relationship (Veith *et al.*, 1979):

$$\log BCF = 0.76 \log K_{ow} - 0.23$$

Mammalian toxicity

Since humans may be exposed to chemicals via drinking water, Red List 2 considers indicators of mammalian toxicity, namely the lowest reported acute oral toxicity value (milligrams per kilogram) for the rat or another mammal. The use of an acute oral LD_{50} is not ideal, because it requires extrapolation to provide an indication of effects on humans (Toft & Hickman, 1990), but no other

more suitable value is consistently reported in the literature.

In addition the EC classification as a carcinogen, mutagen or teratogen (COM, 1990) is considered by Red List 2, along with the exposure route(s) by which the substance has these effects on humans.

Dealing with gaps in the data

An important consideration in priority setting is how to handle missing data: if no value is available for a parameter should a substituted value be included, and should this be set at the high or low threshold limit or somewhere in between? One option that has been used (e.g. Fingas *et al.*, 1991) is to reject chemicals from the analysis if there is no information on them. Such a scheme is acceptable only if rejected chemicals are allocated to a separate list for further study. Although no approach would be universally considered appropriate it is sensible that the precautionary principle should be applied and that a worst-case situation should be assumed (de Nijs *et al.*, 1988). An acceptable alternative is to 'adopt' an equivalent value for a similar chemical which has been thoroughly investigated and has reliable data for the missing parameter (Freitag, 1987). In all cases chemicals that are classified on the basis of some sort of default data should be readily identifiable, and an effort should be made to obtain or generate the required data to allow a proper assessment.

In the Red List 2 scheme all missing values result in a worst-case default value being assigned, and all default values are flagged such that if the chemical becomes assigned to a priority list the use of a default value is clear.

16.3 PRIORITY POLLUTANT LISTS

As discussed earlier, there are numerous lists of priority chemicals applicable to different situations in different countries, and it is not feasible to consider them all here. Instead this section concentrates on a few specific lists in order to identify some of the chemicals that are considered as priorities, and to give an idea of those that appear most frequently on priority lists.

Table 16.4 The Oslo Convention black and grey lists

Black
Cadmium and its compounds
Mercury and its compounds
Organohalogen compounds and their precursors
Persistent plastics
Persistent synthetic materials
Substances agreed as likely to be carcinogenic

Grey
Arsenic and its compounds
Chromium and its compounds
Copper and its compounds
Lead and its compounds
Nickel and its compounds
Non-persistent oils and hydrocarbons
Phosphorus and its organic compounds
Organosilicon compounds
Organotins
Tainting substances (in sea-foods)
Zinc and its compounds

Radioactive substances were recognized as a concern, but were the remit of a number of other organizations.

Table 16.4 shows the Oslo Convention black and grey list, the precursors of many of today's priority lists (Bjerre & Hayward, 1984). Many of the priorities identified were generic groups of chemicals, such as 'organohalogens' and 'non-persistent oils and hydrocarbons'. When these lists were drawn up there were no widely accepted prioritization schemes available, and substances were suggested as priorities on the basis of known problems with certain chemicals. Groups were put forward because of the likelihood of other similar chemicals in the group having similar properties and similar hazard potentials.

All priority lists should be dynamic, as more data become available and as more chemicals are considered as candidates for study. For this reason the black and grey lists are now outdated, and more recent priorities have generally been identified with considerably more precision. Several lists are given in Table 16.5, which includes the following.
1 The Canadian Priority Substances List, applicable to all environmental partitions and published in 1989 under the Environmental Protection Act (here taken from Toft & Hickman, 1990).
2 The US Environmental Protection Agency's List of 129 priority pollutants, also applicable to all environmental compartments (here taken from Keith & Telliard, 1979).
3 The European Community's (EC) list of Potential List I substances, relating to the aquatic environment only (CEC, 1976; Mestres, 1979).
4 The UK list of potential prescribed substances, for which Her Majesty's Inspectorate of Pollution investigate the need for control of releases to all environmental compartments (DoE, 1991; Powlesland, HMIP, personal communication).

In each case candidate substances were selected from a combination of known priorities (such as other countries' priority lists) and chemicals identified as potentially harmful on the basis of their significant toxicity, bioaccumulation, persistence and production volumes.

Examination of Table 16.5 demonstrates the different results of different schemes, even if their aim was the same. For example, considering only the North American situation, the lists for Canada and the USA have only 24 chemicals (including polycyclic aromatic hydrocarbons (PAH) as a group) in common, even though both schemes apply to all environmental compartments and relate to adjacent geographical regions. Comparing lists derived for different environmental compartments, the UK list (all compartments) and the EC list (water only) have 38 substances (including PAH) in common, i.e. more than the two North American lists, even though their aims are different.

Considering all four of these priority lists together, nine chemicals (plus PAH) are common to them all (Table 16.6). The suggestion is that there is a 'hard core' of chemicals that are considered to be particularly hazardous in most or all situations. Other chemicals may be priorities in one case and not in another, because of more specific concerns; perhaps being used in large quantities only in certain countries, not persisting in one part of the environment, or only being toxic to certain types of organisms.

A notable feature of this list is that of the seven specific compounds five are chlorinated. It is thought that naturally occurring chlorinated hydrocarbons are rare or non-existent, and thus it is perhaps not surprising that many of them are particularly hazardous because there are few environmental pathways for their degradation and detoxification. Heavy metals are also prominent on the various priority lists. Even though these commonly occur naturally they tend to be present only at trace levels, except where influenced by humans, and therefore most living organisms are not able to tolerate elevated concentrations.

For comparison, Table 16.7 shows those chemicals that most frequently appear on chemical spills priority lists in North America, rather than general priority lists as considered so far. Fingas *et al.* (1991) considered more than 1300 substances and their inclusion on 19 spill lists in the USA and Canada, each based on a combination of aquatic and/or mammalian toxicity, production volume, and historic spill frequency and volume. Table 16.7 lists 43 chemicals that appeared on 12 or more (an arbitrary number) of these lists. Only three of these (benzene, trichloroethene and toluene) are common to the US EPA list and to the Canadian Priority Substances List, even though they apply to the same geographical area. The reason for the difference is simply that the different prioritization schemes had a different emphasis, with spill lists being much more specifically targeted than other priority lists. For example, their emphasis is on liquids and on end-products that are transported or stored (rather than chemical intermediates). It is important to bear specific issues of this sort in mind at all stages when developing prioritization schemes, as well as when interpreting their results.

16.4 EXISTING STANDARDS FOR PRIORITY CHEMICALS

16.4.1 Types of standards

Prioritization of substances will not alone provide any protection: regulatory standards are also needed, to provide a means of measuring compliance with regulatory actions and thus to help control the release of chemicals. Standards are usually expressed as some sort of permitted concentration (possibly zero), and there are three types.

1 Environmental quality standards specify the maximum level (concentration and frequency) of a contaminant that is considered acceptable in the receiving environment. They can be applied on any scale, from locally to internationally.
2 Emission standards specify the maximum amount or concentration of a chemical that can be released, with a specified frequency. They are applied on both a chemical- and an industry-specific basis.
3 Product standards specify limits for the level of contamination in marketable products, such as food.

Both emission and product standards are, in effect, extrapolated from some sort of environmental quality standard because in each case an 'acceptable' level needs to be determined for the receiving environment (whether a river, the atmosphere, a human consumer or whatever) so that the amounts permissible in emissions or products can be established.

In many instances there has been a deliberate policy to set paired standards. For example, there may be a guideline and a mandatory value set for the same parameter, with the mandatory value being somewhat more relaxed, as in the EC Surface Water Directive (CEC, 1975). Alternatively a temporary standard may be established that may be more relaxed than a defined future standard, as in the EC Cadmium Directive (CEC, 1983). This approach recognizes that it may take considerable time and investment to be able to achieve the more stringent standards, which are set as ultimate targets.

16.4.2 Necessary assumptions in standards

Each type of standard assumes that the environment has some capacity to receive chemical contaminants without incurring unacceptable damage, and that the specified chemicals will act in isolation. Clearly neither will always be true, but with the limited data available it is not possible to establish standards with complete certainty of their appropriateness. Most are established from data provided by laboratory investigations, and as such they must incorporate some sort of extrapolation factor to predict effects in the environment from effects in the laboratory.

Table 16.5 Priority lists for Canada, the USA, the EC (EU) and the UK

Chemical name	Priority chemical?			
	Canada	USA	EC	UK
Acenaphthene	Y[a]	Y	Y[a]	Y[a]
Acenaphthylene	Y[a]	Y	Y[a]	Y[a]
Acrolein		Y		
Acrylonitrile		Y		Y
Aldrin		Y	Y	Y
2-Amino-4-chlorophenol			Y	
Ammonia				Y
Aniline	Y			
Anthracene	Y[a]	Y	Y	Y[a]
Antimony		Y		Y
Arsenic and its compounds	Y	Y	Y	Y
Asbestos		Y		Y
Mineral fibres	Y			Y
Glass fibres				Y
Atrazine			Y	Y
Azides				Y
Azinphos-ethyl			Y	
Azinphos-methyl			Y	Y
Bentazone			Y	
Benzene	Y	Y	Y	Y
Benzidine	Y	Y	Y	
Benzo(*a*)anthracene	Y[a]	Y	Y[a]	Y[a]
Benzo(*b*)fluoranthene	Y[a]	Y	Y[a]	Y[a]
Benzo(*k*)fluoranthene	Y[a]	Y	Y[a]	Y[a]
Benzo(*g,h,i*)perylene	Y[a]	Y	Y[a]	Y[a]
Benzo(*a*)pyrene	Y[a]	Y	Y[a]	Y[a]
Benzyl chloride			Y	
Benzylidene chloride			Y	
Beryllium		Y		Y
BHC (all isomers, including lindane)		Y		
Biphenyl			Y	
Bleach pulp mill effluents	Y			
Boron				Y
Bromine				Y
Bromoform (tribromomethane)		Y		
Bromomethane		Y		
4-Bromophenyl phenyl ether		Y		
1,3-Butadiene				Y
Butyl benzyl phthalate		Y		
Cadmium and its compounds	Y	Y	Y	Y
Carbon dioxide				Y
Carbon disulphide				Y
Carbon monoxide				Y
Carbon tetrachloride		Y	Y	Y
Chloral hydrate			Y	
Chlordane		Y	Y	
Chlorinated paraffin waxes	Y			
Chlorinated wastewater effluents	Y			
Chlorine				Y
Chloroacetic acid			Y	
2-Chloroaniline			Y	
3-Chloroaniline			Y	

Continued

Table 16.5 *continued*

4-Chloroaniline			Y	
Chlorobenzene	Y	Y	Y	
p-Chloro-m-cresol		Y		
Chlorodibromomethane		Y		
1-Chloro-2,4,-dinitrobenzene			Y	
Chloroethane		Y		
2-Chloroethanol			Y	
bis(2-Chloroethoxy) methane		Y		
bis-(2-Chloroethyl) ether	Y	Y		
2-Chloroethyl vinyl ether		Y		
Chloroform		Y	Y	Y
bis-(2-Chloroisopropyl) ether		Y		
Chloromethane		Y		
bis-(2-Chloromethyl) ether	Y	Y		
Chloromethyl methyl ether	Y			
4-Chloro-3-methylphenol			Y	
Chloronaphthalenes (technical mixture)			Y	
1-Chloronaphthalene			Y	
2-Chloronaphthalene		Y		
4-Chloro-2-nitroaniline			Y	
1-Chloro-2-nitrobenzene			Y	
1-Chloro-3-nitrobenzene			Y	
1-Chloro-4-nitrobenzene			Y	
Chloronitrotoluenes			Y	
4-Chloro-2-nitrotoluene			Y	
2-Chlorophenol		Y	Y	
3-Chlorophenol			Y	
4-Chlorophenol			Y	
4-Chlorophenyl phenyl ether		Y		
Chloroprene			Y	
2-Chlorotoluene			Y	
3-Chlorotoluene			Y	
4-Chlorotoluene			Y	
Chlorotoluidines			Y	
2-Chloro-p-toluidine			Y	
Chromium and its compounds	Y	Y		Y
Chrysene	Y[a]	Y	Y[a]	Y[a]
Copper		Y		Y
Coumaphos			Y	
Creosote	Y			
Cyanides (total)		Y		
Cyanuric chloride			Y	
Cyfluthrin				Y
2,4-D			Y	
DDD		Y	Y	
DDE		Y	Y	
DDT		Y	Y	Y
Demeton			Y	
Dibenzo(a,h)anthracene	Y[a]	Y	Y[a]	Y[a]
1,2-Dibromomethane			Y	
Dibutyl phthalate	Y	Y		
Dibutyltin compounds	Y[d]		Y	
Dichloroanilines			Y	
1,2-Dichlorobenzene	Y	Y	Y	
1,3-Dichlorobenzene		Y	Y	

Continued on page 380

Table 16.5 *continued*

Chemical name	Priority chemical?			
	Canada	USA	EC	UK
1,4-Dichlorobenzene	Y	Y	Y	
Dichlorobenzidines			Y	
3,3'-Dichlorobenzidine	Y	Y		
Dichlorobromomethane		Y		
Dichlorodifluoromethane		Y		
Dichlorodiisopropyl oxide			Y	
1,1-Dichloroethane		Y	Y	
1,2-Dichloroethane	Y	Y	Y	Y
1,1-Dichloroethane		Y	Y	
1,2-Dichloroethane		Y	Y	
Dichloromethane	Y	Y	Y	Y
Dichloronitrobenzenes			Y	
2,4-Dichlorophenol		Y	Y	
1,2-Dichloropropane		Y	Y	
1,3-Dichloropropane-2-ol			Y	
1,3-Dichloropropene		Y	Y	
2,3-Dichloropropene			Y	
Dichlorprop			Y	
Dichlorvos			Y	Y
Dieldrin		Y	Y	Y
Diethylamine			Y	
Diethyl phthalate		Y		
Dimethoate			Y	
Dimethylamine			Y	
3,5-Dimethylaniline	Y			
2,4-Dimethylphenol		Y		
Dimethyl phthalate		Y		
4,6-Dinitro-o-cresol		Y		
2,4-Dinitrophenol		Y		
2,4-Dinitrotoluene		Y		
2,6-Dinitrotoluene		Y		
Di-n-octyl phthalate	Y	Y		
1,2-Diphenylhydrazine		Y		
Disulfoton			Y	
Endosulfan		Y	Y	Y
Endosulfan sulfate		Y		
Endrin		Y	Y	Y
Endrin aldehyde		Y		
Epichlorohydrin			Y	
Ethylbenzene		Y	Y	Y
bis-(2-Ethylhexyl) phthalate	Y	Y		
Fenitrothion			Y	Y
Fenthion			Y	
Flucofuran				Y
Fluoranthene	Y[a]	Y	Y[a]	Y[a]
Fluorene	Y[a]	Y	Y[a]	Y[a]
Fluorides (inorganic)	Y			Y
Formaldehyde				Y
Gallium				Y
Heptachlor		Y	Y	
Heptachlor epoxide		Y	Y	
Hexachlorobenzene	Y	Y	Y	Y

Continued

Table 16.5 *continued*

Hexachlorobutadiene (HCBD)		Y	Y	Y
Hexachlorocyclohexane (HCH, lindane)			Y	Y
Hexachlorocyclopentadiene		Y		
Hexachloroethane		Y	Y	
n-Hexane				Y
Hydrogen fluoride				Y
Indeno(1,2,3-c,d)pyrene	Y[a]	Y	Y[a]	Y[a]
Indium				Y
Iodine				Y
Isodrin			Y	
Isophorone		Y		
Isopropylbenzene			Y	
Lead and its compounds		Y		Y
Linuron			Y	
Malathion			Y	Y
MCPA			Y	
Mecoprop			Y	
Mercury		Y	Y	Y
Methamidophos			Y	
Methane				Y
Methyl ethyl ketone				Y
Methyl isobutyl ketone				Y
Methyl methacrylate	Y			
Methyl tertiary-butyl ether (MTBE)	Y			
Mevinphos			Y	
Monolinuron			Y	
Naphthalene	Y[a]	Y	Y	Y[a]
Nickel and its compounds	Y	Y		Y
Nitric acid				Y
Nitrobenzene		Y		
2-Nitrophenol		Y		
4-Nitrophenol		Y		
n-Nitrosodimethylamine		Y		
n-Nitrosodiphenylamine		Y		
n-Nitrosodi-n-propylamine		Y		
NO_x				Y
Omethoate			Y	
Organic solvents				Y
Oxydemeton methyl			Y	
Palladium				Y
Parathion			Y	
Particulate matter				Y
PCSDs				Y
Pentachlorobenzene	Y			
Pentachlorophenol (PCP)	Y	Y	Y	Y
Phenanthrene	Y[a]	Y	Y[a]	Y[a]
Phenol		Y		
Phenols (total)		Y		
Phosphorus and its compounds				Y
Phoxim			Y	
Platinum				Y
Polyaromatic nitro compounds				Y
Polybrominated biphenyl esters				Y
Polychlorinated biphenyls (PCBs)		Y[b]	Y	Y
Polychlorinated dibenzodioxins (PCDDs)	Y	Y[c]		Y

Continued on page 382

Table 16.5 *continued*

Chemical name	Priority chemical?			
	Canada	USA	EC	UK
Polychlorinated dibenzofurans (PCDFs)	Y			Y
Polyhalogenated biphenyls				Y
Polyhalogenated naphthalenes				Y
Polyhalogenated terphenyls				Y
Propanil			Y	
Pyrazon			Y	
Pyrene	Y[a]	Y	Y[a]	Y[a]
Selenium		Y		Y
Silver		Y		
Simazine			Y	Y
SO_x				Y
Styrene	Y			Y
Sulcofuran				Y
2,4,5-T			Y	
Tellurium				Y
Tetrabutyltin	Y[d]		Y	
Tetrachlorobenzenes	Y		Y[e]	
1,1,2,2-Tetrachloroethane	Y	Y	Y	
Tetrachloroethylene		Y	Y	Y
Thallium		Y		Y
Toluene	Y	Y	Y	Y
Toxaphene		Y		
Triazophos			Y	
Tributyl phosphate			Y	
Tributyltin oxides	Y[d]		Y	Y
Trichlorphon			Y	
Trichlorobenzenes	Y		Y	Y
1,2,4-Trichlorobenzene		Y	Y	Y
1,1,1-Trichloroethane	Y	Y	Y	Y
1,1,2-Trichlroethane		Y	Y	Y
Trichloroethene (trichloroethylene)	Y	Y	Y	Y
Trichlorofluoromethane		Y		
Trichloromethane				Y
Trichlorophenols			Y	
2,4,6-Trichlorophenol		Y		
1,1,2-Trichlorotrifluoroethane			Y	
Trifluralin			Y	Y
Triphenyltin acetate	Y[d]		Y	Y
Triphenyltin chloride	Y[d]		Y	Y
Triphenyltin hydroxide	Y[d]		Y	Y
Vanadium				Y
Vinyl chloride		Y	Y	Y
Waste crankcase oils	Y			
Xylenes	Y		Y	Y
Zinc		Y		Y

[a] Not named specifically, but included under 'polycyclic aromatic hydrocarbons (PAH)'.
[b] Aroclors 1016, 1221, 1232, 1242, 1248, 1254, 1260 only.
[c] 2,3,7,8-Tetrachlorodibenzo-*p*-dioxin (TCDD) only.
[d] Specified as 'organotins (non-pesticide use)'.
[e] 1,2,4,5-Tetrachlorobenzene only.

Table 16.6 Chemicals common to priority lists for Canada, the USA, the UK and the European Community

Metals	Aromatic hydrocarbons	Aliphatic hydrocarbons
Arsenic	Benzene	1,2-Dichloroethane
Cadmium	Hexachlorobenzene	Dichloromethane
	(PAH)	1,1,1-Trichloroethane
	Toluene	Trichloroethene

Table 16.7 Chemicals most commonly featuring on spills priority lists in North America

Acetaldehyde	Methanol
Acetone	Methyl ethyl ketone
Acrolein	Methyl methacrylate
Acrylonitrile	Naphthalene
Aldrin	Nitric acid
Ammonia	Nitrobenzene
Aniline	Parathion
Benzene	Pentachlorophenol
Carbon disulphide	Phenol
Carbon tetrachloride	Phosgene
Chlordane	Phosphoric acid
Chlorine	Phosphorus
Chloroform	Propylene oxide
Chloromethane	Sodium cyanide
Cyclohexane	Sodium hydroxide
Dichloromethane	Styrene
Ethylbenzene	Sulphuric acid
Ethylene oxide	Tetraethyl lead
Formaldehyde	Toluene
Hydrochloric acid	1,1,1-Trichloroethylene
Hydrocyanic acid	Vinyl acetate
Hydrofluoric acid	

These weaknesses must be recognized when applying standards, which must constantly be reviewed and questioned (Chapman, 1991).

16.4.3 Standards for priority chemicals

Tables 16.8 and 16.9 summarize some of the environmental quality standards and emission standards, respectively, that are applied by Canada, the USA, the UK and the European Community to the nine specific chemicals present on all of their priority lists, as identified in Section 16.3. These standards are summarized where appropriate for water, land and air.

The most obvious feature of these two tables is that some of the nine chemicals have more standards applied to them than others, depending on their patterns of use and release, and the extent to which they have been considered priorities in the past. The tables also show that different standards designed to protect the same environmental compartment actually vary widely. The distinction between guideline and mandatory values has already been discussed, but other variables include compliance criteria, dependence on water hardness, and the actual standards themselves. The way a standard is expressed (i.e. based on dissolved or total chemical, hardness-dependent, and so on) and the actual value apportioned to it depend on the method used to derive it, and this is too large a consideration to be dealt with here. Differences in compliance criteria are considered below.

Compliance criteria

The character of a standard is dependent on the degree of compliance required. Compliance criteria usually take the form of a percentile; typically 100 (equivalent to a maximum acceptable concentration), 99, 95 or 50 (equivalent to an annual average concentration). The US EPA set standards which should not be exceeded more than once every three years, which is assumed to be a suitable recovery period should there be any adverse effects on aquatic communities (EPA, 1985). Although apparently quite a stringent approach, there is limited evidence supporting this recovery period.

Extreme percentiles such as 100 and 99 are not ideal: sampling and analysis errors mean that 100% compliance cannot be proven (Agg & Zabel, 1989); and although 99 percentile suggests a high level of confidence, in reality it is a statistic that cannot be estimated precisely (Berthouex & Hau, 1991). Use of lower percentiles such as 95 or 50 allows for greater precision of the estimate. For substances whose levels fluctuate in the long term but may remain relatively constant for extended periods, 95 percentile is more suitable, because compliance will ensure that high concentrations will not be sustained. For substances whose levels fluctuate in the short term, as long as there is a reasonably large margin between acute and

Table 16.8 Environmental quality standards for nine priority chemicals

					References
Arsenic					
Water	50–100	µg/l	95%	EC mandatory, potable abstraction, depending on level of treatment	1
	50	µg/l	AA	UK mandatory, freshwaters	2
	25	µg/l	AA	UK mandatory, saltwaters	2
	190	µg/l	4 days	USA mandatory, freshwaters	8
	360	µg/l	1 h	USA mandatory, freshwaters	8
	36	µg/l	4 days	USA mandatory, saltwaters	8
	69	µg/l	1 h	USA mandatory, saltwaters	8
	50	µg/l	MAC?	Canadian guideline, freshwaters	9
Land	50	mg/kg	MAC	UK mandatory, agricultural soil (dry weight)	3
Benzene					
Water	50	µg/l	AA	UK guideline, freshwaters and saltwaters	a
	300	µg/l	MAC?	Canadian guideline, freshwaters	9
Cadmium					
Water	5	µg/l	95%	EC mandatory, potable abstraction	1
	5	µg/l	AA	EC mandatory, freshwaters	4
	2.5	µg/l	AA	EC mandatory, estuaries	4
	2.5	µg/l	AA	EC mandatory, saltwaters	4
	0.66	µg/l	4 days	USA mandatory, freshwater hardness 50 mg $CaCO_3$/l	8
	2	µg/l	4 days	USA mandatory, freshwater hardness 200 mg $CaCO_3$/l	8
	1.8	µg/l	1 h	USA mandatory, freshwater hardness 50 mg $CaCO_3$/l	8
	8.6	µg/l	1 h	USA mandatory, freshwater hardness 200 mg $CaCO_3$/l	8
	9.3	µg/l	4 days	USA mandatory, saltwaters	8
	43	µg/l	1 h	USA mandatory, saltwaters	8
	0.2	µg/l	MAC?	Canadian guideline, freshwater hardness <60 mg $CaCO_3$/l	9
	1.8	µg/l	MAC?	Canadian guideline, freshwater hardness >180 mg $CaCO_3$/l	9
Land	3	mg/kg	MAC	UK mandatory, agricultural soil (dry weight)	3
1,2-Dichloroethane					
Water	10	µg/l	AA	EC mandatory, freshwaters and saltwaters	5
Dichloromethane					
Hexachlorobenzene					
Water	30	ng/l	AA	EC mandatory, freshwaters and saltwaters	6
	6.5	ng/l	MAC?	Canadian guideline, freshwaters	9

Continued

Table 16.8 *continued*

Toluene					
Water	40	µg/l	AA	UK guideline, freshwaters and saltwaters	a
	300	µg/l	MAC?	Canadian guideline, freshwaters	9
Air	750	mg/m^3	MAC	USA mandatory, occupational exposure	7
1,1,1-Trichloroethane					
Trichloroethene					
Water	10	µg/l	AA	EC mandatory, freshwaters and saltwaters	5

AA, Annual average concentration; MAC, maximum acceptable concentration; MAC?, unspecified, but thought to be MAC; 95%, 95 percentile concentration; 4 days, 4-day average not to be exceeded more than once every 3 years; 1 h, 1-hour average not to be exceeded more than once every 3 years.
References: 1, CEC (1975); 2, DoE (1989b); 3, HMSO (1989a); 4, CEC (1983); 5, CEC (1990); 6, CEC (1988); 7, Hooftman & Janssen (1985); 8, EPA (1986); 9, CCREM (1991); a, unpublished provisional standard.

chronic effects, 50 percentile compliance is adequate.

As noted in Table 16.8, environmental quality standards specified by CCREM (1991) do not have the attendant compliance criteria reported with them. They all appear to be maximum allowable concentrations (unverified personal communication), but by not specifying this in the implementation document there is much potential for confusion and problems associated with monitoring. It is important that all standards are properly defined, so that both the regulator and those being regulated know what is expected of them.

16.4.4 Problems associated with monitoring and analysis for compliance

Monitoring for compliance

The implementation of effective and affordable environmental monitoring is not simple. Sampling strategies need to account for different environmental media that may be contaminated and different patterns of contamination. For example, in surface waters there is the water column, suspended solids, sediments and various biota to consider, and contaminants may enter from diffuse or point sources on a continuous or intermittent basis. One approach is to predict the environmental compartment in which the contaminant is most likely to be a problem and then

selectively monitor this. Thus, considering our nine priority chemicals in surface waters, As, Cd and hexachlorobenzene are most likely to be a problem in sediment and biota, benzene and toluene in sediment, and the aliphatics in the water column (Chapman *et al.*, 1982). This approach should be used with caution because of the possibility of getting the prediction wrong and obtaining misleading monitoring data. Thus it should only be used as a framework for monitoring programmes, and should always be backed up by periodically taking some samples from all partitions.

Taking account of these problems, there should be three phases to any monitoring programme for priority chemicals (Keith & Telliard, 1979). The initial screening phase need only be semi-quantitative as it is only necessary to give an indication of the level (if any) of contamination by individual priority chemicals. Screening allows the identification of particular discharges or sites which apparently have significant levels of priority contaminants, and which therefore require further investigation. More precise analysis is then needed in the confirmation phase to establish a baseline level of contamination against which future changes can be measured. Once this has been carried out, actual monitoring for changes in contaminant levels can begin. It is important that all through this process some relatively clean emissions or uncontaminated sites are also included, to provide a baseline or control level.

Table 16.9 Emission-standards for nine priority chemicals

					References
Arsenic					
Water	1	mg/l	Monthly	EC guideline	1
Land	2	mg/kg		UK guideline, sludge (dry weight) applied to land	2
Air	1	mg/m^3		EC mandatory, new incinerators	3
Benzene					
Cadmium					
Water	0.2	mg/l	Monthly	EC mandatory, various industries	4
Land	20	mg/kg		EC sludge, applied to land	5
	1	mg/kg		EC sludge (dry weight) applied to land	6
Air	0.2	mg/m^3		EC mandatory, new incinerators	3
	0.012–0.07	g/m^3	MAC	UK guideline value	7
1,2-Dichloroethane					
Water	0.1–6	mg/l	Monthly	EC mandatory, industry-specific	8
	0.2–12	mg/l	Daily	EC mandatory, industry-specific	8
Dichloromethane					
Hexachlorobenzene					
Water	1–1.5	mg/l	Monthly	EC mandatory	9
	2–3	mg/l	Daily	EC mandatory, industry-specific	9
Toluene					
1,1,1-Trichloroethane					
Trichloroethene					
Water	0.1–2	mg/l	Monthly	EC mandatory, industry-specific	8
	0.2–4	mg/l	Daily	EC mandatory, industry-specific	8

MAC, Maximum acceptable concentration; Monthly, monthly average concentration; Daily, daily average concentration.
References: 1, de Bruin (1980); 2, HMSO (1989a); 3, CEC (1989a,b); 4, CEC (1983); 5, CEC (1986) and COM (1988); 6, HMSO (1989b); 7, HMSO (1988); 8, CEC (1990); 9, CEC (1988).

Analysing for compliance

Analysis of samples can also be problematic, with different methods needed for the range of widely variable chemicals being measured and many analyses being time consuming and expensive. Particular problems are posed when generic groups of chemicals, such as PAHs, are identified as priorities because of the large number of chemicals in such groups (Keith & Telliard, 1979). This is one reason why imprecise priority lists containing such groupings (such as the Oslo Commission 'black' list) can only be used effectively as precursors for more specific priority lists.

For certain chemicals the standards set are below currently achievable detection limits. This problem is exacerbated in heavily contaminated water where more compounds introduce more analytical interference, and for some priority chemicals compliance with standards is difficult or impossible to prove. In such cases, rather than setting standards that are environmentally meaningful but cannot be monitored, should we simply set standards at the limit of detection? In reality we currently have no choice but to accept this position, but we also should be striving to improve our analytical capability so that we can monitor the environmentally desirable standards. In essence we are again adopting the dual standard approach outlined in Section 16.4.1, with somewhat relaxed standards being necessary at present but more stringent ones being used to add impetus to technological (in this case analytical) improvement.

Table 16.10 lists some available analytical techniques for our nine priority chemicals, again concentrating on water. For other phases, particularly solids, identification and quantification of chemicals are similar but the sensitivity is dependent on the sample extraction efficiency. Arsenic, for example, can be analysed to 0.5 mg/kg in solids by atomic absorption spectrophotometry of ashed samples (HMSO, 1987), but clearly such a method could not be used for organics. The US EPA has specified analytical techniques for each of the nine chemicals in water, but where these are not sufficiently sensitive to monitor all the standards summarized in Tables 16.8 and 16.9 then alternative methods are included.

Inorganics

EPA method 200.7 is suitable for the determination of dissolved, suspended or total arsenic and cadmium in drinking water, surface water and wastewaters (EPA, 1984). Briefly, it involves appropriate preparation of the sample (depending on what is being measured) followed by inductively coupled plasma–atomic emission spectrometry for detection. However, as noted in Table 16.10, the limits of detection for arsenic and cadmium are higher than some of the existing quality standards.

For cadmium a more sensitive technique has been reported for fresh and potable waters, involving direct analysis by electrothermal atomization–atomic absorption spectrophotometry (HMSO, 1986). The only pretreatment required is filtration of raw water samples, addition of a lanthanum salt to minimize interference, and acidification of the sample. This technique has a lower detection limit of $0.1-0.2\,\mu g/l$, and even this is only just capable of monitoring the lowest reported standard in Table 16.8, namely the Canadian guideline value for soft waters of $0.2\,\mu g/l$. In reality the errors expected around the lower limit of detection will mean that compliance with the Canadian standard could not be proven with this method.

Atomic absorption spectrophotometry is also applicable to the analysis of arsenic in water: inorganic arsenic V is reduced to arsine, which is then either trapped and converted to arseno-molybdenum blue or decomposed to elemental

Table 16.10 Analytical techniques for nine priority chemicals

	EPA method[a]	Limit of detection ($\mu g/l$)	Alternative method	Limit of detection ($\mu g/l$)
Arsenic	200.7	53[b]	HMSO (1978)	0.19[c]
Benzene	602	0.2		
Cadmium	200.7	4[b]	HMSO (1986)	0.1–0.2
1,2-Dichloroethane	601	0.03		
Dichloromethane	601	0.25		
Hexachlorobenzene	612	0.05[b]	WRc, unpublished[d]	0.002
Toluene	602	0.2		
1,1,1-Trichloroethane	601	0.03		
Trichloroethene	601	0.12		

[a] EPA (1984).
[b] Inadequate for monitoring the standards summarized in Table 16.8.
[c] For seawater.
[d] Based on HMSO (1984).

arsenic, either of which can then be measured spectrophotometrically (HMSO, 1978, 1982). With detection limits of 0.19 and 0.055 µg/l reported for seawater and potable water, respectively, this method appears to be adequate for monitoring all the standards for arsenic reported in Table 16.8.

Organics

EPA methods 601 and 602 are appropriate for purgeable halogenated hydrocarbons and purgeable aromatics, respectively, in clean or dirty water (EPA, 1984). Chemicals are purged by an inert gas bubbled through the sample, trapped in a sorbent trap and desorbed onto a gas chromatograph (GC) column for quantification; method 601 employs a halide-specific detector and method 602 a photoionization detector. Method 624 is similar but specifies GC-mass spectrometry (MS), while method 1624 involves inclusion of deuterium-labelled analogues as internal standards. Method 612 for chlorinated hydrocarbons is also a GC method, but uses solvent extraction rather than a purge-and-trap technique, and an electron-capture detector for quantification. Methods 625 and 1625 are the analogous GC-MS and radiolabel techniques, respectively.

Since their publication these EPA '600 series' methods have been optimized to give up to 100% recovery from samples (Michael *et al.*, 1988). Although some of the details of these analyses are slightly outdated, for example using packed columns rather than capillary columns (which give better resolution), the basic methods are still appropriate. As shown in Table 16.10, they are adequate for monitoring all the reported standards except for those for hexachlorobenzene, where the limit of detection is above both the Canadian and EC standards. For hexachlorobenzene a detection limit of 2 ng/l is achievable in 'clean' water, using a method based on the standard UK analysis (HMSO, 1984; WRc, unpublished). This involves liquid–liquid extraction and analysis by capillary column GC using an electron-capture detector. This method is therefore adequate for monitoring the standards for hexachlorobenzene reported in Table 16.10.

16.5 ISSUES FOR THE FUTURE

16.5.1 Prioritization, standards and control

Prioritization serves only to identify those chemicals that pose a particular threat to the environment and is pointless if no, or inadequate, measures are taken to control their release. Although some progress has been made (such as widespread restrictions on the use and disposal of polychlorinated biphenyls – PCBs), in many instances it would appear that controls have been too lax, too slow being implemented, or simply ignored. We must ensure that priority lists are not developed for their own sake, but are followed up with appropriate control measures both now and in the future.

16.5.2 Prioritization and data availability

Prioritization schemes, whether simple or complex, are only as good as the data they use, and for most chemicals these are inadequate for a reliable assessment (Agg & Zabel, 1990). Some workers have attempted to overcome this by improving the ability to predict missing data, such as environmental concentrations. Many advanced models are available, for example based on Mackay's fugacity and equivalence concepts (Chapter 15), but these still need validation in the environment before they can be used with confidence, and this has so far only been carried out on an *ad-hoc* basis. Additionally, models or predicted data should be used with caution, and only be applied to the situation they were originally intended for (Rohleder *et al.*, 1985).

It has long been recognized that we need to establish testing priorities for generating further data (Cairns *et al.*, 1979), but progress in this area remains slow. An important step in this process is to overcome the negative defensive attitude of industry, as they will then be more inclined to release existing data for use in prioritization and standards setting. We also need to be sure that new data are reliable by continuing to develop validated methodologies for generating (eco)toxicity and environmental fate data (Branson, 1980), in particular for the atmosphere and soil for which no established tests exist (Vosser & Wilson, UK

DoE, personal communication). Additionally, we need to better understand the implications associated with chemical interactions and degradation products in the environment before we can fully assess chemicals' hazard potentials.

It will never be possible to develop a perfect prioritization procedure, but if existing priority schemes could be operated on a good set of data their results could be compared and priorities established by commonality of the results.

Standards established for priority chemicals are very variable, depending primarily on the responsible authority's policy for setting them. We must be sure that however they are expressed, they are as environmentally meaningful and as accurate as possible. At present there is much uncertainty associated with standards: for example, environmental quality standards for water should be based on a concentration known to have no adverse effects on a range of aquatic ecosystems, but in reality are extrapolated from laboratory studies on a limited range of species tested in isolation. The basic problem is an inadequacy of relevant data, and again we should not expend effort refining our procedures for establishing standards at the expense of obtaining these data.

16.5.3 Monitoring and analytical capability

As has been demonstrated, our ability to analyse for chemicals in the environment is not always adequate for monitoring standards that may be set, and therefore we should be striving to improve it. This includes a consideration of cost-effectiveness, because as more chemicals are identified as priorities, monitoring costs will increase dramatically. This problem could be addressed by identifying monitoring sites that are representative of overall patterns of release of priority chemicals, and also account for a known, high, proportion of the total loadings (Zabel, 1988).

16.5.4 Treatment technology

To be able to meet the ever-increasing number of standards we must continually strive to improve our waste-treatment technology for existing industry, such as through increased recycling, and to develop cleaner technology in new areas. In many cases one industry's waste could be another's resource, but without co-operation will remain waste. Improvements have been facilitated by the recent application of ideas such as the 'polluter pays principle' (OECD, 1975), integrated pollution control using best available techniques (DoE, 1990; HMSO, 1990) and agencies co-ordinating waste re-use and recycling (e.g. the commercial Waste Exchange Services, based in Cleveland, UK). These and similar policies need to be adopted on a wider scale to assist both industry and the regulatory authorities to minimize the risk of adverse environmental effects.

16.5.5 Harmonization and education

Information sharing and harmonization of priority schemes and standards setting have been carried out only on a limited basis to date. If increased, it could reduce duplication of effort, reduce costs and accelerate the implementation of controls. Each of the issues considered in this chapter (data generation, prioritization schemes, standards setting, analytical techniques, and treatment and control technology), tends to be handled separately not only by different countries but by different authorities in the same country. Some advances have been made, with the revision of the World Health Organization's drinking water guidelines being a good example of international co-operation (Fawell, 1991), but the involvement of international agencies and communication between authorities need to be increased, and national approaches to chemical control need to reflect the needs of international harmonization (Schmidt-Bleek & Hamann, 1985). Much effort is currently being expended on refining classification schemes, making them more complex so as to better represent the environmental situation. In contrast, a harmonized classification scheme would need to be as simple as possible, so as to be widely accepted and universally applicable (Bro-Rasmussen, 1985). Furthermore, it is more likely that a simple scheme could be operated with the relatively limited data currently available.

There is a related need to ensure that developing nations are not exploited as markets for chemicals we no longer consider suitable for our own use, and are assisted with their own chemical prioritization and control. The Canadian Environmental Protection Act bans the export of any substances prohibited in Canada, and those that are severely restricted can be exported only on agreement of both governments involved (Toft & Hickman, 1990). This policy should be adopted in all developed countries.

16.5.6 Handling uncertainty and predicting problems

The identity, quantity, patterns of release and distribution of chemicals are poorly documented, leading to uncertainty in hazard assessment and prioritization (Branson, 1980), and we need to understand and reduce this uncertainty.

Rippen, Frank & Zietz (1985) reported that a chemical's use type could be predicted entirely from the chemical's functional groups (e.g. intermediates have reactive functional groups such as $-NH_2$ or $-Cl$, dyes have conjugated double-bond systems, and so on). Each use type could then be associated with a use pattern, each having a known likelihood of release (assuming a worst-case scenario):

Destructive (e.g. fuels)	1–10%
Contained (e.g. catalysts)	1–10%
Open non-dispersive (e.g. cutting fluids)	10–100%
Open dispersive (e.g. plasticizers)	100%
Direct environment (e.g. pesticides)	100%

Thus chemicals for which there were no relevant data could be provisionally assessed for their likely hazard in order to identify whether they warrant further study.

The concept of ecological or biological modelling has been developed over a number of years, but few good models exist. If reliable, accurate models could be developed, they could be used to identify the particular relationships between the properties, distribution and effects of chemicals that make them particularly hazardous. This could then be used for screening both existing and new chemicals, in order to identify those warranting further study.

Methods such as these need to be evaluated and developed as a means of predicting potential hazards, but should only be a temporary substitute for deriving actual data for candidate substances, because there will always be uncertainty associated with them.

16.6 SUMMARY

Future prioritization and standard setting needs the following essential actions to be reliable and valid.

1 Identify characteristics of hazardous chemicals.
2 Screen using existing or surrogate data to identify potential priorities.
3 Intensify efforts to obtain essential data.
4 Establish priorities by using a number of schemes or decision criteria and comparing the results.
5 Determine environmentally meaningful standards.
6 Implement controls to achieve the standards, and ensure compliance by improving on current technology where necessary.

REFERENCES

Agg, A.R. & Zabel, T.F. (1989) Environmental quality objectives and effluent control. *Chem. Ind.* 17 July 1989, 443–447.

Agg, A.R. & Zabel, T.F. (1990) Red-List substances: selection and monitoring. *J. Inst. Water Environ. Manage.* **4**(1), 44–50.

BCW (1989) A system of criteria for the designation of substances from the black list. Working Group on Criteria for the Assessment of water pollution (BCW), Ministries of Housing, Physical Planning and the Environment (VROM) and of Transport and Public Works (V and M), the Netherlands.

Berthouex, P.M. & Hau, I. (1991) Difficulties related to using extreme percentiles for water quality regulations. *Res. J. Water Pollut. Contr. Fed.* **63**(6), 873–879.

Bjerre, F. & Hayward, P.A. (1984) The role and activities of the Oslo and Paris Commissions. In: *Environmental Protection: Standards, Compliance and Costs* (Ed. T.J. Lack), pp. 142–157. Ellis Horwood, Chichester.

BMU (1987) Directives for applications for the classification of water-endangering substances in the meaning of Sub-section 19g of the Water Conservation Law (WHG). Federal Office for the Environment.

Branson, D.R. (1980) Prioritization of chemicals according to the degree of hazard in the aquatic environment. *Environ. Health Perspect.* **34**, 133–138.

Bro-Rasmussen, E. (1985) A proposal for an EEC-scheme for definition and classification of substances as dangerous for the environment. In: *Environmental Modelling for Priority Setting among Existing Chemicals.* pp. 437–445. Proceedings of workshop, 11–13 November. München-Neuherberg.

de Bruin, J. (1980) Elimination of the pollution of the aquatic environment by the carcinogenic substances benzidine, arsenic and inorganic compounds of arsenic. CEC report no. ENV/84/81.

Cairns, J.C. (1986) The myth of the most sensitive species; multispecies testing can provide valuable evidence for protecting the environment. *BioScience* **36**(10), 670–672.

Cairns, J.C., Dickson, K.L. & Maki, A.W. (1979) Estimating the hazard of chemical substances to aquatic life. *Hydrobiologia* **64**(2), 157–166.

CCREM (1991) *Canadian Water Quality Guidelines*, 3rd update. Environment Canada, Ottawa.

CEC (1975) Directive concerning the quality required of surface water intended for the abstraction of drinking water in the Member States (75/440/EEC). *Offic. J.* **L194/39**, 25 July.

CEC (1976) Council Directive on pollution caused by certain dangerous substances discharged into the aquatic environment of the Community. (76/464/EEC.) *Offic. J.* **L129**, 18 May.

CEC (1983) Directive on limit values and quality objectives for cadmium discharges (83/513/EEC). *Offic. J.* **L291/1**, 24 October.

CEC (1986) Directive on the protection of the environment, and in particular the soil, when sewage sludge is used in agriculture (86/278/EEC). *Offic. J.* **L181**, 4 July.

CEC (1988) Directive amending annex II to Directive 86/280/EEC on limit values and quality objectives for discharges of dangerous substances included in List I of the annex to Directive 76/464/EEC (88/347/EEC). *Offic. J.* **L158/35**, 25 June.

CEC (1989a) Directive on the prevention of air pollution from new municipal waste incineration plants (89/369/EEC). *Offic. J.* **L163**, 14 June.

CEC (1989b) Directive on the reduction of air pollution from existing municipal waste incineration plants (89/429/EEC). *Offic. J.* **L203**, 15 July.

CEC (1990) Directive amending annex II to Directive 86/280/EEC on the limit values and quality objectives for discharges of certain dangerous substances included in List I of the annex to Directive 76/464/

EEC (90/415/EEC). *Offic. J.* **L219**, 14 August.

CEESC (1986) Methode guide de selection des substances a inclure en list I, dans le cadre de la directive 76/464/CEE. Direction de la Prevention des Pollutions, Mission du Controle des Produits.

CEFIC (1987) Position paper on the criteria for the selection of substances as candidates for list I under Community Directive 76/464/EEC. CEFIC, Brussels.

Chapman, P.M. (1991) Environmental quality criteria: What type should we be developing? *Environ. Sci. Technol.* **25**(8), 1353–1359.

Chapman, P.M., Romberg, G.P. & Vigers, G.A. (1982) Design of monitoring studies for priority pollutants. *J. Water Pollut. Contr. Fed.* **54**(3), 292–197.

COM (1988) Proposal for a Directive modifying Directive 86/278/EEC on the protection of the environment, and in particular the soil, when sewage sludge is used in agriculture. COM(88) 624 final. *Offic. J.* **C319**, 2 December.

COM (1990) List of classified carcinogens, suspected carcinogens under discussion, suspected carcinogens on priority list as at 15 February 1990. Working Group 'Dangerous Substances', report number XI/138/90.

Crookes, M.J. & Nielsen, I.R. (1989) A comparative study of priority setting schemes for existing chemicals. Building Research Establishment of the DoE, report number BRE 125/3/2.

Dieter, H.H., Kaiser, U. & Kerndorff, H. (1990) Proposal on a standardized toxicological evaluation of chemicals from contaminated sites. *Chemosphere* **20**(1–2), 75–90.

DoE (1986) Selection of substances discharged to water as candidates for List I status. Water Quality Division, WQD41P385-IRH7.

DoE (1988) Inputs of dangerous substances to water: proposals for a unified system of control. The Government's consultative proposals for tighter controls over the most dangerous substances entering the aquatic environment ('the Red List'). DoE/Welsh Office, July.

DoE (1989a) The red list selection procedure. August.

DoE (1989b) Water and the environment. The implementation of European Community directives on pollution caused by certain dangerous substances discharged into the aquatic environment. Circular 7/89 (DoE) 16/89 (Welsh Office), 30 March.

DoE (1990) Integrated pollution control—a practical guide. Guidance issued by the Department of the Environment and the Welsh Office.

DoE (1991) The environmental protection (prescribed processes and substances) regulations 1991. Statutory Instrument no. 472.

Environment Ontario (1989) The effluent monitoring priority pollutants list, 1988 update. Hazardous Contaminants Co-Ordination Branch, Environment

Ontario.

EPA (1984) Guidelines establishing test procedures for the analysis of pollutants under the Clean Water Act; final rule and interim final rule and proposed rule. *Fed. Reg.* **49**(209), 43234–43442.

EPA (1985) Guidelines for deriving numerical national water quality criteria for the protection of aquatic organisms and their uses. PB85–227049, US EPA, Washington, January.

EPA (1986) Quality criteria for water 1986. EPA report no. 440/5–86–001.

Fawell, J.K. (1991) Developments in health-related quality standards for chemicals in drinking water. *J. Inst. Water Environ. Manage.* **5**(5), 562–565.

Fingas, M., Laroche, N., Sergy, G., Mansfield, B., Cloutier, G. & Mazerolle, P. (1991) A new chemical spill priority list. In: *Proceedings of the 8th Technical Seminar on Chemical Spills,* pp. 223–332. Vancouver, British Columbia, 10–11 June 1991. Minister of Supply and Services, Canada.

Freitag, D. (1987) Environmental hazard profile of organic chemicals. An experimental method for the assessment of the behaviour of organic chemicals in the environment. *Chemosphere* **16**(2–3), 589–598.

Hedgecott, S. & Cooper, M. (1991) Priority setting scheme for potentially hazardous substances in the aquatic environment in the UK. WRc report number DoE 2665-M/1.

HMSO (1978) Arsenic in potable and sea water by spectrophotometry (arsenomolybdenum blue procedure) 1978. Methods for the examination of waters and associated materials. HMSO, London.

HMSO (1982) Arsenic in potable waters by absorption spectrophotometry (semi-automatic method) 1982. Methods for the examination of waters and associated materials. HMSO, London.

HMSO (1984) The determination of organochlorine insecticides and polychlorinated biphenyls in sewages, sludges, muds and fish 1978. Organochlorine insecticides and polychlorinated biphenyls in water, an addition 1984. Methods for the examination of waters and associated materials. HMSO, London.

HMSO (1986) Lead and cadmium in fresh waters by atomic absorption spectrophotometry, 2nd edn. A general introduction to electrothermal atomization atomic absorption spectrophotometry 1986. Methods for the examination of waters and associated materials. HMSO, London.

HMSO (1987) Selenium in waters 1984. Selenium and arsenic in sludges, soils and related materials 1985. A note on the use of hydride generator kits 1987. Methods for the examination of waters and associated materials. HMSO, London.

HMSO (1988) Best practicable means: general principles and practice. Notes on best practicable means BPM1, January 1988. HMSO, London.

HMSO (1989a) Sludge (use in agriculture) regulations 1989. Statutory Instrument No. 1263. HMSO, London.

HMSO (1989b) Code of practice for the agricultural use of sewage sludge. HMSO, London.

HMSO (1990) Environmental Protection Act 1990. Chapter 43. HMSO, London.

Hooftman, R.N. & Janssen, J.M.A. (1985) Evaluation of the impact of toluene on the aquatic environment. TNO report no. R 86/185a; CEC report no. XI/538/87.

Jackson, J. & Peterson, P.J. (1989) Evaluation of selection schemes for identifying priority aquatic pollutants. Final Report-Study Contract No. B 6612/290/89 For the EEC Commission/DG XI, M.A.R.C., Campden Hill, London W8 7AD.

Keith, L.H. & Telliard, W.A. (1979) Priority pollutants. I—A perspective view. *Environ. Sci. Technol.* **13**(4), 416–423.

Klein, W., Geyer, H., Freitag, D. & Rohleder, H. (1984) Sensitivity of schemes for ecotoxicological hazard ranking of chemicals. *Chemosphere* **13**(1), 203–211.

Mestres, R. (1979) Rapport sur la classification pratique des substances de la Liste I (Directive 76/464/CEE). Laboratoire de Chimie appliquée à l'expertise Faculté de Pharmacie de Montpellier.

Michael, L.C., Pellizzari, E.D. & Wiseman, R.W. (1988) Development and evaluation of a procedure for determining volatile organics in water. *Environ. Sci. Technol.* **22**, 565–570.

de Nijs, A.C.M., Knoop, J.M. & Vermeire, T.G. (1988) Risk assessment of new chemical substances. System realisation and validation. National Institute of Public Health and Environmental Protection, Bilthoven, Netherlands, report no. 718703001.

OECD (1975) *The Polluter Pays Principle.* OECD, Paris.

Rippen, G., Frank, R. & Zietz, E. (1985) Priority setting among existing chemicals by means of use pattern and production volume. In: *Environmental Modelling for Priority Setting among Existing Chemicals,* pp. 482–490. Proceedings of workshop, 11–13 November, München-Neuherberg.

Rohleder, H., Münzer, B. & Voigt, K. (1985) E4CHEM (Exposure and Ecotoxicity Estimation for Environmental CHEMicals)—a computerized aid for priority setting. In: *Environmental Modelling for Priority Setting among Existing Chemicals,* pp. 491–525. Proceedings of workshop, 11–13 November, München-Neuherberg.

Sampaolo, A. & Binetti, R. (1989) Improvement of a practical method for priority selection and risk assessment among existing chemicals. *Regul. Toxicol. Pharmacol.* **10**, 183–195.

Schmidt-Bleek, F. & Haberland, W. (1980) The yardstick concept for the hazard evaluation of substances. *Ecotoxicol. Environ. Safety* **4**, 455–465.

Schmidt-Bleek, F. & Hamann, H.-J. (1985) Priority setting among existing chemicals for early warning. In: *Environmental Modelling for Priority Setting among Existing Chemicals*, pp. 455–464. Proceedings of workshop, 11–13 November, München-Neuherberg.

Schulze, H. & Mücke, W. (1985) Recent approaches to selecting chemicals for priority lists. In: *Environmental Modelling for Priority Setting among Existing Chemicals*, pp. 446–449. Proceedings of workshop, 11–13 November, München-Neuherberg.

Second International Conference on the Protection of the North Sea (1987) Ministerial Declaration, London, 24–25 November.

Slob, W. & de Nijs, A.C.M. (1989) Risk assessment of new chemical substances. Quantification of uncertainty in estimated PEC-values of aquatic systems. National Institute of Public Health and Environmental Protection, Bilthoven, Netherlands, report no. 958804001.

Smith, P.J. & McVeagh, M. (1991) Widespread organotin pollution in New Zealand coastal waters as indicated by imposex in dogwhelks. *Mar. Pollut. Bull.* **22**(8), 409–413.

SRI International (1980) Elaboration of a method for evaluating the hazard to the aquatic environment caused by substances of list I of Council Directive 76/464/EEC and preparation of a list of dangerous substances to be studied by priority. Commission of the European Communities report number ENV/786/80, September.

Third International Conference on the Protection of the North Sea (1990) Final declaration.

Toft, P. & Hickman, J.R. (1990) Canadian approach to chemical safety. *Proceedings of the 4th Conference on Toxic Substances*, 4–5 April Montreal, Quebec.

Veith, G.D., Austin, N.M. & Morris, R.T. (1979) A rapid method for estimating log P for organic chemicals. *Water Res.* **13**, 43–47.

Zabel, T.F. (1988) Cost estimate for monitoring river and direct sewage discharges to UK estuarine and coastal waters. WRc report number PRS 1760-M/1.

Glossary

This glossary defines key terms, abbreviations, acronyms and symbols that occur in the text. Terms that are used exclusively in a particular chapter and defined there are not generally included. Neither, in general, are terms already covered in the glossary of Volume 1 (pp. 461–465).

AAS	Atomic absorption spectrophotometry
AChE	Acetylcholinesterase
AE	Alcohol ethoxylates
AES	Atomic emission spectrophotometry. Also alkyl ether sulphates or alkyl (or alcohol) exthoxy sulphates
AHH	Aryl hydrocarbon hydroxylase
Alkyl	Monovalent hydrocarbon group, i.e. $-C_nH_{2n+1}$
AnS	Anionic sulphonates
AOS	Alpha-olefine sulphonates
AP	Alkylphenol
APDC	Ammonium pyrrolidine dithiocarbonate
APE	Alkyphenol ethoxylates; 1EO, monoethoxylate; 2EO, diethoxylate
APHA	American Public Health Association
Application factor	see Safety factor
Aryl	Organic group produced by removing one hydrogen atom from an aromatic hydrocarbon
AS	Alkyl (or alcohol) sulphates
ASV	Air saturation value for oxygen. Also anodic stripping voltammetry
ATMAB	Alkyltrimethyl ammonium bromides
ATMAC	Alkyltrimethyl ammonium chlorides
AVS	Acid volatile sulphide
BAF	Bioaccumulation factor
BCF	Bioconcentration factor
BCR	Community Bureau of Reference of the EC
BCW	Dutch Working Group on Criteria for the Assessment of Water Pollution
BHC	Benzene hexachloride. Also 1,2,3,4,5,6-hexachlorocyclohexane
BiAS	Bismuth iodide active substances
BMU	Bundesministerium für Umwelt, Naturshutz und Reaktorsicherheits (Federal Ministry for the Environment, Nature Protection and Reactor Safety)
BTEX	Gasoline components: benzene, toluene, ethyl benzene and xylenes
C	Concentration
CAS	Continuous activated sludge test (see also SCAS)
CCREM	Canadian Council of Resource and Environment Ministers
CEC	Commission of the European Community. Also used for Coordinating European Council for the Development of Performance Tests for Lubricants and Engine Fuels (restricted to Chapter 10)
CEESC	Commission d'Evaluation de l'Ecotoxicité des Substances Chimiques (Commission for the Evaluation of Ecotoxicity of Chemicals)
CEFIC	Conseil European des Federations de l'Industrie Chimique (European Council of Chemical Manufacturers' Federations)
ChE	Cholinesterase
Chlp	Chlorinated hydrocarbon pesticides
C_m	Dipole moment
CMT	Carcinogenic, mutagenic, teratogenic
COD	Chemical oxygen demand
Com	Document/memorandum from European Commission; published in Official Journal (OJ)
CONCAWE	Conservation of Clean Air and Water in Europe; oil companies' organization

CS	Cationic surfactants	FAB	Fast atom bombardment
CSV	Cathodic stripping voltammetry	FIFRA	Federal Insecticide, Fungicide and
CTAS	Cobalt thiocyanate active substances		Rodenticide Act (USA)
CWA	Clean Water Act (US)	FPC	Flame photometric detector
D	Transport and transformation rate	GC	Gas chromatography
	coefficients or discharge	GESAMP	Group of Experts on the Scientific
2,4-D	2,4-Dichlorophenoxyacetic acid		Aspects of Marine Pollution
DBAS	Disulphine blue active substances	GFAAS	Graphite furnace AAS
DCPA	Dimethyl tetrachloroterephthalate	Grignard	Has general formula RMgX, where R
DDD	Dichlorodiphenyldichloroethane	Reagent	is alky or aryl group; X is halogen
DDE	Dichlorodiphenyldichloroethylene	H	Henry's law constant (see also K^H)
DDT	Dichlorodiphenyltrichloroethane	HCB	Hexachlorobenzene
Depuration	Process that leads to elimination of a	HMIP	Her Majesty's Inspectorate of
	substance from an organism		Pollution (UK)
DIN	Deutsche Industrie Norm	HMSO	Her Majesty's Stationery Office
DO	Dissolved oxygen	HPLC	High pressure (sometimes precision)
DOE	Department of the Environment (UK		liquid chromatography
	Government)	HRMS	Double focusing sector high
DTDMAC	Ditallow dimethyl ammonium		resolution spectrometer
	chloride	I	Hydrogen bonding indicator. Also
EAAS	Electrothermal atomic adsorption		intensity of transmitted light
	spectrophotometry	IAEA	International Atomic Energy Agency
EC	European Community. Also	IC	Ion chromatography
	European Economic Community	ICES	The International Council for the
	(EEC) and European Union (EU). The		Exploration of the Seas
	latter was established under the	ICPMS	Inductively coupled plasma mass
	Treaty of Union in 1993. The EC		spectrometry
	remains as one of the 'pillars' of the	Imposex	Imposition of male sexual organs on
	EU, so both descriptions are		female
	technically correct. EU is likely to	IQAMS	Imidazolium quaternary ammonium
	increasingly replace EC. EC is also		methylsulphate
	used to refer to European	IR	Infrared
	Commission, one of the institutions	ISO	International Standards Organization
	of the EU. Also effect concentration.	IUPAC	International Union of Pure and
ECD	Electron capture detection		Applied Chemistry
EDTA	Ethylenediaminetetracetic acid	IW	Interstitial water
EDXRF	Energy-dissipative X-ray fluorescence	IWC	Instream water concentration
eec	Estimated environmental	K	Rate constant
	concentration	K_B	Bioconcentration factor (see also BCF)
EIFAC	European Inland Fisheries Advisory	K_{bw}	Biota−water partition coefficient
	Council	K_{cw}	Cuticle−water partition coefficient
ELA	Experimental Lakes Area (Canada)	K_d	Distribution coefficient
ENDS	Environmental Data Services	K^H	Henry's law constant (see also H)
EO	Ethoxylate	K_{oc}	Sediment to water partition
EPA	Environmental Protection Agency		coefficient
	(usually refers to USA)	K_{pw}	Particle−water and also plant−water
EPN	0-Ethyl 0 (4-nitrophenyl)		partition coefficient
	phenylphosphonothioate	LAE	Linear alcohol ethoxylates
EQT	Environmental quality target	LAS	Linear alkylbenzene sulphonates
E_s	Steric substituent constant	LC/MS	Liquid chromatography/mass
f	Fragmental constant. Also fugacity of		spectrometry
	chemical: chemical's partial pressure.	LRMS	Low resolution single quadrupole
	Latter can be viewed as an 'escaping		resolution mass spectrometer
	pressure or tendency'	MAFF	Ministry of Agriculture, Fisheries and
FA	Fatty alcohol		Food (UK Government)
FAAS	Flame AAS	MBAS	Methylene blue active substances

Metalloid	Element having both metallic and nonmetallic properties or a nonmetallic element that can combine with a metal to produce an alloy	Pinocytosis	Nonspecific uptake of fluid (cf. phagocytosis, which is uptake of solid particles) by cell, via small vesicles
MIBK	Methyl isobutyl ketone	PNEC	Predicted no effect concentration
mp	Melting point	polar	cf. nonpolar
MS	Mass spectrometry	POTW	Public owned treatment works
MSD	Mass selective detector	PP	Property−property relationship
MS/MS	Triple stage quadrupole tandem or hybrid mass spectrometer	PPN	Peroxypropionyl nitrate
		PS	Property−structure relationship
NAA	Neutron activation analysis	PTFE	Polytetrafluorethylene
NIES	National Institute for Environmental Studies (Japan)	PVC	Polyvinyl chloride
		QA	Quality assurance (see also QC)
NIST	National Institute of Standards and Technology (USA)	QAC	Quaternary ammonium compounds
		QC	Quality control (see also QA)
NMR	Nuclear magnetic resonance	QSAR	Quantitative structure−activity relationship
NOAA	National Oceanic and Atmospheric Administration (USA)	QSPR	Quantitative structure−property relationship
NOD	Nitrogenous oxygen uptake (demand)	QSRNS	Quality Status Report on the North Sea
Nonpolar	Molecules with no permanent electric moment, having atoms that are bonded by paired electrons, nonelectrolytic in solution and nonionizing in water (cf. polar)	R	Gas constant. Also reaction
		RC	Primary biodegradability
		S	Aqueous solubility
		SABS	South African Bureau of Standards
NP	Nonylphenol	Safety factor	Factor by which presumed no effect levels are reduced (usually by division) to obtain safe levels, to take into account uncertainties in extrapolating from acute to chronic, single species to multispecies and laboratory to field situations. Size depends upon amount of data available. Some would say that because we can never be sure of safety, the name is misleading. Also called application factors
NPE	Nonylphenol ethoxylates: 1EO, monoethoxylate; 2EO, diethoxylate		
NRCC	National Research Council of Canada		
NS	Nonionic surfactants		
NTA	Nitrilotriacetic acid		
NURP	National Urban Runoff Program (USA)		
NWC	National Water Council (UK). No longer exists		
Ombrotrophic	Fed by rain		
OW	Overlying water	SAR	Structure−activity relationship
[P]	Parachor, measure of molar volume	SAS	Alkane sulphonates or alkyl paraffin sulphonates
P_a	Pressure		
P(AA)	Polymers of acrylic acid (followed by molecular weight)	SBR(s)	Structure−biodegradability relationship(s)
P(AA-MA)	Co-polymers of acrylic acid and maleic acid (followed by molecular weight)	SCA	Standing Committee of Analysts
		SCAS	Semicontinuous activated sludge test (see also CAS)
PAH	Polycyclic aromatic hydrocarbons	SD	Standard deviation
2-PAM	Pyridine-2-aldoximemethochloride	SDA (American)	Soap and Detergents Association
PAN	Peroxyacetyl nitrate		
PAR	Property−activity relationship	S_E	Electrophilic superdelocalizability
PBN	Peroxybutyl nitrate	SFD	Silicon furnace detector
PCB(s)	Polychlorinated biphenyl(s)	SPE	Solid phase extraction
PCDD(s)	Polychlorinated dibenzo-*p*-dioxin(s)	SS	Suspended solids
PCDF(s)	Polychlorinated dibenzofuran(s)	STP	As used here (domestic) sewage treatment plant
PCDO	Polychlorinated dibenzodioxin		
PCP	Pentachlorophenol	Stratosphere	Region of upper atmosphere
PEC	Predicted environmental concentration	STW	Sewage treatment works
		T	Absolute temperature

2,4,5 T	2,4,5 Trichlorophenoxyacetic acid	UOD	Ultimate oxygen demand
TAC	Toxic air contaminants	USGS	US Geological Survey
TAP	Toxic air pollutants	V	Volume. Also solute molar volume
TBT	Tributyltin	VOC	Volatile organic chemicals
TCDD(s)	Tetrachlorodibenzo-*p*-dioxin(s)	WAF	Water-accommodated fraction
TEL	Tetraethyl lead	WSF	Water-soluble fraction
ThCO$_2$	Theoretical CO$_2$ production	Z	Fugacity capacity
Thiol	Any of a group of organic compounds resembling alcohols but with oxygen of the hydroxyl group replaced by sulphur	Zeolite	Porous material, natural or synthetic, consisting of hydrated silicates of aluminium and sodium or calcium. Used as a catalyst in 'cracking' complex organics such as petroleum
ThOD	Theoretical oxygen demand		
TIE(s)	Toxicity identification evaluation(s)		
TLC	Thin layer chromatography		
TMAB	Lauryl trimethyl ammonium bromide		
TMAC	Lauryl trimethyl ammonium chloride	**Greek symbols**	
TML	Tetramethyl lead		
TRE(s)	Toxicity reduction evaluation(s)	α	Solvatochromic parameter
Troposphere	Lowest layer of the atmosphere, where the weather conditions exist and nearly all cloud formations occur	β	Solvatochromic parameter
		χ	Conductivity index
		$x_\chi v$	x-order valence connectivity index
TSA	Total surface area	ε	Dielectric constant
TSCA	Toxic Substances Control Act (USA)	Φ	Polarizability parameter
TSS	Total suspended solids	γ	Electronic parameter
TU	Toxic unit	π^*	Solvatochromic parameter
UCM	Unresolved complex mixture	π_x	Substituent constant
ULV	Ultra-low volume	σ	Electronic substituent constant
UNEP	United Nations' Education Programme	Σ	Extinction coefficient
		ζ	Path length

Chemical Index

Page numbers in *italic* type refer to figures; page numbers in **bold** type refer to tables.

Subject Index

Page numbers in *italic* type refer to figures; page numbers in **bold** type refer to tables.